Bridging the Gap Between AI, Cognitive Science, and Narratology With Narrative Generation

Takashi Ogata
Iwate Prefectural University, Japan

Jumpei Ono
Aomori University, Japan

IGI Global
PUBLISHER of TIMELY KNOWLEDGE

A volume in the Advances in Human and Social Aspects of Technology (AHSAT) Book Series

Published in the United States of America by
IGI Global
Information Science Reference (an imprint of IGI Global)
701 E. Chocolate Avenue
Hershey PA, USA 17033
Tel: 717-533-8845
Fax: 717-533-8661
E-mail: cust@igi-global.com
Web site: http://www.igi-global.com

Library of Congress Cataloging-in-Publication Data

Names: Ogata, Takashi, 1958- editor. | Ono, Jumpei, 1987- editor.
Title: Bridging the gap between AI, cognitive science, and narratology with narrative generation / Takashi Ogata and Jumpei Ono, editors.
Description: Hershey PA : Information Science Reference, [2021] | Includes bibliographical references and index. | Summary: "This book combines traditional theories of narratology with current narrative generation practices using artificial intelligence"-- Provided by publisher.
Identifiers: LCCN 2020010862 (print) | LCCN 2020010863 (ebook) | ISBN 9781799848646 (h/c) | ISBN 9781799856245 (s/c) | ISBN 9781799848653 (ebook)
Subjects: LCSH: Rhetoric--Data processing. | Discourse analysis, Narrative--Data processing. | Cognitive science. | Artificial intelligence.
Classification: LCC P301.5.D37 B75 2021 (print) | LCC P301.5.D37 (ebook) | DDC 808/.036--dc23
LC record available at https://lccn.loc.gov/2020010862
LC ebook record available at https://lccn.loc.gov/2020010863

This book is published in the IGI Global book series Advances in Human and Social Aspects of Technology (AHSAT) (ISSN: 2328-1316; eISSN: 2328-1324)

British Cataloguing in Publication Data
A Cataloguing in Publication record for this book is available from the British Library.

Advances in Human and Social Aspects of Technology (AHSAT) Book Series

Ashish Dwivedi
The University of Hull, UK

ISSN:2328-1316
EISSN:2328-1324

Mission

In recent years, the societal impact of technology has been noted as we become increasingly more connected and are presented with more digital tools and devices. With the popularity of digital devices such as cell phones and tablets, it is crucial to consider the implications of our digital dependence and the presence of technology in our everyday lives.

The **Advances in Human and Social Aspects of Technology (AHSAT) Book Series** seeks to explore the ways in which society and human beings have been affected by technology and how the technological revolution has changed the way we conduct our lives as well as our behavior. The AHSAT book series aims to publish the most cutting-edge research on human behavior and interaction with technology and the ways in which the digital age is changing society.

Coverage

- Cyber Behavior
- Gender and Technology
- Human-Computer Interaction
- Technoself
- Human Rights and Digitization
- Cyber Bullying
- Human Development and Technology
- Technology Adoption
- ICTs and social change
- Information ethics

IGI Global is currently accepting manuscripts for publication within this series. To submit a proposal for a volume in this series, please contact our Acquisition Editors at Acquisitions@igi-global.com or visit: http://www.igi-global.com/publish/.

The Advances in Human and Social Aspects of Technology (AHSAT) Book Series (ISSN 2328-1316) is published by IGI Global, 701 E. Chocolate Avenue, Hershey, PA 17033-1240, USA, www.igi-global.com. This series is composed of titles available for purchase individually; each title is edited to be contextually exclusive from any other title within the series. For pricing and ordering information please visit http://www.igi-global.com/book-series/advances-human-social-aspects-technology/37145. Postmaster: Send all address changes to above address.

Titles in this Series

For a list of additional titles in this series, please visit:
www.igi-global.com/book-series/advances-human-social-aspects-technology/37145

Dyslexia and Accessibility in the Modern Era Emerging Research and Opportunities
Kamila Balharová (Pedagogical and Psychological Counseling Center, Brno, Czech Republic) Jakub Balhar (Gisat s.r.o., Czech Republic) and Věra Vojtová (Masaryk University, Czech Republic)
Information Science Reference • © 2021 • 279pp • H/C (ISBN: 9781799842675) • US $165.00

Understanding the Role of Artificial Intelligence and Its Future Social Impact
Salim Sheikh (Saïd Business School, University of Oxford, UK)
Engineering Science Reference • © 2021 • 284pp • H/C (ISBN: 9781799846079) • US $200.00

Overcoming Barriers for Women of Color in STEM Fields Emerging Research and Opportunities
Pamela M. Leggett-Robinson (PLR Consulting, USA) and Brandi Campbell Villa (Belay Consulting, USA)
Information Science Reference • © 2020 • 199pp • H/C (ISBN: 9781799848585) • US $145.00

Maintaining Social Well-Being and Meaningful Work in a Highly Automated Job Market
Shalin Hai-Jew (Kansas State University, USA)
Business Science Reference • © 2020 • 333pp • H/C (ISBN: 9781799825098) • US $215.00

ICTs and Innovation for Didactics of Social Sciences
Emilio José Delgado-Algarra (University of Huelva, Spain)
Information Science Reference • © 2020 • 295pp • H/C (ISBN: 9781799828822) • US $175.00

Civic Engagement Frameworks and Strategic Leadership Practices for Organization Development
Susheel Chhabra (Periyar Management and Computer College, India) and Muneesh Kumar (Maharaja Agrasen Institute of Technology (MAIT), Delhi, India)
Information Science Reference • © 2020 • 314pp • H/C (ISBN: 9781799823728) • US $195.00

Civic Engagement in Social and Political Constructs
Susheel Chhabra (Periyar Management and Computer College, India)
Information Science Reference • © 2020 • 285pp • H/C (ISBN: 9781799823643) • US $195.00

Recent Advances in Digital Media Impacts on Identity, Sexuality, and Relationships
Michelle F. Wright (Penn State University, USA)
Information Science Reference • © 2020 • 347pp • H/C (ISBN: 9781799810636) • US $195.00

IGI Global
PUBLISHER of TIMELY KNOWLEDGE

701 East Chocolate Avenue, Hershey, PA 17033, USA
Tel: 717-533-8845 x100 • Fax: 717-533-8661
E-Mail: cust@igi-global.com • www.igi-global.com

Editorial Advisory Board

Table of Contents

Detailed Table of Contents

This chapter surveys, discusses, and explores the entire concept of visual narrative structure, cognition, and generation from continuity-based to discontinuity-based perspectives. The model of visual narrative structure, including presentation and meaning, is expanded to explain the cognitive transition based on the rhetorical transition techniques and the rhetorical cutting techniques. The classification of the visual narrative structure, including rhetorical transition techniques and rhetorical cutting techniques, is useful for narrative simulation for discussing and exploring the entire visual narrative concept and generation. The rhetorical cutting techniques and the rhetorical transition techniques can reveal various cognitive effects such as reality effects and nostalgia effects, including difficulty. The determinacy-based narrative, the indeterminacy, the diversity, and the ambiguity on narrative can bridge the gaps between cognitive science, artificial intelligence, and narratology.

Video game non-player characters (NPCs) are a type of agent that often inherits emotion models and functions from ancestor virtual agents. Few emotion models have been designed for NPCs explicitly, and therefore do not approach the expressive possibilities available to live-action performing actors nor hand-crafted animated characters. With distinct perspectives on emotion generation from multiple fields within narratology and computational cognitive psychology, the architecture of NPC emotion systems can reflect the theories and practices of performing artists. This chapter argues that the deployment of virtual agent emotion models applied to NPCs can constrain the performative aesthetic properties of NPCs. An actor-centric emotion model can accommodate creative processes for actors and may reveal what features emotion model architectures should have that are most useful for contemporary game production of photorealistic NPCs that achieve cinematic acting styles and robust narrative design.

Chapter 3

The purpose of this chapter is to examine the psychological approach from a narrative viewpoint. In addition, the authors discuss the possibility of collaboration with artificial intelligence (AI). They reviewed the narrative approaches in psychology. At result, they showed how the narrative approach in clinical domain was expanded. The following possibilities exist regarding a narrative generative system in a narrow sense: (1) It is the construction of a fluid system. (2) The systems generate a tale communally. (3) There are no clear beginnings and endings. (4) It is a narrative generation through a dialogue. Next, the following possibilities exist regarding a narrative generative system in broad sense: (1) It is the modeling of the stance of the narrative. (2) It is the modeling of narrative generation by the community. (3) It is the modeling of the process. (4) It is the study of the interlocutory peculiarity.

Chapter 4

In this chapter, the author attempts to define the verbs in the description of Japanese sake taste by employing 1) a usage-based approach, 2) "encyclopedic semantics" rather than a "dictionary view," and 3) sense-making theory, drawing on data from a "sake tasting description corpus" (approximately 120,000 words). The chapter selects eight verbs of high frequency (e.g., hirogaru 'spread') and defines their sense(s) in a bottom-up and abductive fashion, based on a score indicating the strength of co-occurrence between terms. In this study, the authors deal with the verbs for "Understanding" or "Interpretation "; it means, verbs that contribute to narrating the personal, individual story (contents) of the tasters. This study suggests the verbs for understanding have senses related to [Timeline] and [Space]. On the other hand, verbs do not tend to collocate with [Movement] and interestingly, the [Structure], as same as the tendency of adjectival-nouns.

Chapter 5

This chapter investigates the threshold of context memory on numbers 300 through 500 regarding perspective with assumption experiment and theories of cognitive semantics and short-term memory. Context generation became possible with text database with lexical engineering system on artificial intelligence (AI). It was realized an infinite and continuous presentation of context. Several context presentations realized the limitation of context memory on the process of continuous context presentation on numbers 500 and 2,000, as well as the representation process with judgement for memory. The first report studied the context generating machine for Japanese pun generation. This was termed "Class B engine" or "Dajare Machine." This study focused on the examination of the context threshold using an experiment of human brain mapping on an auditory evoked magnetic field (AEF) of a magnetic encephalogram. It also focused on the analysis of the theory of cognitive semantics and folk biology, using a theory of short-term memory of the "magical number seven."

Preface

Narrative generation studies include challenging themes and topics covering artificial intelligence (AI) and cognitive science. Recently, many studies have been conducted on various aspects of narrative generation, including story, plot, character, and language and visual representations. However, few narrative generation studies have systematically inherited and used knowledge and techniques from narratology, literary theories, and other related areas of human and social sciences.

On the other hand, the editors of this book have presented the concept of "post-narratology" to approach their narrative generation study. Post-narratology refers to the narratology expanded by AI, cognitive science, and related researches, and furthered by the unique narratologies and literary theories that have been developed through each culture. One of the editors (Ogata) has presented many ideas, concepts, and systems related to the topic in "Informatics of Narratology" (Ogata & Kanai, 2010), "Computational and Cognitive Approaches to Narratology" (Ogata & Akimoto, 2016), "Content Generation Through Narrative Communication and Simulation" (Ogata & Asakawa, 2018), "Informational Narratology" (Ogata, Kawamura, & Kanai, 2018), "Post-Narratology Through Computational and Cognitive Approaches" (Ogata & Akimoto, 2019), and other books and papers. Moreover, recently, Ogata (2020a, 2020b) presents a research system that consistently collects many related areas from the viewpoint of post-narratology. These books formalize, model, and systematize previous narratology, literary theories, and relating theories, methods, and studies from the perspective of narrative generation (and other content generation) to establish bases and foundations of narratology and literary theories in narrative generation studies. In particular, the last three books (Ogata & Akimoto, 2019; Ogata, 2020a, 2020b) reveal a newly developed stage of the authors' narrative generation study based on post-narratology.

As previously described, while narrative generation is one of the challenging research themes in AI and cognitive science, many narrative generation researches do not efficiently, systematically, and essentially use the knowledge and techniques that have been developed through previous narratology and literary theories. This indicates that bridging both fields is very difficult. In such an academic situation, the objective of this book is to bridge the gap between AI, cognitive science, and narratology with narrative generation in a broad sense, including other content generation, such as a novel, poem, movie, computer game, advertisement. To present an organic, systematic, and integrated combination of both fields to develop a new research area, namely post-narratology, this book plays an important role in the creation of a new research area and has an impact on narrative generation studies including AI and cognitive science, as well as narrative studies including narratology and literary theories.

The target audience of this book includes researchers and university students from fields of information (such as AI, cognitive science, natural language semantics, and content generation) and literature (such as narratology, literary theories, and related human and social sciences). Recently, the research and development of automated content generation has come into practice as business services of automated document generation for newspapers and various types of reports. In contrast, this book proposes the concrete description of automated generation techniques and technologies of narratives that are more difficult to formalize, model, and systematize than the aforementioned general documents. Therefore, its potential users include enterprise practitioners, engineers, and creators of diverse content generation fields, such as advertising production, computer game creation, comic and *manga* writing, and movie production.

TOWARD A POST-NARRATOLOGY OF NARRATIVE GENERATION THROUGH BRIDGING THE GAP BETWEEN NARRATOLOGICAL STUDIES AND AI/COGNITIVE STUDIES

The editors would like to concretely supply the basic framework of this book's approach in Figure 1.

Figure 1 indicates that there is a gap between narratology (or narratological studies related to it, from a broader perspective), and AI and cognitive science (and related research fields). This book aims to reduce the gap through the medium of narrative generation. However, the narrative generation in this case

Figure 1. The basic framework of the whole book's approach

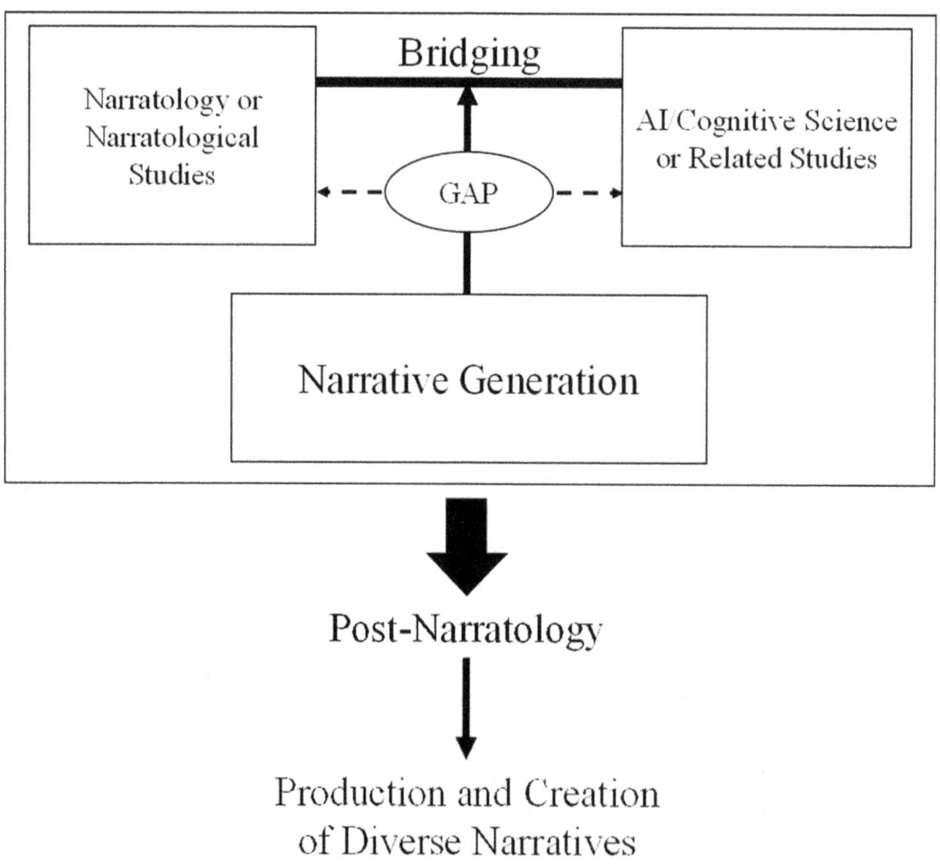

is not limited to the range of narratives in the narrow sense, such as novels and artistic works. Regarding this aspect, the "theory of narrative genre system" by Ogata (2020b) views narratives in the wide range including narratives as literary and artistic works, narratives as practical texts, narratives as social events, and psychological and mental phenomena in the narrow sense. Ogata called his approach—that aims to solve the gap between narratological studies and AI/cognitive science through narrative generation, comprehended in the broad sense—post-narratology in his recent aforementioned books (Ogata & Akimoto, 2019; Ogata, 2020a, 2020b). The editors view the future direction of post-narratology that will produce and create diverse objects or works and impact literature, arts, and societies.

Hence, Figure 1 and the previous explanation provide the basic framework of the entire post-narratology research by the editors, and this edited book is located as a result to post-narratology. On the other hand, Table 1 shows a panoptic list that includes narratological studies, AI and cognitive studies, and targets, specifically, systems and models for bridging between the narratological and AI/cognitive studies by all chapters. (The terms used in this table are frequently rewritten as different terms from the words that the authors originally described.)

Explaining Table 1, first, this whole book adopts theories and methods of narratological studies, including narratology, in addition to marketing and management theories. Although marketing and management theories themselves are not included in narratology in a narrow sense, narratological perspectives to those academic fields will also be able to exist in the editors' post-narratology approach. On the other hand, the theories and methods tied to researches related to AI and cognitive sciences cover many fields such as AI-based narrative generation techniques and cognitive approaches to various areas including image recognition and social cognition. Based on these, this book's chapters deal with narrative (generation) systems and models from literature, arts, and entertainment to management, psychological, and mental phenomena for image, virtual character with emotion, narration by disabled, gustatory sense representation, puns (*dajare*), novels (narratives), *haiku*, poems, video commercials, narration based on a sense of values, storytelling of a new product, service development, and management innovation. In this manner, all the chapters directly and indirectly treat the theme of narrative generation in a broad sense.

As previously stated, this book provides an approach to the systematic post-narratology that the editors have schemed and developed. In this sense, several previous books should be referenced to obtain a better understanding of this book.

BOOK ORGANIZATION

This book contains 12 chapters in all and is divided into three sections, including "Cognitive Approaches to Narratology and Narrative Generation," "Narrative Generation Systems and Narrative Content Generation," and "Social Perspectives of Narratological and Narrative Generation Methods." The theme of each section and outlines of the contained chapters are described below.

SECTION 1: COGNITIVE APPROACHES TO NARRATOLOGY AND NARRATIVE GENERATION

This section includes the following five chapters that explore narratological approaches to cognitive models and systems.

Table 1. Main concepts related to the framework of the book used in all the chapters

Chapter [Author(s)]	Narratological Studies	AI and Cognitive studies	Target (System or Model)
Chapter 1 [Akihito Kanai]	Narratology/Film theory	Cognitive film theories/ Cognitive transition and cutting techniques/Narrative image simulation	Narrative film simulation system
Chapter 2 [Sheldon Schiffer]	Narratology/Film theory	Computational and cognitive psychology/AI-based agent models/Computer game technologies/Emotion	The facial expression system with emotion of NPCs in computer games
Chapter 3 [Kai Seino and Shun Ishizaki]	Narratology/Post-narratology/ Psychological narratology/Clinical psychology	Text mining/ Psychological narratology	Clinical narrative generation system accompanying linguistic analysis
Chapter 4 [Hiroki Fukushima]	Sense-making theory/Narratology	Text mining/The cognition of the sense of taste	Linguistic description of the taste of Japanese *sake*
Chapter 5 [Nobuyoshi Harada]		Cognitive semantics/Human memory/Brain function analysis/Japanese pun generation (*dajare* machine)	Memory model in literary work recognition
Chapter 6 [Takashi Ogata]	Narratology/Post-narratology/ *Kabuki*-oriented narratology/ Creating narratives	Narrative generation system/ *Geinō* information system	Narrative generation systems/*Watakushi Monogatar*i
Chapter 7 [Jumpei Ono and Takashi Ogata]	*Haiku* theory/Narratology/Post-narratology	*Haiku* generation system/ Narrative generation system	*Haiku* generation system/Narrative generation system
Chapter 8 [Akinori Abe]	Narratology/Intertextuality	Abduction/Literary work generation systems	Literary work generation systems, including poems and short songs
Chapter 9 [Yoji Kawamura]	Marketing theory/Advertising theory/Film theory	Image processing techniques/ Cognitive effects of image representation	Video commercial generation system by still images
Chapter 10 [Yasuo Tanida]	Human sense of values/Narratology	Quantitative stochastic model of language/Text mining/Narrative generation techniques	Rewriting system of stories
Chapter 11 [Jun Nakamura]	Business studies/Narrative studies of management fields	Abduction/Storytelling modeling	Storytelling model of new product and service creation
Chapter 12 [Akimitsu Hirota]	Marketing theory/Business studies/ Signature story in management fields	Data mining/Storytelling modeling	Innovative storytelling model of market-creating products

Chapter 1, "Cognitive Transition and Cutting Techniques for Narrative Film Rhetoric Simulation," by Akihito Kanai aims to present an entire concept of visual narrative structure based on the theory of cognitive transition and cutting techniques that Kanai has been studying. Depending on Kanai's cognitive image theory, the rhetorical cutting and transition techniques can reveal various cognitive effects, including reality effects and nostalgia effects, with the feeling of difficulty of image recognition by audiences. An important aspect of Kanai's work is that he emphasizes audiences' difficult rather than easy understanding of image in narrative films. His goal in narrative film rhetoric is to model, systematize, and create narrative film techniques for aiming at difficulty and delay of the understanding by the audiences. This chapter expands his previous model regarding visual narrative structures including presentations and meanings, and shows various cognitive effects, such as reality effects and nostalgia effects, including difficulty in the rhetorical cutting techniques and rhetorical transition techniques. His study is related to narratology or narratological studies in the themes of film or visual narratology and defamiliarization-related thought or theory. Moreover, although his unique cognitive approach to narrative film rhetoric directly aims to present a cognitive model, one of his future goals is to apply his presented cognitive model to actual narrative film rhetoric simulation.

In Chapter 2, "Multi-Disciplinary Paths to Actor-Centric Non-Player Character Emotion Models," Sheldon Schiffer presents a generic emotion model in the basic components of an NPC (non-player character) in computer games as an entertaining virtual agent. The emotion-response algorithms of NPCs dealt with in this chapter need to receive relevant data from the player and game play space and elicit expressive data back into the game play environment for the player to observe and respond. In the broad perspective, as Schiffer regards the world of a computer game as a kind of narrative in which characters—including NPCs—act in narrative spaces and times, he introduces narratological studies, including Chatman who deals with linguistic and image narratives to consider the problems of camera-work of human faces. On the other hand, the author refers to several computational psychological studies that overlapped with AI and cognitive science to consider in detail a virtual agent with emotion and its action. The author considers that cinematic acting styles provide valuable methods for facial expression in video games as they connect the emotional source of actor expression to NPCs that move narratives forward and the new type of NPC model can contribute to robust narrative design for contemporary game production. This chapter describes diverse theories and techniques including through narratological studies and computational and cognitive studies to aim at realization of a facial expression model with emotion in NPCs in the narrative genre of computer game. This chapter especially focuses on the theme of facial expression of a NPC associated to emotion. Such function can have a delicate impact on the continuous narrative flow and the relationship with other game characters and human players. (For example, from a narrative field of the editors of this book, in *kabuki*-dance *Kyōganoko Musume Dōjōji*, the change of the instantaneous facial expression of the heroin, particularly, from the beautiful and gentle face of a girl to dreadful face of a ghost or departed spirit, reveals the most important depth of the narrative). Exploring the problems of facial expression with emotion in a new narrative genre of computer games is an interesting challenge because computer games are a different and unique type of narrative from many previous narrative genres such as novels, poems, dramas, and movies.

In Chapter 3, "Extension of Clinical/Psychological Approach Using Post Narratology: The Proposal of a Psychological Narratology and Review for Systematization," Kai Seino and Shun Ishizaki present the concept of psychological narratology and methods to apply it to their computational linguistic analysis that they have been studying to provide the possibility of bridging the gap between AI-based approaches and narratological studies. Another characteristic of this chapter is that they deal with the clinical domain for supporting people with disabilities such as autism spectrum disorder (ASD) and their AI-based linguistic analysis using a data mining technique is also utilized to support these people. Therefore, their study introduced in this chapter probably shows a possible direction to AI-based narrative analysis from AI-based linguistic analysis. In this regard, this chapter accurately reflects the concept of the whole book, bridging the gap between narratology, AI, and cognitive science with narrative generation. The comparison between literary narratology or previous standard narratology and psychological narratology is also interesting from the theoretical angle or the standpoint of the book's editors. Actually, previously many narratological researches have been conducted in the contents of AI, cognitive science, and psychology. However, narratology, narratological studies, or narratives in such contexts have been regarded as a means or example for serving main academic fields. From the viewpoint of this book's concept, narratology, narratological studies, further narratives should be focused on as primary concepts to be considered. In particular, Seino and Ishizaki explicitly focus on the psychological aspect in narratology.

Chapter 4, "Defining the Verbs for 'Understanding and Interpretation' of Japanese Sake," by Hiroki Fukushima intends to define the verbs for linguistically describing and representing the taste of Japanese *sake*. (Although Fukushima does not refer to it, *Bungakuron* by Natsume Soseki (1997) introduced the sense of taste as an element for forming the content in literary works, such as novels, poems, and scripts.) A series of Fukushima's studies that aim to linguistically describe and represent human vague senses of taste of Japanese *sake* are closely connected to the problems of narratology. The linguistic description and representation of the senses of taste of *sake* will be associated with problems regarding prominently personal and individual stories and narratives, and their generation or creation. In particular, depending on "a usage-based approach," "encyclopedic semantics," and "a sense-making theory," this chapter selects eight verbs that are frequently used in the taste description of Japanese *sake* and defines their senses in a bottom-up way and abduction method based on a score indicating the strength of co-occurrence between terms. Furthermore, this chapter clarifies the definition of the verbs dependent on the terms "timeline," "space," "movement," and "structure."

In Chapter 5, "300–500 Threshold of Context Memory and Assumption Experiment," Nobuyoshi Harada presents an attempt of the problem regarding the threshold of context memory on numbers 300 through 500. In particular, this study focuses on the examination of the context threshold through an experiment of human brain mapping on an "auditory evoked magnetic field (AEF)" of a magnetic encephalogram. Harada introduces, as a hypothesis for the future discussion, several context presentations that realize the limitation of context memory in the process of continuous context presentation on numbers 500 and 2,000. As materials for the experiment, Harada uses many puns (Japanese *dajare*s) generated by a computer program, and this research is related to the problems of human memory processing in cases where many *dajare*s are continuously inputted into a human brain. In this regard, this hypothetic research may contribute to the consideration of a problem concerning narratological studies and cognitive science, especially brain science.

SECTION 2: NARRATIVE GENERATION SYSTEMS AND NARRATIVE CONTENT GENERATION

Four chapters included in this section relatively directly focus on narrative generation systems and narrative content generation for various narrative genres, such as narrative, *haiku*, poem, and advertising.

Chapter 6, "Bridging the Gap Between Narrative Generation Systems and Narrative Contents With *Kabuki*-Oriented Narratology and *Watakushi Monogatari*," by Takashi Ogata presents an idea that bridges the gap between narrative generation based on a synthetic system framework including two types of subsystems—Integrated Narrative Generation System (INGS) and *Geinō* Information System (GIS), and narrative contents known as *Watakushi Monogatari*. *Kabuki*-oriented narratology is also used to bridge them by introducing four *kabuki* theories through the classical and the contemporary. This chapter comprises of the following three major themes to complete the aforementioned idea. The first theme deals with narrative phenomena such as the integration of INGS and GIS. This theme is connected to the second theme, which is presenting the narrative generation of *kabuki* or *kabuki*-oriented narratology, and the third theme, particularly, thoughts and design underlying *Watakushi Monogatari* as a collection of narrative content to be created by the author in the context of internal and external narrative generation to create and distribute narrative contents. INGS and GIS in the main topics of the first theme are respectively located as a single narrative generation and reception mechanism and a multiple narrative production and consumption mechanism. The framework proposed in this chapter aims to perform the creation of actual narrative work of *Watakushi Monogatari* showing the aspect of internal narrative generation and social development of *Watakushi Monogatari* corresponding to the aspect of external narrative generation through a kind of collaborative work of INGS and GIS, and further "I" (*watakushi*) as the author himself.

The purpose of Chapter 7, "Haiku Generation From Narratological Perspective: A Circulation Between Haikus and Stories," by Jumpei Ono and Takashi Ogata introduces the haiku generation method into our narrative generation. There are two typical haiku definitions: the sentence is composed of 17 mora and consists of three phrases, though there are other haikus not included in the definition. This chapter presents an experimental system that can generate haikus that have the aforementioned typical definitions. However, the extreme goal of the study is not presenting a program that can make haiku generation possibility to be compared to the human haiku creation ability, but incorporating such ability into a narrative generation system, specifically, the INGS introduced in Chapter 6. Concretely, the authors discuss a kind of circulative system between haikus and narratives (stories or plots), through which one or more haiku(s) are generated in accordance with a generated narrative and one or more narratives from the generated haiku(s). The continuous execution results in the circulative generation of haiku(s) and narrative(s) and incremental narrative change through haiku generation. Hence, the proposed system framework is regarded as an alternative of narrative generation in the authors' narrative generation study.

In Chapter 8, "On Logical Literary Work Generation and More," Akinori Abe describes the possibility and problems of computational literary work generation, such as *waka*, a type of Japanese classical short song with fixed form of 5, 7, 5, 7, 7 (characters), poems, and stories, largely based on the concept of intertextuality by Julia Kristeva who is a French narratologist. Although Abe's previous approach to narrative generation was based on a relatively simple method of using a random and disconnected combination, so-called "cut-up," of selected words, phrases, and letters, this chapter presents a more intelligent AI-based method. In particular, Abe introduces techniques of abduction into logical literary work generation to maintain the logical consistency and coherency of generated works to aim to bridge the narratological method to narrative generation systems in a broad sense. In contrast with this abduction-based logical approach, Abe indicates important factors in narrative or literary works of very cognitive aspects, such as beautiful, sentimental, and valuable representation, and discusses how to add such cognitive aspects to the logical generation of literary works. From the perspective of the editors of this book, this chapter by Abe will contribute to support actual narrative or literary works through various operations for narrative generation functions.

Chapter 9, "An Analysis of Cognitive Elements and Effects of Video Commercials: Toward Effective Indexing and Video Production," by Yoji Kawamura is a continued report of the advertising system that generates the story-based sequential collection according to still images using a prepared database of fragmentary still images of a specific product type. This chapter analyzes cognitive elements and effects in advertising commercials to produce rich and attractive advertising works. In particular, Kawamura proposes a method to organically combine among video shots based on still images in the database to provide consistency or coherence to the collection using a kind of semantic network. The image database allows the attaching of records of the time and place of shooting and features of the persons as indexes. From the theme of this entire book, first, advertising works also correspond to a type of narrative genre, and this study deals with advertising narrative stories referring to management and marketing theories and image theories as an interdisciplinary narrative study or narrative generation study. Furthermore, this chapter focuses on the problem of still image indexing based on cognitive scientific effectiveness for advertising story generation.

SECTION 3: SOCIAL PERSPECTIVES OF NARRATOLOGICAL AND NARRATIVE GENERATION METHODS

This section includes three chapters on the relationship between social perspectives of narratology or narratological methods and narrative generation.

Chapter 10, "Narrative Generation Using Psychological Value Variables: Probabilistic Model of Language Expressions and Values," by Yasuo Tanida designs a systematic method to rewrite a story generated by a story generation system based on "psychological value variables" for readers' positive evaluation of the future implementation of a system for realizing the proposed design. The concept of psychological value variables is an original idea by the author, and each person comprehends and interprets the world and acts based on the comprehension and interpretation according to the difference in psychological values. The extreme goal by Tanida is probably to utilize the theory of psychological value variables to support and accelerate marketing activities of companies. The proposed study of this chapter is located in the flow to this goal and extends his previous linguistic analysis based on words of people or customers to narratological analysis, and further toward the analysis through narrative generation. Regarding AI-based methods, Tanida mainly adopts a probabilistic language expression model similar to the language analysis appearing in Chapter 3.

Through Chapter 11, "Effect of External Activate Factors Serving as Clue: Creating New Products or Services Ideas With Storytelling," Jun Nakamura is interested in the problem of the effect of storytelling for the process of creating new products or services in the context of companies. To approach to this problem, Nakamura demonstrates an experiment to compare the effect of external activate factors such as objective advice and clues by consultants and outside directors at a graduate lecture. In particular, this experiment clarifies the effect of storytelling by the external activate factor for the emergence of abduction and new rules. This chapter's implication to the concept of this book will be stated with Chapter 12 in the next explanation.

Chapter 12, "Design of Narrative Creation in Innovation: 'Signature Story' and Two Types of Pivots," by Akimitsu Hirota treats the topic of narrative creation in innovation in business organizations. Hirota previously analyzed storytelling through trial and error in the development of market-creating products. However, this chapter further pursues the stories of trial and error of the developer and marketer during the development of a market-creating product, and in it, he especially labels such stories "innovation narratives." In particular, this chapter clarifies the creating process of an innovation narrative from the angle of an organizational dialog based on two types of pivots, including open pivots and closed pivots. The importance of stories and narratives has been pointed out in the world of business administration studies, including marketing, from the perspective that they supplement and structuralize the knowledge type of analysis and measuring. Chapters 11 and 12 commonly approach "organizational narratology" as a narratological genre that aims to deal with a variety of narratological phenomena seen in social organizations as narratives or stories, or to interpret diverse organizational phenomena as narratives or stories. These chapters indicate that organizational narratives and stories play important roles for the development of new products and services and enhance organizational innovation as they function as a kind of control knowledge for actually creating new rules and innovative stories. The concept of this book also discusses the connection between management knowledge and narratological knowledge in Chapters 11 and 12.

Takashi Ogata
Iwate Prefectural University, Japan

Jumpei Ono
Aomori University, Japan

REFERENCES

Natsume, S. (2007). *Bungaku ron (I, II)* [Literary theory (I, II)]. Iwanami Shoten. (Original work published 1907)

Ogata, T. (2020a). *Toward an integrated approach to narrative generation: Emerging research and opportunities*. IGI Global. doi:10.4018/978-1-5225-9693-6

Ogata, T. (2020b). *Internal and external narrative generation based on post-narratology: Emerging research and opportunities*. IGI Global. doi:10.4018/978-1-5225-9943-2

Ogata, T., & Akimoto, T. (Eds.). (2016). *Computational and cognitive approaches to narratology*. IGI Global. doi:10.4018/978-1-5225-0432-0

Ogata, T., & Akimoto, T. (Eds.). (2019). *Post-narratology through computational and cognitive approaches*. IGI Global. doi:10.4018/978-1-5225-7979-3

Ogata, T., & Asakawa, S. (Eds.). (2018). *Content generation through narrative communication and simulation*. IGI Global. doi:10.4018/978-1-5225-4775-4

Ogata, T., & Kanai, A. (2010). *Monogatariron no jōhōgaku josetsu: Monogatari seisei no shisō to gijutsu wo megutte* [An introduction to informatics of narratology: Around the thoughts and technologies of narrative generation]. Gakubunsha.

Ogata, T., Kawamura, Y., & Kanai, A. (2018). *Jōhō monogatariron: Jinkōchinō, ninchi, shakai-katei to monogatari seisei* [Informational narratology: Artificial intelligence/cognition/social process and narrative generation]. Hakutō Shobō.

Section 1

Cognitive Approaches to Narratology and Narrative Generation

Chapter 1
Cognitive Transition and Cutting Techniques for Narrative Film Rhetoric Simulation

Akihito Kanai
Hosei University, Japan

ABSTRACT

This chapter surveys, discusses, and explores the entire concept of visual narrative structure, cognition, and generation from continuity-based to discontinuity-based perspectives. The model of visual narrative structure, including presentation and meaning, is expanded to explain the cognitive transition based on the rhetorical transition techniques and the rhetorical cutting techniques. The classification of the visual narrative structure, including rhetorical transition techniques and rhetorical cutting techniques, is useful for narrative simulation for discussing and exploring the entire visual narrative concept and generation. The rhetorical cutting techniques and the rhetorical transition techniques can reveal various cognitive effects such as reality effects and nostalgia effects, including difficulty. The determinacy-based narrative, the indeterminacy, the diversity, and the ambiguity on narrative can bridge the gaps between cognitive science, artificial intelligence, and narratology.

INTRODUCTION

This chapter surveys, discusses and explores the entire concept of visual narrative structure, cognition and generation from continuity based to discontinuity based. Visual narratives such as films are analyzed in terms of the *cognitive transition*, the cognitive processing change of a narrative focus, in this chapter.

The question of how people comprehend visual narratives such as films is difficult and complicated (Loschky et al., 2020). A viewer tends to recognize a film as a visual narrative to understand the story (Cutting, Brunick, & Canden, 2012; Cutting & Iricinschi, 2014). Although viewers' scene perception and event comprehension based visual narrative comprehension process and generation mechanism

DOI: 10.4018/978-1-7998-4864-6.ch001

are important questions for the field of cognitive science or artificial intelligence, the other aspects of narrative such as the mental difficulties on cognition are also important (Corner, 2018). The mental difficulties are intentionally present, caused by a visual narrative's core feature. For example, the mental difficulty cognition from massive indeterminacy of place and time caused by the discontinuity editing can enhance the viewpoint diversity and the cognitive transition of the viewer (Kanai, 2018). Besides the determinacy and the continuity based narrative, the indeterminacy, the diversity, the ambiguity, and the discontinuity of narratives are important questions for the field of narratology. For example, Bakhtin (1984) argued the polyphonic theory of novels. The polyphonic theory emphasized the indeterminacy, the divergence, and the ambiguity. In recent years, Hasumi (2014) argued the indeterminacy and the ambiguity of novels such as *Madame Bovary*.

Narrative is mainly discussed in the field of narratology. Theories on narratology make a strong distinction between *story* (*fabula*) and *discourse* (*syuzhet*) (e.g., Genette, 1972; Shklovsky, 1965). Story comprehension and meaning are not as important as discourse and rhetoric themselves in theories of narratology (Chatman, 1990). For the study of film, shots, editing, and film structures themselves are equally important. A film has not only a story and meaning structure but also visual and audio structures. Story comprehension is affected by a film's structure. Furthermore, the cognition difficulties caused by story comprehension are also affected by the structure. The cognition difficulties resulting from the structure are related to the quality and the artistry of the film.

Kanai (2016, 2019) defined the *rhetorical cutting technique* as rhetorical techniques for creating irrational cuts in a film based on discontinuity editing. Rhetorical cutting techniques generate irrational relationships or unfamiliar situations in film and create an alienation effect (Brecht & Willett, 1964). Rhetorical cutting techniques for narrative space and time in a film can cause a viewer's cognitive transition from being event-based and story-driven to being presentation-based and rhetoric-driven processing without cognitive effects from the story.

Furthermore, this chapter discusses *rhetorical transition techniques*, rhetorical techniques for creating plurality and diversity in a film with some story and continuity. Rhetorical transition techniques for narrative space and time in the film can cause the viewer's cognitive transition from event-based story-driven to presentation-based and rhetoric-driven processing with cognitive effects from the story.

Rhetorical transition techniques and cutting techniques can cause a viewer's cognitive transition. The transition and the interaction between event-based, story-driven processing and presentation-based, rhetoric-driven processing are among the most important aspects of film cognition. Only the cognitive science approach with narratology can adequately analyze the transition and the interaction. Without a cognitive science approach, narratology cannot address them. In addition, a cognitive science approach without narratology cannot address cognition difficulties in the transition and the interaction. Furthermore, narratology with cognitive science based artificial intelligence can reach the transition and the interaction, including cognition difficulties, and can do several narrative simulations based on cognitive effects for content generation.

Researchers such as Ogata and Asakawa (2018) have argued for techniques such as visual narrative generation through narrative simulation using artificial intelligence techniques. *Narrative simulation* can be a process of composing and designing various kinds of narrative rhetoric and of testing the cognitive effects of that narrative (Kanai, 2018). Many cognitive effects from a narrative result from the effects of the determinacy and the indeterminacy on the interaction between cognitive process, story, discourse, and rhetoric themselves. For example, the indeterminacy emerging from rhetorical cutting techniques in narrative space and time in a narrative film can cause a viewer to experience cognitive transition and intensive cognitive effects. Based on the rhetorical cutting techniques used to generate indeterminacy with regards to time, space, and characters' actions in continuous shots, viewers can feel less constrained with regards to story comprehension and can reset their viewpoints to subtle elements, such as details of individual shots. The cognitive transition caused by the indeterminacy of narrative film rhetoric can cause cognitive effects, such as reality or nostalgia effects relating to the actual world. Those effects are discussed in a later section of this chapter. Moreover, based on the rhetorical transition techniques used to generate the plurality, the diversity, and the ambiguity with regards to time, space, and characters' actions in continuous shots, a viewer can still feel constrained with regards to story comprehension but can reset the viewpoint to subtle elements. Based on the indeterminacy, the diversity, and the ambiguity, the narrative simulation and generation mechanism can emphasize the whole rhetorical narrative aspects of films.

In this chapter, visual narrative is analyzed and surveyed, especially from the perspective of the cognitive transition relate to the narrative story and rhetorical presentation, including both continuity and discontinuity. A model of visual narrative structure, including presentation and meaning provided by Cohn & Magliano (2020), is expanded to explain the cognitive transition based on the rhetorical transition techniques and the rhetorical cutting techniques. The classification of the visual narrative structure, including rhetorical transition techniques and rhetorical cutting techniques, is useful for narrative simulation for discussing and exploring the entire visual narrative concept and generation.

BACKGROUND

Visual narrative cognition involves the simultaneous processing of many factors. Generally, a story and its cognition are important factors to narrative cognition. The viewer can comprehend the totality meaning and the story from continuous events (e.g., Avrahami & Kareev, 1994). The many previous approaches primarily focused on characters' goal-directed actions (e.g., Jhale & Young, 2010). Recent focus has turned to the visual narrative structure (e.g., Cohn, 2013, 2014; Cohn & Magliano, 2020).

Although theories of story grammar inspired by Propp's (1968) theory (e.g., Rumelhart, 1975; Thorndyke, 1977) blurred the distinction between narrative (i.e., presentation) and events (meaning), Cohn & Magliano (2020) provided a model of visual narrative structure that divides visual narrative systems across a modality of expression (i.e., graphics), organized using combinatorial systems (grammar), which express conceptual information (meaning), and it does so across both the unit and sequence levels (Figure 1). Aspects of modality involve both the unit (i.e., graphic structure) and sequence (layout) levels. The combinatorial systems organize form and meaning for both units (morphological structure) and sequences (narrative structure). Aspects of the meaning involve conceptual and event structures. Contrary to the myth of transparency based on the ecological theory of perception such as Gibson (1979), that visual narrative requires basic perceptual or event cognition alone, visual narrative systems differ in many ways and have opaque aspects, even when setting aside the modality differences of filmic and drawn narratives. For example, comics and illustrated stories vary in many conventions (such as layout, morphology, and narrative structure), despite both conveying meaning across sequential images. Figure 1 mainly features drawn visual narratives, but dynamic visual narratives appear as film, which uses actual percepts instead of drawings.

In addition to the determinacy or the transparency created by the stories' or characters' goal-directed actions and conceptual or event meanings, the difficulties or the opaqueness from the indeterminacy or

Figure 1. A model of the structures involved in visual narratives from Cohn & Magliano (2020)

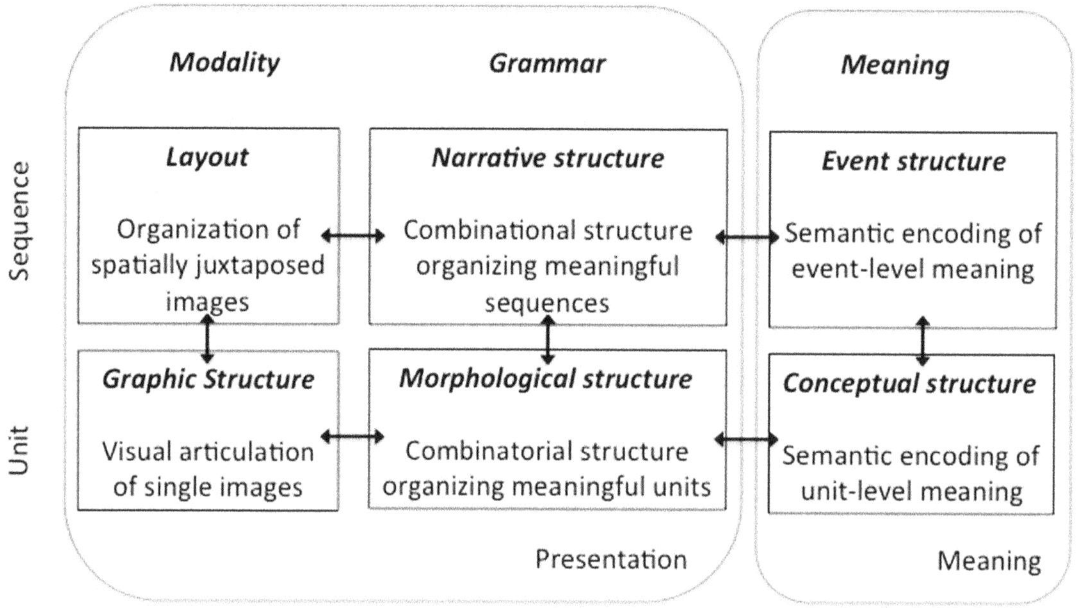

the plurality created by the visual elements of rhetoric are essential issues of film cognition (Kanai, 2017). Based on the indeterminacy or the plurality, the film can emphasize the rhetorical aspects of presentation. The aspects of rhetoric can cause cognitive effects that do not arise only through comprehending a story and an overall meaning based on continuities but which instead arise through the audiovisual situation or the cognitive transition itself.

Previous studies argued for various continuities and cognition (e.g., Berliner & Cohen 2011; Magliano & Zacks, 2011; Smith 2012). The *continuity* means that the flow from shot to shot in the film is smooth. The continuity is an illusion created in the mind of a viewer (Smith, 2011). Thanks to spatial continuity, a viewer can understand characters' locations. Moreover, continuity editing supports and sustains temporal manipulation as well. Film continuity is managed in the service of telling a story and thus creating a sense of continuity and discontinuity of events within the story world. In this perspective, film discontinuity is regarded as an obstacle to story comprehension. Scene perception and event comprehension theory is applied to continuity based visual narratives (Loschky et al., 2020). In the other perspective, the discontinuities or the plurality in a film themselves are important for the quality and artistry. Those aspects can cause intensive cognitive effects from the difficulties cognition.

Previous studies of film cognition showed how viewers understood event-based stories, because the rhetoric of a film is generally accomplished by telling some story through explicit or implicit events. However, a film is not always created with the purpose of representing a story's events: the rhetoric of the film may have a purpose that is not to represent an event-based story. Philosopher Deleuze (1990) said that after World War II, films had stopped being "movement-images" and instead had become "time-images." There are films for which it is necessary to realize a film's structure is independent of the story's events. In such a film, a purely audiovisual situation emerges, and the point where force is applied is placed in the description itself, in addition to the actions and goals of the characters in events. Here, a film does not always have an overall meaning; and the subtle element of each shot becomes important. With regards to the purpose of the rhetoric of the film, there is a difference between the movement-image and the time-image, as Deleuze (1990) said. Then, the parts where emphasis is placed on the elements of the film structure are different. Deleuze (1985) argued that irrational cuts in the film are related to the time-image.

Furthermore, documentary film is not accomplished by the purpose of telling some story. For example, films directed by Frederick Wiseman have no main characters or events. Films such as *Ex Libris: The New York Public Library* (2017) directed by Wiseman consists of everything about the library. In this case, event structures and overall story are not so important. *Ex Libris: The New York Public Library* has a consistent narrative based on the film's associations between the objects, the moods, and the events rather than on the continuity of the events.

In the other case, art film is not accomplished by the determinacy of the event or conceptual structure. For example, films directed by Jean-Luc Godard use visual effects for unclear meanings. Films such as *The Image Book* (2018) directed by Godard consist of images emphasizing indeterminacy and plurality.

Slow cinema or contemplative cinema using recent digital cameras such as *The Woman Who Left* (2016) directed by Lav Diaz and *Stray Dogs* (2013) directed by Tsai Ming-liang feature long continuous shots. Çağlayan (2016) discussed that slow cinema can transform boredom into an aesthetically rewarding experience through its celebration of dead time and foregrounding dramatic ambiguity. The long take performs productive functions of boredom and mind wandering by dedramatizing narrative structures. The long, continuous shot can amplify the ambiguity of the meaning of presentation.

In order to discuss and explore the entire visual narrative concept such as film, not only the viewers' scene perception and event comprehension based meaning creation mechanism, the difficulty, the indeterminacy, the diversity, and the ambiguity aspects of visual narrative caused by the cutting and the rhetorical transition techniques for the cognitive transition are also to be considered.

RHETORICAL CUTTING TECHNIQUES

Based on the rhetorical cutting techniques to make the indeterminacy from the irrational relationships of time or character in continuous shots, viewers can relax constraints on story comprehension based on the difficulty of cognition and reset their viewpoints to better view subtle elements, such as individual shots details.

The rhetorical cutting techniques for narrative space and time in the film can cause difficulty cognition and the viewer's cognitive transition from event-based, story-driven determinacy to presentation-based, rhetoric-driven indeterminacy. Based on the cognitive transition, cognitive effects such as intensive reality effects can emerge.

Cutting techniques can be used to create indeterminacy between (a) the rhetorical elements of two shots (Cutting 1); (b) events and images (Cutting 2); or (c) sounds and images (Cutting 3; Kanai & Ogata, 2004a). The indeterminacy created through the cutting techniques described above is used in many artistic films, television advertisements, and music videos.

This chapter primarily discusses visual aspects such as Cutting 1 and Cutting 2. For many films, Cutting 1 can make the indeterminacy between two events merge in continuous shots. Cutting 2 can make indeterminacy between presentation and meaning. Figure 2 is derived from Figure 1; it illustrates the role of the cutting techniques. Film directors such as Wiseman only use Cutting 1. On the other hand, film directors such as Godard use both Cutting 1 and Cutting 2. Additionally, the sudden appearance of the long continuous shot can involve Cutting 2.

Figure 2. A model of the structures involved in visual narratives with the cutting techniques

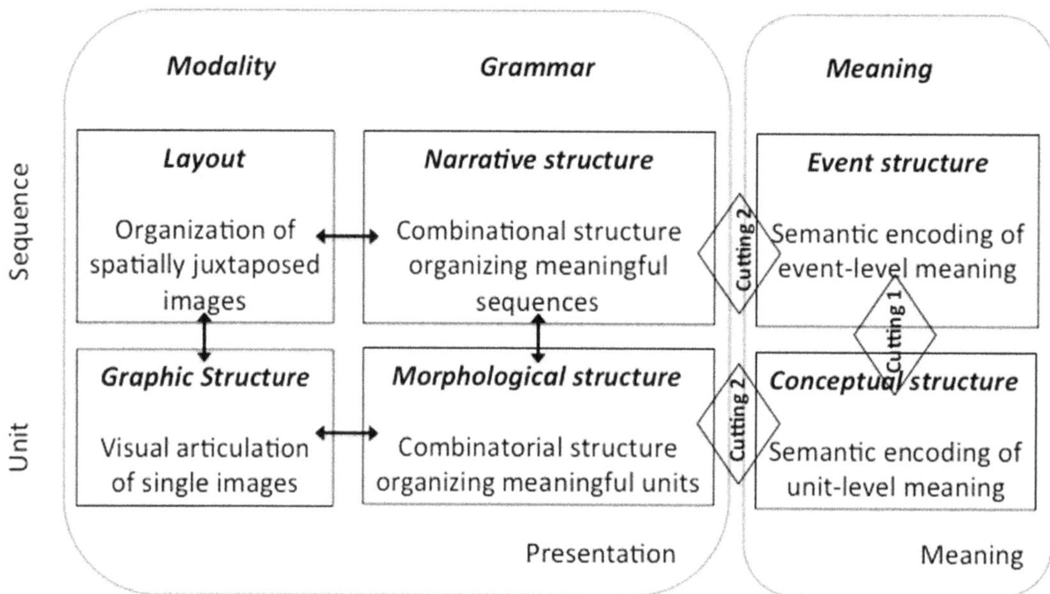

For example, *The Image Book* (2018) directed by Godard used various cutting techniques. Some continuous shots of the film had no event relationship (Cutting 1). *The Image Book* consisted of many old films' retouched shots, which created a discontinuity or an ambiguity regarding the events and images (Cutting 2).

To cause intensive cognitive effects, the discontinuity of events must be emphasized in the rhetoric. A mix of identical and different elements in continuous shots using the rhetorical cutting techniques is important for the cognitive effects of the rhetoric of the film, such as the difficulty.

Rhetorical cutting techniques resemble memories of the past (Kanai, 2019). Humans cannot remember one consistent story from the past. Especially, disaster memories may consist of many fragments like the rhetorical cutting techniques with mental difficulty. Films about disasters also may consist of many fragments with the rhetorical cutting techniques for the intentional mental difficulty. Disaster narratives not only consist of consistent stories, memories, and meanings. Conversely, in order to appropriately relate to the disaster narrative film with the rhetorical cutting techniques, viewers must relax their constraints in terms of the need to comprehend continuous events and can feel the strength and the pain of the disaster.

RHETORICAL TRANSITION TECHNIQUES

For the rhetorical cutting techniques, the discontinuity of events is emphasized. On the contrary, the rhetorical transition techniques keep the continuity of events. The rhetorical transition technique is based on the continuity editing, including the plurality, the diversity, and the ambiguity other than the meaning of consistent events.

Rhetorical transition techniques can change the flow of cognitive processing for narrative. The changing flow can be reduced to two categories. For many films, Transition 1 can make the plurality, the diversity, and the ambiguity between two events emerged in continuous shots. Transition 2 can make the plurality, the diversity, and the ambiguity between presentation and meaning. Figure 3 is derived from Figure 1; it illustrates the role of the transition techniques. The rhetorical transition techniques for narrative space and time in the film can cause the viewer's cognitive transition from event-based and story-driven to presentation-based and rhetoric-driven processing with cognitive effects from the story. In this case, event-based, story-driven processing is not emphasized.

Figure 3. A model of the structures involved in visual narratives with the transition techniques

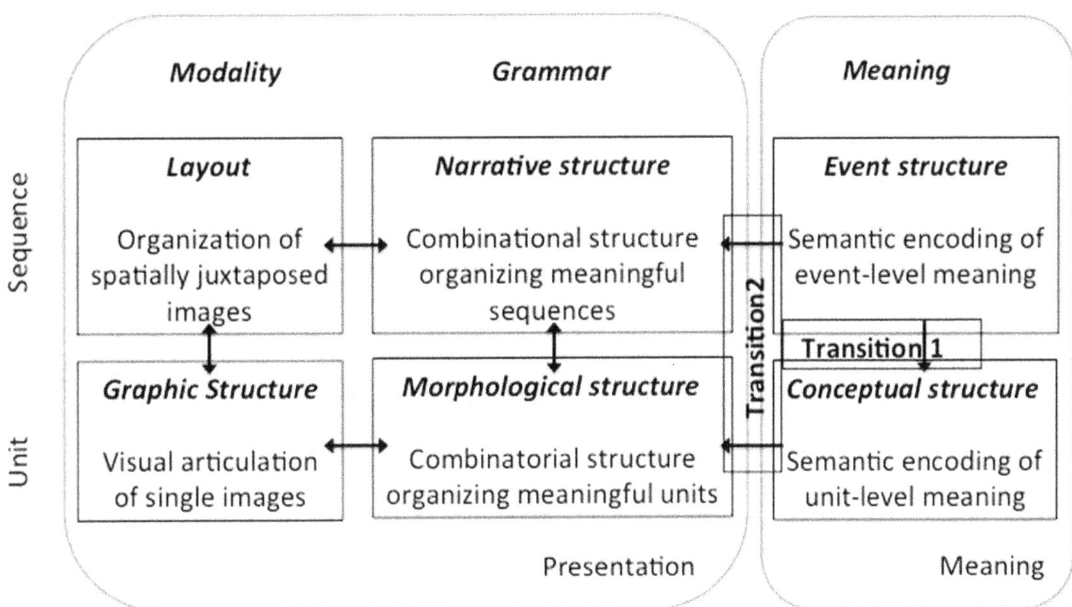

Characters' details, environment, nearby objects, elements, cognition, and rhetorical explanation can be the rhetorical transition techniques, which can generate the plurality and the diversity with regards to time, space, characters' actions, and meaning in continuous shots. For example, the opening scene from *Touch of Evil* (1958) directed by Orson Wells consists of a single continuous long shot lasting three minutes and twelve seconds. Hutson et al. (2017) investigated the levels of attentional synchrony were less in the continuous long shot of *Touch of Evil* than in the highly edited sequence. *Touch of Evil*

opens on a close-up of someone setting a time bomb. The time bomb is then placed into the trunk of a car, after which a couple unknowingly gets into the car and drives off, as the camera follows them. Then, a second couple walking down the street is introduced, and the camera begins to follow them with the car always lurking around them. The relationship between the time bomb and the second couple can change the flow of cognitive processing for narrative structure (Transition 1). *Touch of Evil* is lacking cuts but it includes numerous objects from a Mexico–US border town. Characters' environments and nearby objects are transition techniques to reset viewers' viewpoints from event understanding to the plurality, the diversity, and the ambiguity of presentation (Transition 2). Generally, the heavy-handed context manipulation can change the nature of viewers' event models of the film. Despite large differences in viewers' understanding and event models of the film, there were small differences in attentional selection. Using the attentional selection nature, the transition techniques can reset the viewpoint to subtle elements. For cognitive science, attentional synchrony itself is an important matter. The diverse attention can cause the viewer's cognitive transition from story-driven to rhetoric-driven processing with cognitive effects from the story, which is an important matter for narratology.

For another example, *Pickpocket* (1959) directed by Robert Bresson frequently used close-up shots of hands picking pockets. The close-up shots of hands can be the rhetorical transition technique. First, using the attentional selection nature, the close-up shots of hands can reset the viewers' viewpoints from the whole event structure to picking pockets events only (Transition 1). Second, the close-up shots of hands can reset the viewers' viewpoints from pocket-picking events to presenting the characters' hands themselves (Transition 2). Bordwell (2007) discussed two forms of editing: analytical editing and constructive editing. *Analytical editing* is based on the continuity of events. On the other hand, Bordwell (2012) argued for constructing editing based on *Pickpocket*. Bresson adhered to the idea that a film is only a sum of the elements that are necessary for visual narrative, such as the shots of the hands. Rhetorical transition techniques can be the components of constructive editing. Other than the close-up shots of hands, the close-up shots of paper money or a fake bill used in *Money* (1983) directed by Bresson also can be rhetorical transition techniques.

COGNITIVE EFFECTS FROM RHETORICAL CUTTING AND TRANSITION TECHNIQUES

Rhetorical cutting techniques and rhetorical transition techniques can emerge various cognitive effects such as reality effects and nostalgia effects, including the difficulty, which can bridge the gaps between cognitive science, artificial intelligence, and narratology.

Previous studies on film cognition primarily discussed the first type of reality effects related to the condition of being natural and less deviated with regard to movements (Visch & Tan, 2009; Rooney et al., 2012). The first type of reality effects are especially realized by Transition 2. The second type of reality effects, related to the unfamiliar situations, were argued by the Russian Formalists (Shklovsky, 1965). The indeterminacy from the rhetorical cutting techniques, especially Cutting 2, caused the second type of reality effects (Kanai, 2006).

The indeterminacy, the plurality, and diversity derived from the rhetorical cutting or transition techniques creates new realities, nostalgia effects, or meanings related to unknown memories and affects that do not arise through event comprehension alone (Kanai, 2018). Particularly, the indeterminacy can emerge through the sudden appearance of an event unit (Cutting 1). In this case, a viewer sometimes cannot apply the determinacy from the consistent events. The scene perception and event comprehension theory cannot be applied to indeterminacy based visual narratives. Therefore, attention to the audiovisual situation itself or the subtle elements is enhanced (Cutting 2). This occurs because the indeterminacy generates new cognitive effects related to unknown memories and affects that do not arise through event comprehension and meaning.

For visual narratives, nostalgia effects are especially important. The connections to the past, the memories, and the present always involve the indeterminacy regarding the event's meaning and the ambiguity regarding the place and time. The rhetorical cutting and the transition techniques and nostalgia effects are more closely related to each other. In general, nostalgia refers to a longing for the past. Past images in a film not associated with a specific event can evoke many kinds of nostalgia based on the viewer's cognition. Nostalgia has mainly been argued by researchers like Stern (1992) as a longing for a good but lost personal and historical past. Holbrook and Schindler (1991) defined it as a preference for objects (people, places, or things) that were more common when one was younger. Nostalgia can emerge from various different types of presentation. Some nostalgia is evoked from the loss of a place's good past. Other nostalgia is evoked from the unforgettably bad past of a place. Furthermore, nostalgia may emerge suddenly with the determinacy of the place and the time, and the indeterminacy of rhetorical elements such as irrational shots. Indeterminacy based nostalgia does not always follow comfortable cognition.

To manage various kinds of nostalgia, Naito and Kanai (2012a, 2012b) redefined *nostalgia* as emotions arising from the rise and fall in other times in the place. Kanai (2016, 2019), classified the rhetoric of a film into four kinds of nostalgia—including non-nostalgia. Kanai focuses on three kinds of nostalgias regarding a certain place, as suggested by Davis (1979): "simple nostalgia," "reflexive nostalgia," and "interpreted nostalgia." Simple nostalgia emerges from the good past with the rhetorical transition techniques. *Reflexive nostalgia* emerges from both the good and bad past, with the rhetorical cutting and transition techniques. *Interpreted nosta*lgia emerges from the reality or the uncertainty of the past with the rhetorical cutting techniques.

Simple nostalgia emerges from ambiguous affection for the past of a place relating to some events based on the rhetorical transition techniques (Transition 1 and Transition 2). By contrast, reflexive and interpreted nostalgia emerge from the reality and the uncertainty about the past of a place as related to unfamiliar situations, with the cognitive transition and alienation effects based on the rhetorical cutting techniques (Cutting 1 and Cutting 2). The editing, recording, and photography of a film are not always subordinate to the determinacy of the overall events and meanings.

Nostalgia can emerge from every film. Nonetheless, some film directors, whose purpose is to create timeless films located in unknown places, do not want to generate nostalgia effects for viewers. Non-nostalgia based narratives can be composed using the strategy for the nostalgia effects. Moreover, every film has not only past aspects but also present aspects, as the film may be viewed in the present time. To emphasize the present aspects of films, the strategy for the non-nostalgia cognitive effects must be involved.

As stated above, the focuses of the rhetorical cutting and the transition techniques are on presentation of the narrative, the mood and the object associations, or the reality related to the unfamiliar situations; rather than on the events and the meaning in the story. The indeterminacy and the ambiguity of such associations and situations cause cognitive effects that do not arise in the process of comprehending continuous events only. Kanai (2001a, 2001b) indicated that when a viewer can relax his or her constraint with regard to the need to comprehend a story, this strengthens those cognitive effects, which do not arise in the process of event comprehension.

NARRATIVE SIMULATION

Narrative simulation, including the strategy for the rhetorical cutting and the transition techniques, can broaden the potential for narrative generation—especially from the perspective of the indeterminacy and the ambiguity. Although event-based plots and characters' goal-directed actions are essential for visual narrative cognition and generation, events comprehension and meanings can also constrain this (Ogata & Kanai, 2010). The rhetorical cutting and the transition techniques, which can relax constraint with regards to the event comprehension and meaning, can be the main components of visual narrative simulation.

The author had proposed a film rhetoric composition system based narrative simulation (Kanai & Ogata, 2004b; Kanai, 2018). The system can be used for narrative simulation with the rhetorical cutting techniques. The authors revised the system, which can compose nostalgia-based narrative rhetoric with the cutting and transition techniques. The user can decide which the rhetorical cutting and transition techniques to use for making discontinuities and continuities. The system then retrieves the shots from the database that fits the chosen strategy: Cutting 1, Cutting 2, Transition 1, and Transition 2. The shots are edited as a sequence of shots and outputted. By using this computational approach, we could make many different films from the same shots and use them to study the nature of the visual narrative. Computational narrative simulation can be argued as being the choice of the cutting and the transition techniques used as a strategy in order for the viewer's cognitive processes. The generated films can reveal both the event-based story and the broad presentation aspects of the rhetoric, narrative, and cognition of the past and the film. Without using event structures and goal-oriented characters, the nostalgia-based visual narrative could be mainly simulated.

The montage theory of Eisenstein (1942, 1949) can be used as a tool for more controlled and expanded narrative simulation with the transition techniques. Kanai & Kodama (2010) created 16 kinds of one-minute films using 90-second sequences of 18 shots each from *The Passion of Joan of Arc* (1928) directed by Carl Th. Dreyer, using the narrative simulation mechanism, and investigated the difference between the rhetorical transition aspect and the rhetorical cutting aspect in the films by use of a psychological experiment. *The Passion of Joan of Arc* is one of the most famous of art films containing no spoken dialogue. The shots from *The Passion of Joan of Arc* are famous for their usage of close-up face shots. The close-up face shots of Joan of Arc themselves can be a transition technique (Transition 1), so that quality makes them suitable for serving as the source of narrative simulation. In this case, the close-up face cuts from the sequence can make the continuity and the determinacy of the character (i.e., Joan) and the location (the cell; Transition 2). By contrast, the irrationally cut films from the sequence can make the discontinuity for the indeterminacy of the various characters or the various locations (Cutting 1). Also, some characters' shots with a slanted angle can make the indeterminacy between presentation and meaning (Cutting 2). The Kanai & Kodama conducted the experiment based on the hypothesis "the rhetorical cutting aspect can enhance the viewpoint diversity of the viewer" and "the rhetorical transition aspect can enhance the viewer's immersion into the story, and the viewer's impression of character." The results somewhat proved the hypothesis. The cutting aspect for the indeterminacy-based montage can enhance viewer viewpoint diversity. Additionally, the transition aspect for the determinacy-based montage can enhance the viewer's immersion into continuous events.

FUTURE RESEARCH DIRECTIONS

In this chapter, a model of visual narrative structure provided by Cohn & Magliano (2020) is expanded for explaining the cognitive transition caused by the rhetorical transition techniques and the rhetorical cutting techniques. An expanded visual narrative structure can be used for doing narrative simulation. Future research directions of the main themes in this chapter would be as follows.

One hypothesis to be discussed more is that many of the cognitive processes that support understanding in literature, also support the processing of visual narratives such as film. Only a few studies have directly explored the extent to which processes are common across textual, filmic, or drawn narratives. Filmic narrative only is mainly discussed in this chapter. The rhetorical transition techniques can be the components of constructive editing. The details of the relationship between rhetorical transition techniques and constructive editing can be discussed more.

This chapter does not discuss the difference between fiction and reality, for example, the documentary. All visual narratives such as films include both fiction and reality aspects. Film continuity for both fiction and reality are illusions created in viewers' minds. Both fiction and reality-based visual narratives can evoke nostalgia effects. The question of the difference between fiction and reality effects is important for both cognitive science and narratology.

Emotional and affective aspects other than nostalgia and difficulty are also important. Emotional and affective cognition emerge from both event-based, story-driven processing and presentation-based, rhetoric-driven processing. For example, the beauty and ugliness effects of visual narrative mainly emerged from presentation-based, rhetoric-driven processing. Also, the beauty and ugliness effects of visual narrative affect and relax event-based, story-driven processing.

Moreover, the beauty and ugliness of the visual narrative can involve reverse-transition techniques, which have not been discussed in this chapter. Viewers' scene perception and event comprehension based visual narrative comprehension process are still important matters for cognitive science and narratology. The rhetorical reverse-transition techniques for narrative space and time in the film can cause the viewer's cognitive transition from being event-based and story-driven through being presentation-based and rhetoric-driven back to being event-based story processing again. In this cycle based on rhetorical reverse-transition techniques, viewers' scene perception and event comprehension-based visual narrative comprehension process are enhanced. The rhetorical reverse-transition techniques can be a key concept for how people comprehend visual narrative.

Finally, auditory information is vital to film cognition and making (Bordwell, 1985). Cutting and transition aspects of auditory information can be focused more.

CONCLUSION

In order to discuss and explore the entire narrative concept and generation, not only the viewers' scene perception and event comprehension based visual narrative comprehension and generation mechanism, the other aspects caused by the rhetorical cutting techniques and the rhetorical transition techniques are also to be considered.

The transition and the interaction between event-based, story-driven processing and presentation-based, rhetoric-driven processing with the rhetorical cutting techniques and the rhetorical transition techniques are one of the most important aspects of film cognition. The transition and the interaction are only analyzed by a cognitive science approach with narratology.

Narratology without a cognitive science approach cannot address the transition and interaction. In addition, a cognitive science approach without narratology cannot reach the difficulties cognition in the transition and the interaction. Furthermore, narratology with a cognitive science based artificial intelligence can only address the transition and the interaction, including cognition difficulties, and can do several narrative simulations based on cognitive effects for content generation.

Although the scene perception and event comprehension based visual narrative comprehension is an important question for the field of cognitive science or artificial intelligence, the other aspects of narrative are also important. The rhetorical cutting techniques and the rhetorical transition techniques can emerge various cognitive effects such as reality effects and nostalgia effects, including difficulty. Not only the determinacy based narrative, the indeterminacy, the diversity, and the ambiguity on narrative can bridge the gaps between cognitive science, artificial intelligence, and narratology.

REFERENCES

Avrahami, J., & Kareev, Y. (1994). The emergence of events. *Cognition, 53*(3), 239–261. doi:10.1016/0010-0277(94)90050-7 PMID:7842635

Bakhtin, M. (1984). *Problems of Dostoevsky's poetics* (C. Emerson, Trans.). University of Minnesota Press. (Original work published 1963) doi:10.5749/j.ctt22727z1

Berliner, T., & Cohen, D. J. (2011). The illusion of continuity: Active perception and the classical editing system. *Journal of Film and Video, 631*(1), 44–63. doi:10.5406/jfilmvideo.63.1.0044

Bordwell, D. (1985). *Narration in the fiction film*. University of Wisconsin Press.

Bordwell, D. (2007). *Poetics of cinema*. Routledge.

Bordwell, D. (2012). *Constructive editing in Robert Bresson's Pickpocket* [Video]. Vimeo. https://vimeo.com/52312154

Brecht, B., & Willett, J. (1964). *Brecht on theatre: The development of an aesthetic*. Hill and Wang.

Bresson, R. (1959). *Pickpocket* [Film]. A. Delahaie (Producer). France: Compagnie Cinématographique de France.

Bresson, R. (1983). *Money* [Film]. J. M. Henchoz & D.T. Plantier (Producer).

Çağlayan, E. (2016). The aesthetics of boredom: Slow cinema and the virtues of the long take in once upon a time in Anatolia. *Projections, 10*(1), 63–85. doi:10.3167/proj.2016.100108

Chatman, S. (1990). *Coming to terms: The rhetoric of narrative in fiction and film*. Cornell University Press.

Cohn, N. (2013). Visual narrative structure. *Cognitive Science, 34*(3), 413–452. doi:10.1111/cogs.12016 PMID:23163777

Cohn, N. (2014). You're a good structure, Charlie Brown: The distribution of narrative categories in comic strips. *Cognitive Science*, *38*(7), 1317–1359. doi:10.1111/cogs.12116 PMID:24646175

Cohn, N., & Magliano, J. P. (2020). Editors' introduction and review: visual narrative research: an emerging field in cognitive science. *Topics in Cognitive Science*, *12*(1), 197–223. doi:10.1111/tops.12473 PMID:31865641

Corner, J. (2018). Documentary spectatorship and the navigation of "difficulty.". In C. Brylla & M. Kramer (Eds.), *Cognitive theory and documentary film* (pp. 59–73). Palgrave Macmillan. doi:10.1007/978-3-319-90332-3_4

Cutting, J. E., Brunick, K. L., & Canden, A. (2012). Perceiving event dynamics and parsing Hollywood films. *Journal of Experimental Psychology*, *38*(6), 1476–1490. PMID:22449126

Cutting, J. E., & Iricinschi, C. (2014). Re-presentations of space in Hollywood movies: An event-indexing analysis. *Cognitive Science*, *35*, 1–23. PMID:25087776

Davis, F. (1979). *Yearning for yesterday: A sociology of nostalgia*. Free Press.

Deleuze, G. (1985). *L' Image-temps*. Les Editions de Minuit.

Deleuze, G. (1990). *Pourparlers*. Les Editions de Minuit.

Diaz, L. (2016). *The woman who left* [Film]. L. Diaz (Producer).

Dreyer, C. T. (1928). *The passion of Joan of Arc* [Film]. France: Société générale des films.

Eisenstein, S. M. (1942). *The film sense* (J. Leyda, Trans.). Harcourt Brace.

Eisenstein, S. M. (1949). *Film form* (J. Leyda, Trans.). Harcourt Brace.

Genette, G. (1972). Discourse du recit: Essai de method. [Narrative discourse: An essay in method] In *Figures III*. Seuil.

Gibson, J. J. (1979). *The ecological approach to visual perception*. Houghton Mifflin.

Godard, J. L. (2018). *The image book* [Film].

Hasumi, S. (2014). *Bovary fujin ron* [A Theory of "Madame Bovary"]. Chikumashobō.

Holbrook, M. B., & Schindler, R. M. (1991). Echoes of the dear departed past: Some work in progress on nostalgia. In R. H. Holman & M. R. Solomon (Eds.), *Advances in Consumer Research, 18* (pp. 330–333). Association for Consumer Research.

Hutson, J. P., Smith, T. J., Magliano, J. P., & Loschky, L. C. (2017). What is the role of the film viewer? The effects of narrative comprehension and viewing task on gaze control in film. *Cognitive Research: Principles and Implications*, *2*(1), 46. doi:10.118641235-017-0080-5 PMID:29214207

Jhale, A., & Young, M. R. (2010). Cinematic visual discourse: Representation, generation, and evaluation. *IEEE Transactions on Computational Intelligence and AI in Games*, *2*(2), 69–81. doi:10.1109/TCIAIG.2010.2046486

Kanai, A. (2001a). The story as a constraint on film. *Proceedings of the Third International Conference on Cognitive Science*, 503–506.

Kanai, A. (2001b). Cognitive process model toward the rhetoric of the film. *Cognitive Studies*, *8*(2), 139–150.

Kanai, A. (2006). Reality of unfamiliar situations on moving image. *Proceedings of 19th Congress of the International Association of Empirical Aesthetics*, 800–803.

Kanai, A. (2016). Non-story, nostalgia, and film cognition: Nostalgia-based narrative rhetoric composition. In T. Ogata & T. Akimoto (Eds.), Computational and Cognitive Approaches to Narratology (pp. 376–390). Hershey, PA: Information Science Reference (IGI Global).

Kanai, A. (2018). Narrative simulation for film rhetoric composition with or without story and nostalgia effects. In T. Ogata & S. Asakawa (Eds.), Content generation through narrative communication and simulation (pp. 148–161). Hershey, PA: Information Science Reference (IGI Global). doi:10.4018/978-1-5225-4775-4.ch003

Kanai, A. (2019). Past, non-story narrative film, and nostalgia. In T. Ogata, & T. Akimoto (Eds.), Post-narratology through computational and cognitive approaches (pp. 277–291). Hershey, PA: Information Science Reference (IGI Global). doi:10.4018/978-1-5225-7979-3.ch006

Kanai, A., & Kodama, K. (2010). Film edit design: Story and irrational cutting. *Cognitive Studies*, *17*(3), 444–458.

Kanai, A., & Ogata, T. (2004a). Non-story processing on the film rhetoric composition system. *Proceedings of 18th Congress of the International Association of Empirical Aesthetics*, 433–436.

Kanai, A., & Ogata, T. (2004b). Aspect of non-story processing and film rhetoric composition in the narrative generation mechanism. *Proceedings of the Ninth International Symposium on Artificial Life and Robotics*, 162–165.

Loschky, L. C., Larson, A. M., Smith, J. T., & Magliano, J. P. (2020). The scene perception and event comprehension theory (SPECT) applied to visual narratives. *Topics in Cognitive Science*, *12*(1), 311–351. doi:10.1111/tops.12455 PMID:31486277

Magliano, J. P., & Zacks, J. M. (2011). The impact of continuity editing in narrative film on event segmentation. *Cognitive Science*, *35*(8), 1489–1517. doi:10.1111/j.1551-6709.2011.01202.x PMID:21972849

Naito, Y., & Kanai, A. (2012a). The narrative structure of nostalgia cognition and film. In *Proceedings of the International Conference on Cognitive Science '12*. Academic Press.

Naito, Y., & Kanai, A. (2012b). Nostalgia and film creation. In *Proceedings of the 26th Annual Conference of the Japanese Society for Artificial Intelligence (1N1-OS-1a-2)*. Yamaguchi, Japan: Academic Press.

Ogata, T., & Asakawa, S. (Eds.). (2018). Content generation through narrative communication and simulation. Hershey, PA: Information Science Reference (IGI Global). doi:10.4018/978-1-5225-4775-4

Ogata, T., & Kanai, A. (2010). *Monogatariron no jyohogaku jyosetsu: Monogatariseisei no shiso to gijyutsu wo megutte* [An introduction to informatics of narratology: Towards the thoughts and technologies of narrative generation]. Gakubun-sha.

Propp, V. Y. (1968). *Morphology of the Folktale* (L. Scott, Trans.). University of Texas Press. (Original work published 1928)

Rooney, R., Benson, C., & Hennessy, E. (2012). The apparent reality of movies and emotional arousal: A study using physiological and self-report measures. *Poetics*, *40*(5), 405–422. doi:10.1016/j.poetic.2012.07.004

Rumelhart, D. E. (1975). Notes on a schema for stories. In D. G. Bobrow & A. Collins (Eds.), *Representation and understanding: Studies in cognitive science* (pp. 211–236). Academic Press. doi:10.1016/B978-0-12-108550-6.50013-6

Shklovsky, V. (1965). Art as technique. In *Russian formalist criticism: Four essays* (L. T. Lemon & M. J. Reis, Eds. & Trans.). University of Nebraska Press. (Original work published 1917)

Smith, T. J. (2012). The attentional theory of cinematic continuity. *Projections*, *6*(1), 1–27. doi:10.3167/proj.2012.060102

Stern, B. B. (1992). Historical and personal nostalgia in advertising text: The fin de siècle effect. *Journal of Advertising*, *21*(4), 11–22. doi:10.1080/00913367.1992.10673382

Thorndyke, P. W. (1977). Cognitive structures in comprehension and memory of narrative discourse. *Cognitive Psychology*, *9*(1), 77–110. doi:10.1016/0010-0285(77)90005-6

Tsai, M. (2013). *Stray Dogs* [Film]. V. Wang (Producer).

Visch, V. T., & Tan, E. S. (2009). Categorizing moving objects into film genres: The effect of animacy attribution, emotional response, and the deviation from non-fiction. *Cognition*, *110*(2), 265–272. doi:10.1016/j.cognition.2008.10.018 PMID:19118823

Welles, O. (1958). *Touch of evil* [Film]. A. Zugsmith (Producer). USA: Universal Pictures.

Wiseman, F. (2017). *Ex libris: The New York public library* [Film].

KEY TERMS AND DEFINITIONS

Cognitive Transition: The cognitive processing change of a narrative focus, especially the transition from story-based to non-story-based forms of narrative.

Discontinuity Editing: The irrational flow from shot to shot in a film.

Indeterminacy: The irrational relationships between time and character or between place and time within continuous shots.

Mental Difficulty Cognition: Hard cognition caused by the indeterminacy or the divergence from the presentation.

Narrative Simulation: A process of composing various kinds of narrative rhetoric and of testing the cognitive effects of a narrative.

Rhetorical Cutting Technique: A technique used to irrationally cut film with the intent of creating an irrational relationship between adjacent segments.

Rhetorical Transition Technique: A technique used to generate the plurality, the diversity, and the ambiguity with regards to time, space, or characters' actions in continuous shots.

Chapter 2
Multi-Disciplinary Paths to Actor-Centric Non-Player Character Emotion Models

Sheldon Schiffer
Georgia State University, USA

ABSTRACT

Video game non-player characters (NPCs) are a type of agent that often inherits emotion models and functions from ancestor virtual agents. Few emotion models have been designed for NPCs explicitly, and therefore do not approach the expressive possibilities available to live-action performing actors nor hand-crafted animated characters. With distinct perspectives on emotion generation from multiple fields within narratology and computational cognitive psychology, the architecture of NPC emotion systems can reflect the theories and practices of performing artists. This chapter argues that the deployment of virtual agent emotion models applied to NPCs can constrain the performative aesthetic properties of NPCs. An actor-centric emotion model can accommodate creative processes for actors and may reveal what features emotion model architectures should have that are most useful for contemporary game production of photorealistic NPCs that achieve cinematic acting styles and robust narrative design.

INTRODUCTION

"Constructing characters" is a phrase that infers distinct meanings for two participants in the creative process of computer-based media. On one hand, for the narrative architect of video games or other kinds of narrative computational media, it is a semiotic process of fiction authoring where the character designer provides a written personal history. Three-dimensional or two-dimensional models of anthropomorphic shape combined with voice can also suggest agency in a game space. A weapon-wielding muscular humanoid with big bright eyes is well equipped for video game combat and all of the emotional expression players associate with fighting. Once audio-visual elements are programmed to react to user input, these elements can signify to the observant player an imagined persona. On the other hand, for the developer of video game computer code, "constructing characters" is a process of designing a system that uses

DOI: 10.4018/978-1-7998-4864-6.ch002

quantitative data derived from the game program, the computer operating system or from player input data to control or trigger character animation and voicing such that the player experiences the presence of a seemingly intelligent cohesive character identity. The result of the work of the designer and the developer is a composite signified that evolves in the player's mind over the time of game play.

The manner of construction for narrative architects of video games depends on mental processes of player participation. Over the course of game play time, the player may observe actions and behaviors of Non-Player Characters (NPCs) so that a pre-game play biography and an in-game "alterbiography" (Calleja 2009) combines the NPC's pre-game past with the evolving NPC actions the player witnesses or learns through game interaction since the start of the game. The manner of character construction for video game developers depends on computer languages whose frameworks contain data structures (primitives, classes, objects) and data behaviors (methods, functions) that can trigger and manipulate unique animations of the three-dimensional mesh model and its sound emanations (usually a voice) in ways that resemble the player's understanding of human and animal emotional expressions. These NPC animations and sounds must be recognizable by the player as specific to the NPC's type as a fictional narrative agent (human or non-human) and consistent with the character's role within the world of the game.

The study of character construction coincides with the related research in other disciplines. The late twentieth century coincidentally delivered strains of thought from disciplines that sought to provide taxonomies of two human endeavors – storytelling and emotion expression. Narratology addressed the former and evolved from literary theory and folklore studies to describe systematically the human perception and representation of stories in various media. The categorization of characters within stories based on a typology of roles and emotion sets afforded to those roles is one specialization within narratology. Cognitive psychology and its subdiscipline, Computational Cognitive Psychology, evolved as a reaction to Behaviorism and as an alternative explanation to the mental and emotional processes that drive human behavior. The categorization of human emotions as well as their neurological processes is one subdiscipline that Cognitive psychologist frequently consider. The two disciplines converge in computer game design because game character designers and game code developers both use models from which characters can be efficiently produced. These characters and their behaviors can be embodied as preconfigured audio-visual animation and sound synthesis systems for "static" characters or can be used in-game to spawn procedurally generated characters or behaviors.

This research focuses on NPC modeling rather than Player Character modeling for the benefit of taking a simpler problem before tackling the more complex Player Character with its additional set of variables that describe each player's unique persona and context. Daniel Vella initialized a discussion of the Player Character's implications on game narrative that can apply to related studies of NPCs (Vella 2014). A discussion of NPC modeling benefits from the character theories of Narratology and the emotion theories of Computational Cognitive Psychology because both examine the behavior of human or human-like agency. Narratology considers how foregrounding particular NPC behaviors in the context of an audio-visual story system forms the role of a character. These behaviors are a functional necessity for story development and an elaboration in the player's mind. Computational Cognitive Psychology considers how to represent in computational form, a simulation of the internal processing machinery of the emotive part of the human mind so that when one implements and embeds an emotion model in an agent such as an NPC, the sensory input will yield an expressive output as an appropriate behavior that simulates a coherent human-like character. To design an NPC model of emotion for video games, one must consider how emotions in characters are useful in the elaboration of narrative, and how they are generated by actors for use in animated characters.

BACKGROUND

Whether one considers video games as narratives or narrative systems, the fact a player *moves* through the time and space of a game world provides sufficient conditions to apply the frameworks of narratology. For games with NPCs, the player experiences over time the construction of character within her mind. Each interaction with an NPC is a momentary witnessing of in-game action with all of its semiotic unwrapping over the duration of gameplay. Cognitive Psychology recognizes that expression on the body, and in particular as seen on the face, belies the mental properties and emotional states of a person (Ekman 1997). It also communicates emotional meanings from the face of one person into the eyes of another. The latter function of emotion serves narrative expression in the visual arts through the agency of character. The degree to which facial expression effects game narrative depends on the complexity of expression, and therefore the degree of immersion and believability a player may experience. Thus, an effective NPC emotion model must strive for complex expressions.

Narrative Expansion Through Facial Expression Complexity

Seymour Chatman gives a comparative study of Maupassant's short story *Une Partie de campagne* (*A Country Excursion*) (Maupassant 2017) with Jean Renoir's rendition of the story in the film *A Day in the Country* (Braunberger & Renoir 1936). Chatman provides persuasive analyses of how emotions within characters as described by words in a literary narrative are interpreted through cinematography and editing to convey similar ideas in a film adaptation (Chatman 1980). Many of these emotional ideas that Chatman identifies depend on the framing of shots in the film where actors' faces communicate their emotional states. But cinema is not video game. In Renoir's film we are forced to see faces within a frame and in the order of a sequence of images. The dialectical process of montage allows the meanings of faces to be refined and elaborated by the image that precedes a facial closeup and by the image that follows. In video games, only cutscenes lock the player into seeing specifically framed and ordered images of faces. When the camera is released to the volition of the player, there are few if any cuts to other images. Once the camera is free to move in response to the player's will, there is no way for the game designer to be sure that the player sees anything in a particular order. An NPCs facial expression must convey its ideas independent of other objects in the viewing screen. The need for the player to maneuver and see an NPCs face must be embedded in the mechanic of the gameplay. While lighting, music and voice can reinforce the emotional state of an NPC as well as the mood of the game space, the facial expression itself must provide the emotional clarity the game designer intends. If the goals of characters steer their action, then the emotional expression on the face of NPCs reveal how the designer wants the player to understand the character's attitude toward their goals and the consequence of their action at a particular instance. (Later in this discussion, there will be a more robust elaboration of how Computational Cognitive Psychology informs the design of the internal workings of NPCs.)

But is there enough information from the face of an NPC to "move" the narrative and cause the player to decide something significant within a game? The answer depends on the resolution of the animation. A higher resolution facial animation provides the character designer more expressive possibilities. In the past, the concept of visual resolution has described the density of pixels for a given square of screen surface. But *animation resolution* describes the number of vertex groups of an NPCs facial mesh that move distinctly and simultaneously for a unique emotion expression. While this description does not yet provide a quantifiable ratio of dependent and independent variables (like pixels per inch, for example), it does communicate a problem in game animation that effects the player's experience. Game animation must render each frame in real time (usually 60 frames each second). A character with complicated expressions simultaneously moving many parts of a mesh, can demand a lot of parallel calculations of a graphics process, thus challenging the graphics processor to complete all necessary calculations within 1/60th of a second. Cinematic acting for NPCs depends on such computational power. But with complex animated facial expressions, players can observe more detailed expression; they can read more nuanced emotional information when observing NPCs up close. Chatman's close reading examines the polysemic phrasings of Maupassant's story with the polysemic cinematic expression of Renoir's filmed rendition. High resolution animation allows a similarly comparison between a character in a work of literature and its representation in a video game adaptation. Complex facial expression in game characters can allow the player to perceive NPC emotions in ways similar to human actors. NPC facial expression can then be used to effect narrative generation.

Augmenting the expressive capabilities of NPCs has the side effect of augmenting narrative expressivity of video games. Narratologist Uri Margolin defines *characterization* as a "human or human-like individual… capable of fulfilling the argument position in the propositional form DO(X) – that is, a Narrative Agent (=NA), to whom inner states, mental properties (traits, features) or complexes of such properties (personality models) can be ascribed on the basis of textual data" (Margolin 1986). The definition is applicable especially when examining the detailed form of a game character as seen in a single frame of video game play, in a model sheet typically produced in the preproduction stages of character design, or in a written biography of a video game character. In these video game text documents, signification occurs by inference. The character's behaviors are imagined by what its body looks like statically, and what has been reported about its past in the written form. Margolin continues by describing *character-building* in a narrative, which is an "accumulation of a number of traits" or trait clusters "from several successive acts of the NA [Narrative Agent]…" A built character is "a unified stable constellation" of traits experienced in narrative time. While Margolin had in mind primarily literary characters, the player likewise experiences NPCs during gameplay over time and observes their actions in much the same way, gradually building an idea of a character in mind.

When one thinks of actions in video game play, often the idea conjures dramatic movements that propel the body of the character or impact objects the character touches. These actions reveal internal and external states and features of NPCs through body gestures, postures and facial expressions. Facial expressions tell much about the internal mental states and actions of NPCs. Literary representations of emotion take the form either of direct predicate statements where the emotional state becomes the predicate or adverb (e.g. It is angry, charmed or It acts lovingly, scornfully) or are presented indirectly as actions or reactions in the form of a verb (e.g. It *flees* from fear, It *flirts* from arousal). Visual representations of character must use other means. Posture, gesture and facial expressions of emotion are the primary means of presenting emotion in NPCs. Early video games, much like early animation in the cinema were limited to a smaller vocabulary of emotional expressions because the processor speeds and workflow did not easily allow for much variation of expression. But more powerful graphics processors and modularized workflow allow for complex groups of simultaneous animations of different parts of an NPC face to form many more combinations and variations of emotional expression. By expanding the vocabulary of a character's emotive expression, the possible meanings derived by the player from an NPC's expressive system also expand. With more possible meanings entering the space of a narrative, the count of possible player experiences and interpretations grow as well. Thus as NPC emotional expressivity increases in combinatorial complexity with an expanding vocabulary of movements and gestures,

NPCs have the ability to evolve from expressing simpler cartoon-like acting styles where performative meanings are simple, discrete and the expressive variables show few dimensions, into a more robust cinematic acting style where expression is polysemic, often ambiguous and the expressive variables of the face have many more dimensions.

Agency and Acting

The "construction" or "building" of Non-Player Characters by the player within a video game borrows heavily from experiences with other non-ludic art forms, such as literary fiction and films, as well as perceived experiences of simulation systems, such as software. Theatrical and cinematic narrative depend on the appearance of willful actions of characters in a story world. *Agency* in drama is the apparent freedom of a character to act, react, not act, or sublimate desire to achieve a goal within a fictional world. Aristotle and many modern theorists of agency attribute voluntary goal acquisition as an underlying cause of human action (Charles 2017). For NPCs and Player Characters, Janet Murray translates the concept effectively. "Agency is the satisfying power to take meaningful action and see the results of our decisions and choices" (Murray, 1997). The definition of agency has extended itself into software engineering.

As NPCs in video games developed, likewise *agents* were run-time entities within software. The concept of *agency* in software engineering emerged commercially in the mid-1990s when software required run-time decisions to manage its states autonomously. Developers realized that software systems need agents to assess the state and behavior of a system's environment to autonomously manage the system in the background so that desired measurable states are maintained. Agents in software are distinct from other software components because they do not need to wait for a user to tell them what to do. Agents can have goals that they remember pervasively, and they can be provided readable and writable access to execute methods on other software objects to accomplish those goals (Maes, 1997).

Eventually, agents within software surfaced to interact with users directly using text, speech or character animation. They helped users accomplish software dependent tasks. These have been called *virtual* agents. The primary functional tradition of servitude for virtual agents limits their autonomy and their range of emotional expressivity. Virtual agents were not designed to entertain. But as computer games began to use facial animation for emotion elicitation of NPCs, an emotion model architecture was required. Animators provided basic iconic facial expressions of emotions and software developers integrated morph animations driven by emotion models adopted from virtual agents.

Virtual agents, however, were not originally embedded in artfully crafted interactive narratives, but were relegated to software assistance for users. As emotion models of Cognitive Computational Psychology were implemented into virtual agents, validation of virtual agent efficacy affirmed their use for pedagogical applications (Rickel and Johnson 1997, Lane et al. 2013). The implementation of the same virtual agent technology for video games was an obvious temptation. But the limitations of graphics processors, central processors and memory required a low bar of aesthetic expectations for players as demonstrated in the crude character designs of the first three decades of computer games. Player aesthetic expectations of autonomously animated NPC "acting" had to be simpler to adapt to technology limitations. Following comic book artists and non-photoreal animators before them, game character artists and developers more often mimicked the expressions in the cannon of Delsartean-based typology of facial and body models to create automated animations that could express basic recognizable emotions (Nixon 2010). Virtual agent emotion models could at best validate their ground-truth assumptions based on a player's recognition of an NPC emotion, an NPCs believability of the authenticity of expression, or the

NPCs social appropriateness. These three assumptions reflect the narrow range of expressivity a virtual agent could emote given the technological constraints on real-time 3D animation. These methods of emotion expression in agents do not consider the concept of complexity of animation expression, which can be defined as the number of simultaneous facial animations an NPC may elicit in response to stimuli as either an active or reactive expression. Simultaneous animations with higher resolution animation allow developers and animators to collaborate with actors and realize more complex emotion elicitation.

Behave Like a Human, Think Like a Machine

MIT Media Lab researcher Patti Maes in 1997 wanted agents of "intelligent software" to "sense the current state of its environment and act independently to make progress toward its goal" (Maes, 1997). While describing agents, Maes uses one of the underlying concepts of artificial intelligence laid out forty years before by Alan Turing in his notable 1950 essay "Computing Machinery and Intelligence" (Turing, 1950). Turing proposed that machines could think autonomously and learn from their environment and past actions. But four decades earlier, Turing went further than Maes. He proposed that machines could respond in human-like ways to elicit enough rapport to convince a user that the machine was human. Without stating so directly, his rebuttals redrafted the idea of what "thought" was, by allowing that the production of a thought can include distinctly "mechanical" methods. The notable proposal of a continuous state machine suggests that he at least intuited some types of "thought" were not finite states that depended on a combination of true (1) or false (0) relations but were gradients that allowed for a continuous range [0,1], much as basic emotional expression states are currently given normalized values, and much as emotion expression morphs are blended between each other (e.g. blendshapes). Turing anticipated that a machine may have to perform machine-like thinking to come to human-like results. "May not machines carry out something which ought to be described as thinking but which is very different from what a man does" (Turing, 1950)? Virtual agents, as Turing described in his essay, may process "thought" like a machine, but still elicit behavior like a human. Virtual agents are programmed to elicit emotional reactions like humans while processing input data mathematically, probabilistically and algorithmically.

Long before computers existed, French philosopher and mathematician, René Descartes reasoned how a machine could never think like a human (Descartes, 2006). Descartes decried in 1668 the "impossibility" of sentient AI. "...It is for all practical purposes impossible for a machine to have enough different organs to make it act in all the contingencies of life in the way in which our reason makes us act" (Descartes, 2006). The reasons he gave described the challenge of computing semantic correctness and social appropriateness with limited computing power and memory to process a wide range of meaningful expression combinations. The basic needs for semantic correctness and social appropriateness accomplish two things that humans do not always do well: *teaching* and *entertaining*. Both human activities require sensitive and dynamic awareness of the emotional state of the user. A detailed psychological model of the user or player is essential to direct the actions of an agent precisely and to measure the agent's effectiveness. A virtual agent with eyes must know where on the user's body of the elicitor it should look and gather input. It should also be able to interpret what that input means. Gestural and facial expression, with its potential for producing ambiguous meanings that trigger emotions, is what Descartes intuited as a combinatorial morass.

PEDAGOGIC AGENTS AND EMOTIONAL BEHAVIOR

Descartes was correct. Logical processes are not effective to make decisions if available information is not complete enough to find a rational conclusion, or if premises are contradictory. Cognitive psychologists and neuroscientists contend that "emotional intelligence" allows the mind to fall back on another process that is based on probabilistic reasoning (Damasio, 1998; Johnson-Laird & Oatley, 1992). The brain contains neural networks that use pattern recognition to provide probabilistic answers to situations that rational thought cannot adequately answer. By applying emotional reasoning methods in a virtual agent, it can "predict" or recognize patterns from machine-learned (ML) instances to make behavioral decisions.

"Socially Appropriate" Behavior

Rosalind Picard, a research pioneer in the field of Affective Computing, describes an emotionally intelligent agent as possessing "abilities to recognize, express, and have emotions, coupled with the ability to regulate these emotions, harness them for constructive purposes, and skillfully handles the emotions of others" (Picard, 2000). In the context of an animated video interaction, a three-dimensionally animated agent often moves and speaks in ways perceived as "socially appropriate" for the context of the user-agent relationship. It should demonstrate emotionally logical behavior, follow an expected cause-and-effect sequence, that would respond to the sentience of the user. The term socially appropriate (which is sometimes called "behaviorally appropriate") describes behaviors that fall within a normal range of variance of behaviors given a specific social situation with a known set of persons in specific types of relationships to a user (Price & Bouffard, 1974). The range of variance is affected by an expectation of satisfaction from a user and the culturally determined role the agent plays in the relationship with the user. For example, a virtual agent classmate in a learning application who assumes the role of a *peer* will likely behave with a broader range of appropriate behaviors than a virtual agent that is an *instructor* who assumes a role of pedagogic authority. Behavioral variance in virtual agents can constrain or expand user input as users adapt their decision-making variance to the variance expressed by agents (Yin & Sun, 2015). Additionally, some feature values of the human user are considered that can affect an agent's behavior – physical location, language, culture, gender, class, education level, age, sexual orientation, country of origin and ethnicity (Gratch, Okhmatovskaia & Duncan, 2006). A virtual agent can be designed to choose for vocalization variables: diction, prosody, volume, pitch and backchannel utterances or for facial expression variables: movement direction, velocity and acceleration so that it can reflect and respond in socially appropriate ways to the user. A designer's intention is for a virtual agent's elicitation parameters to effectively tune to the user's familiar behaviors so that it will maximize rapport during the interaction (Zhao, Papangelis & Cassell, 2014).

Virtual agents that teach or entertain, challenge the engineer to model its algorithms to accomplish a task that has no finite outcome of "correctness". A virtual agent with affective abilities should at least be able to fulfill each of Picard's criteria for emotional intelligence – emotion recognition, elicitation, and regulation – to accomplish its goal. For example, a virtual piano teacher must demonstrate a nomenclature in musical theory, but also must sense (by facial or vocal expression) if the student "feels" she is able to understand more complex or simple examples, or no examples at all (Picard & Rosalind, 2000). The virtual agent must decide if its own gestural and aural expression should be firmer and more resolute in posture and tone, or gentler and more accepting. A hospital-based virtual agent, called a

Hospital Buddy, provides companionship and reduces psychiatric side-effects of long-term patients. The Hospital Buddy must sense a level of trust to inquire from the patient perceptions of his hospital experience (Bickmore, Bukhari, Vardoulakis, Paasche-Orlow & Shanahan, 2012). Virtual immigration interviewers in the E.U. and the U.S. have been developed and tested for border crossings (Dormehl, 2018). They are designed to flag possible deception by visa applicants using affective biometric data, including facial micro-expressions, during inquiries for facts collected from an interview. The agent must choose questions that intend to trigger emotional responses in relation to known facts and to notice an absence of emotional response (Deahl, 2018). In all these examples of pedagogic virtual agents, they must sense the emotional state of the user and express itself in a socially appropriate way so that its tasks and manner prioritize the productive flow of information between computer to human, and back from human to computer.

Elicitation-by-Design

For this discussion here forward, the term *pedagogic* agent shall refer to didactic, interrogative, informative and simulative virtual agents (all characteristics of a teacher). Pedagogic agents are generally designed to elicit "controlled" emotional responses so that the messages they express produce a predictable experience from the user and for the developer. For pedagogic agents, emotional expression must carry an intended message and outcome so that the user's behavior leads to an interactive experience that enables the agent to gather the emotional information from the user as intended – surprises can defeat the software's purpose. Pedagogic agents are not free to express the full capacity of a human personality any more than a human could doing a similar pedagogic task. To accomplish this emotion-elicitation-by-design, the author of the agent must program to specification by biasing the emotion that the agent should demonstrate during an instance of human interaction. Pedagogic agents for the most part substitute a human service that the user is presumed to need. Once the user ceases to need the service, then a pedagogic agent can become an intolerable sycophantic nuisance. The "coached" emotion-elicitation-by-design approach often fails to convey authenticity because the emotional "results" performed in a scripted interaction are intended to synchronize with a developer-permitted pedagogic task to fulfill an anticipated user need rather than respond to the spontaneous emotion elicitation of the user.

Negative Personality Traits

Unpredictability, Immorality, Deceptiveness, Self-Loathing, Flippancy, Distractedness and Forgetfulness. The seven words are negative personality traits discouraged for any person assigned to a pedagogic role. Teachers, counselors, judges, doctors would not likely gain more clients if user reviews included descriptions with these negative characteristics. However, they are all attractive characteristics for fictional NPCs because they create inter-personal conflicts that test a player's beliefs and assumptions about the game world (and perhaps the player's world). While these characteristics may not be desired in a non-NPC emotion model (such as that of a virtual agent), each of them makes useful functional components in an NPC emotion model as it decides either what emotion its appraisal will select, or what elicitation it should animate.

Neuroticism

Sometimes the user appreciates hints of awareness of the negative characteristics in human behavior, or what is typically referred to in personality dimensions as "neurotic behavior" (Trull, 2013). Neurotic behavior from pedagogic agents (virtual or real) distract from the disingenuousness of a commercial-cum-social interaction. (Consider the cynical humor of flight attendants who joke about passenger attention during pre-flight announcements.) But neuroticism tolerance varies among users and contexts; it may contradict the social "believability" of the virtual agent (Malatesta, Caridakis, Raouzaiou, & Karpouzis, 2007). While humor for the user can be the result of an agent's neuroticism, it can be socially inappropriate for a user-agent relationship if the agent is pedagogic. Additionally, if the pedagogic agent has little or no awareness of human misdeeds or maladaptation, then a neurotic response would be implausible. Moreover, a pedagogic agent must have a narrow range of expressive variability to keep the user serviced and focused on the tasks of the software.

DISTINCT EMOTIONAL FEATURES OF NPCs

An entertaining agent gathers, processes, and elicits sentient information with greater freedom of emotional expression than a pedagogic one, much as an actor has more emotionally expressive liberty than a teacher. *Entertaining* virtual agents are found in games and interactive media where the agent is personified to express the features of its personality that serve the game world. For clarity, this author refers to entertaining virtual agents as NPCs. Their emotion-response algorithms receive relevant data from the player and game play space and elicit expressive data back into the game play environment for the player to observe and respond. A cycle of data exchange can create rapport or dissonance between player and NPC as the game evolves and as the NPCs role requires. The narrative "facts" revealed in this exchange can be distorted or omitted by the NPCs subjective filters, much as any fictional character does in other media. Furthermore, neuroticism may be welcome insofar as it fits the role required for the world of the game. NPCs are often designed to elicit from the player a wider range of emotional responses than pedagogical agents. Similarly, live action actors and comedians or animated cartoon characters provoke emotions that are not permitted from instructors or public spokespersons. An actor, through his character, can elicit rage, fear and disgust in "socially inappropriate" ways when playing a conflicted character. Pedagogic agents such as piano teachers or public health officials usually cannot; the roles of pedagogic agents rarely allow neurotic behavior. Spectacular neuroticism as a performative aesthetic of cinematic acting is a feature of cinematic acting insofar as game players also watch movies and expect a similar degree of performative complexity, emotional authenticity and celebrity attraction (Morgans, 2018; Stuart, 2016).

Cinematic Acting Style

The trend in the last half-decade has demonstrated strong interest in cinematic acting styles as games increase their ability to use higher resolution graphics, motion, and complex facial expression animation (Torres, 2014). Cinematic acting differs from theatrical acting in the production method. While performance preparation methods are very similar, cinematic acting requires the actor to deconstruct a performance into smaller units for each shot. Acting for animation of video game NPCs require actors to decompose further – down to the gesture, posture, or facial expression. To capture motion data for high resolution animation, increased computational power is needed to recognize the face, track its inflection, and predict its movement over time. Contemporary graphics processors allow for the capture of more complex and simultaneous emotion expressions. High-resolution animations support the aesthetic of cinematic acting styles in video games.

What are cinematic acting styles? Cinematic acting inherited many methods for training from theater. There are two dominant currents in cinematic acting styles. One works from a collective external vocabulary of postures, gestures and facial expressions that are loosely assigned to roles within a theatrical narrative. The task of the actor is to shape his own variation of a role as a set of expressions that convey the emotional expressions of a character for the specific narrative. This approach, often called the outside-in approach and is associated with performance theorist Vsevelod Meyorhold and Francois Delsarte, is highly adaptable between cultures and requires a studious actor to learn the emotive vocabulary of the target audience. The outside-in approach affiliates well with animation and its tradition of referencing a Delsartean taxonomy of physical expression, much like trained and practiced for silent film acting (Hart 2005). The other major approach requires a process of self-examination to identify existing emotional impulses from past lived experience so that when an actor in-character focuses on an action, the resulting emotion from obstructions or assistance are personally authentic. This inside-out approach has been affiliated with Konstantin Stanislavski and his Method approach as well as Sanford Meisner. In animated or live-action films, both currents of cinematic acting deploy complex simultaneous movements motivated by inferred psychological states that are used by actors to create character facial expression for motion capture or voice recording for video games.

For this research, both cinematic acting styles provide valuable methods for facial expression in video games as they connect the emotional source of actor expression to NPCs that move narratives forward. Games that aspire to cinematic high resolutions of image and movement, such as *Beyond Two Souls* (Quantic Dream 2013) and *The Last of Us* (Naughty Dog 2013), have shown substantial consideration of the aesthetics of cinematic acting through complex animations of the face. Therefore, the methods and techniques of cinematic acting, character development and performance deserve consideration as video games embrace NPCs with cinematic acting styles.

Responsiveness to Multiple Simultaneous Stimuli

The game player, attracted to cinematic acting, expects more complex animation to enrich the perceived emotional and intellectual life of the character. To develop cinematic acting for NPCs, its animation must move in complex ways that are consistent with the design of a complex screen performance; this means moving groups of vertices of an NPCs mesh in ways that represent a response from simultaneous stimuli from both external-physical and internal-mental sources. When we closely examine well-prepared actors that move in any fully developed performance style, such as naturalism or physical comedy, what makes a performance complex (and possibly "good") is when an actor responds to multiple stimuli almost simultaneously, from sources that are external-physical and also from sources that are internal-mental. Acting Theorist John Lutterbie draws from Cognitive Science as he describes the "acting instrument" as a Dynamic System Theory model that listens inside and outside the body, filters decisions through memory, then responds with complex movement, language and gesture (Lutterbie, 2011). Complexity increases when an actor listens, acknowledges and responds with "executive control" to multiple stimuli with an intensity that seems appropriate to the scene, role, its story, and its world.

Often attention to stimuli in the external-physical space appears to the player with a subjective internal-mental amplification or diminishment of intensity. As game designers deploy cinematic acting styles, the Player observes stimuli generated from the NPC that seems internal-mental and occurs simultaneously with stimuli from the external-physical space. That disparity of intensity creates a competition of focus for the NPC which results in polysemic expression. The player notices this distinction and imagines, often with pleasurable uncertainty and insatiable curiosity, what the actor is thinking or feeling. Often the most pleasure-inducing experience is an ambiguous facial expression, and the audience or player invents or projects her own meanings that caused the expression. This ambiguity of focus can be built into the NPC and is fundamental to how cinematic acting styles contribute to the expansion of the narrative space. No longer is the game limited to a visual physical space. Cinematic acting styles open the narrative to the mental space of the NPC. To achieve cinematic acting, NPCs require uniquely subjective filters and algorithms to idiosyncratically appraise and elicit reactions to the emotion-evoking entities it perceives.

Cinematic Mediation of Emotion Expression

Principal game characters are presented to the player base with spectacular trailers and posters, much modeled on theatrical film trailers. These advertising media reinforce character archetypes and stereotypes, and also create a set of behavioral expectations in the player before ever playing the game. Additionally, in-game NPCs are experienced by players as mediated persona through game video screens and speakers, much like film and television characters. Audio-visual moods and cues reinforce mediated emotional expressions in a game much as one observes in film and television (Smith, 2003; Wiley, 2003). The construction of mediated emotional expressions results from already familiar production methods of actor training, casting, rehearsal, recording and reperformance technologies that games use for their production. A facial expression on an NPC after a kiss or bullet to the back triggers an expected category of emotional response that can stand alone or come accompanied with sympathetic or dissonant music, sound effects or dialog. The combined facial expression and sound can affect the perceived meaning of the response and negate or affirm the expected response. These expectations do not precisely resemble real life but are a synthesis of memories of watching previous mediated experiences of similar events combined with real life memories. Actors are trained to express real emotions and to synthesize them with culturally preconceived expressions of real emotions.

Performative Mediation of Emotion Expression

Mediation is not only determined by industrial or societal norms through cinema. Mediation can also be the result of an actor's own imaginary construction of the character's mental processes as she believes the audience will perceive them. As each elicitation of an actor or NPC is presumably the result of the flow of emotion-evoking stimuli, it is an actor's detailed design of those mental processes for the character that provides actor-centric mediation of elicitation. Thus, what has here been defined as cinematic acting style for NPCs, with its complex mesh animations, is the result of an actor's invention of an internal emotion model that filters sensations, identifies them, prioritizes them, associates them with memories, maps sensations to affect definitions, and selects an elicitation in response that appears as an emotional facial expression.

EMOTION MODELING OF COGNITIVE COMPUTATIONAL PYSCHOLOGY

Performance theories that are foundational for cinematic acting share some concepts that describe the derivation of human emotion with those defined in the literature of Computational Cognitive Psychology. Psychologists Ortony, Clore and Collins were among the first to propose a human emotion model for simulation on the computer. While theirs was not intended for game development, it is easy to imagine that its deployment would sense emotion triggers in the game play space. The model would process the sensed data through its own character-specific filters, choose an emotion that describes the state of the NPC, and then elicit that emotion through an animatable face, body or synthetic voice. During the last thirty years, emotion models were designed to test the human affect processing system by observing them embedded in virtual agents more than NPCs. Computational Cognition Theories of emotion are collectively known as Appraisal Theories. To understand the rationale for an Actor-Centric emotion model for NPCs, a fundamental understanding of an Appraisal Theory framework for emotion models is helpful.

Appraisal Theory for NPCs

Computational Cognitive Psychologists for the most part agree that emotions are generated in response to an "appraisal" of the environment and its condition to affirm one's *beliefs*, provide access to fulfill *desires* and enable one's *intentions* (BDI). A person's environment is often laden with entities to appraise in ways that concern a person's physical and social survival and quality of existence. Appraisal occurs initially as a person assesses the immediate situation and its immediate impact (Ortony, Clore & Collins, 1990). And for some theorists, there is a *secondary* appraisal (reappraisal) where one considers how to cope long-term by accepting the situation as it is and adjust one's attitude about the situation. A person might then take action to change the situation in response (Lazarus, 1991).

Entities that have emotional effects on a person are either inanimate *objects*, animate *agents,* or consequential *events*. A person will judge an object's *likeability*, an event's desirability (of consequences), or an agent's (past, current or future) *praiseworthiness* (of actions) in terms that either favor positively or negatively one's own BDIs. Observing and evaluating an entity (object, event or agent) and its feature values registers an emotional appraisal along one or more dimensions of measurement. The values of these measurements are stored as *appraisal variables*. A fundamental scale of measurement is *pleasure* and *displeasure* within a range of [-1,1]. This dimension has also been interpreted as *aversion* and *attraction.* As a two-dimensional appraisal scale, *pleasure-displeasure* runs on the X-axis and rises with *arousal (intensity)* on the Y-axis within a range of [-1,1]. Originally known as the Circumplex Model of Emotion (Russell, 1980), it has been further refined as a 12-emotion unit circle with evenly divided expression positions called the Core Affect (Yik, Russel and Steiger, 2011), as shown in Figure 1. Core affect prioritizes the "basic" emotion positions in the unit circle. A three-dimensional scale variation includes *dominance* on the Z-axis within a range of [-1, 1] (Mehrabian, 1996) where dominance represents the degree to which a person can control the emotion from triggering impulsive elicitations that could be anti-social, unwelcomed or inappropriate, as shown in Figure 2. Most Computational Cognitive Models based on Appraisal Theories use either two- or three-dimensional emotion mapping, as found in Embodied Conversational Agent development (Malatesta, Raouzaiou, Karpouzis & Kollias, 2009). Using two- and three-dimensional emotion maps is known to provide a system modeled on closely observed human behavior. Emotion maps allow researchers to test a virtual agent accurately in relation to a human reference (Lisetti & Hudlicka, 2015; Gratch, Marsella, Wang, Stankovic, 2009).

Figure 1. Yik, Russel and Steiger (2011) represent Core Affect in a two-dimensional emotion map that evenly divides a unit circle into 12 sections

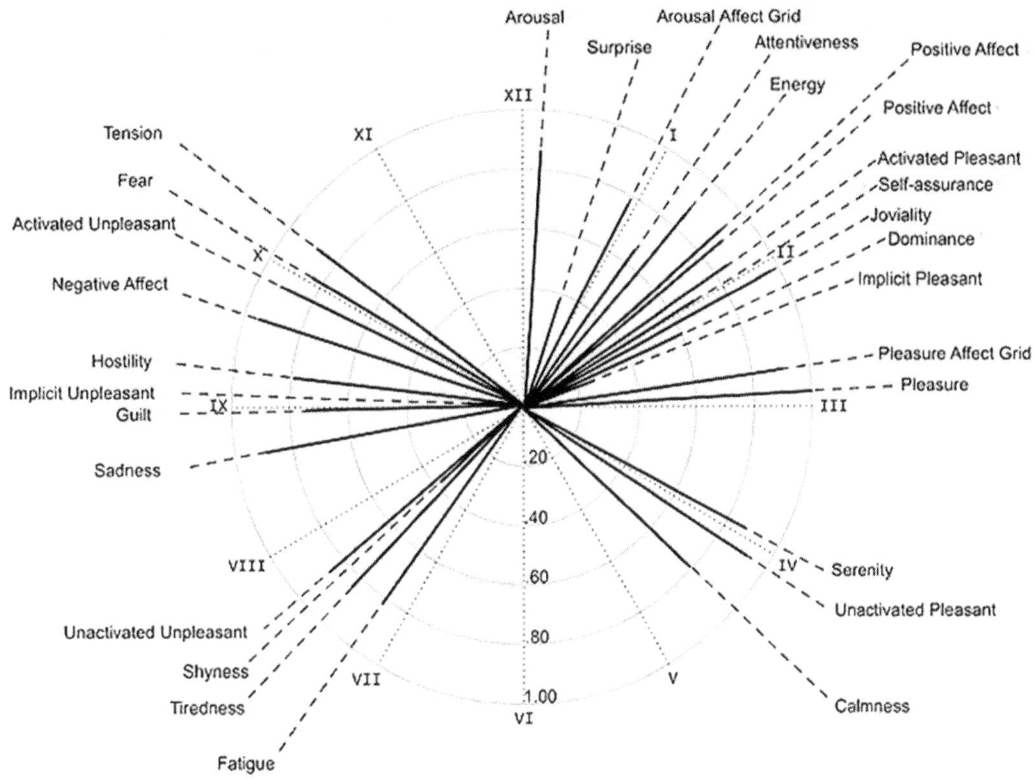

NPCs perceive entities with subjectively determined values of their feature variables based on their character-specific BDIs. For every entity that triggers an emotion, an algorithm must translate and map all the entity's feature values in a two-dimensional Circumplex, as shown in Figure 3, or a three-dimensional volume, as shown in Figure 4, to a coordinate point positioned by its appraisal variable values -- Pleasure, Arousal and (if in three-dimensions) Dominance (PAD). These two or three dimensional values are normalized within a [-1, 1] range of a unit circle or sphere.

Much of the distinction between appraisal theories is the structure of the emotion model, and therefore the path of stimuli signal through the model. Some theories emphasize memory and pattern recognition components in relation to goals and praiseworthiness, others emphasize reappraisal and changes of emotion response, while others emphasize the effect of stimuli on planning for goal acquisition. The resulting PA or PAD values may change as the appraisal process cycles through each appraisal or reappraisal over time. PA and PAD values infer a vector with a coordinate point at its tip. With each reappraisal cycle, the vector can move to a new position as the emotional state of the agent changes.

A coordinate point within a unit circle or sphere falls into a region or volume that is bounded by an emotion label. The idea of the emotion label is controversial. Some Computational Cognitive Psychologist prefer to abandon labels for emotions. The trepidation to work with emotion labels stems from the subjective quality of assigning a culturally determined word to an object (an emotion) only known by

Figure 2. Lance and Marsella (2010) use Mehrabian's three-dimensional emotion map PAD to show example emotional categories

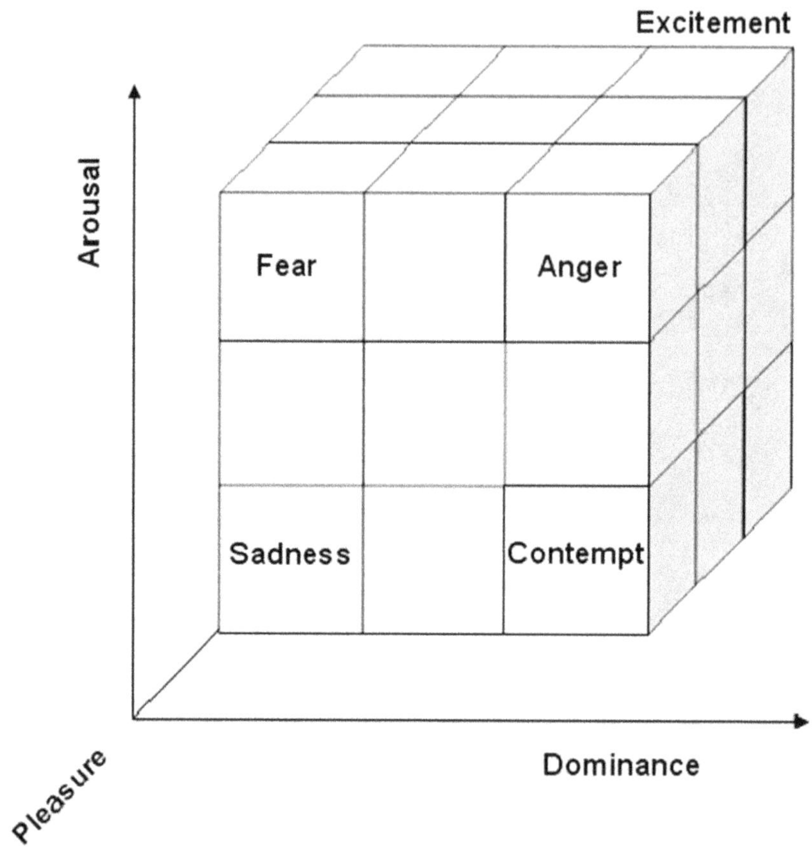

elicited effects in the body, and often repressed by the vagaries of culture (Russell & Feldman Barrett, 1999). Labeling an emotion is a linguistic problem rife with semantic contradictions of inclusion and biases. However, since NPC design depends on collaboration among creative and technical professionals, labels must be used to communicate software development processes. This author recommends the following considerations to avoid misunderstanding labeling emotion expression of the face. First, a developer must consider if in the culture of the player the face is where a specific feeling is expressed. Second, a developer must decide if the system for measuring elicitation (usually ML-based) can detect that labeled emotion on the face. Third, is the emotion word a categorical label that includes other more specific labels or is it a more detailed and descriptive word that does not contain other labels. Labels that are at the top of an inclusion hierarchy tend to draw more agreement of meaning.

At the top of a hierarchy of emotion labels are "primary" or "basic" emotions. These are allegedly triggered by biological instinct and evolutionary necessity (Izard, 1992). If the label is not a primary emotion, then it is a secondary emotion. Some Cognitive Psychologist and Constructivist Neuroscientists believe that many emotions are contextually constructed, categorized and classified as expression

Figure 3. Yik, Russell & Steiger (2010) use a 12-Point Affect Circumplex (12-PAC) model of Core Affect to plot the vectors of 30 mood words

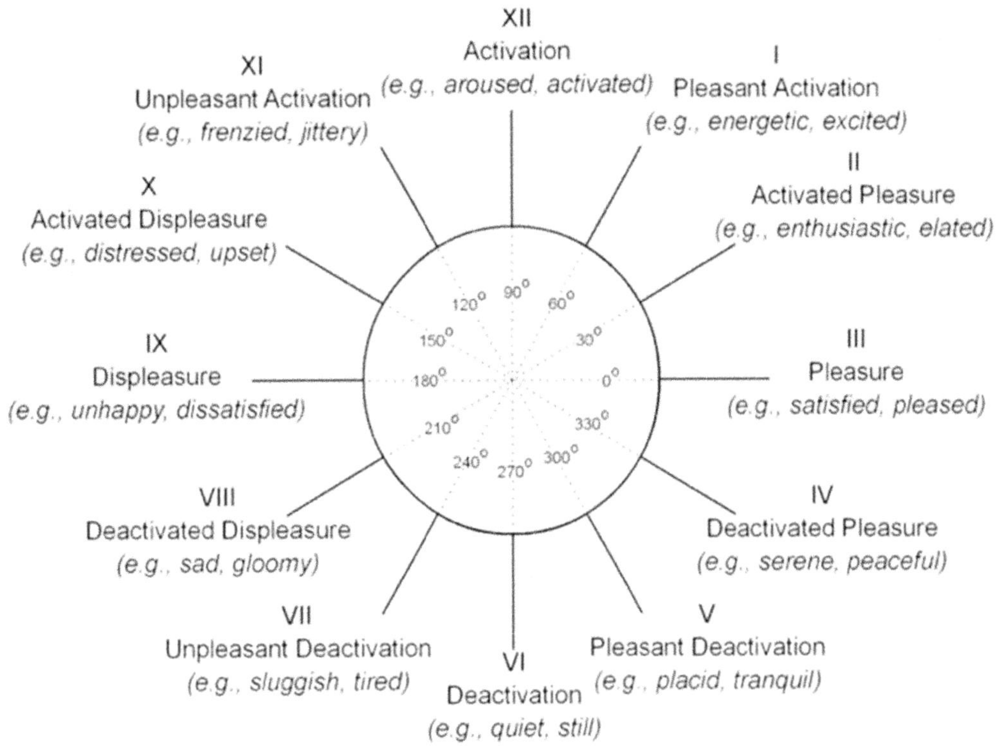

concepts. Constructivists contend humans have a proclivity to express many emotions only if learned and permitted, and that some expressive characteristics may not be innate at birth (Quartz, 1999; Barrett & Satpute, 2013; Barrett, 2017). The labeling of patterns "nurtured" in context, explains the differences in emotion label meanings from one culture or social group to another.

The components described in the various Appraisal Theory-based models are designed to best attain the principal objective or action of a person. Typically, object-action pairings are ascribed to an NPC. They could for example be assist player-character to attain goal, obstruct player-character to attain goal, obtain a game object or manage a game environment. If a virtual agent or NPC is intended to simulate the full psychological profile of a human, then a very complex model will be necessary to accommodate types of memory, learning, adaptation, natural and social reasoning, language acquisition and expression, motor-skill acquisition and adjustment, and other influences. Additionally, since reappraisal of an entity can be triggered by internal stimuli from memory, a scheduling algorithm must control the bussing of data through the components, thus allowing for more than one path through the system from sensation to elicitation, and allowing for interruption by cycles of signal between components.

Figure 4. Zhang, S., Wu, Z., Meng, H. M., and Cai, L. (2010) use Mehrabian's three-dimensional emotion map to plot facial expressions of a talking avatar from an expression database

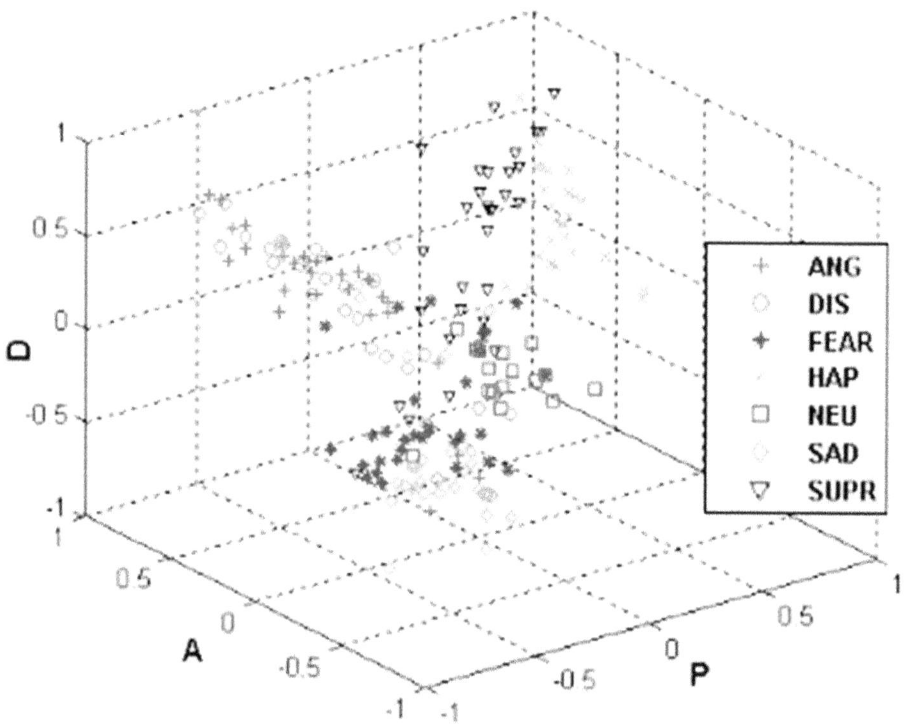

SOLUTIONS AND RECOMMENDATIONS

For an NPC emotion model to accommodate the complexity of perceiving and processing multiple emotional entity categories of stimuli simultaneously, components that sense and process emotion-evoking entities must be built into the emotion model for sensing, evaluating, prioritizing and selecting affect. Ultimately, the path of either a sensation signal or a processed affect selection is controlled by algorithms that schedule the flow of data from sensation to appraisal to elicitation so that behavior is synchronized to receive new sensations and to reconsider past sensations (reappraisal) held in memory. Present sensations and memories compete for attention. Simultaneous appraisal and elicitation with reappraisal require parallel processing and scheduling to manage the complexity of computational processes. To achieve cinematic action and therefore high-resolution animated emotion expression, complexity in an emotion model must increase with the growth of more specific components and more non-linear scheduling of affect selection and elicitation.

An Actor-Centric Emotion Model

Extracting the principal points identified thus far, an actor-centric emotion model should have components that allow for the NPC performative characteristics. Arrays, structs or vectors serve to store immutable ordered data that describe the implicit biases of an agent. A Feed-Forward Neural Network (FFNN) provides the structure to algorithmically process input from multiple handlers-listeners or sensors into a gradient output that will result into an animated elicitation. These structures are used in Table 1 to describe how an emotional model may deploy the critical performative characteristics of an Actor-Centric emotion model.

Table 1. Data Structures and Algorithms for Performative Characteristics

Performative Characteristic	Data Structure(s)	Algorithmic Method(s)
Expression of Negative Traits	Arrays or vectors that store coefficients to alter values of post-appraisal dimensional vector	Receives post-appraisal signal as dimensional vector. Each dimensional value is multiplied by a coefficient C Î ¡
Multiple Sensations Simultaneous Sensations	Multiple FFNNs or structs with object handlers-listeners and feature recognition array	FFNNs or structs receive signals from environment during same frame (appraisal instance).
Internal Stimuli	Array 1 stores feature values of significant objects from previous appraisals (memories). Array 2 stores relevant structures indexed by object type whose methods can trigger internal sensations.	If external stimuli are recognized as similar to significant objects in array 1, then method in structure of matching object type in array 2 triggers internal sensation.
External Stimuli	*Objects in game space:* Struct with vector representing feature values. Objects in player space: FFNN recognizes object and converts to vector representing feature values.	External stimuli are recognized as similar to objects in array 1, then method in structure of matching object type in array 2 triggers internal sensation.
Self-Oriented Goals	Data types that permit gradient values, such as 3D vectors. Or, any variable if the goal is not an emotional state or change in emotional state.	Method must check goal value and current state value. Difference may trigger new appraisal. Method might check player variables to decide to forgo action to serve self and instead engage player.
Modularity	Affect Derivation Component object or struct contains arrays or links to FFNNs. More modules increase ability to recognize more emotional entities.	Method switch can enable or disable modules as needed to minimize computational expense as agent changes state or space.
Scalability	Within Affect Derivation Component, increasing Affect Components allows for more granular sensitivity to variations of entity feature variables.	Method switch can enable or disable components as needed to minimize computational expense as agent requires greater sensitivity to entity feature variables.
Affective Loop	The entire emotion model runs as a process. It contains two cycles. *Cycle 1:* entirely internal and allows reappraisals. *Cycle 2:* partially external to the agent, allows the agent to send out elicitation and receive response from other agents or the player.	Scheduling of signal vector must allow for each component to complete its task. Must store in memory significant appraisals for subsequent recognition or reappraisal. Must regulate and prioritize competing signals to determine which should occupy appraisal bus: new signal or reappraisal.

A generic emotion model as shown below in Figure 5, presents the basic components of an NPC. This generic model was synthesized from the MAMID model developed by Hudlicka (2003), from the FeelMe model developed by Broekens and DeGroot (2004), and from the FAtiMA model developed by

Dias, Mascarenhas and Paiva (2014). The resulting model satisfies Picard's componential requirements. Table 2 elaborates on the function of each component.

Inside of the Affect Derivation component in Figure 5 is what is depicted in Figure 6. The generic emotion model uses FAtiMA's modularity feature. The Appraisal Components provide the Appraisal Frame the current values of each Appraisal Component's predictive result. This author proposes that each component consist of an FFNN that recognizes the entity and its emotion relevant feature variables and values to predict appraisal variable value for each affect from [0, 1]. For a facial emotion expression, that would be the gradient value of an emotion expression of which there is usually at least six expressions plus neutral.

Figure 5. A generic emotion model that combines features from MAMID (Hudlicka, 2003), FeelMe (Broekens and DeGroot, 2004) and FAtiMA (Dias, Mascarenhas and Paiva, 2014)

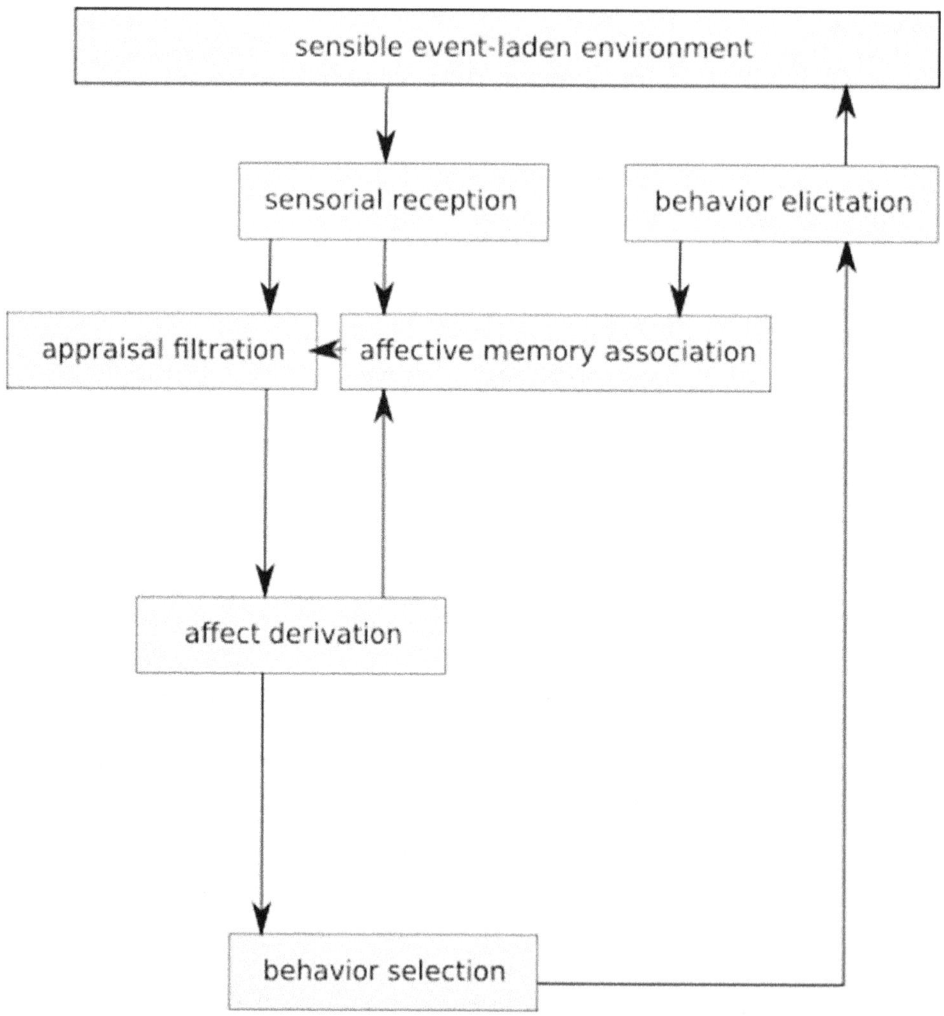

Each Affect Derivation Component consists of a data structure that receives input from an Appraisal Filter. The Appraisal Filter receives an appraisal signal that recognizes each emotion expression for each respective Appraisal Component. The filter will likely implement an object recognition neural network or an array of feature value ranges – both will process a vector whose elements and their values are within range to process and predict a meaning. For a facial emotion expression system, the Appraisal Filter will look for facial features in a video stream that fit the parameter values of an emotional expression on a human face. Then the signal must be sent to the appropriate Appraisal Component within the Affect Derivation unit. An Appraisal Component will evaluate a feature of the face that best indicates a value of a specific emotion expression type (happiness, anger, etc.). Then the Appraisal Component must use an emotion recognition FFNN to predict the expressive meaning of the signal. The resulting gradient value will reside in the Appraisal Frame. All Appraisal Frames will be evaluated to choose a final Affect Derivation for the appraisal instance. The resulting derivation will be the Affective State, which will make a Behavior Selection, and in turn trigger a Behavior Elicitation.

Applying the Affective Loop for Emotion Model Dynamics

A consistent theme in the discussion of emotion models is the idea that affectively relevant sense data runs a cycle between an agent and its environment. Sundström calls this cycle the Affective Loop (2005). The Affective Loop has three stages familiar to actors from improvisation and rehearsal exercises: 1) a human user elicits an emotional expression, 2) a system responds with an appropriate emotional expression, and 3) the human user internalizes and interprets the affective meaning of the eliciting system, and 4) the loop repeats. A similar concept is found in the Repetition Exercise of renowned theater director and acting teacher Sanford Meisner (Meisner, 1987). The three stages of the Affective Loop have parallel stages in the process of developing an actor's skill at finding authentic emotional responses to interactions with other characters, objects and events observed in a scene. The Affective Loop expands human-to-human interaction, as practiced in the dramatic arts, to a "design principle" of the Affective Computing domain that implements motion-capture-performer-to-computer and computer-to-player principles.

The Affective Loop design principles are interpreted for player to NPC interaction as follows from Sundström (2005). *Embodiment* is enhanced as the player's body expresses its affect through the game controller or other sensing interface. Some part of the player's elicitation data becomes input that the NPC's emotion system can sense. Likewise, the player can sense the NPC sensing the player, and sensing the NPC eliciting its organismic idiosyncrasies through its own body.

Flow is another attribute where, as the player interacts, there is a sense that there is always potentially a new sensation coming from which to repeat the Affective Loop. Or, an NPC can react to internal stimuli from memory. The emotion model not only repeats as loops, but it is always open for new stimuli from internal or external sources that allow each loop iteration to elicit distinct values.

Another Affective Loop attribute occurs when NPCs are designed with elicitation *ambiguity*. When there is more than one meaning and possibly more than one stimulus attributed to an NPC elicitation, the uncertainty can create interest in subsequent sense information with the belief that clarity of meaning will be revealed over time. Ambiguity is accomplished with facial expressions by always allowing multiple simultaneous stimuli sensations and appraisals causing distinct affect selections that can blend in the facial elicitation.

Lastly, *Natural-But-Designed Expressions* are characteristics that value elicitation forms found in nature, such as shapes of the face (eyes, mouth, ears) and their movements that resemble shapes and

Table 2. Component descriptions of the generic emotion model in Figure 5

Component Name	Description
Sensible event-laden environment	Container of all events, objects and agents that can be sensed by emotional agent.
Sensorial reception	Events that are accepted by sensors.
Appraisal filtration	Filters perception of objects, agents, events in terms of beliefs, desires (goals) and intentions (BDI).
Affective memory association	Storage of perceptions, affects and elicitations.
Affect Derivation	Pleasure-Arousal-Dominance mapped emotion vector algorithm.
Behavior selection	Resulting animation of face indicating current emotional state.
Affective memory association	Storage of perceptions, affects and elicitations change the "rules" or "nets" (and don't maintain a memory of events.
Behavioral elicitation	The physical movement of the mesh as gesture or facial expression.

Figure 6. Detail of Affect Derivation Component from FAtiMA component (Dias, Mascarenhas, & Paiva, 2014)

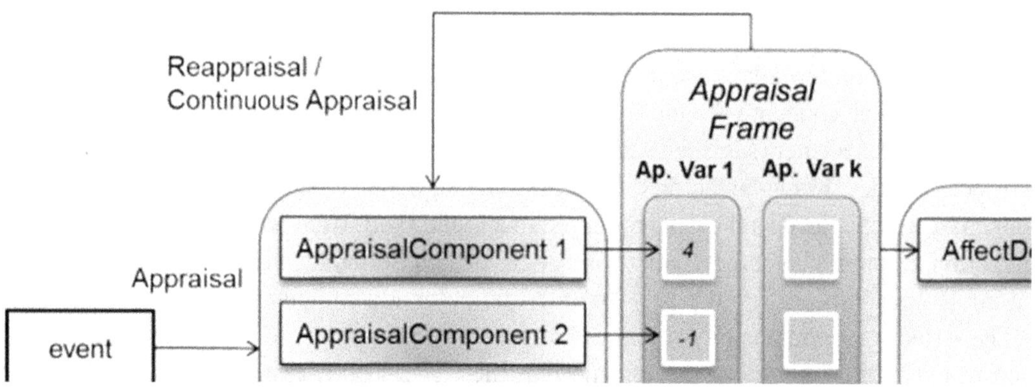

movements of recognizable natural objects. For example, when eyebrows rise in surprise or rage, they take the shape of volcanos or other swollen vessels filled with energy.

With its flexible deployment, the Affective Loop as applied within a system provides a guide that reinforces an NPCs resemblance to intelligent behavior and appearance of living entities. The Affective Loop intends continuous engagement by providing the player with familiar organismic processes.

Subjectively Filtered Imperfect Perception and Memory

NPC emotion models should also allow for occluded, skewed or false perceptions of sensed data by deploying subjectively tuned feature value distorting filters consistent with the NPCs idiosyncratic personality. Subjectivity in virtual agents is a known topic in software engineering, and it is used to localize agent behaviors to restricted domains, even when the agents are NPCs (Omicini & Ossowski, 2003). Designing subjectivity in an NPC then allows for building "imperfect", non-machine-like or organismic features that simulate human constraints on intelligence. Thus, rationality can be limited as the NPCs character design requires. Similarly, an NPCs memory can also be designed as "imperfect". Memories can be subject to disappearance, erosion and distortion over time. For an NPC to show signs of organismic entropy is for its memory to simulate decay and self-doubt as the game play proceeds. Furthering organismic limitations, an NPC may not always respond too quickly, nor should it always respond without the delaying force of anticipatory doubt, remorse or regret. Similarly, NPCs can reflect before choosing a course of action. They can likewise elicit a choice with temerity or tentativeness. People often mumble and stumble because they doubt their choices, forget their memories, become distracted by competing but pertinent stimuli, or fail to figure out a course of action in due time. Occasionally unclear elicitation is the result of an overloaded overwhelmed system. Such "imperfections" built into an emotion model make for NPCs that cause curiosity in players and reinforce the lifelikeness of the world that spawned them.

Modularity and Scaling of Emotion Resolution

Less an organismic and more a mechanical concern, the most adaptable emotion models are those that allow for modular components that increase the variety of sensible entities and the resolution of sensing input data. Modularized emotion models can allow an increase or decrease in the number of feature variables an NPC can sense, and an increase in the complexity of elicitation. Modularity gives a designer the ability to use the same model with fewer components for simple mass-produced NPCs, while for more complex supporting characters, and for very complex antagonists or "buddy" characters, more modules can expand the character's emotion complexity. Modularity also allows for real time scaling up or down, where if a complex character has a brief and simple appearance, its model can retain less data with fewer components, or vice-versa. Thus, when designing NPC emotion models, scale of entity detection and emotion elicitation may vary enormously, depending on the importance of a character in a game.

Like the limitations of scale, games provide circumscribed interactions with NPCs that reveal an incomplete view of a character. We get to know an NPC over time through a mediated context that serves the whole game. There is no need to model more than what the game design will allow the player to sense during anticipated game play time. A model needs to show only the emotions that the game mechanic allows. Like many forms of humanistic expression that comprise an art form, players must infer what they do not sense. And they cannot sense everything at once.

FUTURE RESEARCH DIRECTIONS

With an emotion model that is modular, it becomes conceivable to implement a procedurally generated emotion as part of a procedurally generated NPC and thus within a procedurally generated narrative system. Procedurally generated game components can become principal features of adaptive games that reshape themselves to the variables of the player. Adaptivity would require some mechanism that can categorically predict the features of the player so that generated game components would spawn in variations that best suit the designer's prediction of what the player needs to experience. A predictive system would require a robust machine learning algorithm that enables adaptivity in ways productive to the aesthetic of the game and player-designer relationship. From the perspective of the actor who must perform the character motion, it remains to be seen if the workflow proposed in this research is more or less consistent with the performance preparation process. This question needs further research and experimentation. Future work described is viable, reproduceable and could provide a new paradigm for production of photorealistic NPCs and cinematic acting in video games.

CONCLUSION

Enabled by processor innovations and the techniques of machine learning, photorealistic NPC design and development can adopt the performative aesthetics of motion picture acting with its complex facial animations. The challenge that lies ahead for developers is to design an emotion model and working process that adapts to existing entertainment methods and techniques. Past practices that relied on hand-crafted morph animations have provided remarkable accomplishments at creating life-like characters for computer games. However, animator-centric NPC emotion modeling based on morph animations is a cost prohibitive method for most game developers and designers. This research promotes a synthesis of the structures of emotion developed by Computational Cognitive Psychology with the most influential Performance Theories and acting training methods and practice toward the service of procedurally generated character and narrative. Future work will reveal if putting a human actor's creative process at the center of emotion modeling and the workflow of character design will make NPCs more enjoyable and closer to cinematic style acting for computer-based games.

REFERENCES

Bickmore, T., Bukhari, L., Vardoulakis, L., Paasche-Orlow, M. & Shanahan, C. (2012). *Hospital buddy: a persistent emotional support companion agent for hospital patients*. Academic Press.

Braunberger, P. (Producer), & Renoir, J. (Director). (1936). *A Day in the Country* [Motion picture]. France: Panthéon Productions.

Broekens, J., & DeGroot, D. (2004). Scalable and flexible appraisal models for virtual agents. In *Proceedings of the Fifth Game-on International Conference* (pp. 208-215). Academic Press.

Calleja, G. (2009). Experiential narrative in game environments in breaking new ground: innovation in games, play, practice and theory. *Proceedings of DiGRA 2009*.

Chatman, S. (1980). What novels can do that films can't (and vice versa). *Critical Inquiry*, *7*(1), 121–140. doi:10.1086/448091

Damasio, A. R. (1998). Emotion in the perspective of an integrated nervous system. *Brain Research. Brain Research Reviews*, *26*(2-3), 83–86. doi:10.1016/S0165-0173(97)00064-7 PMID:9651488

de Maupassant, G. (2007). *The Short Stories of Guy de Maupassant*. Black's Readers Service.

Deahl, D. (2018), The EU plans to test an AI lie detector at border points. *The Verge*. https://www.theverge.com/2018/10/31/18049906/eu-artificial-intelligence-ai-lie-detector-border-points-immigration

Descartes, R. (2006), A discourse on the method of correctly conducting one's reason and seeking truth in the sciences (I. MacLean, Trans.). Oxford University Press.

Dias, J., Mascarenhas, S., & Paiva, A. (2014). Fatima modular: Towards an agent architecture with a generic appraisal framework. In *Emotion Modeling* (pp. 44–56). Springer. doi:10.1007/978-3-319-12973-0_3

Dog, N. (2013). *The Last of Us* [Computer video game]. Sony Interactive Entertainment.

Dormehl, L. (2018), A.I. border agents could use machine smarts to tell if travelers are lying. *Digital Trends*. https://www.digitaltrends.com/cool-tech/ai-border-airport-virtual-agent/

Dream, Q. (2013). *Beyond Two Souls* [Computer video game]. Sony Interactive Entertainment.

Ekman, P. (1997). *What the face reveals: Basic and applied studies of spontaneous expression using the Facial Action Coding System (FACS)*. Oxford University Press.

Gratch, J., Marsella, S., Wang, N., & Stankovic, B. (2009). Assessing the validity of appraisal-based models of emotion. In *3rd International Conference on Affective Computing and Intelligent Interaction and Workshops* (pp. 1-8). IEEE. 10.1109/ACII.2009.5349443

Gratch, J., Okhmatovskaia, A., & Duncan, S. (2006, November). Virtual humans for the study of rapport in cross cultural settings. *25th Army Science Conference*, 27-30.

Hart, H. (2005). Do You See What I See? The Impact of Delsarte on Silent Film Acting. *Mime Journal*, *23*(1), 184–199. doi:10.5642/mimejournal.20052301.11

Hudlicka, E. (2003). Modeling effects of behavior moderators on performance: Evaluation of the MAMID methodology and architecture. *Proceedings of BRIMS*, 12.

Lane, H. C., Cahill, C., Foutz, S., Auerbach, D., Noren, D., Lussenhop, C., & Swartout, W. (2013). The Effects of a Pedagogical Agent for Informal Science Education on Learner Behaviors and Self-efficacy. In Artificial Intelligence in Education. AIED 2013. Lecture Notes in Computer Science, (vol. 7926). Springer.

Lazarus, R. S. (1991). Cognition and motivation in emotion. *The American Psychologist*, *46*(4), 352–367. doi:10.1037/0003-066X.46.4.352 PMID:2048794

Lisetti, C., & Hudlicka, E. (2015). Why and how to build emotion-based agent architectures. In R. A. Calvo, S. K. D'Mello, J. Gratch, & A. Kappas (Eds.), *Oxford Library of Psychology. The Oxford Handbook of Affective Computing* (pp. 94–109). Oxford University Press.

Lutterbie, J. (2011). *Towards a general theory of acting: Cognitive science and performance*. Palgrave Macmillan. doi:10.1057/9780230119468

Maes, P. (1997). Intelligent Software. *IUI '97 Proceedings of the 2nd International Conference on Intelligent User Interfaces*, 41-43.

Malatesta, L., Caridakis, G., Raouzaiou, A., & Karpouzis, K. (2007). Agent personality traits in virtual environments based on appraisal theory predictions. *Artificial and ambient intelligence, language, speech and gesture for expressive characters*, 7.

Malatesta, L., Raouzaiou, A., Karpouzis, K., & Kollias, S. (2009). Towards modeling embodied conversational agent character profiles using appraisal theory predictions in expression synthesis. *Applied Intelligence*, *30*(1), 58–64. doi:10.100710489-007-0076-9

Margolin, U. (1986). The doer and the deed: Action as a basis for characterization in narrative. *Poetics Today*, *7*(2), 205–225. doi:10.2307/1772759

Mehrabian, A. (1996). Pleasure-arousal-dominance: A general framework for describing and measuring individual differences in temperament. *Current Psychology (New Brunswick, N.J.)*, *14*(4), 261–292. doi:10.1007/BF02686918

Meisner, S. (1987). Sanford Meisner on acting. Vintage Random House.

Morgans, J. (2018). How video games cast actors just like movies do. *Vice*. https://www.vice.com/en_uk/article/gymde7/how-video-games-cast-actors-just-like-movies-do

Murray, J. H. (1997). *Hamlet on the Holodeck: The future of narrative in cyberspace*. MIT Press.

Nixon, M., Pasquier, P., & El-Nasr, M. S. (2010). DelsArtMap: Applying delsarte's aesthetic system to virtual agents. In *International Conference on Intelligent Virtual Agents* (pp. 139-145). Springer. 10.1007/978-3-642-15892-6_15

Omicini, A., & Ossowski, S. (2003). Objective versus subjective coordination in the engineering of agent systems. In *Intelligent Information Agents* (pp. 179–202). Springer. doi:10.1007/3-540-36561-3_9

Ortony, A., Clore, G. L., & Collins, A. (1990). *The cognitive structure of emotions*. Cambridge University Press.

Picard, R. W., & Rosalind, W. (2000). Toward agents that recognize emotion. *Vivek-Bombay*, *13*(1), 3–13.

Price, R. H., & Bouffard, D. L. (1974). Behavioral appropriateness and situational constraint as dimensions of social behavior. *Journal of Personality and Social Psychology*, *30*(4), 579–586. doi:10.1037/h0037037

Rickel, J., & Johnson, W. L. (1997). Integrating pedagogical capabilities in a virtual environment agent. In *Proceedings of the First International Conference on Autonomous Agents* (pp. 30-38). 10.1145/267658.267664

Russell, J. A. (1980). A circumplex model of affect. *Journal of Personality and Social Psychology*, *39*(6), 1161–1178. doi:10.1037/h0077714

Smith, G. M. (2003). *Film structure and the emotion system*. Cambridge University Press. doi:10.1017/CBO9780511497759

Stuart, K. (2016). Video games where people matter? The strange future of emotional AI. *The Guardian*. https://www.theguardian.com/technology/2016/oct/12/video-game-characters-emotional-ai-developers

Sundström, P. (2005). *Exploring the affective*. University of Sweden. http://www.diva-portal.org/smash/record.jsf?pid=diva2:1041047

Torres, S. (2014), Video game characters modeled after real people. *Venture Beat*. https://venturebeat.com/2014/06/28/video-game-characters-modeled-after-real-people/

Trull, T. J., & Widiger, T. A. (2013). Dimensional models of personality: The five-factor model and the DSM-5. *Dialogues in Clinical Neuroscience*, *15*(2), 135. PMID:24174888

Turing, A. (1950, October). Computing machinery and intelligence. *Mind*, *59*(236), 433–460. doi:10.1093/mind/LIX.236.433

Vella, D. (2014). Modeling the semiotic structure of game characters. *Proceedings of DiGRA 2014.*

Wiley, N. (2003). Emotion and film theory. *Studies in Symbolic Interaction*, *26*, 169–187. doi:10.1016/S0163-2396(02)26012-3

Yik, M., Russell, J. A., & Steiger, J. H. (2011). A 12-point circumplex structure of core affect. *Emotion (Washington, D.C.)*, *11*(4), 705–731. doi:10.1037/a0023980 PMID:21707162

Yin, M., & Sun, Y. (2015). Human behavior models for virtual agents in repeated decision making under uncertainty. *Proceedings of the 14th International Conference on Autonomous Agents and Multiagent Systems.*

Zhao, R., Papangelis, A., & Cassell, J. (2014). Towards a dyadic computational model of rapport management for human-virtual agent interaction. In Intelligent Virtual Agents. IVA 2014. Lecture Notes in Computer Science, (vol. 8637). Springer. doi:10.1007/978-3-319-09767-1_62

ADDITIONAL READING

Feldman Barrett, L. (2017). *How emotions are made: The secret life of the brain.* Houghton Mifflin Harcourt.

Hudlicka, E. (2008, August). Affective computing for game design. In Proceedings of the 4th Intl. North American Conference on Intelligent Games and Simulation (pp. 5-12). McGill University Montreal, Canada.

Hudlicka, E., & Broekens, J. (2009, September). Foundations for modelling emotions in game characters: Modelling emotion effects on cognition. In 2009 3rd International Conference on Affective Computing and Intelligent Interaction and Workshops (pp. 1-6). IEEE.

Marsella, S. C., Carnicke, S. M., Gratch, J., Okhmatovskaia, A., & Rizzo, A. (2006, August). An exploration of delsarte's structural acting system. In *International Workshop on Intelligent Virtual Agents* (pp. 80-92). Springer, Berlin, Heidelberg. 10.1007/11821830_7

Popescu, A., Broekens, J., & Van Someren, M. (2013). Gamygdala: An emotion engine for games. *IEEE Transactions on Affective Computing*, *5*(1), 32–44. doi:10.1109/T-AFFC.2013.24

Scherer, K. R., Bänziger, T., & Roesch, E. (Eds.). (2010). *A Blueprint for Affective Computing: A sourcebook and manual.* Oxford University Press.

Vick, E. (2009). *Emotion notions: Modeling personality in game character AI.* Cengage Learning.

Warpefelt, H., & Verhagen, H. (2015). Towards an updated typology of non-player character roles. In *Proceedings of the International Conference on Game and Entertainment Technologies.*

KEY TERMS AND DEFINITIONS

Affective Loop: A cycle of sensing and reacting between an agent and its environment. It models the process of human interaction with things external to the body.

Agency: The apparent freedom of a character to act, react, not act, or sublimate desire to achieve a goal within a fictional world.

Agent: (In software development) An artificially intelligent software object that can monitor a program's state and execute commands autonomously to achieve those goals. (In literary and performance theory) an object or being that appears to behave with an intention to achieve a goal. Actions taken toward goal attainment foreground the agency of the object or being.

Appraisal Variables: A set of values categorically assigned to objects that an agent senses that can affect the emotional state of the agent.

Computational Cognitive Psychology: A subdiscipline of Cognitive Psychology and Cognitive Science that draws from Neuroscience and Communication Theories to develop models of cognitive experience that can be programmed as computer-based application and tested with data collected from human subjects.

Emotion Model: An abstract representation of the components that comprise the process for an agent of sensing appraisal variables in objects, identifying objects, processing objects in relation to beliefs, desires and intentions, and eliciting emotions.

High Resolution Animation: An animation method, often assisted by motion capture technology, that uses very detailed movement to animate a photorealistic game character.

Non-Player Character (NPC): Characters in a video game, excluding the one performed by the player during gameplay.

Performative Mediation: The process of integrating information outside the domain of the game design for the construction of a character. An actor's personal experiences, an awareness of the distorting features of camera optics. These are external information that often is considered in an actor's performance preparation for a character and its role.

Player Character: The character in a game performed by the player.

Virtual Agents: An artificially intelligent software object programmed with an emotion model with the intent of teaching or assisting a user complete a task or learn a process.

Chapter 3
Extension of Clinical/ Psychological Approach Using Post Narratology:
The Proposal of a Psychological Narratology and Review for Systematization

Kai Seino
National Rehabilitation Center for Persons With Disabilities, Japan

Shun Ishizaki
Keio University, Japan

ABSTRACT

The purpose of this chapter is to examine the psychological approach from a narrative viewpoint. In addition, the authors discuss the possibility of collaboration with artificial intelligence (AI). They reviewed the narrative approaches in psychology. At result, they showed how the narrative approach in clinical domain was expanded. The following possibilities exist regarding a narrative generative system in a narrow sense: (1) It is the construction of a fluid system. (2) The systems generate a tale communally. (3) There are no clear beginnings and endings. (4) It is a narrative generation through a dialogue. Next, the following possibilities exist regarding a narrative generative system in broad sense: (1) It is the modeling of the stance of the narrative. (2) It is the modeling of narrative generation by the community. (3) It is the modeling of the process. (4) It is the study of the interlocutory peculiarity.

DOI: 10.4018/978-1-7998-4864-6.ch003

INTRODUCTION

In this chapter, we examine the potential of combining the narrative as a problem-solving method with computational techniques such as artificial intelligence (A.I.), and the applicability of the integrated approach in a clinical environment. As a general term, "narrative" is understood as the "act of discussing and being discussed using language (Spoken Language and written language, etc)" (Yamada, 2007b). "Narrative" can also be defined more narrowly as in the term "narratology," which refers to the reproduction of real or fictional events and is always accompanied by time constraints (Prince, 2003).

Bruner (1987) suggested that the narrative provides individuals with a method for understanding and thinking about their everyday lives and interactions with others. In recent years, the narrative approach has been utilized in clinical domains such as medicine, nursing, psychology, and welfare. The narrative may be thought of as an evidence base containing conventional, objective data. It may also be conceptualized as a narrative base, wherein individual subjectivity and originality are considered important, and this conceptualization of narrative leads to narrative-based medicine and narrative-based practice. Furthermore, an evidence base is considered objective science and is based on modernism, whereas a narrative base is relative, post-modernist, and post-structuralist.

In the clinical domain, the approach is key to any narrative, and is called the narrative approach. For example, representation is a narrative therapy in psychotherapy. However, the conceptualization of a narrative in recent years has not been limited to the narrative approach and is considered to have a wider extent and possibility. One key to understanding the wider possibility of a narrative is the term "narrative generative system," which is advocated by Ogata, the author and editor of this book. The narrative generative system conceived by Ogata not only considers the structure or form of a story, but also incorporates the process by which the story is constituted and received. In addition, "computational narratology," one of the underlying theories of the narrative generative system, is a concept resulting from the fusion of computational technology and story. Authors have divided narratology into literary narratology and psychological narratology from the viewpoint of a narrative generation system (Seino 2016). A literary narratology considers the contents and form of a story, whereas psychological narratology is called a story treatment in the clinical domain and is a narrative approach.

Against the backdrop of Ogata's theory, Seino et al (2016, 2018, 2019) proposed a new technique for data mining designed to provide support for persons with disabilities. The technique provides a method for the analysis and generation of a story, through a combination of text-mining technology and the viewpoint of literary narratology. The results demonstrated that a theory could be established based on a narrative generative system. However, to investigate the practical implications of the system for the theory and method of a psychological narratology, a discussion of technology (e.g., computer, software, coding) is omitted. Therefore, the psychological approach is presented in this chapter from the viewpoint of a narrative, and aims at clarifying the possibility of a psychological narratology.

As aforementioned, the purpose of this chapter is to explore the theory and the method of using narrative in the psychological approach. Medical treatment and support for any illness that exists in the clinical domain is defined as a psychological approach. In this chapter, the possibility of extending the psychological approach through collaboration with AI or cognitive science has been examined. Specifically, we review the research and practice of psychology, psychiatry, and a proximity domain in recent years and their use of narrative. Our discussion is broadly focused on the approach and practice relevant not only to narrative therapy but also to a narrative, or to narratology.

A DEFINITION AND THEORY OF A NARRATIVE

Definition of "Narrative"

A narrative has been defined in multiple ways, and there are also many theoretical positions on the topic (Riessman, 1991). For example, one definition of narrative is that it is an explanation of the experience of a continuous story or people (Bloor et all, 2006). A narrative may also be communicated using several methods (Clandinin, 2006). For example, it is possible for a narrative to originate from a document, such as an interview, a diary, a biography, or an ethnography (Liebich, Tuval-Mashiach et all, 1998). In this study, three types of narrative are defined: first, a literary narrative is a story found in literature, including mythology and folk tales; second, a personal narrative is a story of an individual and family; third, a social narrative is a social story about a group of people. In addition, narratology, which is the theory of a narrative, is defined in the narrowest sense in the literary theory of "literary narratology." Moreover, a narrative may be used outside of the field of literature, as in a clinical domain such as psychology, medicine, nursing, and social work that provides services to support persons. The combination of narratology and the use of a narrative in a clinical domain is called "psychological narratology." Both literary narratology and psychological narratology are explained in detail later in this chapter.

Literary Narratology as a Basic Theory

The narrowest definition of narratology is regarded as a literary theory, which is the study of the nature, form, and function of narrative (Prince, 2003). This section gives an overview of the theories of narratology proposed by Propp (1968) and Genette (1980). Ogata (2010) organized narratology into five categories that establish the theoretical underpinnings of narratology: (1) "The Poetics" of Aristotle (1997); (2) structuralism, which developed based on the ideas of Saussure; (3) the literary art movement of Russian Formalism (Kuwano & Oishi, 1988); (4) British and American literary theory (Yamoto, 1974); and (5) the literary theory of structuralism (Tsuchida, Aoyagi, & Ito, 1996).

Narrative comprises "story" and "discourse." According to narratology (Prince, 2003), the story is the content of the narrative while the discourse concerns the expressive side. Narratology employs an analysis of both story and discourse, and the relation between the story and discourse has become the main theme of narratology. To illustrate, Russian Formalism, which is one of the theoretical underpinnings for narratology, regarded the components of narrative as being fabula (corresponding to the story) and sjuzet (corresponding to the discourse). Forster (1927) defined the components of narrative as "story" and "plot" (i.e., discourse). He observed that "story talks about a case and an event according to progress of time," whereas the plot focuses on the causal relation of incidents and events. Propp (1968) described the fundamental components of any Russian folktale. In particular, he defined the narratological notions of "function" and characters' "roles." "Function" is a character's act defined from the viewpoint of its significance in the story, and Propp described 31 types of "functions" that compose Russian folktales. Characters' "roles" are classified based on their range of performable "functions" and, as a result, he described the following seven roles": (1) villain, (2) dispatcher, (3) helper, (4) princess or prize and her father, (5) donor, (6) hero, and (7) false hero. Genette (1980) classified narrative into three components: story, discourse, and narrating. Furthermore, Genette (1980) divided discourse into three categories: (1) "tense," which is the temporal relation between discourse and story; (2) "mood," which is the reproduction of the story through the discourse; and (3) "voice," which is the relationship between the narrating and the narrated and the relationship between the narrating and text. Tense, mood, and voice are further divided into several sub-items.

In applied studies of narratology, some researchers analyze and generate stories based on the theories constructed by Propp and Genette (Akimoto & Ogata, 2013; Ogata, 2011; Klein et al, 1974). For example, in the study of the narrative generation system, Ogata (2015) devised a conceptual dictionary that can provide the basic components of narrative; specifically, character, objects, and place. Regarding these applied studies, Ogata (1999) suggested a study framework called "computational narratology." This research offered a literary study frame that fused an understanding of narratology with the generation of narrative through the use of AI and cognitive science, thereby broadening the concept of computational narratology.

Interdisciplinary Narrative Generative System

In this section, "interdisciplinary narrative generative system," which is derived from Ogata's narrative generation system, is outlined. Considerable research on story generation has been conducted in the domains of cognitive science and AI. Ogata and Kanai (2010) have completed advanced research on the interdisciplinary approach for the narrative generative system, which introduced narratology and literary theory. In addition, Ogata (2011) proposed the concept for the narrative generative system, which continues to be developed at two levels: at a broad level, the narrative generative system constitutes a method for symbolically understanding a human being and society in aggregate; and at a narrow level, a narrative generative system consists of a computer program. Recently, research in the clinical domain, which is based on Ogata's concepts, has been progressing. Specifically, the research is on disease and disability, and the results will be used for clinical support (Aoki et al, 2017; Seino et al, 2016, 2018). This research realizes the concept of Ogata's narrative generative system at a narrow and broad level. Two underlying concepts in Ogata's narrative generative system are extended literary theory and multiplex tale structure (Ogata, 2001). Extended literary theories unite various disciplines and include the dynamic state analysis of a phenomenon, the experimental creation of a new literary genre, and a concept that aims at deployment (Ogata, 2001). The descriptive method of an integrative meta-level is theorized as a model of a literary phenomenon. A multiplex story structure encompasses the idea of integrating the understanding and explanation of a literary phenomenon (Ogata 2000), and it is performed by multiplexing the process of generation-acceptance.

The underlying idea of Ogata's narrative generative system is the method of the meta-level which unifies a discipline. It is a model which aims at analysis and a certain creation of a phenomenon. In addition, with multiplex story structure, a phenomenon can be realized as a model that is understood and explained by the multiplexing of a generation-acceptance process. Authors argue about the kind of suggestion this idea may provide to practice and research in the clinical domain in the "Discussion" section.

The Research Framework for Narrative

Several studies have suggested how to treat a narrative. The relationship between the concept of the narrative and the usage of the narrative varies (Clandinin, 2006). The content of the narrative fluctuates as well (Mishler, 1995). In a macroscopic position, the framework of a narrative includes all life activities. In a micro position, a narrative (story) expresses the causalities of an event (Riessman, 1993). In addition, the narrative approach is defined as "the method of making the concept of a narrative key and

Table 1. A humanities and social science classification system, and the contained discipline

	Classification	The contained discipline
1	Humanities / social science research that treats narrative directly	History, the philosophy of history, literary criticism, psychology
2	Humanities social science research relevant to narrative	European philosophy, hermeneutics
3	Practice research that utilizes narrative	Clinical psychology and sociology, risk psychology, economy and business administration, folklore, engineering works and city planning

presenting a certain phenomenon" (Noguchi, 2009). Conversely, researchers have criticized research on the narrative. Nelson (2011) noted that qualitative research based on the narrative was not fully discerned and the meaning of the research was not fully explained to the reader.

Numerous disciplines study narratives. Fujii et al (2001) classified the research in connection with the story of the discipline of humanities and social science into three categories (Table 1). In addition, with regard to the entire discipline, Ogata (2018) classified the division between a scientific domain and a story into three more categories (Table 2).

Table 2. The division between a discipline and narrative

	Classification	The contained discipline
1	Symbol level [Literature, art, and an entertainments domain]	Narratology, poetics, rhetoric, literary criticism, linguistics, folklore, folktale study, and an old tale, entertainment research, Noh play theory, kabuki theory etc., cultural anthropology and folklore, philosophy and thought, movie theory and cinematic review
2	A brain, a nerve, and a psychological level [Psychiatry and a psychological domain]	Psychology, psychiatry and psychopathology, psychoanalysis, tale treatment, neuro-psychoanalysis, brain science (The author added a postscript.)
3	Social level [Social domain]	Sociology, history, business administration, economics, jurisprudence, marketing theory, and advertising theory

Moreover, Mishler (1995) classified research into three categories as Models of Narrative Analysis (Table 3).

THE NARRATIVE IN CLINICAL / PSYCHOLOGICAL APPROACH

A Clinical Narrative Turn and Narrative-Based Medicine (NBM)

The purpose of this research is to contribute to problem-solving for clinical issues. Therefore, we examined psychological narratology from the viewpoint of a narrative, and discussed its extension and the possibility of applying it to other areas of research.

The conventional clinical domain attaches importance to evidence and the basis for demonstrating that the evidence is objective. The concept of an evidence-based approach originated from evidence-based

Table 3. Models of narrative analysis

	Category	Subcategory
1	Reference and temporal order: The "telling" and the "told"	1. Recapitulating the told in the telling 2. Reconstructing the told from the telling 3. Imposing a told on the telling 4. Making a telling from the told
2	Textual coherence and structure: Narrative strategies	1. Textual poetics: Figuration, tropes, and style 2. Discourse linguistics: Oral narratives
3	Narrative functions: Contexts and consequences	1. Narrativization of experience: Cognition, memory, self 2. Narrative and culture: Myths, rituals, performance 3. Storytelling in interactional and institutional contexts 4. The politics of narrative: Power, conflict, and resistance

Source: Mishler (1995)

medicine (EBM) in the 1990s (Guyatt, 1991; Sackett, Richardson, Straus, Rosenberg, & Haynes, 1997), and there is objective evidence in this regard. Sackett et al. (1997) define EBM as the "the conscientious, explicit, and judicious use of current best evidence in making decisions about the care of the individual patient." After the introduction of EBM, the concept was extended to other fields such as nursing, welfare, psychology, and education. Conversely, the idea of "evidence-based" has been reasonably criticized and the following criticisms have been made: (1) it downplays qualitative studies on the hierarchy of evidence (Concato, 2004; Manuel, Fang, Bellamy, & Bledsoe, 2008); (2) in the framework of objective fact, it is all but impossible to acquire the opinions of socially-oppressed minorities, and therefore, this method fails to capture a pluralistic reality (Gergen, 1994a); and (3) it is also impossible to imitate the medical EBM model in a practice domain based on social sciences because the EBM model assumes only natural sciences as a research foundation (Yamabe, 2012). Based on the limits of objective facts, social constructivism, which provides the relativist conception of reality, has come to be recognized (Gergen, 1994b, 1999). Against the backdrop of social constructivism, a trend that allows a reconsideration of EBM from the perspective of the narrative is Narrative Based Medicine (NBM) (Greenhalgh & Hurwits, 1998). Saito (2003) identified the following five points as elements of the NBM approach: (1) illness as a story; (2) the patient as a storyteller; (3) the allowance of ambiguity of phenomena; (4) the insignificance of linear cause and effect (it is not considered important that a specific cause and effect should be joined together linearly); and (5) dialogue as a treatment. The narrative approach has paid close attention to the narrative understanding of NBM. Since the 1990s, the narrative approach has attracted greater attention in interpersonal help domains, such as medical care, nursing, and psychology. The development of epistemology and the methodology that attached importance to narrative is known as the "Narrative Turn." (Denzin & Lincoln, 2000). Noguchi defined this narrative approach as "going closer to reality with the clues in the forms of narratives" (Noguchi, 2005).

From this background, we may be able to develop an effective method if evidence and a narrative can be unified by the integration of quantitative and qualitative analysis in a clinical context. In this chapter, it may become a help of integration of evidence and a narrative to examine integration of psychological approach and computational technology by making a narrative into a key.

The Ideological Background of Narrative Approach

The narrative approach is set against the context of the philosophy of post-modern thought and post-structuralism, and the theory of social constructionism. Post-modern philosophy originates in semiology or literary criticism from the first. Therefore, post-modern ideology has a tale and a deep relation. In addition, social constructionism was greatly impacted by the study of the concept of the text theory by Derrida (1978), and the power of Foucault (1975). The social constructionism way of thinking can be summarized in the following five points. (1) Reality is socially constituted by language. (2) It is social practice to begin to develop a meaningful language, and existence of the individual mind is not regarded as evident. (3) Meaningful discourse does not exist without a "relation." If meaningful discourse does not exist, the "object" and the "act" cannot occur. (4) Feelings such as "It is right" and "this is good" are produced out of a community. However, there are many instances in which "right" and "good" were limited to the community. (5) Any enforcements must be open to evaluation and criticism (Gergen, 1994). In other words, social constructionism brought about the paradigm shift that converted the method of actual recognition into relativism from conventional absolutism. These theories are used to analyze personal narratives and social stories in the context of society and culture. Lyotard (1979) described a category of classification of "grand narratives" that can be used to justify various supporting stories that are seen as "little narratives." Foucault (1980) accounted for a classification of "a dominant story," which influences a certain situation, and "an alternative story," which appears when the dominant story is doubted. Yoshimoto's (1968) community illusion may also be connected to dominant and alternative story. Therefore, it was claimed that a tale is an indispensable concept for the interpretation of man's world or its historical recognition (Danto 1965).

REVIEW FOR SYSTEMATIZATION OF A PSYCHOLOGICAL NARRATOLOGY

The system of a psychological narratology is proposed in this section, which shows that the clinical / psychological approach is extensible through post-narratology. Post-narratology is a narrative generative system that includes story generation by A.I. In order to actuate story generation, a literature review is performed with a clinical/ mental approach considering that the extension to post-narratology is possible. The system then constructs the story through AI and authors argue about the ability of a bridge to be carried out and whether by re-extending the clinical/mental approaches through story generation, a clinical contribution is possible. These approaches are concretely aimed at the approach (research and practice) in clinical domains, such as psychology. Moreover, what can be called a narrative approach is dealt with from the viewpoint of post-narratology, in which the possibility of extension of the present psychological narratology is proposed. A psychological narratology is the theory of the clinical/ psychological approach that is relevant to the narrative in the broadest sense that the author has proposed. Therefore, the following psychological narratologies are outlined and considered.

- Narrative therapy: A type of psychotherapy and family therapy/ The client pays to talk to the clinician.
- Self-help group: A type of group treatment in which one receives support from peers (companions or the parties concerned)
- Open Dialogue: A therapeutic intervention for an acute term mental disease, in which a dialogue is considered important.

- Tohjisha-Kenkyu (In Japanese) (Self-directed studies): An empowerment approach produced out of the life movement of the affected party.

Psychotherapy

Outline

In clinical psychology or the counseling field, a "narrative" has come to be associated with the narrative therapy of psychoanalysis, analytic psychology, a gestalt therapy, and family therapy since the 1980s. In a narrow sense, psychological narratology is narrative therapy in psychotherapy and family therapy. Narrative therapy was advocated by White and Epston (1990) and it developed through the context of family therapy in the 1980s. The core belief was an awareness of the importance of how narration and the tale of the client are respected. White and Epston (1990) applied a classification of Foucault to narrative therapy, aiming to excavate the new "alternative story" from the "dominant story" that restricts life. This excavation is conducted by "externalizing" and seeking a "unique resolution" for the problem.

"Formation of outside," "discovery of unique outcome," "re-writing," a "remembering," a "definition festival," "outsider witness," and a "therapeutic letter" are among the techniques used in this process (White & Epston 1990). In addition, Spence (1983) described the occurrence in which a narrator is in treatment, and a hearer hears the talk of a narrator as a tale. In analytic psychology, as with traditional clinic psychology, the role of a "tale," like the myth, is touched upon by Kawai (2001), who presupposes that the process of psychotherapy is a tale. In addition, in Gestalt therapy, Polster (1987) presupposed that the recovery of harmony of the heart requires a dialog, and proposed the method of making achieving this harmony is by having a real person sit on an "empty chair" that is assumed to be there and engage in a dialog.

Consideration of a proposal

Narrative therapy creates a narrative that provides the client with a tool for problem solving. A typical narrative approach heads toward the solution to a question as the client completes a tale. We propose the following possibility from the viewpoint of post narratology: what is treated is the discovery or description of a "dominant story" and an "alternative story," and the technique of externalization. There are two implications of this suggestion. First, the "dominant story" and "alternative story" may be identified in order to construct the theory of a narrative generative system. Second, computational technology may be able to realize two stories, and if so, a narrative generative system may be expanded.

Other Psychological Narratologies

There are multiple other psychological narratologies that have been advanced in the last 30 years. Elliott (2005) divided narratives into two classifications: a "first order narrative" and "second order narrative." The former focuses on individual's discussion of their own experience, whereas the latter centers on researchers' discussion about a personal quest to understand the social world. McAdams and Aubin (1998) analyzed a recited life story and, through the classifications of "a story of recovery" and "a story of the turning point," identified fixed forms, such as "sacred stories."

Moreover, there is a theory known as "literary narratology about story and discourse," that includes psychological narratology. Bruner (1986) classified forms of narrative into the (1) "narrative mode," which includes the time-linking information about an event, and (2) "paradigmatic mode," which includes information about the necessary relations between multiple events as well as causal relations. Morgan (2000) views a story (corresponding to the narrative) as "an event connected on temporal axes in succession which becomes a plot (corresponding to the discourse)." Adam (1999) contended that a narrative is composed of time, characters, predicates, and events that can change the predicates. Yamada (2007a) distinguished four levels of narrative: (1) real, (2) mutual action, (3) text, and (4) model. In line with Yamada's classification, Shojima (2008) classified narrative into four acts: (1) the side of mutual acts, (2) the side of the social act, or social historic linkup, (3) something that cannot be discussed because it exists outside of the narrative, and (4) multiple lives or voices surrounding the life of the narrator.

Some studies have examined the narrative approaches of persons with disabilities. For example, Tagaki's (2008) study examined the speech of persons with disabilities, whereas others have focused on the psychological details of schizophrenia (Komori & Yamada, 2001), narrative approaches to mental disorders (Buchanan-Barker & Barker, 2008; Vuokila-Oikkonen, Janhonen, & Vaisanen, 2004), the speech of aphasic persons (Nouchi, 2008), and the life of a small child undergoing treatment for cancer (Saiki, 2008). Studies of these narrative approaches have thus far used psychological narratology. However, these earlier studies, except for the current authors' research, are yet to closely examine the term "narrative" or to adopt a literary narratology method.

Psychological Narratologies Other Than Narrative Therapy

Additional clinical approaches are discussed below that are not narrative therapy, but nevertheless contain what the authors view as psychological narratology. These approaches are related to a tale and are considered useful for the extension of a narratology and a study of a narrative generative system.

Self-help Group

Outline

A self-help group is practiced among those who suffer from the same illness and offers support through peer relations. For example, self-help groups may be formed for individuals addicted to alcohol, drugs/psychotropic substances, or gambling, or who are dealing with the same life obstacle or chronic malady. In recent years, self-help group have come to be viewed as a form of a narrative approach (Noguchi 2002). "Alcoholics Anonymous (AA)" was the first self-help group, which was developed in the United States in 1935, and is for recovery from alcohol addiction. AA was effective for treating alcohol addiction, and over time, many other groups adopted the method used by AA to deal not only with addiction or dependence but also for other problems. The typical definition of a self-help group is that of Kats &E. I. Bender (1976), who defined a self-help group as a small group of companions (peers) who provide mutual aid for the achievement of a specific purpose. AA meetings are regularly scheduled and the members who gather at each meeting talk in turn about various topics. The talk may be of experiences from one's past or the utterance of a single word may constitute a speaking turn (Noguchi 1996). The self-group does not have a technical definition (Ito, 2000).

Little consideration

The acts that a self-help group perform are the primary methods of a group. With the aim of aiding recovery, the talk among members is considered a narrative approach. Sakiyama (2018) indicated that self-help group research was a domain where research was advanced from the viewpoint of narratology. In addition, Sakiyama (2018) arranged the self group as it is what constructs "alternative story" ; it compared with "dominant story" which a professional makes.

The first-line specialty of support of a self-help group is "experiential knowledge." Experiential knowledge is the knowledge based on a member's experience, and serves as a source of support for a self-help group. The knowledge of members is more practical and more comprehensive than the knowledge of professionals (Borkman 1976). Each member's "experiential knowledge" is discussed and the collective knowledge if the group is accumulated and utilized as one tale knowledge from a group. We can interpret sharing of knowledge as the process of making one big narrative from two or more members' narratives. The process may be adopted into theory as a narrative generative system and may be realized as a system. For example, studies have proposed collect many texts that were spoken and generating the tale based on some typical elements. In addition, Ito (2000) examined a self-help group from the perspective of an individual tale. The results of the study indicated that the tale should have been regarded as an object with the effect of showing not the collective knowledge system of a group, but of a participant's self. In other words, participants produced or consumed "the good talk," through which each participant realized a sense of togetherness with self-excellence and the group. To determine the degree of change for each participant, it was presupposed that it was important to determine the standard by which a participant made a tale "the good talk." In other words, it was important for a tale or suggestion to achieve a set basis for valuation. An evaluation of the tale could be conducted based on how the group absorbed the tale and how it affected the production and consumption of future tales.

Open Dialogue

Outline

In recent years, Open Dialogue (OD) is another psychological approach that has attracted attention in Japanese psychiatry. OD started in Finland in the 1980s and is a method of therapeutic intervention for an acute-term mental disease. However, an open dialog is not a mere technique or system. Seikkula, a central figure who developed the open dialog emphasizes that OD is a "technique," "philosophy," and a "way of thinking," but not a "medical treatment process." In OD, two or more staff conduct a crisis intervention within 24 hours of a request. Patients and families practice OD every day by sitting in a circle and performing "OD." OD seldom uses a drug but makes a patient's critical situation cease (Saio 2014). All matters are discussed and determined as a characteristic portion of OD at the members' meeting place. In addition, an expert team discusses a patient and a family (exchange) in their presence, as they reflect on what is being shared with them. Olson, Seikkula, and Ziedonis (2014) summarized the following key OD items:

The Twelve Key Elements of Fidelity to Dialogic Practice in Open Dialogue (Olson, Seikkula, and Ziedonis, 2014)

1. Two (or More) Therapists in the Team Meeting

2. Participation of Family and Network
3. Using Open-Ended Questions
4. Responding to Clients' Utterances
5. Emphasizing the Present Moment
6. Eliciting Multiple Viewpoints
7. Use of a Relational Focus in the Dialogue
8. Responding to Problem Discourse or Behavior in a Matter-of-Fact Style That is Attentive to Meanings
9. Emphasizing the Clients' Own Words and Stories, Not Symptoms
10. Conversation Amongst Professionals (Reflections) in the Treatment Meetings
11. Being Transparent
12. Tolerating Uncertainty

Little Consideration

OD is considered to be a narrative approach given that a conversation called a dialog becomes a key medical treatment. OD contains narrative therapy and a common feature of OD is that it has an ideological background. According to Saito (2014), OD has various philosophical bases. It is a basedof Bateson's double bind theory, Bakhtin's polyphony theory of a dialog principle, and Vygotsky's theory of psychology. In addition, according to Saito (2014), OD is perceived of as a cure within the context of critical psychiatry or post psychiatry (Thomas & Braken, 2004). Post psychiatry considers the technique of scientific psychiatry and seeks new moral medical treatment, given that the present psychiatry is criticized. In addition, it is also a common feature with narrative therapy to use reflecting. The 12 elements of OD consider tale generation and its effect, and the process is thought provoking. A dialogue is performed by two or more members of various positions. In addition, open-ended questions are used as an interlocutory trigger. OD may be able to be realized as an automatic dialogue system or a narrative system by speech recognition. By pulling out two or more viewpoints and observing relationships, an automatic dialogue system may be able to return utilize an extraction method to return an element as the narrative is being generated.

Tohjisha-Kenkyu (Self-directed studies)

Outline

In recent years, the "Tohjisha-Kenkyu" approach that originated in Japan has attracted significant international attention. Tohjisha-Kenkyu may be introduced as an open dialog in Japan (Saito 2016) and is an approach commonly used with patients with schizophrenia. In recent years, it has also been developed as a treatment for persons with developmental disabilities. Tohjisha-Kenkyu is an empowerment approach, and it happened among the activity and liveing of the person with disabilities who held schizophrenia etc (Mukaiyachi, 2013).

There are seven main characteristics of Tohjisha-Kenkyu (Mukaiyachi, 2013). (1) Tohjisha-Kenkyu is a self-help program for the parties concerned, such as individuals with schizophrenia who live in the local region. The party concerned becomes "a hero of difficulties." (2) The wisdom that dissolves difficulty is in one's experiences of the experiences of one's companions. (3) The areas of exploration are the difficulties of life faced by the concerned party. (4) The party concerned with a disability investigates his symptoms with companions. They cooperate with all involved parties and talk with one another to explore difficulties in an open-minded manner without judgment. Occasionally they develop a figure (picture) and action plan. An important discovery is the identification of a pattern and the mechanisms by which a difficult even occurs and the meaning and possibility that exist in a difficult context. (5) While repeating positive trial and error, a unique understanding and idea that are produced improvisatorially may lead to an important discovery of "how to help oneself." The means for helping oneself are found there, in a unique understanding of oneself. The participant values sharing and implementing their discoveries in life and sharing with a friend. (6) It is a method of mere "problem solving." It changes one's attitude toward facing the occurrence considered to be a problem, and determines how to resolve it. It examines and cherishes the "language" that related the experience in which a person has lived, and lives now. It can be called a "program of language" that transforms "being connected" with detailed "acting" into an actual life scene. (7) The Tohjisha-Kenkyu program is necessary and effective for a supporter.

Little consideration

Tohjisha-Kenkyu provides suggestions for examining a narrative generative system and how to connect narrative generation for problem solving. In Tohjisha-Kenkyu, it is assumed that the party concerned will themselves become a hero of difficulties. In addition, Tohjisha-Kenkyu allows the narrative to pass and has suggested that it is a narrative approach tied to the problem solving of the party concerned. In addition, Noguchi (2002) reported that Tohjisha-Kenkyu is "a community of narration" that makes it "the community of the tale." Many audiences are in a "narrative community," and new narration becomes more certain with an existing audience. A narrative community is the space wherein new narration is shared and established. In addition, the aspects of changing attitudes, and how to identify and position attitudes Tohjisha-Kenkyuare are also common in narrative therapy. Moreover, the conceptualization of one's own experience and that of a friend is similar to the notion of "experiential knowledge" in a self-help group. In addition, when considering tale generation, an improvisatorial, unique understanding or an idea is important and shows how to utilize the result, which creates a "relation" as a "behavior" in a scene received by the narrative generative system.

Approaches relevant to other narratives

The approaches relevant to other tales are enumerated as a reference. Arai (2014) identified narrative therapy, collaborative therapy, and reflection as narrative approaches from the perspective of social work. Arai (2014) also reported that narration of the illness, the self-help group Tohjisha-Kenkyu, the human library, and the narrative interview are approaches related to the narrative approach. Other than this, authors show "the social story for an autistic child (Gray 2000) ", and "the Wellness Recovery Action Plan for a person with disabilities (Copeland 2002)" as approach relevant to a narrative. In addition, Kato (2010) reported that an item on narrative was considered in preparation of ICD-11 (the International Statistical Classification of Diseases and Related Health) by the WHO (the World Health Organization). However, the item on narrative was not adopted into ICD-11, which was released in 2018.

PSYCHOLOGICAL APPROACH AND NARRATIVE GENERATION BY A.I.

In this section, two studies of the clinical/ psychological approach and the possibility of narrative generation by A.I. are introduced.

Narrative Generation For Employment Support of Persons With Disabilities

In this section, the authors' research is introduced, in which we extracted text elements and employed a system of story generation using A.I. Specifically, we analyzed a disabled person's free descriptive answer to an interview question using the method of "informatics tale analysis" to extract the elements of the story. This method combines the technology of text mining, and some theories of a literary narratology. Text mining involves the quantitative analysis of text data based on natural language processing or data mining. After we extracted the element of the story from the individual's text response, the system proposed creating a collectivestory.

An example of the narrative is shown below. Please note that to protect respondents' privacy, some answers are altered, but the basic meanings remain unchanged. In the authors' study, both the wider and narrow senses of narrative were used depending on the particular contexts of the analysis.

[Concerning work description] In response to the question: "Describe your job in detail."

Example 1: Since my legs were injured, I have been assigned paperwork. I am now engaged in computer work. (Physical disability)

[Concerning care and support at the workplace] In response to the question: "Specify the care and support provided in your workplace as a means of assisting your career."

Example 2: At my workplace, employees often get acquainted with coworkers at informal meetings, such as drinks parties. I did not attend these parties because of my disease, so I had no opportunity to get to know them. The other day, I took the opportunity of participating in one, and it worked much more effectively than I had expected. (Mental disorder)

Example 3: I first wanted to work without mentioning my disease. The other day, a relative of a person who had supported me asked me to work in their office. When I began working there, I soon found that this relative did not care about my condition, so it was a disastrous year. From this experience, I realized that I should conceal my disease because people consider me to be inferior to healthy people. (Intractable disease)

The authors' study analyzed the descriptive texts that reflected the personal stories of persons with disabilities, and clarified social stories of disability based on the results of multivariate analysis involving text mining. Note that the descriptive texts analyzed in this study were narrative in the narrow sense while others were narratives in the wider sense.

This is an example from our research entitled "Employment Support and Story Generation for Students with Developmental Disabilities." The results indicated that story generation tends to materialize through the combining of existing elements with those from previous research (Seino, Enomoto, & Miyazawa, 2018). The figure below shows a story generative system.

The analysis of job descriptions analysis results were used to create the image. For cases in which this system was created, it was important to consider how elements were combined. The probability of a story being generated is attained by calculating the probability that it may happen and the strength of the relationship between the text elements. In addition, in order to utilize the result of the support generation simulation, it is important to understand what kind of elements have been collected. For example,

Figure 1. The image of the job description story generative system

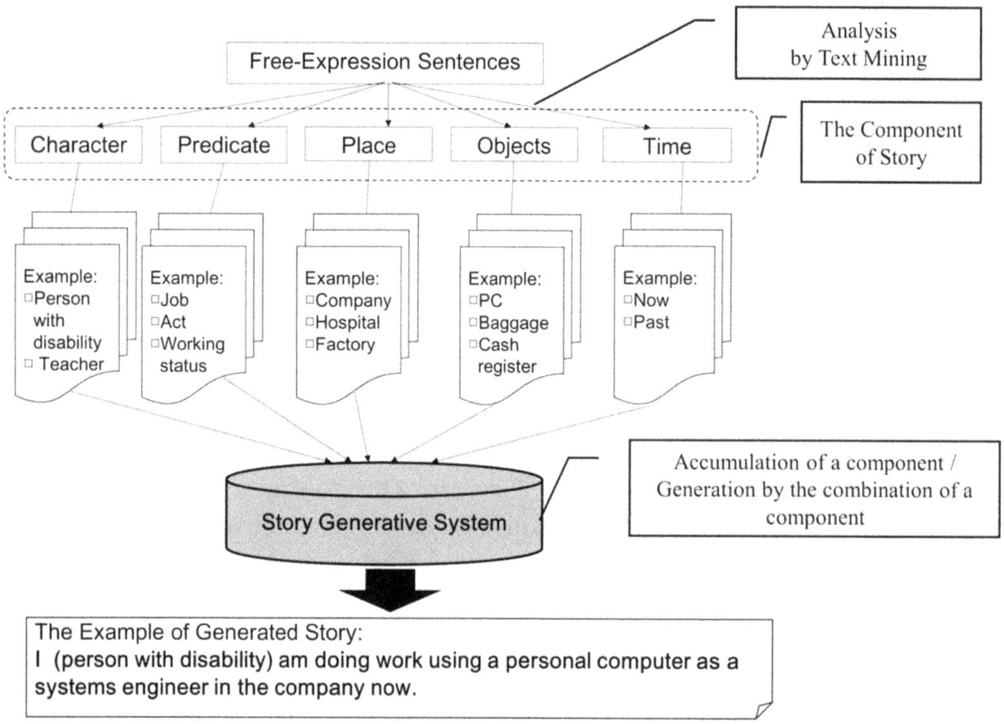

the elements of support needs may be collected based on responses to a specific problem, and it may be possible to examine what kind of problem and support needs occur at what time, thus enabling the preparation of appropriate support.

Narrative Generation for Paper Creation Support of a Student With a Developmental Disease Tendency

Joint research by Ogata and Aoki et al. (Aoki, et al. 2018) proposed students' paper creation support and narrative generative system based on the cognitive tendencies and action patterns of individuals with autism spectrum disorders. Aoki et al (2018) suggested that a paper creation might address the problem of a student with autism spectrum disorder. they explained why the problem occurs based on the cognitive tendencies and the action patterns of people with autism spectrum disorder, who may be sensitive to "a surprise and a gap" in the "discontinuous nature" of a story. Therefore, a story (paper creation), after making it continuous, stops the reaction of a student with autism spectrum disorder. The evaluation mechanism was set up and support was proposed for advancing a story by issuing directions on a macro and a micro level.

DISCUSSION

This section has following discussions. First, literary narratology as basic theory is compared to psychological narratology as a practice or application theory. Second, the extension and possibility of a psychological narratology are considered.

Comparison of Two Narratologies

Literary narratology focuses on a literary narrative while a psychological story centers on social and personal stories. Both literary and psychological narratologies share some characteristics. For example, Russian Formalism, Forster and Bruner, and Morgan all reached the consensus that narrative can be classified into two elements: (1) those aspects of the narrative that align along the event in a time lapse (corresponding to the story) and (2) those aspects of the narrative that align along the event in a causal relation (corresponding to the discourse). Additionally, an interest in the components of the story is seen commonly in Propp (1968), Greimas (1966), Ogata (2015), and Adam (1999). The story consists of characters (including their roles and actions), functions, time, objects, places, predicates, and events that can affect the predicates. Additionally, Genette's (1980) classification of "voice" is divided into the time for narrating, a standard, and the persona of the narrator. The words and narrative have similar elements as the classification by Elliot (2005), Yamada (2007a), and Shojima (2008). In other words, literary narratology and psychological narratology may have a common viewpoint. Conversely, there is a distinct difference between literary and psychological narratologies. Unlike literary narratology, psychological narratology analyzes social and personal narratives, allowing the relationship between individuals, their culture, and society to be clarified.

Extension and Possibility of a Psychological Narratology

In this chapter, the psychological narratology was divided into some items and reviewed. The authors showed that the narrative approach in clinical domain, defined as psychological narratology, went beyond the narrative therapy in psychotherapy and was expanded to various experiments and research. Moreover, is thought that four common features are among the psychological narratologies reviewed in this chapter. As the first, many practice is not simple technology and method but stance, and a way of thinking. It is thought that this is related to narrative approach of the background of philosophy or an ideology. The approach will be fixed supposing if it is technology and a method. However, in fact, since it has narrative approach against the background of social constructionism, fixation is avoided and relativized. As the second, the narrative can point out being made communally. . Moreover, if it is a specialist and the concerned party, the concerned party will be the hero of the tale, and will become a narrator. In narrative therapy, the alternative story of a client is reconstructed in the separation between a therapist and a client. In addition, in a self-help group, an open dialog, and in Tohjisha-Kenkyu, a community is a group completes a tale. The third point is that it is considered to be an effective form of treatment that provides an account of its support effect. In narrative therapy, there is temporary directivity of an alternative story of being strong and carrying on. However, in all practice, the contents or forms which are told are not uniquely absolute and are relative. It is thought that the process of the practice has produced an effect. The practice is free dialogue among members of the community and the concerned party. The fourth point may be obvious if it seen from the method of psychological approach, but all the tales are built through a dialog. They were not monologues and were constructed through colloquial expressions.

Based on these points, the narrative generative system is considered in both a narrow sense and a broad sense. The following possibilities exist regarding the automatic tale generation machine realized as a narrative generative system in a narrow sense, i.e., a computer program etc. (1) It is the construction of a fluid system. For example, a random element is taken in accepted. This system carries out *narrative* generation by combining elements. (2)Two or more persons are the systems that generate a tale communally. (3) There are no clear beginnings and endings. For example, it increases the flexibility of an item or an element. Since the generation of a *narrative*, it can be conceived so that the system aims at the process of generation itself. (4) It is a system of the tale generation through simple dialogue but not an input.

In addition, the following is mentioned as a suggestion for the practice of this research. A psychological narratology is not merely a methodology but is also the mechanism for converting of a relative sense of values. It is given significance with conversion (or improvement) of the worth of the tale of the party concerned and the party concerned. The following specific suggestions are based on such a system. (1) Consider the community itself which generates a tale and utilizes it to be a system and as well as its construction and practical use. (2) The idea of a system and the realization that this idea places more value on the process rather than the contents of tale generation. (3) Get participation and collaboration of the concerned parties towards realization and practical use of a system.

FUTURE RESEARCH DIRECTIONS

This chapter reviewed research on the narrative approach of a clinical domain, and aimed to utilize the actual extension of a narrative generative system. Finally, it did not show a clear objective conclusion like the target narrative approach. The following suggestions should be used to guide future work in this area:

1. Examine the ideology and the philosophical background of a psychological narratology.
2. It is treating overall an ideology and philosophy, and a literary narratology and a psychological narratology.
3. Consider research of tales from the humanities and social science system as a reference for psychological narratology.
4. Consider the narrative approach of a clinical domain more broadly as a psychological narratology. In addition, the following are suggested as future research possibilities:
    a. Modeling and implementation of a system as a narrative generative system in a narrow sense (system which produces the detailed contents of language) from suggestions in this chapter.
    b. Carry out modeling or elaboration of the existing system based on suggestions from this research as a narrative generative system (system which recognizes an actual phenomenon and computes a meaning) in a broad sense.

CONCLUSION

This research defines the narrative approach of a clinical domain as a psychological narratology, and examines its extension and possibility. As a result of the review, psychological narratology was confirmed to have widely expanded beyond the so-called narrative therapy of psychotherapy and family therapy. In particular, the self-help group, the open dialog which attracted attention in recent years, and Tohjisha-Kenkyu also have many common features with narrative therapy, and are seen to be using the same narrative approach. In addition, suggestions from these practices are thought-provoking and also utilize expansion, realization, and the system of the narrative generative system in a narrow sense and a broad sense for actual problem solving. It is meaningful to consider the practice of a clinical domain from the viewpoint of a narrative, as well as in respect to clinical reduction. I expect that expansion of new discovery and practice will be performed because the party concerned, a supporter, the clinical man of deeds, and a researcher in connection with a theme, and the researcher of related domains, such as A.I., computing science, cognitive science, literary theory, and philosophy, will argue and collaborate in the future.

REFERENCES

Adam, J. M. (1999). *Le récit* [The narrative]. Presses Universitaires de France.

Akimoto, T., & Ogata, T. (2013). Towards a discourse mechanism in narrative generation system: Proposal of a system introducing narrative discourse theory and reception theory. *Cognitive Studies*, *20*(4), 396–420.

Aoki, S., Ogata, T., & Ono, J. (2018). Narrative generation related to cognitive patterns seen in ASD -From the perspective of "Surprise". In *Proceedings of the 35th Annual Meeting of the Japanese Cognitive Science Society* (pp. 652-660). Academic Press.

Aristotle. (1997). *Poetics* (N. Matsumoto & M. Oka, Trans.). Iwanami Shoten. (Original work published 335 B.C.)

Borkman, T. (1976). Experiential knowledge: A new concept for the analysis of self-help groups. *The Social Service Review*, *50*(3), 445–456. doi:10.1086/643401

Bruner, J. (1986). *Actual minds, possible worlds*. Harvard University Press.

Buchanan-Barker, P., & Barker, P. J. (2008). The Tidal Commitments: Extending the value base of mental health recovery. *Journal of Psychiatric and Mental Health Nursing*, *15*(2), 93–100. doi:10.1111/j.1365-2850.2007.01209.x PMID:18211556

Clandinin, D. J. (Ed.). (2007). *Handbook of narrative inquiry: Mapping a methodology*. Sage Publications Inc. doi:10.4135/9781452226552

Copeland, M. E. (2002). *Wellness recovery action plan* (2nd ed.). Peach Press. doi:10.1300/J004v17n03_09

Danto, A. C. (1965). *Analytical philosophy of history*. Cambridge University Press.

Denzin, N. K., & Lincoln, Y. S. (Eds.). (2000). *Handbook of qualitative research*. Sage Publications.

Derrida, J. (1978). Writing and difference. London: Routledge & Kegan. (Original work published 1967)

Elliott, J. (2005). *Using narrative in social research: Qualitative and quantitative approaches.* Sage Publications. doi:10.4135/9780857020246

Forster, E. M. (1927). *Aspects of the novel.* Edward Arnold.

Foucault, M. (1980). *Power/knowledge* (C. Gordon, L. Marshall, J. Mepham, & K. Soper, Trans.). Pantheon Books. (Original work published 1972, 1975, 1976, 1977)

Genette, G. (1980). *Narrative discourse: An essay in method* (J. E. Lewin, Trans.). Cornell University Press. (Original work published 1972)

Gergen, K. J. (1994a). *Toward transformation in social knowledge.* Sage Publications.

Gergen, K. J. (1994b). *Realities and relationships.* Harvard University Press.

Gergen, K. J. (1999). *An invitation to social construction.* Sage Publications.

Gray, C. A. (2000). *The new social story book: Illustrated edition.* Future Horizons Inc.

Greenhalgh, T., & Hurwitz, B. (Eds.). (1998). *Narrative based medicine.* BMJ Books.

Guyatt, G. H. (1991). Evidence-based medicine. *American College of Physicians Journal Club, 114*, A-16.

Ito, T. (2000). Serufu herupu gurupu to kojin no monogatari [Self-help groups and personal stories]. *Japanese Sociological Review, 51*(1), 88–103.

Kato, S. (2010). Gendai seishinigaku niokeru rejiriansu no gainen no igi [Significance of the concept of the resilience in present psychiatry]. In Resilience: The new paradigm of present (pp. 1-23). Kanehara & Co., ltd.

Katz, A. H., & Bender, E. I. (1976). Self-help groups in western society: History and prospects. *Journal of Behavioural Science, 12*(3), 265–282. doi:10.1177/002188637601200302

Kawai, H. (2001). Shinriryoho niokeru Monogatari no Igi [Significance of the "narrative" in psychotherapy]. *Japanese Journal of Psychotherapy, 27*, 3–7.

Komori, Y., & Yamada, M. (2001). Seishinbunretsubyou no kazokushinrikyouiku niokeru naratibu apurochi [Narrative approach in family psychology education of schizophrenia]. *Family Therapy Research, 18*(2), 143–150.

Lyotard, J. F. (1979). *La condition postmoderne* [The postmodern condition]. Minuit.

Manuel, J., Fang, L., Bellamy, J. L., & Bledsoe, S. E. (2008). Evaluating evidence. In R. M. Grinnell & Y. A. Unrau (Eds.), *Social work research and evaluation: Foundation of evidence-based practice* (8th ed., pp. 482–495). Oxford University Press.

McAdams, D. P., & De St. Aubin, E. (Eds.) (1998). Generativity and adult development: How and why we care for the next generation. American Psychological Association.

Mishler, E. G. (1995). Models of narrative analysis: A typology. *Journal of Narrative and Life History, 5*(2), 87–123. doi:10.1075/jnlh.5.2.01mod

Morgan, A. (2000). *What is narrative therapy? An easy-to-read introduction*. Dulwich Center Publication.

Mukaiyachi, I. (2013) *Tojisha-kenkyu toha: Tojisha-kenkyu no rinen to kousei* [What is a Tojisha-Kenkyu? The concept and structure of a Tojisha-Kenkyu]. Tojisya-Kenkyu network (https://toukennet.jp/?page_id=56)

Nelson, C. D. (2011). Narratives of classroom life: Changing conceptions of knowledge. *TESOL Quarterly*, *45*(3), 463–485. doi:10.5054/tq.2011.256799

Noguchi, Y. (Ed.). (2005). Naratibu no rinshoshakaigaku [Clinical sociology of narrative]. Keiso shobo.

Ogata, T. (1999). Proposal of computational narratology. In *Proceedings of IPSJ SIG Computers and the Humanities* (pp. 53-60). Academic Press.

Ogata, T. (2010). "Monogatari seisei sisutemu" no haikei oyobi monogatari to bungaku no aida [Backgrounds of the "narrative generation system" and "relations between narrative and literature"]. In T. Ogata & A. Kanai (Eds.), *Monogatariron no jouhougaku josetsu: monogatari seisei no shisou to gijutsu wo megutte* [An introduction to the informatics of narratology: about the thought and technologies of narrative generation] (pp. 186–258). Gakubunsha.

Ogata, T. (2011). Narrative generation system as the practice of "informatics of narratology.". *Journal of Japan Society for Fuzzy Theory and Intelligent Informatics*, *23*(5), 676–685. doi:10.3156/jsoft.23.5_676

Olson, M. E., Seikkula, J., & Ziedonis, D. (2014). *The key elements of dialogic practice in open dialogue: Fidelity criteria*. https://www.umassmed.edu/globalassets/psychiatry/open-dialogue/keyelementsv1.109022014.pdf

Polster, E. (1987). *Every person's life is worth a novel.* W. E. Norton & Company Inc.

Prince, G. (2003). *A dictionary of narratology* (Revised ed.). University of Nebraska Press.

Propp, V. Y. (1968). *Morphology of the folktale* (L. Scott, Trans.). University of Texas Press. (Original work published 1928)

Riessman, C. K. (1991). Beyond reductionism: Narrative genres in divorce accounts. *Journal of Narrative and Life History*, *1*(1), 41–68. doi:10.1075/jnlh.1.1.04bey

Riessman, C. K. (1993). *Narrative analysis*. Sage.

Saiki, S. (2008). Shounigan no kodomo no toubyoutaiken [Narrative based medicine and clinical intellect? Narrative of a pediatric cancer patient]. In Y. Yamada (Ed.), *Jinsei to yamai no katari* [Narrative of life and illness] (pp. 103–132). University of Tokyo Press.

Saio, T. (2014). Open dialogue approach to acute psychosis: Is its effectiveness well-established? *Clinical Evaluation*, *42*(2), 531–537.

Saito, S. (2003). Naratibubeisudomedisun to rinshochi [Narrative based medicine and clinical intellect]. In Y. Yamada (Ed.), *Jinsei to yamai no katari* [Narrative of the life and the illness] (pp. 133–163). University of Tokyo Press.

Seino, K., Enomoto, Y., & Miyazawa, S. (2018). Narrative analysis of employment support for students with developmental disabilities: Using an objective analysis of free-expression answers. In T. Ogata & S. Asakawa (Eds.), Content generation through narrative communication and simulation (pp. 341-357). Information Science Reference (IGI Global).

Seino, K., Haruna, Y., & Ishizaki, S. (2016). Employment status and support needs of persons with disabilities in Japan: An analysis of narrative using narratology and text mining on a national survey. In T. Ogata, & T. Akimoto (Eds.), Computational and cognitive approaches to narratology (pp. 245-275). Information Science Reference (IGI Global).

Shojima, S. (2008). Multilayered contexts in which narrative is generated and interpreted: Toward the next narrative based inquiry. *Kyoto University Research Studies in Education*, *54*, 138–151.

Spence, C. D. (1983). Narrative persuasion. *Psychoanalysis and Contemporary Thought*, *6*, 457–468.

Tagaki, M. (2008). Experiences of people with acquired long-term physical disability: Life stories of persons with spinal cord injuries in Japan. *Qualitative Psychology Nexus*, *7*, 51–79.

Thomas, P., & Braken, P. (2004). Critical psychiatry in practice. *Advances in Psychiatric Treatment*, *10*(5), 361–370. doi:10.1192/apt.10.5.361

Vuokila-Oikkonen, P., Janhonen, S., & Vaisanen, L. (2004). "Shared-rhythm cooperation" in cooperative team meetings in acute psychiatric inpatient care. *Journal of Psychiatric and Mental Health Nursing*, *11*(2), 129–140. doi:10.1111/j.1365-2850.2003.00679.x PMID:15009487

Yamabe, S. (2012). Sosharuwaku niokeru ebidensu tohananika: Purorogu [What is the evidence in the social work?: Prologue]. In Japanese Society for the Study of Social Welfare (Ed.), Taironshakaihukushigaku [Discussion social welfare studies] (pp. 85-87). Chuohoki Publishing.

Yamada, Y. (2007a). Shitsutekikenkyuu niokeru taiwatekimoderukouseihou [Interactive model construction method in qualitative research]. *Japanese Journal of Qualitative Psychology*, *6*, 174–194.

Yamada, Y. (2007b). *Shitsutekishinrigaku no houhou: Katari o kiku* [Method of qualitative psychology: Listening to narrative]. Shinyousha.

Yoshimoto, T. (1968). *Kyodo Gensoron* [The theory of communal fantasy]. Kawade Shobo Shinsha.

KEY TERMS AND DEFINITIONS

Literary Narrative: A narrative found in literature, including mythology and folktales.

Literary Narratology: The theory about story and discourse in the narrow sense.

Narrative in the Narrow Sense: A reproduction of a real or fictional event that is always accompanied by time restraints.

Narrative in the Wider Sense: The individual's act of narrating using his or her own language.

Psychological Narratology: The theory of the clinical/ psychological approach that is relevant to the narrative in the broadest sense.

Chapter 4
Defining the Verbs for "Understanding and Interpretation" of Japanese Sake

Hiroki Fukushima
Kyushu Women's University, Japan

ABSTRACT

In this chapter, the author attempts to define the verbs in the description of Japanese sake taste by employing 1) a usage-based approach, 2) "encyclopedic semantics" rather than a "dictionary view," and 3) sense-making theory, drawing on data from a "sake tasting description corpus" (approximately 120,000 words). The chapter selects eight verbs of high frequency (e.g., hirogaru 'spread') and defines their sense(s) in a bottom-up and abductive fashion, based on a score indicating the strength of co-occurrence between terms. In this study, the authors deal with the verbs for "Understanding" or "Interpretation "; it means, verbs that contribute to narrating the personal, individual story (contents) of the tasters. This study suggests the verbs for understanding have senses related to [Timeline] and [Space]. On the other hand, verbs do not tend to collocate with [Movement] and interestingly, the [Structure], as same as the tendency of adjectival-nouns.

INTRODUCTION

In this study, the author attempts to give encyclopaedic definitions to the verbs in a Sake corpus based on Sake tasting related terminology. This study is in a series of Sake terminology studies. In the previous study (Fukushima 2020b), the author tried to give encycropaedic definitions to adjectives and adjectival-nouns. In this terminology series, the author emphasizes the importance of predicates (verbs, adjectives, and adjectival nouns). They are the starting point of our cognition of taste or flavor as an "event".

As is often the case with wine sommeliers, the expression or description in Sake tasting tends to be regarded as a reductional 'flavor-finding' process. However, in this study series, the author respect Sake tasting expressions as the fruits of the dynamic event construction and sense-making process.

DOI: 10.4018/978-1-7998-4864-6.ch004

There are some Japanese linguistic studies on Sake taste terms (e.g., (Matsuura, 1992; Otsuka, Suwa, & Yamaguchi, 2015; Utsunomiya, Isogai, Iwata, & Nakano, 2006), but the investigation is in its infancy, largely because of the historical context of the Sake brewing industry.

As the flavors of Sake have expanded, more expressions have been required. However, the study of the development of verbal expressions for Sake has been neglected. Technical terms for brewing and descriptive terms used to indicate some of the off-flavors of a *Sake* are emphasized, leaving terms to describe appealing flavors unstudied. In response, Fukushima (2014) compiled a small encyclopedia listing Sake taste words. Considered epoch-making in the Sake industry,[1] this work has inspired other work, leading to the development of a method of defining Sake taste terms.

BACKGROUND

Theoretical Background

The author regards cognitive content generation of taste as a kind of aesthetic appreciation process. Sake, wine, coffee, cigarettes or other items have communities of lovers, from professional tasters to amateur enthusiasts. They express differences in taste or flavor by brands and vintage years, and many comment-supporting tools have been proposed. One typical example is "flavor wheels".

The process of aesthetic appreciation of these items is as follows:

Aesthetic experience ® reflection ® expression (primarily by language)

For this process, the phenomenological background theory, from the aesthetic experience to the reflection and explanation ("emergence-motivated event construction"), is explained in Fukushima (2020a, 2020b).

Sense-making theory

To define the aesthetic Sake taste terms, I draw on "sense-making theory"(Fukaya & Tanaka, 1996; Tanaka & Fukaya, 1998), arguing that the meaning of a word is determined through various levels of interaction. The interaction includes the relationships among words (i.e., co-occurrence), between words and sentence or context, and even between people (i.e., communication level). Fukaya and Tanaka claim the sense of a single word cannot be determined a priori; rather, the sense is cooperatively "made" during the communication process. This chapter concurs with sense-making theory on this dynamic aspect of word sense.

Emergence-motivated event construction

The author proposes to define Sake taste by focusing on adjectivals and verbs, as an alternative to a dominantly used method focusing on nouns. The author calls this latter way of verbalization an "object-motivated event construction" where the experiencer primarily uses nouns to describe the event of tasting. This is commonly found in English tasting comments by wine sommeliers, as in "I feel a note of black cherry, cassis, and the rich flavor of the oak," where the sommelier detects the elements of the flavor and verbalizes them, perhaps selecting the terms from his or her list of tasting words.

This is analogous to an "audio" or a "visual event construction," where an event is reported objectively. As an alternative event construction, Fukushima (2020a, 2020b) proposed the theory of "emergence-motivated event construction."

The way of constructing the event of tasting (or smelling) differs from the process of vision. When we taste something—that is, when we have an event construction of tasting or when we conceptualize what we taste in our mouths—what we feel first is not the element of taste, such as sweetness, acidity, apple flavor, or other flavors in Sake (as expressed by nouns), but the emergence of the tasting event itself.

Supporting emergence-motivated event construction means that adjectives, adjectival nouns, and verbs (but not nouns) take the leading role in the taste description. The recognition of the emergence of an event is primarily expressed by predicates (verbs, adjectives and adjectival nouns).

Linguistic background

This chapter's proposed method of defining taste terms has the following characteristics, in harmony with the themes of cognitive linguistics:

- the dynamic aspect of word senses
- usage-based
- "sencyclopaedic semantics" rather than a "dictionary view"

The meaning of a taste term is often different from the word's general definition. For example, when *tōmei* 'clear' is used to express the taste of Japanese Sake, it could represent (or modify) the lightness of the body, or clean sweetness, or a quick fading of the aftertaste. However, 'light body', 'sweetness', or 'aftertaste' would never be listed in the definition of *tōmei* 'clear' in general dictionaries. As illustrated in this example, I am emphasizing the dynamic aspect of word senses. The "sense of a word" is not fixed and static as in a dictionary. It varies, depending on the context; it is *made* dynamically in communication. This is the dynamic aspect of a word's meaning.

Narratological Background: Narrative Generation and Narrative Analysis

The approaches in narratology are roughly classified into two kinds: narrative generation and narrative analysis. (Ogata, 2020) The primary approach in content generation study is to develop a computational (or computer-assisted) content generation system. In this book, the chapters in Part 2 are illustrative examples.

As an applicational domain of content generation, the author has carried out some studies focusing on "cognitive content" as a story, which is called "Cognitive Content Generation" (Fukushima 2020a). The cognitive content is, in short, the symbolic expression (including non-verbal representations) of a mental image of the five senses.

As a case of the study of the cognitive content, the author has been studying the domain of taste. Over the past few years, the author has conducted studies on cognitive content *generation* of taste. In (Fukushima, 2020a), the author presented a methodology for bridging the gap between the mental representation of taste and the external representation (i.e., words, drawing) as content *generation*.

In this study, the author attempts to *analyze* the cognitive content of taste, with a tasting comments corpus. There are some text types expressing the sense of taste: recipes, essays, advertisements and food packaging, online restaurant reviews, and so on. At this stage, the author does not define how much of these are in the scope of narratology.

Among these genres, this study deals with tasting comments concerning *Sake*. Japanese Sake, which is also called rice wine, has recently become more characterful in its flavor and taste.

MAIN FOCUS OF THE CHAPTER

Verbs for Describing the Taste of Sake

This study aims to define the sense of verbs in Sake tasting description, asking the question, what kind of verbs are used in the Sake corpus?

Appendix 4 shows the top 100 verbs in the Sake corpus. Note that the table does not include a number of functional verbs (e.g., *naru* 'be', *suru* 'do').

Figure 1 shows the hierarchical clustering chart of the top 29 words (word frequency of over 100). From the hierarchy branching, Cluster 01 appears to be independent of other clusters (Clusters 02-05). As a general trend, verbs in Cluster 01 and Clusters 02-05 seem to differ in usage tendency.

Cluster 01 consists of the following verbs: *tsukuru* '(Sake) making,' *kamosu* 'brew,' *hiyasu* 'chill,' *motsu* 'have,' and *tanoshimeru* 'enjoy(able).' These words seem not to be the expressions for the taste or flavor, from the viewpoint of describing the quality of the experience. Here, in order to clarify the role of the verbs, let us take a look at a standard example of Sake reviewing in books (Sentence 1) and co-occurrence networks with nouns (Figure 2).

The "Azuma-ichi" which aims at the optimum quality of Sake during meals uses Yamada-nishiki, cultivated by the brewer himself. **[Ingredients]**

The privately-incubated Kumamoto yeast is used. Azuma-ichi constantly produces the fine junmaiginjo (pure rice premium Sake) by fermenting Kumamoto yeast for a long time at low temperature where the yeast is difficult to be active. **[Brewing Techniques]**

The harmony of calm fragrance and umami, refreshing sourness and light taste is splendid. **[Flavor and Taste]**

The sharpness of the aftertaste and the faint feeling of gas are impressive. **[Flavor and Taste]**

You can enjoy the taste for a long time even after opening. It is a Japanese Sake that goes well with foods. **[How to enjoy]**

For Azuma-ichi, Yamada-Nishiki Junmai-ginjo,
(Sake Competition, 2019); translated by the author, **[Caption]** *added by the author)*

Figure 2 shows the MDS (Multi-Dimensional Scaling) chart (drawing nouns and verbs, word frequency of over 100)

Cluster 01 contains words for the general taste and flavor of Sake. Sake, taste, aroma and other words in Cluster 01 are common words in Sake tasting comments. Some words like umami, balance, *ginjo* and

Figure 1. Hierarchical Clustering Chart of verbs in the Sake description
Note, the bar graph on the left of the figure shows the frequency of word usage.

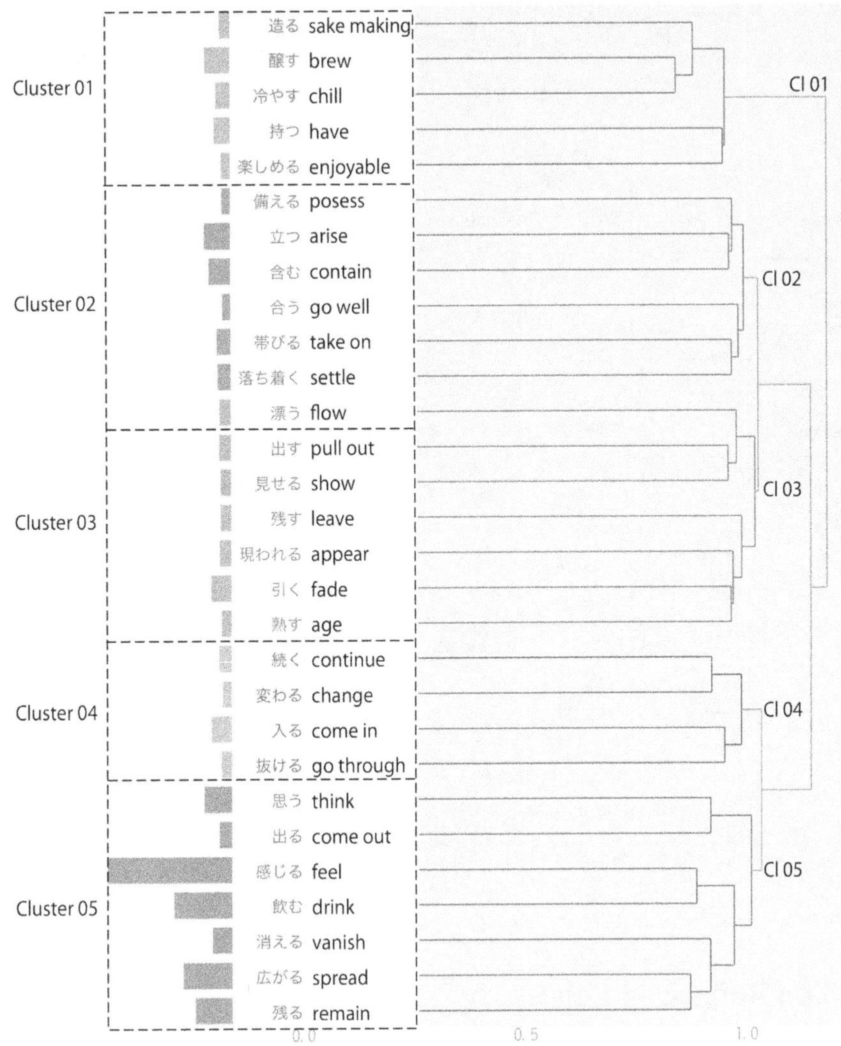

pure-rice (cf. Appendix 1) describe some quality of the taste, but these words can be used in almost any context in Sake descriptions.

Clusters 02 and 03 seem to be clusters for the domain of taste and flavor. Cluster 04 consists of words for the [Space] or [Parts of the mouth] tag, including some words for the [Timeline] tag (e.g., last, after flavor).

Clusters 01 to 04 can be roughly divided into the "taste / flavor" category, which describes the sensory impression in the tasting comments.

On the other hand, Clusters 05 and 06 seem to have another role in the tasting comments (Cluster 07 is for 'other words' or 'un-clustered words.'). Words in Cluster 05 mainly concern Sake brewing (Yamada-Nishiki is a brand of special rice for Sake). And Cluster 06 concerns words of classification (designation) of *Sake* ('dai-ginjo'), drinking situation ('chill'), food pairing ('meal'). Clusters 05 and 06 can be divided into the "brewing" category.

Figure 2. Multi-Dimensional Scaling chart (nouns and verbs, word frequency 100+ times)
Note. Verbs are framed, and the bold frames are the verbs for "understanding" the taste of Sake.

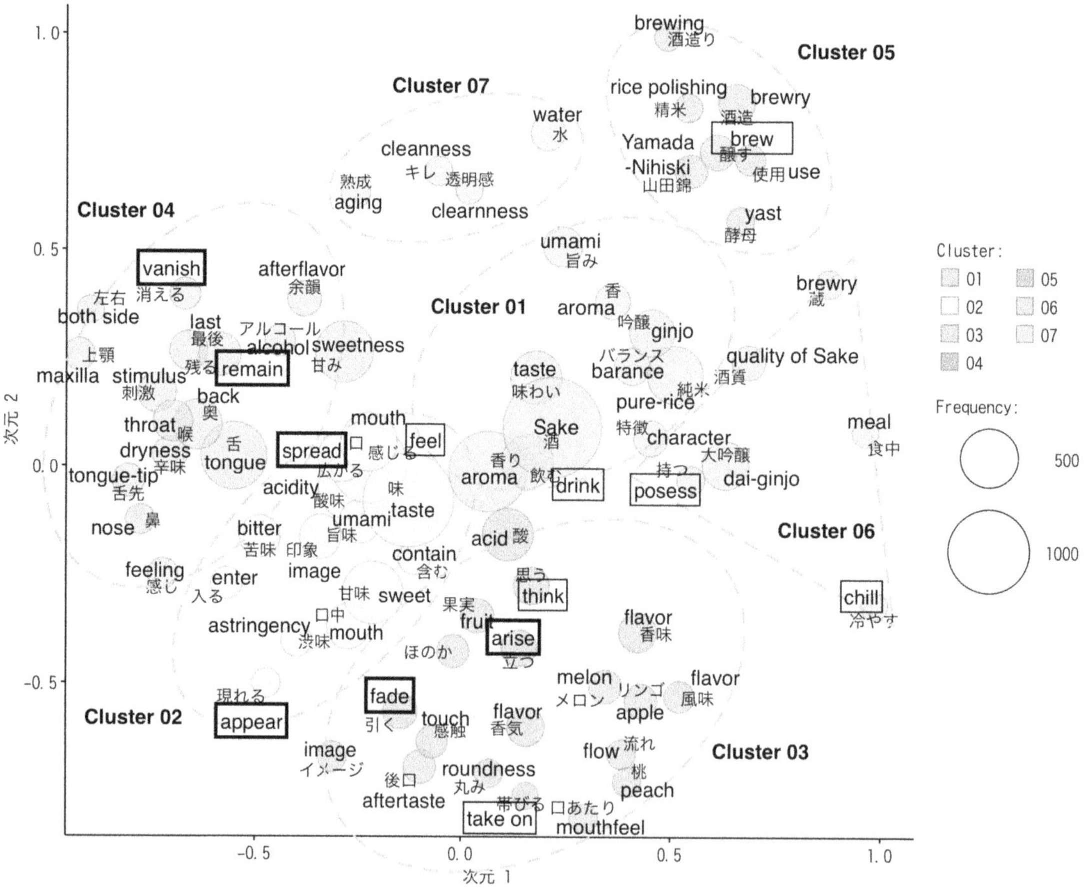

The Verbs to be Defined in this Study

If divided in terms of narratology, verbs in Clusters 05-06 are regarded as verbs for "explaining (Erklaren)", and verbs in Clusters 01-04 can be considered as verbs for "understanding" or "interpretation" (Dilthey, 1900).

In this study, the author focuses on the verbs that contribute to describing the *understanding* of the tasters' experience. The verbs for the understanding are, in other words, verbs for e xpressing some quality of the taste or flavor. The quality of the taste of flavor is personal and is a private sensation felt in the mouth. It cannot be objectively (or scientifically) *explained* with a 'taste-sensor machine'.

More concretely, the verbs to be defined in this study are the following (see also Figure 2.): *hirogaru* 'spread', *nokoru* 'remain', *tatsu* 'arise', *hiku* 'fade', *arawareru* 'appear', and *kieru* 'vanish'. In addition to these verbs from Figure 2, the following frequently used verbs are investigated: *tsuduku* 'continue' and *fukuramu* 'swell'.

METHOD

Corpus and Text Coding (Mining) Tool

CORPUS

For this analysis, the author uses a Sake corpus, a corpus of Japanese Sake tasting expressions primarily taken from Sake-reviewing books and magazines written in Japanese. The Sake corpus also includes tasting comments and expressions provided by six tasters, myself included. Table 2 summarizes the details. Note that the paragraphs refer to the different Sake brand descriptions. In total, the Sake corpus consists of 120,789 words.

Table 1. Details of the Sake corpus

	Details
Tokens	120,789
Types	6,018
Type Token Ratio	20.07
Sentences	5,582
Paragraphs (brands of *Sake*)	2,388
Average Frequency Standard Deviation	10.50 64.55

The Sake corpus draws on data from 14 books and magazines, as well as data from a tasting experiment. At first glance, the number of consulted books may seem limited. However, the books or magazines for Sake tend to introduce the brewing method of Sake brands, not the taste or flavor. Thus, eliciting the "tasting comment data" from published material is a difficult task. Sentence 1 below is a typical example of a flavor description in a Sake magazine. Figure 3 shows an image where such descriptions appear (basic translations are given in the bottom part of the image).

2. a. kajitsukei no hanayaka-na kaori to,

'Fruity, elegant flavor and'

b. pukkuri-to-shita umami ga kōchū de fukurami,

'pulump-umami swells in the mouth,'

c. taoyaka ni nagare-te iku. hōrei na Sake o jikkan.

'then it calmly flows. (I) realized the rich, splendid taste.

Figure 3. A standard style of tasting comment description in Japanese Sake books and magazines. Captured from Sake Catalog 755 (Umehara, 1999)

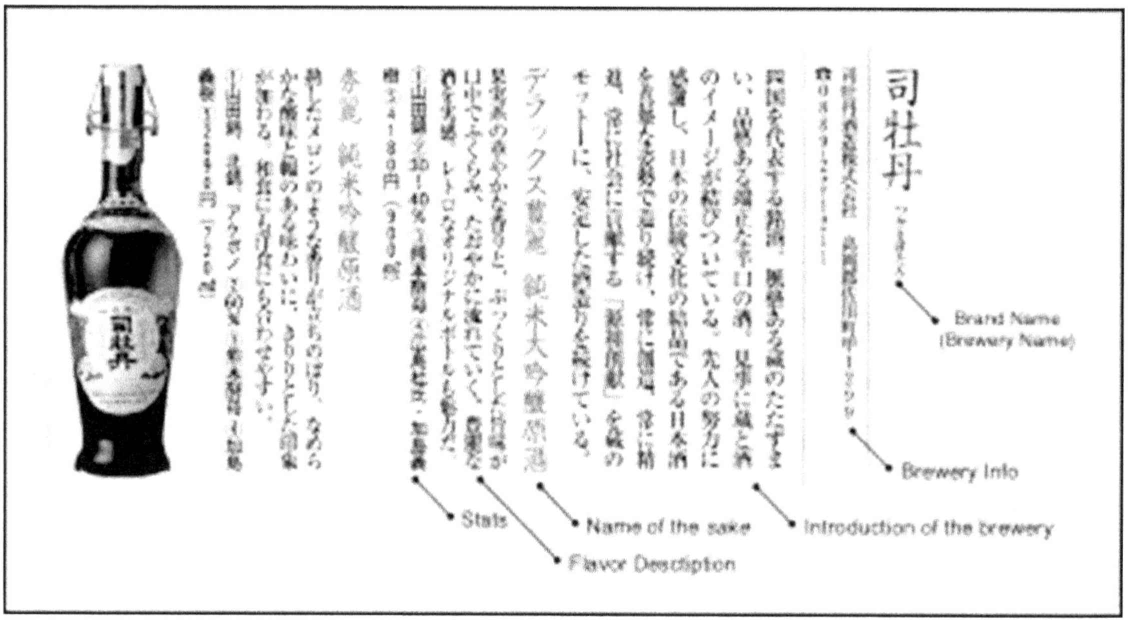

Coding tool

As a text-mining tool, the author used KH Coder (Higuchi, 2004). For the entire process of corpus analysis (i.e., word extraction with ChaSen, PoS tagging, listing word frequency, making KWIC concordance, and drawing word networks), KH Coder was used. KH Coder is free software for quantitative content analysis or text mining. It is also used for computational linguistics and can be used to analyze Catalan, Chinese (simplified), Dutch, English, French, German, Italian, Japanese, Korean, Portuguese, Russian, Slovenian, and Spanish text ("KH Coder Index Page," 22-09-2019).

How to Define Verbs

Calculation of the co-occurrence score with KWIC

Using KWIC, we can calculate the "score" for determining which words most frequently co-occur with the target term. The calculated "score" is used to indicate the strength of co-occurrence relationships. For a detailed explanation of KWIC, see Appendix 2; the protocol for calculating a score is shown in Appendix 3. In this analysis, I considered the words with a score of over 1.000 in defining the target term (i.e., approximately the top 30 words in the concordance list).

Defining the target terms

The terms to be used to define the target term are selected from the list of the top words in the concordance. This task relies on the knowledge and skill of the individual producing the definition. If the person knows very little about Sake or lacks tasting skill or does not understand the meaning of the words in the list, he or she will clearly be unable to define the target term properly. Moreover, people are likely to define the same term differently.[2]

Although the method is not fixed or rigidly defined, I set the following tags (in square brackets and bold text) as the points to be defined.

[Flavor] or [Taste]: [Flavor] or [Taste] terms tend to modify taste or flavor (e.g., Words for [Taste] include: taste, sweetness, bitterness, acidity, etc.; Words for [Flavor] include: flavor, scent, floral, fruity flavor, etc.).

[Dominance]: [Dominance] describes which tastes among the basic tastes (i.e., sweet, acidity, umami, bitter, astringency, and dry[3]) are more likely to co-occur with the target term.

[Structure]: [Structure] terms include words for the structure or physical texture of the Sake (e.g., *bodī* 'body', *waku* / *wakugumi* / *kokkaku* 'frame', *rinkaku* 'contour', *katamari* 'lump', *futoi* 'bold'). Examples of the usage of **[Structure]** terms include instantiations of a "conceptual metaphor" (Lakoff & Johnson, 1980), "TASTE IS BUILDING" or "TASTE IS BODY."

Sentence (3) Shows an example of the two conceptual metaphors.

3. <u>kokkaku</u> ga shikkari-shita, <u>futoi</u> umami

'(I feel) the firm <u>framed</u>, <u>bold</u> umami'

In (3), the first part of the phrase *kokkaku ga shikari-shita* 'firm framed' involves the equation of "building" with "taste", where the elements of "taste" are conceptualized as corresponding to (or "mapped onto") those of "building." Stated differently, the taste is conceptualized as having a frame, just like a building has a frame, and the stability of the "frame" of the taste is expressed by *shikari-shita* 'firm'. Similarly, the latter part of (3) involves the equation of "human body" with "taste", where the elements of "taste" are conceptualized as corresponding to those of the "human body". Because of this correspondence, the taste term *umami* can accompany a modifier *futoi* '(lit.) fat'.

[Organoleptic Feelings]: [Organoleptic Feelings] terms include terms on **[Texture]**. The majority of [Organoleptic Feelings] are words for [Texture] or mouthfeels (e.g., *kanshoku* 'feeling', *tacchi* 'touch', *sofuto* 'soft', *kurīmī* 'creamy'). Words like *marui* 'round', *tsutsumu* 'wrapping', *naderu* 'stroking', and other words for stimulus are not tagged as [Texture], but as [Organoleptic Feelings]. For [Organoleptic Feelings], onomatopoeic terms are often used (e.g., *zara-zara* 'rough texture', *suru-suru* 'smooth').

[Space] or **[Parts of the mouth]:** [Parts of the mouth] terms refer to the words for the parts of mouth used to indicate the point or the place where the taste or flavor is felt (e.g., *kuchi* 'mouth',, *hanasaki* 'nose tip', *shita* 'tongue'). These words can be found in expressions like the one in (4), an instance of *shitasaki* 'tip of the tongue'. [Space] is a tag for the words that indicate the place (but not the part of the mouth), like *oku* 'back (of the mouth)', *sayu* 'right and left', and so on.

4. <u>Shitasaki</u> de amami o kanjiru

'(I feel) sweetness on the <u>tip of the tongue</u>.'

[Timeline]: [Timeline] terms describe the timeline of the tasting experience, from beginning to end (i.e., in my "emergence-motivated event construction" model, from the emergence of the taste to its disappearance). In a tasting comment, describing the way the taste or flavor appears and disappears is very important. Words for appearance are, for example, *saisho* 'beginning', *arawareru* 'appear', *tatsu* 'stand' (cf. Section 4.2.6), and *hanasaki* 'tip of the nose'. Words for disappearance include *kieru* 'fade', *kireru* 'finish', *nokoru* 'remain', and *atokuchi* 'aftertaste'. It is notable that some words for the part of the mouth often (indirectly) refer to the point of the timeline (e.g., 'nose tip' = 'the starting point', or 'back of the mouth' = 'the last point').

[Movement]: [Movement] terms include verbs modified by adjectivals expressing the movement of the taste. Representative examples include: *fukuramu* 'swell', *hirogaru* 'spread', *nagare* 'flow', *tadayō* 'drift', and *osamaru* 'subside'.

[Characteristic Words]: [Characteristic Words] terms include words that supplement the information of the target terms but are important to describe the characteristics of the taste, such as *sukoshi* 'a little'.

[Related Words]: If the PoS of the word is the same as the PoS of the target term (i.e., adjectives or adjectival nouns in the concordance list), and if the word has an intralinguistic relationship with the target term, such as synonymy, antonymy, hyponymy, or gradable antonymy (cf.(Lehrer, 2009)), then the word should be included. For example, if the description includes *yawarakaku marui* 'soft and round', and the target term is *yawarakai* 'soft', *marui* 'round' should be listed as a related word, as it expresses a synonymous meaning as 'soft' in the Sake tasting context.

RESULTS

Hirogaru 'Spread'

As Table 2 shows, hirogaru 'spread' co-occurs with the following words;

[Taste]

Among the five basic tastes, the term *hirogaru* 'spread' often co-occurs with *amami* 'sweetness' [4, 9, 19], umami 'umami' [10, 13], and *sanmi* 'acidity'[14]. As stated above, these three tastes are all the primary tastes for Sake. Thus, *hirogaru* 'spread' describes the essential and desirable taste of Sake.

[Flavor]

In the concordance list, only two words for flavor are listed: *kōmi* and *kaori* 'flavor' [7, 8]. The words *kōmi* and *kaori* are rather abstract words, and no specific flavor words (e.g., apple, melon, etc.) are seen in the concordance list.

[Dominance]

The term *hirogaru* 'spread' is used more frequently to describe "taste" than "flavor," as indicated by the higher token frequency of the taste terms for 'sweetness' [4, 9, 19], 'umami' [10, 13], and 'acidity' [14] versus the flavor terms [7, 8].

Table 2. Concordance list for hirogaru 'spread'

N	Word	Pron.	Eng. Trans.	PoS	Total	LT	RT	Score
1	味	aji	taste	noun	48	29	19	22.57
2	口	kuchi	mouth	noun	52	42	10	16.18
3	口中	kōchū	mouth	noun	37	32	5	14.97
4	甘み	amami	sweetness	noun	33	24	9	12.25
5	舌	shita	tongue	noun	34	16	18	10.13
6	全体	zentai	entire	adv.	23	20	3	9.53
7	香味	kōmi	flavor and taste	noun	17	14	3	9.00
8	香り	kaori	flavor	noun	21	15	6	8.88
9	甘味	amami	sweetness	noun	22	15	7	8.47
10	旨み	umami	umami	noun	15	11	4	6.13
11	優しい	yasashii	tender	adj.	12	9	3	5.73
12	左右	sayū	right and left	noun	15	11	4	5.50
13	旨味	umami	umami	noun	13	8	5	5.48
14	酸味	sanmi	acidity	noun	14	6	8	5.43
15	最後	saigo	last	noun	11	1	10	4.42
16	丸い	marui	round	adj.	6	5	1	3.78
17	上顎	uwaago	maxilla	noun	11	7	4	3.70
18	横	yoko	side	noun	9	9	0	3.57
19	甘い	amai	sweet	adj.	14	11	3	3.57
20	薄い	usui	thin	adj.	8	4	4	3.57
21	感じ	kanjiru	feeling	noun	4	1	3	3.50
22	飲む	nomu	drink	verb	7	0	7	3.33
23	その後	sonoato	later	adv.	7	0	7	3.25
24	奥	oku	back	noun	12	5	7	3.20
25	米	kome	rice	noun	10	4	6	3.20
26	酒	Sake	Sake	noun	9	3	6	3.08
27	きれい	kirei	clean	AN	5	4	1	2.75
28	印象	inshō	impression	noun	8	3	5	2.73
29	味わい	ajiwai	taste	noun	7	5	2	2.65
30	香気	kōki	aroma	noun	6	5	1	2.53

Note. From left to right, the columns in the table give the ranking (1 through 21) of the score (N) of the co-occurring words (Word), the English translation of the co-occurring words (English Trans.), pronunciation of the co-occurring words (pronunciation), the part of speech (PoS) of the words, the total number of co-occurrences (Total), the number of occurrences to the left of the target term (LT), the number of occurrences to the right side of the target term (RT), and the scores.

[Space and Body-part]

Hirogaru 'spread' is originally a word for spatial nuance. Naturally, hirogaru 'spread' frequently collocates with words ofthe [Space] and {Parts of the Mouth] tag, as seen in kuchi 'mouth' [2], kōchū 'mouth' [3] and zentai 'entire' [6], sayū, 'right and left' [12], uwaago, 'maxilla' [17], yoko, 'side' [18], oku, 'back' [24]. These collocating words include almost all areas of the mouth cavity. Note that hirogaru 'spread' can be used for describing aroma diffusing in the nose cavity, though words about 'nose' are not seen in the concordance list.

Table 3. Concordance list for nokoru 'remain'

N	Word	Pron.	Eng. Trans.	PoS	Total	LT	RT	Score
1	味	aji	taste	noun	46	31	15	18.65
2	最後	saigo	last	adv.	27	19	8	11.30
3	舌	shita	tongue	noun	28	15	13	10.15
4	少し	sukoshi	a little	adv.	19	8	11	9.12
5	余韻	yoin	aftertaste	noun	16	11	5	6.33
6	長い	nagai	long	adj.	9	7	2	6.03
7	苦味	nigami	bitterness	noun	12	10	2	5.45
8	印象	inshō	impression	noun	14	12	2	5.23
9	口	kuchi	mouth	noun	15	10	5	5.03
10	薄い	usui	thin	adj.	12	8	4	4.93
11	奥	oku	back	noun	17	12	5	4.70
12	雑味	zatsumi	odd taste	noun	8	6	2	4.50
13	酸味	sanmi	acidity	noun	11	6	5	4.37
14	刺激	shigeki	stimulus	noun	11	11	0	4.27
15	口中	kōchū	mouth	noun	8	3	5	3.83
16	香り	kaori	flavor	noun	11	3	8	3.82
17	甘み	amami	sweetness	noun	12	7	5	3.73
18	酒	Sake	Sake	noun	8	0	8	3.67
19	甘い	amai	sweet	adj.	10	4	6	3.45
20	先	saki	tip	noun	10	8	2	3.35
21	丸い	marui	round	adj.	5	3	2	3.20
22	感じ	kanjiru	feeling	noun	9	5	4	3.15
23	上顎	uwaago	maxilla	noun	9	1	8	3.07
24	甘味	amami	sweetness	noun	6	4	2	3.03
25	その後	sonoato	later	adv.	5	1	4	3.00
26	舌先	shitasaki	tip of tongue	noun	8	3	5	2.83
27	渋味	shibumi	astringency	noun	8	6	2	2.78
28	渋み	shibumi	astringency	noun	6	6	0	2.58
29	味わい	ajiwai	taste	noun	4	1	3	2.45
30	消える	kieru	disappear	verb	7	3	4	2.37

Nokoru 'REMAIN'

[Taste]

In the taste domain, *nigami* 'bitterness' [7], *zatsumi* 'odd taste' [12], and *sanmi* 'acidity' [13] and even *shibumi* 'astringency' [27] are characteristic collocational words for *nokoru* 'remain.' These tastes are not the primary tastes for Sake. Too much astringency, bitterness, and acidity spoil the balance of Sake, but the proper quantity of secondary tastes can give richness to the taste of Sake.

[Dominance]

Though both flavor and taste terms can co-occur with *nokoru* 'remain', from the concordance list, the frequency dominance is clearly given to the domain of taste.

[Timeline]

Nokoru 'remain' is originally a word for describing the last part of the taste. Therefore, some words for indicating the latter period can be seen in the concordance list: *saigo* 'last' [2], *yoin* 'aftertaste' [5, 34], *sonoato* 'later' [25], *nagai* 'long' [6].

[Space]

Nokoru 'remain' seems to often co-occur with the parts of the mouth: *shita* 'tongue' [3], oku 'back of the mouth' [11] *kōchū* 'mouth' [15], *uwaago* 'maxilla' [23], *shitasaki* 'tip of the tongue' [26].

These words have a role in pointing out where the taste "remains." The remaining tastes (i.e., bitterness and odd taste) tend to be felt at the back [11] of the tongue.

[Organoleptic Feelings]

For organoleptic feelings, *shigeki* 'stimulus' [17] is seen in the list. The word 'stimulus' generally means (undesirable) *dryness* of the alcohol.

[Aesthetic Terms]

Yoin 'aftertaste' [5], usui 'thin (weak)' [10], and *marui* 'round' [21]

[Verbs]

For the co-occurring verbs, *kieru* 'vanish' [30] and *hirogaru* 'spread' [31] are listed.

Tatsu 'Arise'

The literal English translation of *tatsu* is 'stand.' However, if the verb *tatsu* is used for describing some quality of taste, it means 'arise' or 'appear.'

Table 4. Concordance list for tatsu 'arise'

N	Word	Pron.	Eng. Trans.	PoS	Total	LT	RT	Score
1	香り	kaori	flavor	noun	61	48	13	31.70
2	昇る	noboru	rise	verb	22	0	22	22.00
3	香気	kōki	aroma	noun	26	24	2	12.58
4	華やか	hanayaka	gorgeous	AN	16	9	7	8.53
5	湧く	waku	well up	verb	7	7	0	7.00
6	口中	kōchū	mouth	noun	15	7	8	6.92
7	酸	san	acid	noun	12	5	7	5.40
8	味	aji	taste	noun	17	4	13	5.38
9	甘い	amai	sweet	adj.	15	9	6	5.23
10	果実香	kamika	fruit fragrance	noun	11	10	1	4.83
11	リンゴ	ringo	apple	noun	8	3	5	3.77
12	酸味	sanmi	acidity	noun	5	1	4	3.20
13	甘味	amami	sweetness	noun	8	5	3	3.03
14	上げる	ageru	raise	verb	3	0	3	3.00
15	クリーミー	kurīmī	creamy	AN	5	1	4	2.95
16	全体	zentai	entire	AN	5	0	5	2.87
17	鼻先	hanasaki	nose tip	noun	5	5	0	2.50
18	香味	kōmi	flavor and taste	noun	6	4	2	2.28
19	爽やか	sawayaka	freshness	AN	8	6	2	2.08
20	ふくよか	fukuyoka	plump	AN	9	3	6	2.05
21	苦味	nigami	bitterness	noun	4	2	2	2.00
22	先	saki	tip	noun	4	4	0	2.00
23	シャープ	shāpu	sharp	adj.	5	2	3	1.83
24	軽妙	keimyō	light and easy	AN	3	0	3	1.83
25	芳香	hōkō	fragrance	noun	4	4	0	1.83
26	軽い	karui	light	adj.	5	2	3	1.75
27	イチゴ	ichigo	strawberry	noun	4	1	3	1.73
28	アルコール	arukōru	alcohol	noun	4	2	2	1.67
29	含む	fukumu	contain	verb	7	2	5	1.60
30	スムーズ	sumūzu	smooth	AN	2	0	2	1.50
31	辛味	karami	dryness	noun	3	2	1	1.50
32	舌先	shitasaki	tip of tongue	noun	3	2	1	1.50
33	桃	momo	peach	noun	2	0	2	1.50

Typical usage of *tatsu* 'arise' is seen in example (5).

5. A citrus type (lime) flavor gently arises.

Raimu kei no kaori ga hikaeme ni tatsu
For Wakaebisu Junmai-Daiginjo, (JS, 2001)

[Flavor]

The verb *tatsu* 'arise' significantly co-occurs with words for flavor. In addition to 'flavor' itself [1, 3, 18], phrases for fruity flavor should be specially mentioned. In the concordance list, many names of fruits [10] are listed; *ringo* 'apple' [11], *ichigo* 'strawberry' [27, *banana* 'banana' [34] and *meron* 'melon' [39].

[Taste]

As well as the flavor words, tastes relating to fruits collocate with *tatsu* 'arise'; *sanmi* 'acidity' [7, 12], *amami* 'sweetness' [13]. Among the primary tastes for Sake (i.e., sweetness, acidity, and umami), umami is not listed in the collocation of *tatsu* 'arise.' This wo_uld be because of the types of Sake. There are mainly four or more types of Sake, including the fruity type and the umami type (others are the clear 'tasteless' type, matured type, and so on). From the concordance list, *tatsu* tends to be used for the fruity type, rather than the umami type of Sake.

For the secondary taste, *nigami* 'bitterness' [21] and *karami* 'driness' [31] are listed, but their rank is rather low.

[Space]

The verb *tatsu* 'arise' often describes the first part of one experience of tasting Sake, but interestingly, there are no words for the [Timeline] tag in the concordance list. Instead, there are some words for the front part of the mouth and nose; *hanasaki* 'nose tip' [17], *shitasaki* 'the tip of the tongue' [32] and simply *saki* 'tip' [22]. Note that some words for the entire space of the mouth are also seen; *kōchū* 'mouth' [6] and *zentai* 'whole' [16].

[Aesthetic Concepts]

Tatsu *'arise'* significantly co-occurs with aesthetic concepts: *hanayaka* 'gorgeous' [4], *kurīmī* 'creamy' [15], *sawayaka* 'freshness' [19], *fukuyoka* 'plump' [20], *shāpu* 'sharp' [23], *keimyō* 'light and easy' [24], *hōkō* 'fragrance' [25], *karui* 'light' [26], *sumūzu* 'smooth' [30], *katai* 'firm' [37], *wakawakashii* 'youthful' [41]

These aesthetic concepts listed above tend to modify the flavor, rather than the taste. Let's take a look at an example (6).

6. An elegant ginjo-flavor (fruity flavor) arises.

Yuuga na ginjō ka ga kaori tatsu

For Shiki-zakura, (SSI, 2019), underlined words indicate the aesthetic concepts.

Tsuduku 'Continue'

[Taste]

In the taste [3] domain, the typical collocation is found in words for the secondary taste; *nigami* 'bitterness' [6, 25], and *san* 'acid' [9]. 'Sweetness' [17, 22] is listed as always, but the rank is relatively low.

[Timeline]

As shown in Table 5, various words for describing the timeline are listed; *saigo* 'last' [7], *sonoato* 'later' [12], *saisho* 'first' [19], *atokuchi* 'aftertaste' [23], *atoni* 'after' [27]. Other words than *saisho* 'first' are words for indicating "later" or "after in the timeline.

[Space]

As well as the timeline domain, words for the space domain also co-occur with the verb *tuzuku* 'continue.' The listed words are: *kuchi* 'mouth' [13, *shitasaki* 'the tip of the tongue' [14], *uwaago* 'maxilla' [16], and *shita* 'tongue' [24].

[Organoleptic Feelings]

The word *shibire* 'numbness' [21] is a unique word for *Tsuzuku* 'continue.' It describes the drying sensation or rough 'mouthfeel'[4].

Numbness has a relation to astringency, as well as the feelings associated with tannin in wines. As for the taste of Sake, the *shibire* 'numbness' can represent the sensation of the astringency, acidity, maybe bitterness, and dryness of the alcohol.

[Characteristic words]

The words *junmai* 'pure rice,' *meigara* 'brand' [8], *kambam* 'signboard' [5] are characteristic. However, these words are not for describing some quality of taste, but for mentioning the background of the Sake being reviewed.

[Aesthetic Terms]

Tsuyoi 'strong' [26] and *yowai* 'weak' [29] are listed.

Futoi 'bold' and *nōmitsu 'dense'* represent the way the thick taste 'continues.'

Table 5. Concordance list for tsuduku 'continue'

N	Word	Pron.	Eng. Trans.	PoS	Total	LT	RT	Score
1	長い	nagai	long	adj.	17	17	0	16.50
2	余韻	yoin	aftertaste	noun	20	16	4	9.58
3	味	aji	taste	noun	6	1	5	3.70
4	感触	kanshoku	feel	noun	5	4	1	2.45
5	看板	kamban	signboard	noun	2	0	2	2.00
6	苦味	nigami	bitterness	noun	3	2	1	2.00
7	最後	saigo	last	noun	4	1	3	2.00
8	銘柄	meigara	brand	noun	3	0	3	2.00
9	酸	san	acid	noun	3	1	2	1.75
10	純米	junmai	pure rice	noun	5	2	3	1.75
11	印象	inshō	impression	noun	5	4	1	1.53
12	その後	sonoato	later	adv.	3	3	0	1.50
13	口	kuchi	mouth	noun	3	1	2	1.50
14	舌先	shitasaki	tip of tongue	noun	3	0	3	1.50
15	流れ	nagare	flow	noun	4	1	3	1.50
16	上顎	uwaago	maxilla	noun	4	2	2	1.40
17	甘み	amami	sweetness	noun	4	3	1	1.33
18	感じる	kanjiru	feel	verb	6	3	3	1.30
19	最初	saisho	first	noun	3	1	2	1.25
20	香り	kaori	flavor	noun	4	2	2	1.20
21	痺れ	shibire	numbness	noun	3	3	0	1.17
22	甘味	amami	sweetness	noun	3	1	2	1.08
23	後口	atokuchi	aftertaste	noun	3	2	1	1.08
24	舌	shita	tongue	noun	3	0	3	1.08
25	苦み	nigami	bitterness	noun	3	1	2	1.03
26	強い	tsuyoi	strong	adj.	1	1	0	1.00
27	後に	atoni	after	adv.	1	1	0	1.00
28	山田錦	Yamada-Nishiki	Yamada-Nishiki	noun	1	0	1	1.00
29	弱い	yowai	weak	adj.	1	1	0	1.00
30	精悍	seikan	virile	AN	1	0	1	1.00
31	創業	sōgyō	establish	noun	3	3	0	1.00
32	太い	futoi	bold	adj.	1	0	1	1.00
33	濃密	nōmitsu	dense	AN	1	0	1	1.00

Fukuramu 'Swell'

Table 6. Concordance list for fukuramu 'swell'

N	Word	Pron.	Eng. Trans.	PoS	Total	LT	RT	Score
1	味	aji	taste	noun	29	23	6	13.08
2	厚い	atsui	thick	adj.	15	7	8	10.83
3	感じる	kanjiru	feel	verb	11	2	9	4.53
4	香り	kaori	flavor	noun	12	9	3	3.72
5	丸い	marui	round	adj.	5	4	1	3.53
6	口中	kōchū	mouth	noun	8	7	1	3.37
7	味わい	ajiwai	taste	noun	8	1	7	3.00
8	吟醸香	ginjō-ka	ginjo-flavor	noun	7	7	0	2.95
9	香味	kōmi	flavor and taste	noun	8	4	4	2.75
10	旨み	umami	umami	noun	6	5	1	2.70
11	優しい	yasashii	tender	adj.	4	4	0	2.53
12	香気	kōki	aroma	noun	6	5	1	2.45
13	広がる	hirogaru	spread	verb	6	3	3	2.33
14	良い	yoi	good	adj.	5	1	4	2.03
15	引き出す	hikidasu	take out	verb	4	1	3	2.00
16	厚み	atsumi	thickness	noun	4	3	1	2.00
17	太い	futoi	thick	adj.	3	1	2	2.00
18	有す	yūsu	possess	verb	4	0	4	2.00
19	きれい	kirei	clean	AN	6	3	3	1.95
20	幅	haba	width	noun	4	2	2	1.83
21	後口	atokuchi	aftertaste	noun	7	1	6	1.77
22	丸み	marumi	roundness	noun	4	3	1	1.70
23	持つ	motsu	have	verb	4	1	3	1.70
24	甘み	amami	sweetness	noun	5	3	2	1.67
25	穏やか	odayaka	calm	AN	3	2	1	1.50
26	保つ	tamotsu	keep	verb	3	0	3	1.50
27	良好	ryōkō	good	AN	3	0	3	1.50
28	立つ	tatsu	stand	verb	5	3	2	1.40
29	純米	junmai	pure rice	noun	6	3	3	1.38
30	張り	hari	tension	noun	3	1	2	1.33

[Taste]

The words for the taste domain are umami [10], sweetness [24], and acidity [33, 41]. These are the primary tastes of Sake and are often listed in the concordance list. Judging from Table 6, the rank of these primary taste words is relatively low.

[Flavor]

The unique words for the flavor domain are *ginjōka* 'ginjō-flavor' [8]. *Ginjōka* is the aroma that is produced by a special brewing process (*ginjō*-brewing, see Appendix 1). *Ginjōka* generally contain fruity flavors, such as melon [36], apple, green apple, pear, or banana.

[Dominance]

Fukuramu 'swell' can co-occur both with the domains of taste and flavor. In comparison to other verbs, the concordance list (Table 6) tells us that the flavor domain [4, 8, 9, 12, 36] relatively dominates over the taste domain [1, 7, 10, 24, 33, 41].

[Timeline]

Words for the timeline domain: *atokuchi* 'aftertaste' [21]

[Space]

Words for the space domain: *kōchū* 'in the mouth' [6]

[Aesthetic Terms]

Atsui/atsumi 'thick(ness)' [2, 16] is a unique term for fukuramu 'swell'. Thickness is the experience of Sake with a rich, full-bodied, or umami taste (or flavor). *Futoi* 'bold' [17] or *haba* 'width' [20] have a similar nuance to *atsui.*

Other aesthetic terms in the concordance list are: *marui/marumi* 'round(ness)' [5, 22], *yasashii* 'tender' [11], *kirei* 'clean' [19], *odayaka* 'calm' [25].

Hiku 'Fade'

The verb *hiku* 'fade' is literally translated as 'pull.' However, in a tasting description, 'fade' or 'go down' would be the better translations.

[Taste]

Words for sweetness [3], umami [4], and acidity [19, 25, 26] are listed. These are the primary tastes for Sake, therefore, *hiku* 'fade' illustrates the manner of fading or the going down of a comfortable and desirable taste.

Table 7. Concordance list for hiku 'fade'

N	Word	Pron.	Eng. Trans.	PoS	Total	LT	RT	Score
1	後	ato	after	noun	36	36	0	17.75
2	尾	o	tail	noun	24	24	0	12
3	甘味	amami	sweetness	noun	23	11	12	10.933
4	旨味	umami	umami	noun	16	5	11	7.367
5	軽い	karui	light	adj.	9	0	9	5.917
6	後口	atokuchi	aftertaste	noun	7	3	4	3.067
7	切れ	kire	clearing	noun	5	1	4	3.033
8	味	aji	taste	noun	8	5	3	3
9	感触	kanshoku	feel	noun	7	2	5	2.7
10	優しい	yasashii	tender	adj.	4	3	1	2.667
11	穏やか	odayaka	calm	AN	5	2	3	2.583
12	きれい	kirei	clean	AN	6	5	1	2.333
13	後に	atoni	after	adv.	5	5	0	2.333
14	印象	inshō	impression	noun	5	1	4	2.233
15	香り	kaori	flavor	noun	8	5	3	2.15
17	浅い	asai	shallow	adj.	2	2	0	2
18	丸み	marumi	roundness	noun	4	1	3	1.833
19	酸	san	acid	noun	6	4	2	1.833
20	奥	oku	back	noun	5	2	3	1.7
21	感じる	kanjiru	feel	verb	7	3	4	1.633
22	香味	kōmi	flavor and taste	noun	3	0	3	1.583
23	なめらか	nameraka	smooth	AN	2	1	1	1.5
24	余韻	yoin	aftertaste	noun	4	2	2	1.5
25	甘酸っぱい	amazuppai	sweet and sour	adj.	3	2	1	1.4
26	酸味	sanmi	acidity	noun	5	2	3	1.35
27	タッチ	tacchi	touch	noun	4	4	0	1.333
28	深い	fukai	deep	adj.	2	0	2	1.333
29	味わい	ajiwai	taste	noun	4	2	2	1.333
30	太い	futoi	thick	adj.	2	0	2	1.25

[Dominance]

We can accept the dominance of [Taste]. As for the [Flavor] category, *kaori* 'flavor' [15] and *kōmi* 'flavor' [24] are listed, but the ranking is low, and no specific flavor names are seen in the list.

[Timeline]

Hiku 'fade' describes a decreasing or vanishing. Consequently, the [Timeline] words, especially words for the latter timing tend to collocate: *atokuchi* 'aftertaste' [6], *kire* 'clearing' [7], *atoni* 'after' [16], *yoin* 'after flavor' [24].

Kieru 'Vanish'

[Taste]

We can recognize [Dominance] in [Taste], rather than [Flavor]. Then, what kind of tastes vanish during Sake tasting? In the concordance list, we can find "stimulating tastes" for Sake; *nigami* 'bitterness' [13], and *shibumi* 'astringency' [25]. These words sometimes appear in other concordance lists, but the frequency is characteristically high in the list for *Kieru* 'vanish'.

Note that too much of these stimulating tastes is generally thought to create "odd tastes" in Sake. However, a moderate amount of bitterness or astringency often gives the feeling of outline or structure to a Sake.

[Space (Parts of the mouth)]

Bitterness and astringency tend to be felt around the root of the tongue or the peripheral area of the mouth cavity (i.e., hard palate, soft palate, or gums). In Table 11, there are some words of the parts of the mouth; *oku* 'back' [7], *shita* 'tongue' [8], *uwaago* 'maxilla' [9], *kōchū* 'mouth' [10], *nodo* 'throat' [12], *shitasaki* 'tip of the tongue' [19], *saki* 'tip' [26].

As shown above, there are a large number of words for the root of the tongue, or the peripheral area of the mouth cavity. This implies that the bitter taste or astringency determines the final impression of the Sake taste.

[Timeline]

Kieru 'vanish' is an event in the last part of one tasting attempt. In the concordance list, we can find some words for describing the last part of the timeline; *yoin* 'aftertaste'[3], *nokoru* 'remain' [14], *atoaji* 'aftertaste' [16] *nokosu* 'leave' [18].

Collocating with the words for [Space] (especially the back area of the mouth), these words are applied for describing the bitter taste and astringency in the Sake.

On the one hand, some words for the first part of the Timeline; *arawareru* 'appear' [23], *saisho* 'first' [24]. *Arawareru* or *saisho* are often collocated with words of the front parts of the mouth (i.e., *shitasaki* 'tip of the tongue' [19], *saki* 'tip' [26]) in the [Space] domain. These words can be regarded as antonyms for *kieru* 'vanish', but they are often collocated together. Take a look at example (7).

7. Chiizu no aji ga kieru koro, umami no yoin ga arawareru

When the taste of cheese disappears, the rich aftertaste appears.

Table 8. Concordance list for kieru 'vanish'

N	Word	Pron.	Eng. Trans.	PoS	Total	LT	RT	Score
1	味	aji	taste	noun	13	12	1	6.25
2	印象	inshō	impression	noun	11	11	0	4.867
3	余韻	yoin	aftertaste	noun	12	6	6	4.633
4	口	kuchi	mouth	noun	10	2	8	3.35
5	甘み	amami	sweetness	noun	8	5	3	3.2
6	感じる	kanjiru	feel	verb	9	8	1	3.117
7	奥	oku	back	noun	10	5	5	2.967
8	舌	shita	tongue	noun	9	0	9	2.8
9	上顎	uwaago	maxilla	noun	7	1	6	2.75
10	口中	kōchū	mouth	noun	8	3	5	2.7
11	スッ	su'	quickly	onoma	5	5	0	2.5
12	喉	nodo	throat	noun	9	2	7	2.433
13	苦味	nigami	bitterness	noun	7	3	4	2.4
14	残る	nokoru	remain	verb	7	4	3	2.367
15	軽い	karui	light	adj.	6	3	3	1.817
16	後味	atoaji	aftertaste	noun	3	0	3	1.667
17	甘味	amami	sweetness	noun	4	3	1	1.583
18	残す	nokosu	leave	verb	4	4	0	1.583
19	舌先	shitasaki	tip of tongue	noun	5	1	4	1.533
20	すーっ	sū'	smoothly	adv.	3	3	0	1.5
21	透明	tōmei	transparent	AN	4	0	4	1.4
22	香り	kaori	flavor	noun	3	2	1	1.333
23	現れる	arawareru	appear	verb	3	2	1	1.25
24	最初	saisho	first	noun	3	0	3	1.25
25	渋み	shibumi	astringency	noun	2	1	1	1.25
26	先	saki	tip	noun	3	1	2	1.25
27	溶ける	tokeru	dissolve	verb	2	2	0	1.2
28	引っかかる	hikkakaru	stuck	verb	4	3	1	1.117
29	感じ	kanjiru	feeling	noun	3	3	0	1.083

This example illustrates the vanishing of one taste, and the contrasting appearance of another taste. In this way, antonyms for *kieru* 'vanish (disappear)' are often collocated in order to give some contrast between the vanishing taste and the appearing or remaining taste.

[Onomatopoeias]

Onomatopoeic expressions are the most distinctive modifiers for vanish. In general, making a taste description can be rephrased to point out the existing taste on the tongue. The way of 'vanishing' or finishing of a taste is hard to recognize and cannot be seen visually.

In a previous study (Fukushima, 2017), the author raised the theory that onomatopoeic expressions in taste descriptions are employed to describe breaking points, turning points, and changing processes of taste, rather than stable states. *Kieru* 'vanish' is a typical example of the "breaking points" or "turning points" of a taste.

Tadayou 'Flow'

[Dominance]

Tadayou 'flow' has an exceptional feature of dominance. While other terms tend to indicate their dominance for Taste, *Tadayou* 'flow' shows its dominance in the Flavor domain. In the concordance list, words for the flavor domain are ranked higher: 'flavor' (*kaori* [1], *kōmi* [6], *fūmi* [10]), *kōki* 'aroma' [3],

[Space]

Except for the [Taste] or [Flavor] words, the major part is in the [Space] domain, especially body parts: *bikū* 'nasal cavity'[9], *oku* 'back'[13], *hanasaki* 'nose tip' [14], *shita* 'tongue' [15], *hana* 'nose' [19].

Within the co-occurrence of *tadayou* 'flow', these body parts terms are used for indicating the place the flavor or taste is felt. The feeling of *tadayou* 'flowing and retaining' is a sensation of gas (not liquid). Thus, 'the nose' or 'the nasal cavity' is seen in the list.

[Aesthetic Terms]

In the concordance list, the following terms are thought of as aesthetic terms for Sake; kirei 'clean', kihin 'elegance,' *sugasugasii* 'refreshing.' These terms have some clean and clear qualities in common. Kihin is one of the best complimentary words for the taste of Sake. It refers to the elegant, polished, and refined quality of Sake.

CONCLUSION AND FUTURE RESEARCH DIRECTIONS

In this study, the author attempted to give encyclopaedic definitions to the verbs in a Sake corpus based on Sake tasting related terminology. In this terminology, the author emphasized the importance of the predicates (verbs, adjectives, and adjectival nouns). They are the starting point of our cognition of taste or flavor as an "event".

Through the terminological studies, the author proposed a defining method using collocation tags (Taste, Flavor, Timeline, and so on). With the collocation tags, we can analyze what types of words tend to co-occur with the target terms.

Table 9. Concordance list for tadayou 'flow'

N	Word	Pron.	Eng. Trans.	PoS	Total	LT	RT	Score
1	香り	kaori	flavor	noun	19	16	3	7.317
2	かすか	kasuka	faint	AN	6	5	1	3
3	香気	kōki	aroma	noun	8	4	4	2.95
4	酒	Sake	Sake	noun	10	6	4	2.933
5	酸味	sanmi	acidity	noun	7	5	2	2.65
6	香味	kōmi	flavor and taste	noun	6	6	0	2.5
7	リンゴ	ringo	apple	noun	4	2	2	2.4
8	きれい	kirei	clean	AN	6	1	5	2.283
9	鼻腔	bikū	nasal cavity	noun	4	3	1	2
10	風味	fūmi	flavor	noun	4	4	0	2
11	気品	kihin	dignity	noun	3	2	1	1.833
12	甘味	amami	sweetness	noun	4	2	2	1.667
13	奥	oku	back	noun	4	3	1	1.5
14	鼻先	hanasaki	nose tip	noun	4	4	0	1.5
15	舌	shita	tongue	noun	5	3	2	1.483
16	味	aji	taste	noun	5	2	3	1.417
17	甘み	amami	sweetness	noun	4	2	2	1.4
18	清々しい	sugasugasii	refreshing	adj.	2	1	1	1.333
19	鼻	hana	nose	noun	4	2	2	1.25
20	ブドウ	budō	grape	noun	2	1	1	1.2
21	酸	san	acid	noun	3	0	3	1.033
22	スマート	sumāto	smart	AN	1	0	1	1
23	プレミアム	puremiamu	premium	AN	1	0	1	1
24	果実	kajitsuka	fruit	noun	1	0	1	1
25	果実香	kamika	fruit fragrance	noun	1	0	1	1
26	感じ	kanjiru	feeling	noun	1	0	1	1
27	甘辛い	amakarai	salty-sweet	adj.	1	0	1	1
28	厚い	atsui	thick	adj.	1	1	0	1
29	香木	kōboku	fragrant wood	noun	1	0	1	1
30	桃	momo	peach	noun	1	0	1	1
31	雰囲気	fun'iki	atmosphere	noun	2	2	0	1
32	余韻	yoin	aftertaste	noun	1	0	1	1

The author examined the meanings of 16 adjectives and adjectival nouns (Fukushima, 2020b)), and eight verbs (this study). In conclusion, this terminology is summarized in the following table.

As a brief summarization of this Sake terminology study, let us compare the overall tendency of each PoS (part of speech, i.e., adjectives, adjective-nouns, and verbs) based on Table 10. Table 10 shows two

Table 10. Tag-collocation with the Target Terms

	Target Terms (Defined Terms)	Taste	Flavor	Dominance	Movement	Timeline	Structure	Space (Parts of the mouth)	Organoleptic Feelings
Adjectives	Yawarakai 'soft'	●	¡	T				●	●
	Karui 'light'	●		T					●
	Katai 'firm'	●		T				●	●
	Marui 'round'	●		T	●	●	●		
	Futoi 'bold'	●		T	●		●		
	Kobashii 'roasted aromatic'	●	●	F/T				●	
	Yasashii 'tender'	●		T					●
	Chikaraduyoi 'powerful'	●		T		●			
Adjective-Nouns	Kirei 'clean'	●	●	F/T	●	●			●
	Odayaka 'calm'	●	●	F/T		●			
	Sawayaka 'fresh'	●	●	F/T					
	Tomei 'clear'	●		T					
	Fukuyoka 'plump'	¡	¡				●		●
	Hanayaka 'gorgeous'		●	F		●			
	Maroyaka 'mellow'	●							
	Fukuzatsu 'complex'	●		T					
Verbs	Hirogaru 'spread'	●	●	F/T				●	
	Nokoru 'remain'	●		T		●		●	●
	Tatsu 'arise'		●	F				●	
	Tsuduku 'continue'	●				●		●	●
	Fukuramu 'plump'	●	●	F		¡		¡	
	Hiku 'fade'	●		T		●			●
	Kieru 'vanish'	●		T		●		●	
	Arawareru 'appear'	●	¡	T		●		●	
	Tadayou 'flow'	¡	●	F				●	

Note. ●: strong collocation, ¡: weak collocation, T: Taste dominant, F: Flavor dominant, F/T: Both Flavor and Taste collocate with the target term. Adjectives and Adjectival-Nouns are studied in (Fukushima, 2020b).

tendencies: first, each PoS has its unique role, as well as each word. Secondly, among one PoS, there can be seen some functional differentiation.

As a general tendency, adjectives can collocate with broad categories. Concerning adjectival-nouns, *kirei* 'clean' collocates with many kinds of categories, but the others' primary function seems to modify the taste or flavor itself. Adjectival nouns seem to hardly collocate with the [Movement] or [Structure] domain. The collocations of the verbs seem to concentrate on the [Timeline] and [Space] domain. On the other hand, verbs do not tend to collocate with the [Movement] or [Structure] domains, the same as the tendency of adjectival nouns.

Adjectives

Common features for adjectives

While they commonly occur with Taste, the co-occurrence frequency with fragrance is low (cf. adjective verbs). On the whole, adjectives seem to co-occur with various domains (i.e., [Tags]), and there is no specific tag specific modification function.

Differentiation among adjectives

There are two groups among adjectives: first, terms that tend to co-occur with the [Space] tag and [Organoleptic Feelings] tag (i.e., *yawarakai* 'soft', *Karui* 'light', *Katai* 'firm'). Second, terms that tend to co-occur with the [Movement], [Timeline], and [Structure] tags (i.e., *marui* 'round', *futoi* 'bold', *kobashii* 'roasted aromatic').

Adjective-Nouns

Common features for adjective-nouns

Adjective-nouns tend to collocate not only with the [Taste] tag but also with [Flavor] tag. Concerning adjectives, the [Movement] tag and the [Timeline] tag often share the target terms. On the one hand, in the case of adjectival-nouns, few terms collocate with Movement but some words collocate with Timeline.

Differentiation among adjectival-nouns

For adjectival-nouns, grouping the terms by tags seems difficult, because of the less frequent collocating between the target terms and tags. But if roughly divided, we can see a group of 'tag-collocating terms' and 'less tag-collocating terms.' Tag-collocating terms include; *kirei* 'clean', *odayaka* 'calm', *fukuyoka* 'plump', and *hanayaka* 'gorgeous'. As with adjectives, these terms work as modifiers of taste or flavor in the collocation with the tags.

'Less tag-collocating terms' (i.e., *sawayaka* 'fresh', *tōmei* 'clear', *maroyaka* 'mellow', and fukuzatsu 'complex') are terms that are less likely to collocate with other tags. The primary function of these terms is not modifying tastes. Rather, these terms express some aesthetic quality by itself. Thus, 'less tag-collocating terms' work as predicates (not modifying words) in taste description sentences.

Verbs

Common features for verbs

The co-occurrence frequency between the listed verbs and the [Movement] tag or the [Structure] tag is relatively low. On the one hand, the [Timeline] and [Space] tags seem to co-occur frequently.

While adjectives and adjectival-nouns indicate a property (or some quality) of the taste, verbs describe the motion of the liquid (or flavor gas), appearance/disappearance, and the changing process of the taste.

Thus, verbs tend to collocate with [Timeline] and [Space] tag words, that are the markers of chronological time flow or the place movement, appearance, or disappearance occurs.

Differentiation among verbs

Similar to adjectival-nouns, there is differentiation in the dominance of Taste and Flavor. Taste-dominant terms are: *nokoru* 'remain', *tsuduku* 'continue', *hiku* 'fade', *kieru* 'vanish', and *arawareru* 'appear'. Flavor-dominant terms are: *tatsu* 'arise' and *tadayou* 'flow'.

Another differentiation among verbs are seen in the collocation with [Organoleptic Feelings] tag. *Nokoru* 'remain', *tsuduku* 'continue', *hiku* 'fade' tend to collocate with [Organoleptic Feelings] tag.

FUTURE STUDIES

In this study series, the author has defined the co-occurrence unit size as one sentence ('sentence co-occurrence'). In future studies, collocational tendencies should be more microscopically investigated from the viewpoint of more close word collocation, such as 'dependency' or 'modification' relationships.

As well the analyzing unit size, the genre or domain of the corpus is important. Since the meanings of the aesthetic terms depend on the domain, the comparison among various domains such as wines, whiskeys, perfumes, or others would be worth investigating. As well as the diversity among domains, the variety among the speaker groups would be impressive. There would be significant term usage variation between professional tasters and novices. Likewise, variations between languages (e.g., comparing the wine descriptions in English and in French), book genre (e.g., comparing the commercial magazines and critique essays), generation, or gender would reveal the novel features of the aesthetic terms. In any case, success or failure largely depends on the quantity and the quality of the corpus.

Another future research direction is on grammar theory. In this study, we analyzed the co-occurrence relationship between [Tags] and each part of speech (PoS), as summarized in figure 10. We found some differences in co-occurrence characteristics among PoS and within PoS, as stated above. However, it should be noted that this analysis employs a particular grammatical theory or PoS classification method. There is an influential grammatical theory called "four major grammar" in Japanese grammar. The classification and the way of the perspective of the PoS are different, respectively. In this study, we could not give a sufficient explanation as to why adjectives are difficult to co-occur with [Flavor] tag, whereas adjective verbs are easy to co-occur with [Flavor] tag. In order to give a clear answer to this question, it may be necessary to review the grammatical theory that we stand on.

REFERENCES

Caballero, R. (2007). Manner-of-motion verbs in wine description. *Journal of Pragmatics*, *39*(12), 2095–2114. doi:10.1016/j.pragma.2007.07.005

Caballero, R. (2017). From the glass through the nose and the mouth: Motion in the description of sensory data about wine in English and Spanish. *Terminology*, *23*(1), 66–88. doi:10.1075/term.23.1.03cab

Dilthey, W. (1900). *Die Entstehung der Hermeneutik*. Retrieved from https://archive.org/details/bub_gb_icZZAAAAcAAJ

Fukaya, M., & Tanaka, S. (1996). *The sense-making theory of <words>(<Kotoba> no Imizuke ron).* Kinokuniya.

Fukushima, H. (2014). *Encyclopedia of Sake Terms.* Keio University.

Fukushima, H. (2020a in press). Cognitive Contents Generation: A Basic Introduction (Ninchiteki monogatari seisei heno shotai). In T. Ogata (Ed.), *Various Aspect of Post Narratology (Post Narratology no shoso).* Shinyosya.

Fukushima, H. (2020b in press). Defining the aesthetic sake taste terms: A usage-based approach. In K. Toratani (Ed.), *The Language of Japanese Food: Cognitive Perspectives and Beyond.* John Benjamins.

Gauntner, J. (2011). Sake Handbook. North Clarendon: Tuttle Publishing.

Lehrer, A. C. N.-P. L. (2009). Wine & conversation (2nd ed). Oxford: Oxford University Press.

López-Arroyo, B., & Roberts, R. P. (2014). English and Spanish descriptors in wine tasting terminology. *Terminology, 20*(1), 25–49. doi:10.1075/term.20.1.02lop

Matsuura, T. (1992). Terms for tasting Sake (Sake wo ajiwau kotoba). In I. Tajima & K. Niwa (Eds.), *Gendai Nihongo no kenkyū.* Izumi Shoin.

Otsuka H. (2004). *Analysis for Evaluated Expressions of Tasting Japanese Sake.* doi:10.11517/pjsai.JSAI04.0.147.0

Otsuka, H., Suwa, M., & Yamaguchi, K. (2015). *Studies of Expressions to Taste Japanese Sake by Creating Onomatopoeia.* Japanese Society for Artificial Intelligence. doi:10.11517/pjsai.JSAI2015.0_2N5OS16b5

Paradis, C., & Eeg-Olofsson, M. (2013). Describing Sensory Experience: The Genre of Wine Reviews. *Metaphor and Symbol, 28*(1), 22–40. doi:10.1080/10926488.2013.742838

Sake Competition. (2019). *The Official Guidbook for Sake Competition 2019.* Pia.

Tanaka, S., & Fukaya, M. (1998). *The Evolvement of the <Sense-Making Theory>(<Imizukeron> no Tenkai).* Kinokuniya.

Utsunomiya, H., Isogai, A., Iwata, H., & Nakano, S. (2006). Flavor Terminology and Reference Standards for Sensory Analysis of Sake. *Journal of the Blewing Society of Japan, 101*(10), 730–739.

KEY TERMS AND DEFINITIONS

Sake: *Sake* (Japanese *Sake*, also spelled *saké*) is a fermented alcoholic beverage made from rice, commonly referred to as Japanese rice wine. For more information, see Appendix 1.

Sense-Making Theory: A semantic theory advocated by Fukaya and Tanaka (Fukaya & Tanaka 1996; Tanaka & Fukaya, 1998). Fukaya and Tanaka claim that the sense of a single word cannot be determined a priori; rather, the sense is cooperatively "made" during the ongoing communication process.

ENDNOTES

1 In 2014, this work received the Good Design Award in Japan.

2 This type of subjective approach may be avoided in some research fields such as experimental psychology and cognitive science. However, the sense of taste itself is firmly in the subjective domain. It cannot be shared directly with another person, nor can it be represented by sensor information from a "tasting machine," as the sense of taste cannot be reduced to the sum of its component elements.

3 Astringency and dryness are not basic tastes in an anatomical context (they are algesthesia), but in this chapter, I regard them as "basic tastes." Note that a salty taste will almost never be noted in the taste of *Sake*.

APPENDIX 1

Sake (Japanese *Sake*, also spelled *saké*) is a fermented alcoholic beverage made from rice, commonly referred to as Japanese rice wine. *Sake* is a fermented food product like soy sauce, miso, and *katsuobushi* 'dried bonito', standard seasonings for Japanese foods. Premium *Sake* uses special rice called *Sake* rice, suited for brewing premium *Sake* (Gauntner, 2011).

There are various types of *Sake*, just as there are various types of wine. Red wine can be roughly broken into three categories: full bodied, medium bodied, and light bodied. For white wine, the gradation from sweet to dry is generally used. In the case of Japanese *Sake*, sweet types (*amakuchi*) and dry types (*karakuchi*) are the most common categories. Light, pale types (*tanrei*), and full, rich types (*nōjun*) also appear. In the 1980s and 1990s, pale and dry (*tanrei-karakuchi*) types of *Sake* were popular (and remain so). However, around 2000, there was a boom in *ginjō-shu* (premium *Sake* made from highly polished rice using a special technique). *Ginjō-shu* has a floral and fruity flavor.

Figure 4. Window showing KWIC concordance results for the word "say" (Higuchi 2017)

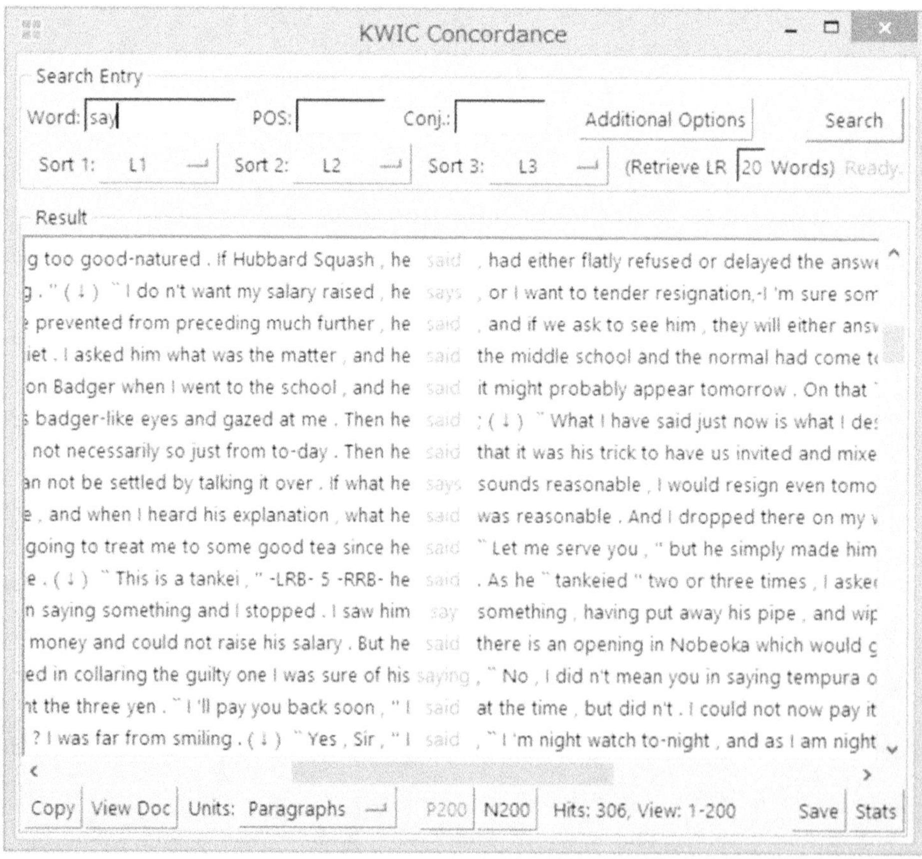

APPENDIX 2

In order to clarify which words co-occur with the target words (the adjectives and adjectival nouns listed above), KWIC (Key Words in Context), or simply "concordance," is used. With KWIC, we can analyze how a target word is used in a corpus. In this appendix, an example of KWIC using KH Coder, a text mining application, is shown. Figure AP-2 illustrates the KWIC for the word "say" in a sample corpus. KH Coder is a very useful text mining tool. It provides statistics, multi-dimensional scale (MDS), word networks, and other helpful visuals.

Using the Collocation Stats window in KH Coder (Figure AP-2), it is easy to determine which words frequently appear before and after the target word (or node word). In Figure AP-2, the statistics show that the word "hear" appears thrice in a position two words before (L2) and twice just before (L1) the node word "say." In addition, we see clearly that words like "Red," "Shirt," and "Porcupine" are often used in association with "say" in this corpus.

(*KH Coder Reference Manual*: Higuchi, 2017, Revised)

APPENDIX 3

Concordance *score* is calculated using the function $f(w)$ shown below, where l_1 is the frequency of a certain word w that appears just before the node word; l_2 is its frequency, two words before the node word; r_1 is its frequency just after the node word; and r_2 is its frequency, two words after the node word. In general, the greater the frequency that a certain word w appears before or after the node word (l_i+ r_i), the larger the value of $f(w)$. In calculating the value of $f(w)$, frequencies (l_i + r_i) are divided by "i,"

Figure 5. Collocation statistics window (Higuchi, 2017)

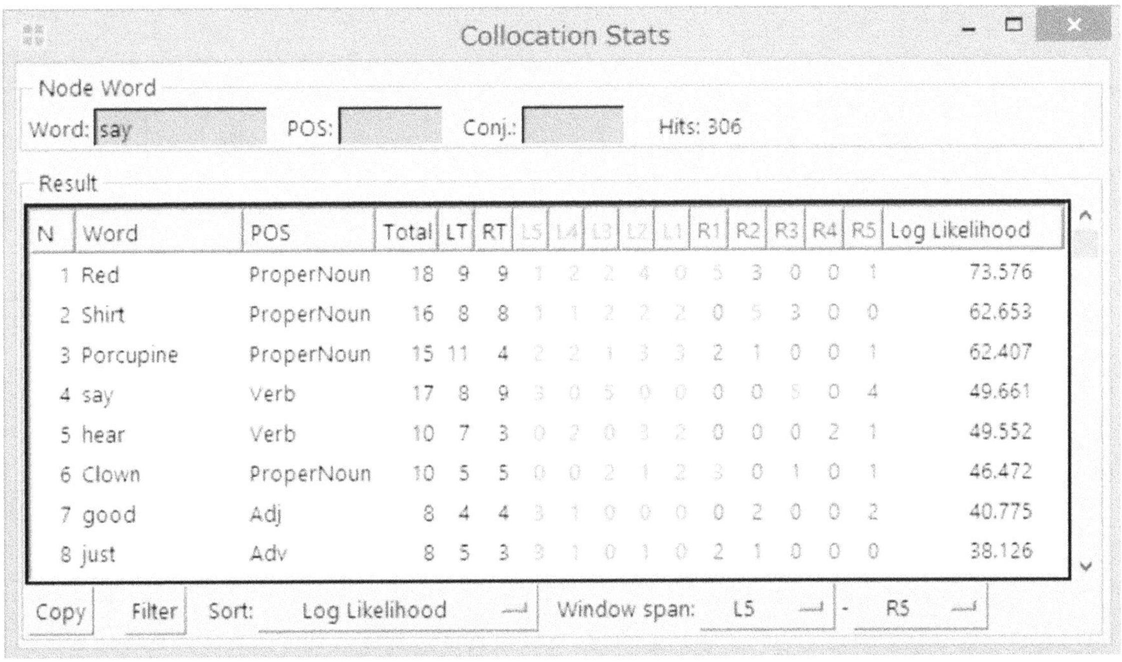

N	Word	POS	Total	LT	RT	L5	L4	L3	L2	L1	R1	R2	R3	R4	R5	Log Likelihood
1	Red	ProperNoun	18	9	9	1	2	2	4	0	5	3	0	0	1	73.576
2	Shirt	ProperNoun	16	8	8	1	1	2	2	2	0	5	3	0	0	62.653
3	Porcupine	ProperNoun	15	11	4	2	2	1	3	3	2	1	0	0	1	62.407
4	say	Verb	17	8	9	3	0	5	0	0	0	0	5	0	4	49.661
5	hear	Verb	10	7	3	0	2	0	3	2	0	0	0	2	1	49.552
6	Clown	ProperNoun	10	5	5	0	0	2	1	2	3	0	1	0	1	46.472
7	good	Adj	8	4	4	3	1	0	0	0	0	2	0	0	2	40.775
8	just	Adv	8	5	3	3	1	0	1	0	2	1	0	0	0	38.126

which weighs the frequencies according to their distance from the node word. Thus, words that appear nearer to the node word (i.e., with a smaller "*i*") have greater weight than those that occur five words before or after the node word. In this formula, the frequencies of words that appear just before and after are simply added, since they are divided by unity.
("KH Coder Reference Manual": Higuchi, 2017)

APPENDIX 4

Table 11. Frequency of the verbs in Sake Corpus

N	word	pron	Eng. Trans	freq.		51	上げる	ageru	raise	41
1	感じる	kanjiru	feel	890		52	目指す	mezasu	aim for	41
2	飲む	nomu	drink	416		53	戻る	modoru	Back	41
3	広がる	hirogaru	spread	352		54	用いる	mochiiru	use	41
4	残る	nokoru	remain	264		55	覆う	oou	cover	40
5	思う	omou	think	199		56	でる	deru	come out	39
6	立つ	tatsu	stand	186		57	押す	osu	push	38
7	醸す	kamosu	brew	178		58	取る	toru	take	38
8	含む	fukumu	include	155		59	取れる	toreru	come off	37
9	ふくらむ	fukuramu	swell	150		60	言う	iu	say	36
10	引く	hiku	draw	145		61	入れる	ireru	put in	36
11	入る	hairu	enter	145		62	引き締める	hikishimeru	tighten	35
12	消える	kieru	vanish	141		63	味わう	ajiwau	taste	34
13	現れる	arawareru	appear	120		64	与える	ataeru	give	34
14	持つ	motsu	have	113		65	向く	muku	face	32
15	冷やす	hiyasu	chill	102		66	混じる	majiru	mingle	32
16	帯びる	obiru	take on	101		67	似る	niru	resemble	32
17	落ち着く	ochituku	settle down	94		68	示す	shimesu	denote	32
18	続く	tsuduku	continue	92		69	膨らむ	fukuramu	swell	32
19	出る	deru	come out	91		70	こだわる	kodawaru	stick to	32
20	出す	dasu	put out	85		71	まとめる	matomeru	put together	32
21	漂う	tadayou	drift	82		72	飲み込む	nomikomu	swallow	31
22	残す	nokosu	leave	77		73	際立つ	kiwadatsu	stand out	31
23	見せる	miseru	show	76		74	知る	shiru	know	31
24	抜ける	nukeru	slip out	75		75	伝わる	tsutawaru	pass	31
25	造る	tsukuru	make	73		76	比べる	kuraberu	compare	31
26	熟す	jukusu	age	72		77	枯れる	kareru	wither	30
27	まとまる	matomaru	come together	70		78	引き締まる	hikishimaru	tighten up	29
28	変わる	kawaru	change	68		79	搾る	shiboru	squeeze	29
29	楽しめる	tanosimeru	enjoyable	65		80	開く	hiraku	open	28
30	備える	sonaeru	equip	62		81	誇る	hokoru	boast	28
31	合う	au	match up	60		82	思える	omoeru	feel	28
32	できる	dekiru	I can.	60		83	置く	oku	place	28
33	引き出す	hikidsasu	pull out	59		84	撫でる	naderu	stroke	28
34	合わせる	awaseru	match	59		85	つける	tsukeru	attach	28
35	進む	susumu	proceed	59		86	押し上げる	oshiageru	push up	27
36	まとう	matou	take on	59		87	伸びる	nobiru	grow	27
37	練れる	nereru	knead	58		88	包む	tsutsumu	wrap	27
38	もつ	motsu	have	57		89	ざらつく	zaratsuku	gritty	27
39	使う	tsukau	use	56		90	引っかかる	hikkakaru	get stuck	26
40	磨く	migaku	polish	53		91	仕上げる	shiageru	finish off	26
41	たたえる	tataeru	praise	53		92	焦がす	kogasu	burn	26
42	楽しむ	tanoshimu	enjoy	52		93	食べる	taberu	eat	26
43	増す	masu	increase	51		94	抑える	osaeru	curb	26
44	流れる	nagareru	flow	50		95	すべる	suberu	slip	26
45	仕込む	shikomu	train	49		96	効く	kiku	do good	25
46	保つ	tamotsu	keep	47		97	仕上がる	shiagaru	finish	25
47	立ち上がる	tachiagaru	stand up	47		98	上がる	agaru	rise	25
48	もたらす	motarasu	bring	45		99	醸し出す	kamoshidasu	brew	25
49	行う	okonau	do	42		100	成す	nasu	make	25
50	伴う	tomonau	come with	42		101	来る	kuru	come	25

APPENDIX 5

Multi-dimensional Scaling

Performing multi-dimensional scaling with many (50 to 100) words results in a plot that may be difficult to interpret. To make interpretation easier, KH Coder can perform a cluster analysis of words and present these clusters using different colors. Given that only 12 colors can be used at a time to identify clusters on the screen, a maximum of 12 clusters can be specified. This function is not available for three-dimensional plots.

If [Adjacent clusters] is checked under the "Multi-Dimensional Scaling Options" area on the right, cluster analysis is performed based on the score obtained in multi-dimensional scaling. This cluster analysis uses the Ward method based on Euclidian distance. Thus, words plotting close to each other are classified in the same cluster. Clearly, clustering can assist interpretation of the results of multi- dimensional scaling. This type of cluster analysis is also applied in "Concept Mapping" proposed by W. M. K. Trochim (1989), and is thought to be an effective method for exploring data with an unknown group or category structure (Afifi & Clark 1996).

Chapter 5
300-500 Threshold of Context Memory and Assumption Experiment

Nobuyoshi Harada
Chiba University, Japan

ABSTRACT

This chapter investigates the threshold of context memory on numbers 300 through 500 regarding perspective with assumption experiment and theories of cognitive semantics and short-term memory. Context generation became possible with text database with lexical engineering system on artificial intelligence (AI). It was realized an infinite and continuous presentation of context. Several context presentations realized the limitation of context memory on the process of continuous context presentation on numbers 500 and 2,000, as well as the representation process with judgement for memory. The first report studied the context generating machine for Japanese pun generation. This was termed "Class B engine" or "Dajare Machine." This study focused on the examination of the context threshold using an experiment of human brain mapping on an auditory evoked magnetic field (AEF) of a magnetic encephalogram. It also focused on the analysis of the theory of cognitive semantics and folk biology, using a theory of short-term memory of the "magical number seven."

INTRODUCTION

There was no discussion on the limitations in the memory of context in the "Poetics" of Aristotle (1949) because it is extremely difficult to generate and expose (present) the context on the over number of limitations of humans. The context-generating system of artificial intelligence (AI) was possible to generate the unlimited number of contexts. One example was the "dajare machine" (Japanese pun-generating machine), which could generate an unlimited number of contexts, with the over number of limitations of the homogeneous context. A qualitative difference was observed between over- and underlimitation on the effect of memory construction (Appendix 1).

DOI: 10.4018/978-1-7998-4864-6.ch005

This chapter describes the quantitative difference in the number of contexts exposed by the assumed experiment of the auditory evoked magnetic field (AEF). The establishment of context memory was investigated with two conditions of the over and under numbers of thresholds of context memory. This chapter also presents the concept of the context memory threshold from the perspective of working memory of magical number seven and the semantic classification of taxonomy on hierarchical structure. The part of the function of the context threshold of taxonomy on hierarchical structure was reflected in the function of the long-term working memory.

The narrative and pun had a functional base of long-term declarative memory. The human memory function was classified into short-term memory and long-term memory on the property of time course of memory preservation. Long-term memory was divided into two types of memory, namely, declarative memory and nondeclarative memory, with the property of recollection for consciousness (Squire & Zola-Morgan, 1985). Tulving (1972) reported the definition of the two types of long-term memory for semantic memory and episodic memory. Semantic memory is the memory of word and word semantics and concepts and the semantics and relationships of perceptual objects, as organized memory. Episodic memory is the memory of events that individuals experience. According to Tulving, semantic memory is constructed by repeated experience, while the memory of the situation of acquisition disappears. Episodic memory is constructed by a single trial experience, while the memory of the situation of acquisition is maintained with the memory of the episode. The memory of the narrative plot and pun is constructed by a single trial experience, while the memory of the situation of acquisition disappears. The memory of the narrative plot and pun is midway between semantic and episodic memory.

The pun was possible to adopt for one narrative division with the definition of the narrative genre (Ogata, 2018), which was included and dominated by narrative factors, as the broadly defined narrative, which did not have independence of artwork and had some property of accessory and method of other artworks.

Adam (1984) described the predictability of narrative and narrative genre (fantastic story, journalism story, and laughter) of discourse. The expectations of narrative and individual narrative genre of discourse were established by the predictability of the organizational form of narrative and the individual discourse of narrative genre, which was supported by the shared knowledge between the narrator and listener and general reasoning.

In this chapter, the authors of this study discuss the threshold of context memory. A difference in the memory context was found in the process of recalling context memory with different contexts of numbers of 500 and 2,000. The process of recalling the sound file of 500 contexts provoked the sense of having already heard these sounds. This was not the same for the process of 2,000 contexts. This result indicated a limitation and threshold of context memory with numbers between 500 and 2,000. The threshold may control the status of the memory and novelty contexts, as it controls the presentation of the number of contexts.

This chapter first discusses the case study and background of the context threshold of 300–500. The case study of memory construction with "dajare machine" is the first case of overexposure (and presentation) of the context memory threshold using an AI system. This chapter presents a case study of 300 thresholds in "Shippaigaku," the 316 cases of counseling in "A Common Casebook," and number of limitations of 300 of Rakugo Stories.

This chapter also discusses the main focus of the assumption experiment. Study 1 shows the background of human brain mapping of the relationship between anticipation and variation of average waveform of an AEF. Study 2 shows the actual case of variation of average waveform with the anticipation of the sound of mora with the status of memory-reflected familiarity. Study 3 shows the effect of memory status as familiarity on stimulus anticipation of four-character Chinese idiom with an auditory evoked magnetic field. The preprocessing of the first sound changed the response of the second sound with the elicitation anticipation for arrival of the second sound. Study 4 shows the design of the assumption experiment. The difference in memory status is discussed from the perspective of an AEF between 500 and 2000 Japanese pun presentation. The difference in memory status made it possible to change the average waveform of the second sound with the preprocessing of the first sound of the elicitation of the anticipation of the second sound.

This chapter finally discusses the concept of the context memory threshold. This discussion is constructed with two perspectives: the hierarchical structure of semantics as cognitive semantics and magical number seven as the function of working memory. The first discussion is constructed with the peats of the hierarchical structure of semantics, folk biology, and layers of semantics. This first discussion presents the perspective of the function of three layers of semantics.

The second discussion presents the perspective of magical number seven as a function of working memory.

The perspective of the first and second discussions is combined and investigated in the last discussion of the hypothesized concept with the semantic layer and magical number seven. This perspective is supported by the hierarchical structure of the long-term working memory.

BACKGROUND AND CASE STUDY

Dajare Machine

The *dajare* machine, also known as a "class B engine," generates Japanese puns, both continuously and infinitely, using a lexical engineering system of AI. This system was developed to present the concept of the language engineering study group, *Kotoba Kougaku* (linking art and engineering), and research results. The *dajare* machine was named as "slightly generating infinite puns" because it generates an infinite number of Japanese puns (Abe, 2011).

The *dajare* machine creates Japanese puns through either a printed or voice output. There are four types of content. The first three patterns are combinations of two similar words, including an operation to insert a sound word, such as the end of a sentence. The fall of the sentence reflects the contemporary social situation.

500 to 2,000 Sound Files

This was the first case reflecting the threshold of context memory using a *dajare* machine. The threshold was found via a process of presenting Japanese puns of sound files generated in a number of contexts.

The 38th Language Engineering Workshop was held on August 26 and 27, 2011, at Kanagawa University Yokohama Campus, Building Number 23. The *dajare* machine was exhibited by Professor Kazumitsu Matsuzawa's laboratory. To simplify the mechanism, the system was displayed to play the sound files. This was generated and recorded using a *dajare* machine.

The author had an opportunity to obtain a comment from Professor Matsumoto regarding the number of contexts of Japanese puns and subjects' responses related to the memory of context.

Figure 1. Class B engine system developed to present the concept of the language engineering study group Kotoba Kougaku and research results

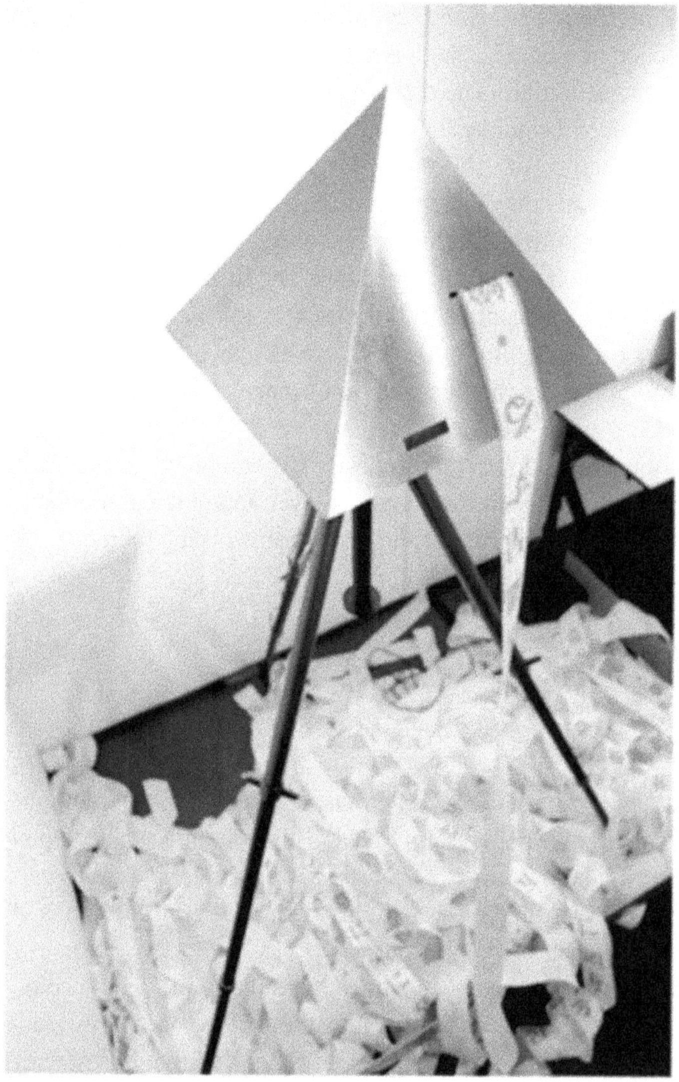

In trials where Japanese puns were regenerated in 500 contexts, participants could recognize regenerated puns the second time they heard the sound files. However, when puns were regenerated in 2,000 contexts, the participants were unable to recognize the regenerated puns upon hearing them the second time. This result reflects the presence of a memory threshold between 500 and 2,000 contexts.

The data indicated a qualitative difference in the context of memory processes when exposed to 500 and 2,000 contexts related to recall. Thus, the result reflects the limitation property of context memory.

Shippaigaku

Shippaigaku (i.e., the Japanese word for "failure analysis") concerns the sampling and analysis of failures to create failure measures and manual operation of failure status (Hatanaka, 2000). Chapter 8 of *Shippaigaku* provides a discussion on the limits of manipulation number failure objects for the construction of measures and manual operations. The number of limitations of manipulation should not exceed 300. If manual chapter numbers exceed 300, an omission of the discovery of the chapter could be provoked over 300 chapters (e.g., 20,000). A knowledge unit of failures and their cause and effect could have limitations structured in groups and hierarchical arrangements. The construction of the structure of manual chapters was required for grouping and hierarchical arrangement to understand the background of failures. Grouping and hierarchical arrangement of chapters were reflected with manual chapter threshold of 300.

An Uncommon Casebook

An Uncommon Casebook is a collection of 316 case studies published by Dr. Milton H. Erickson (O'Hanlon & Hexum, 1990). Erickson grouped these cases into 45 units with nine classified sets: (1) habits and impulses; (2) physical problems and distress; (3) sexual problems; (4) sleep problems; (5) fear and emotional problems; (6) cognitive and communication problems; (7) marriage, family, and human relationship issues; (8) serious disabilities in behavior or cognition; and (9) other issues. Units were grouped as finger sucking and claw biting issues. There were three layers within each group and unit. The number of members in each group and/or unit deviated between 1 and 12 and from 2 to 9. These averages did not deviate significantly from 7. Seven was known as the limitation of short-term memory (Miller, 1956). The group set structure and unit of cases showed a limitation to the number related to recognition, classification, and systematization of counseling cases.

Limitation of Memory Related to *Rakugo*

One chapter of the book *Shippaigaku* described limitations related to human memory. This section presents limitations related to number memory in the Japanese storytelling termed "*rakugo*" (Hatanaka, 2000). Enshou was a famous master of *rakugo* storytelling. A *rakugo* storyteller has approximately 100 stories in his/her life. However, Enshou's 300 stories were relayed with accurate memory. The book *Shippaigaku* noted that 300 was the limitation of the memory of numbers.

MAIN FOCUS 1: ASSUMPTION EXPERIMENT

Observation: 300 to 500 Context Memory

The threshold of context memory assumed an experiment of 300 to 500 threshold of context memory with a fall on a brain mapping experiment using magnetoencephalogram and electroencephalogram based on the difference of memory status reflected with familiarity of context between under and over number of threshold on auditory evoked magnetoencephalographic and electroencephalographic fields.

The threshold of context memory that exists between 500 and 2,000 on context number was reflected with a recognition process for the regeneration of 500 to 2,000 contexts of Japanese puns. The difference in recognition allowed us to reflect the states of degree of memory for words of Japanese puns between 500 and 2,000 contexts. Moreover, different states of the degree of memory of words allowed expression with the degree of familiarity of words. The effect of word familiarity allowed the investigation of the variation in AEF and auditory evoked electrical potential (AP) in human subjects.

It is possible to show the experimental design of the examination of the effect of the 300–500 threshold with the difference in provocation of word familiarity between 500 and 2,000 Japanese puns.

Study 1: Anticipation and AEF

Figure 2. Response anticipation of the averaged waveform on different inter-stimulus intervals and word familiarity

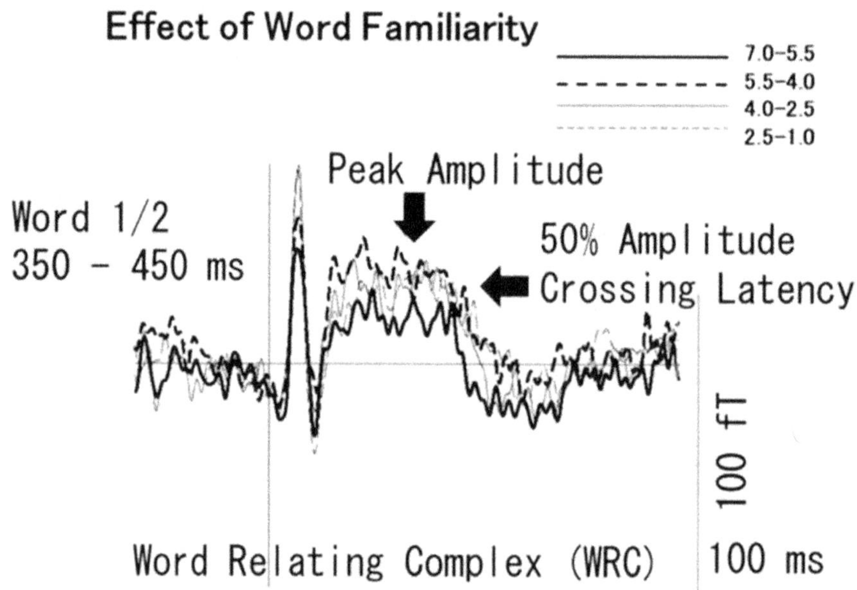

Engstrom, Kelso, and Holroyd (1996) observed that the reaction time for tapping to a metronome sound was a Gaussian distributed relative to the metronome timing for short inter stimulus intervals (ISIs), such as 1 or 2 s. On the contrary, the reaction time for long ISIs, such as 4 or 8 s, was distributed as a large, sharp, and narrow peak almost 200 ms following the metronome. The difference in the response between the short and long ISI conditions was thought to be derived from anticipating the timing of the metronome. The subject could not anticipate the timing of the metronome with a long ISI but could anticipate it with a short ISI.

The averaged waveform of the N100m component with changes in the amplitude and latency were also observed with changes in the length of the ISI in the fixed ISI form, in the short ISI (i.e., 1–2 s) and long ISI (i.e., 4–8 s) conditions. Näätänen (1986) reported a large N100m component in long ISI conditions, such as 4 or 8 s. The amplitude and latency of the N100m component increased with the increment of the ISI interval from 1 to 2 s to 4 to 8 s (Appendix 2).

In a previous study (Amano, 1996), subjects were asked to press a button as soon as they recognized the phoneme /k/ in Japanese spoken words or nonsense words that consisted of four morae. The phoneme was either in the first, second, third, or fourth mora, and its position was always similar in words and nonsense words. The response time was measured under three conditions of different speeds with which the words were played (i.e., word duration): time-compressed (× 0.5), normal (× 1.0), and time-expanded (× 2.0). The response time for phoneme /k/ was faster in words than in nonwords for all three duration conditions but only if the phoneme was in the third or fourth mora. When the phoneme was in the first or second mora, response times for the phoneme were similar in words and nonwords, regardless of the duration condition. Nonsense words are, by definition, completely unfamiliar; therefore, subjects could not anticipate the mora sounds in this condition. Thus, in this study, the familiarity of a word facilitated the detection of a phoneme in the third or fourth mora of 4-mora-long words. Amano (1996) suggested that the brain processes word sounds (phoneme recognition) independently of word duration and that the facilitatory effect of word familiarity on word recognition is independent of word duration.

If Amano's (1996) finding that the reaction time for a particular phoneme does not depend on word duration is a reflection of the general word recognition function of the human brain, it implies that the brain processes word sounds similarly for different word durations. Furthermore, it implies that the effect of word familiarity on word recognition will be similar for different word durations and between continuous word sound streams and clearly separated phoneme sounds. The reaction time was different in the third and fourth morae of words between words and nonwords; the difference in anticipation function appeared on the onset time of the third and fourth morae.

The average waveform of the word sound (speech sound) showed the N100m component at 100 ms of sound onset time and broad N100m complex after the N100m component. The broad N100m complex was considered the sustained field, and the sustained dipolar field pattern was estimated as one single ECD with location similar for N100m (Eulitz, Diesch, Pantev, Hampson, & Elbert, 1995; Hari, 1991). The average wave of the N100m complex for word sounds could be adopted as word-related complex (WRC).

The effect of familiarity was observed on the amplitude and latency of the WRC. The decrease in familiarity was expressed as the increment of amplitude and prolongation of latency of WRC (Harada, Iwaki et al., 2005a).

Study 2: Word Familiarity with AEF

The effect of word familiarity of auditory stimuli on the word-recognizing function of the human brain was investigated (Harada, N., Iwaki, S., Nakagawa, S., Amano, S., Yamaguchi, M., & Tonoike, M. 2005a,b). Word familiarity, characterized by facilitation and accuracy of word recognition, is an index of the relative ease of word perception (Amano, Kondo, & Kakehi, 1995; Connine, Mullennix, Shernoff, & Yele, 1990). The author studied the effect of word familiarity using four morae of Japanese auditory words as auditory stimuli on the elicitation of AEFs with a word-naming task. The author selected words from a database of the lexical properties of Japanese words based on auditory familiarity (Amano & Kondo, 1999).

The author grouped and presented the four morae of Japanese auditory words in four classes of degree of familiarity. The researcher used three sound conditions as the auditory stimuli for recording the AEFs. The first condition was a normal length of a word sound (word 1/1 condition). The second was shortened by 50% for the length of a word sound (word 1/2 condition). The third was a simple pure tone at 1 kHz with 100 ms duration (1 kHz, 100 ms condition). This was the control condition to record the simplest AEF waveform. The researcher observed the waveforms of AEF on channels of the maximum amplitude of N100m components in the left and right hemispheres.

Figure 3. Average waves of auditory evoked magnetic field of the magnetoencephalogram of the four Japanese mora sounds used in four classes related to the degree of familiarity (The thickness and pattern of line indicates familiarity [thick line, 7.0–5.5; thick dotted line, 5.5–4.0; fine line, 4.0–2.5; thin dotted line, 2.5–1.0])
Source: Harada, Iwaki, Nakagawa, Amano, Yamaguchi, & Tonoike, 2005a

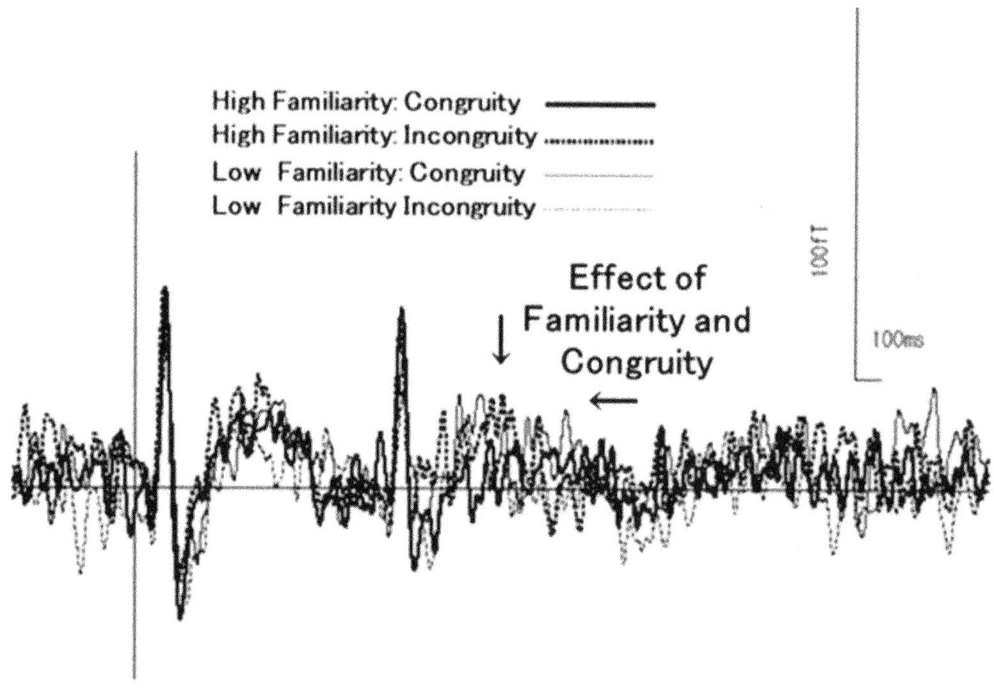

The author observed WRCs after the N100m component as large broad components in the word 1/1 and 1/2 conditions in the left hemisphere channel, rather than the 1 kHz 100 ms condition (Figures 1 and 2). As shown in Figure 3, the peak amplitude of WRCs in the word 1/2 condition significantly decreased ($F(3/27) = 9.44, p = 0.000195$). The 50% amplitude crossing latency of WRCs of the maximum amplitude in the four conditions of familiarities significantly shortened ($F(3/27) = 11.71, p = 0.0000427$) with an increment in the value of familiarity on the decreasing curve of WRCs (Figure 2).

Some scholars reported that the increment in the regularity of interstimulus intervals (with the increment of exponent n of fluctuation $1/f^n$) provoked a decrement in the N100m amplitude and latency shortening with the construction of memory traces of stimuli or short-term memory traces (Harada et al., 2001; Harada, Masuda et al., 2005). The amplitude decrement and 50% amplitude crossing latency shortening of WRCs were provoked by an increment in the value of familiarity. This may be reflected in the internal construction of memory traces for word sounds (long-term memory traces) constructed with experience from exposure to and generation of word sounds in a lifetime (Harada, Iwaki et al., 2005a).

The researcher observed the effect of word familiarity of auditory stimulus of four morae of Japanese auditory words on the amplitude decrement and 50% amplitude crossing latency shortening of WRC in the word 1/2 condition. Long-term memory traces constructed with the experience of word sounds resulted in the WRC waveform in the left hemisphere.

Study 3: Effect of Memory (Familiarity) on Stimulus Anticipation of Four-Chinese-Character Idiom with AEF

The author investigated the effect of word familiarity and incongruity of auditory stimuli with four-Chinese-character idioms (four kaiji idiom) with a difference in familiarity of four-Chinese-character idiom and congruity or incongruity of idiom with exchange of forward and backward idioms with the same familiarity group (Harada, Iwaki, & Tonoike, 2006c). A previous study reported that they expressed the effect of word familiarity on AEF on increment in amplitude and shortening of latency of the N100m component of WRC on AEF (Harada, Iwaki et al., 2005a). The expression of congruity and incongruity of auditory stimuli of four-character Japanese idiom were investigated at different levels of familiarity (i.e., high and low familiarity of Japanese) with or without exchange of the forward and backward parts of the idiom (Appendix 2).

Figure 4. Average waveforms of auditory evoked magnetic fields of magnetoencephalographs elicited by word sounds of congruity and incongruity in a Japanese sentence of four Chinese characters with high (7.0–5.0) and low (2.0–1.0) degrees of familiarity (recorded in channels with maximum amplitude of N100m components of the left hemisphere channel located on the supratemporal auditory cortex)
Source: Harada, Iwaki, Tonoike, 2006c

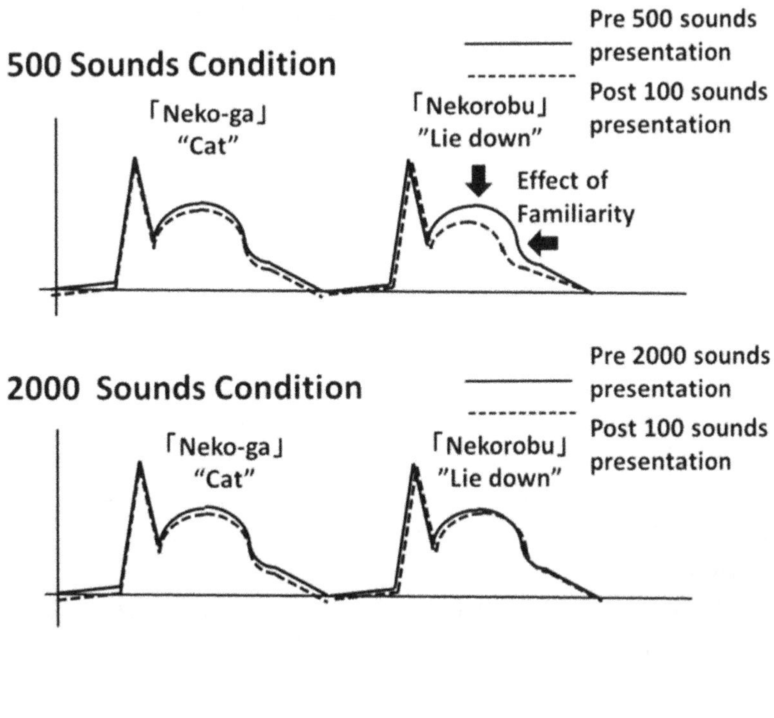

The author grouped and presented the four-character idiom (four kaiji idiom) of Japanese auditory words into two classes (i.e., high, 7.0–5.5, and low, 2.0–1.0) of degree of familiarity. The researcher separated the four-character idiom using the forward and backward parts of the idiom. The author investigated the congruity and incongruity conditions of idioms with and without exchange of forward and backward parts of idiom with the group of idioms with the same familiarity. Time interval of auditory stimuli of the forward and backward parts of the idiom was located at 1000 ms. The researcher observed waveforms of AEF of the forward and backward parts of idiom sounds on channels of the maximum amplitude of N100m components in the left and right hemispheres.

The author observed WRCs after the N100m component in the left hemisphere as large broad components in four conditions of high and low familiarity and with congruity and incongruity on the forward and backward parts of the idiom. The peak amplitude and latency of the WRCs of the forward part had no difference in the four conditions. The peak amplitude and latency of WRCs of the backward part decreased only with high familiarity and no exchange (congruity) condition. There was no difference in the amplitude of WRCs in the backward part of the three other conditions (i.e., high familiarity and exchange, low familiarity and no exchange, and high familiarity and exchange).

Auditory stimuli of the backward part of the idiom with high familiarity and no exchange condition allowed anticipation of the stimulus presentation. Instead, the stimulus presentation of the backward idiom part exchanged with high familiarity condition was not anticipated, even in low familiarity conditions with exchange and no exchange conditions. The researcher observed that the sound set of idioms expressed WRC amplitude decrement and latency shortening of the backward idiom on the condition of anticipating the backward idiom. The anticipation process may reflect the internal construction of memory traces for word sounds (i.e., long-term memory traces) constructed with the experience of exposure to and generation of word sounds. The amplitude decrement and latency shortening of N100m of WRC allowed the investigation of the provocation of anticipation of auditory stimuli with construction of memory of word sounds on constructed familiarity.

Study 4: Design of Assumption Experiment to Investigate for Context Memory Threshold of 300 to 500 Numbers

The sections above illustrate the general structure of the experiment. To examine the difference between the construction of context memory and presentation of Japanese puns with a number between 500 and 2,000, it is necessary to study the difference in amplitude and latency of AEF and AP of WRC of Japanese puns (Harada, 2019). Construction of context memory was different in the presentation of the number of Japanese puns between 500 and 2,000. Memory construction was reflected in anticipation of the sound of context and allowed the examination the provocation auditory response of AEF and AP of WRC. The author investigated memory construction of 500 and 2,000 Japanese puns by provocation of AEF and AP of WRC.

Japanese puns can be prepared using the *dajare* machine for numbers 500 to 2,000, with 400 for the measurement of AEF and AP of WRC. The researcher was also able to sample the Japanese pun from the Web data available on its website. The author sampled the Japanese pun with two criteria of selection: 1) easiness to understand for individuals without knowledge or expertise and 2) clearly distinguishable structures of the forward and backward parts.

The experimental design consisted of two processes: presentation process and measurement process. The author presented Japanese pun groups in the range of 500–2,000. After the presentation of the Japanese pun sound, the listener has a 15-min break; then, the measurement of the auditory response starts. The author analyzed the measurement of AEF and AP of WRC with 100 pun sounds sampled randomly from the range 500–2000 of Japanese pun groups. The author represented the sampled 100 puns with 200 new Japanese puns with random mixed conditions. The researcher asked the listener to assess new or already heard sounds for the presented Japanese pun with two selection buttons. The author separated part of the Japanese pun sounds into the forward and backward parts, with a time interval of 1000 ms. The researcher averaged the auditory responses 100 times for the measurement of AEF and AE of WRC.

The design of experiment was constructed with two processes of pre- and postpresentation of word sound under two conditions of 500 and 2,000 numbers.The 500 and 2,000 sounds of Japanese puns were in the prepresentation process and then the 100 responses of selected word were averaged for AEF. The selected 100 word sounds were presented a postpresentation with 200 new Japanese pun word sounds. The auditory response of 100-word sounds were averaged for AEF. Subsequently, the amplitude and latency of the AEF of WRC were compared with the backward part of the response of WRC between pre- and postpresentation of word sound. The 500-word sound presentation condition decreased the amplitude and shortened the latency on WRC with construction anticipation with the structure of context memory.

Figure 5. Designed experiment of results of auditory evoked magnetic field of 500 or 2,000 word sound presentation of Japanese puns

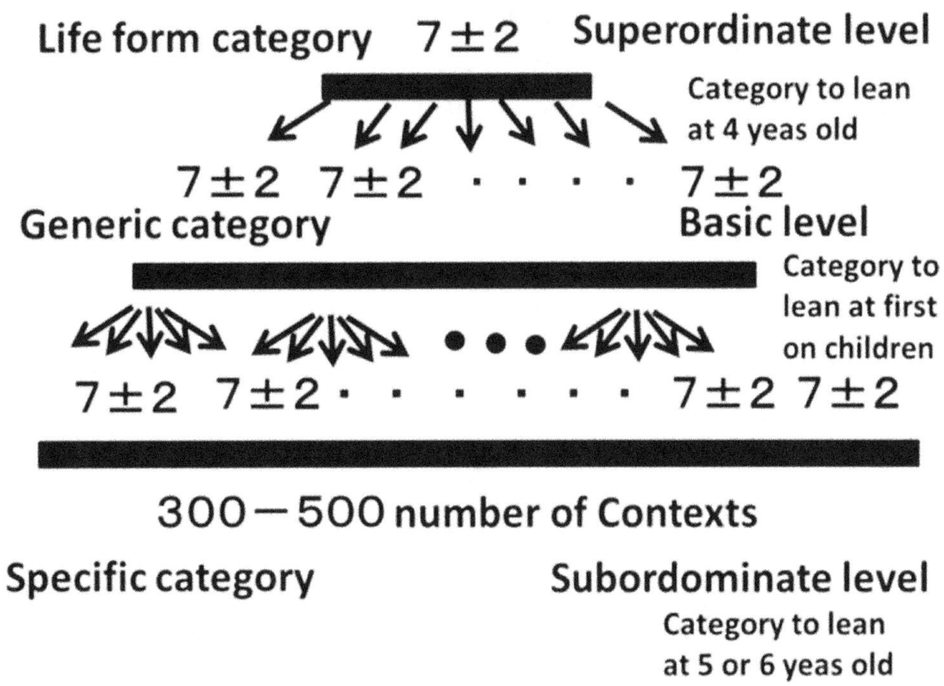

The author studied the difference in amplitude and latency of AEF and AP of WRCs with a backward sound of Japanese puns from a selection of 100 puns from the range of 500–2,000 groups. It was impossible to anticipate the forward part sound of the Japanese pun when the sound of the stimuli arrives. Instead, the researcher could influence the anticipation of the backward part sound with the construction of memory of the context of the Japanese pun. If context memory of the Japanese pun was constructed with the first presentation, the sound memory of the Japanese pun was easily provoked with the forward part of the sound presentation. This arrived with the backward sound, as it easily anticipated it via sound memory. It was possible to produce anticipation of the backward part sound with decrement of amplitude and shortening of latency of the N100m and N1 component of AEF and AP of WRCs. If context memory was not constructed, there was no amplitude decrement or latency shortening with AEF and AP of WRCs.

It was possible to investigate the construction of context memory by the variation of AEF and AP of WRCs. The author was able to observe the amplitude decrement and latency shortening with 500 Japanese pun groups, but not with 2,000 Japanese puns. This resulted in the construction of context memory in the 500 Japanese pun groups, but not in the 2,000 Japanese pun groups.

Figure 6. The detection of context memory threshold 300–500 is shown by the difference in the establishment of memory of the pun presented in the range of 500–2000

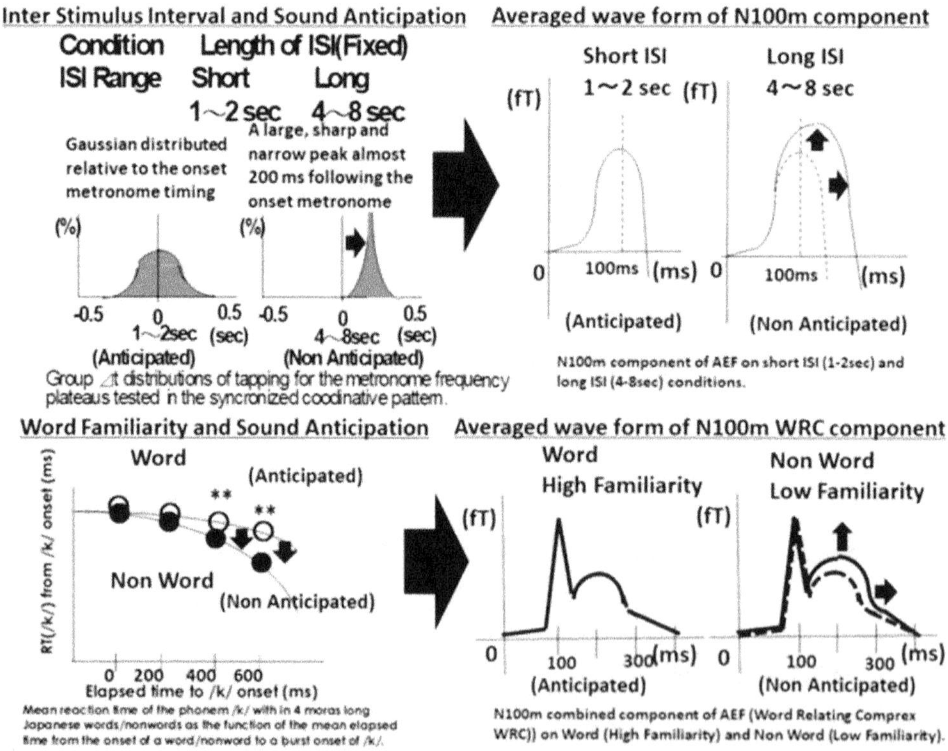

The design of the experimental steps is shown in Figure 6 for the detection of the threshold of context memory with the difference in the status of memory of the pun, which was presented under or over the threshold.

Step (1) consisted of the 500 and 2000 pun presentation. On the adoption of the case study of the dajare machine, the 500 pun presentation established the memory, while the 2000 pun presentation did not. In the first and second phrases, one pun constructed. This is the memory establishment process of a pun.

Step (2) consisted of a memory checking process for two conditions. The 100 puns were corrected from the groups of 500 and 2000 puns. The corrected 100 puns were presented after the 500 and 2000 pun presentations with short break. An AEF of 100 puns was recorded and averaged 100 times for the detection of the average waveform.

In the second pun presentation of 90 times (Step 2-1), the presentation of the first phrase recalled the second phrase and anticipated the second pun. The condition of 500 puns recalled and anticipated the second pun. The condition of 2000 puns did not recall or anticipate the second pun.

At the time before the second phrase presentation, the sound arrival of the second pun was anticipated. In Step (2-2), after the second phrase presentation, the facilitation of auditory informational processing in the auditory cortex occurred. The condition of the 500 pun presentation maintained the anticipation, while the condition of 2000 pun presentation did not.

The result of Step (2-2) appeared on the variation of the average waveform of AEF. The condition of 500 puns allowed the facilitation of informational processing, while the condition of 2000 puns did not. The condition of 500 puns of AEF decreased the peak amplitude and shortened the latency, while the condition of 2000 puns did not.

The difference in AEF between the conditions of 500 and 2000 puns indicated the difference in the establishment of memory for the presented pun and showed the existence of the threshold of context memory.

MAIN FOCUS 2: CONCEPT OF CONTEXT MEMORY THRESHOLD

Hypothesis of Semantic Layer and Magical Number Seven

The author examined the 300–500 threshold with concepts of short-term memory limitation of the magical number seven and semantic layer development of words in early childhood. The number of objects held in short-term memory was 7 ± 2. The researcher constructed a semantic layer of words with the basic level on a fast process. The superordinate level was on the second process (4 years). The subordinate level was in the third process (5 or 6 years).

The researcher expressed the 300–500 threshold functional base with the function of short-term memory of the magical number seven on three semantic layers of word for context memory. The 300–500 threshold can be shown as $7 \times 7 \times 7 = 343$ (memory number of short-term memory process was presented as 7 ± 2) and with the taxonomic function of long-term working memory (Ericsson and Kintsch, 1995).

Structural Model of Context Memory Threshold

Shippaigaku presented a manipulation threshold of 300 in the chapter number of operations manual of machine (Hatanaka, 2000). It was necessary to construct the structure of chapters for their grouping and hierarchical arrangement to understand the background of failure reflected with 300 chapters. The hypothesis of context memory threshold was influenced by memory property of grouping and hierarchical arrangement of memory target of context. The author discussed the memory property of hierarchical arrangement by the theory of the semantic layer; it was rerating for the property of the layer on the memory target. Further, the author illustrated the memory property of grouping by describing the short-term memory concept of the magical number seven (Miller, 1956); it was rerating for property of limitation and operability on memory target. The author discussed the hypothesis of context memory threshold using the semantic layer theory and the theory of short-term memory concept of the magical number seven.

Hierarchical Arrangement of Memory

Hierarchical Structure of Semantics

The author constructed a category from a hierarchical structure in relation to inclusion. The semantic feature theory of structural semantics shows hierarchical structures of semantic relations as hyponyms (i.e., low-layered words). These include the semantic features of hyperonyms (i.e., high-layered words). These belong to the idea of the discrete and binary oppositional feature of the category (Katz & Fodor, 1963; Katz & Postal, 1964).

All category levels of the hierarchy have equal importance in structural semantics. Instead, cognitive semantics attach special importance to a single category level in the hierarchical category. The specific category, or basic level, plays a central role as the structural base for the superordinate

level (i.e., higher layer of the basic level) and subordinate level (i.e., lower layer of the basic level). The basic level depends on the semantic theory of the prototype. Cognitive semantics describes the function of a prototype. The semantic category of words, namely, the prototype and nonprototype, are central members (prototype = robin or sparrow of the bird member). The peripheral members (nonprototype = ostrich or penguin of the bird member) in a semantic category have a subjective judgment of prototypicality (Barsalou, 1985). Members of this category do not have equal importance (Rosch, 1973, 1977). The role of the prototype includes (1) a short response time to discriminate belonging to the members of the category; (2) the facility to image a typical example of a category member; and (3) ease of learning.

The basic level is the prototype of the taxonomic layer in a hierarchical structure. Its functional roles include (1) concentration of number information compared with other layers; (2) first layer of a category for children to learn words; (3) maximum layer of frequency of use; (4) expression with one word; and (5) less difference in interferences between languages.

Folk Biology

All humans have the word taxonomy of animals and plants. Folk biology, a research area in the field of anthropology, is a study on the structure of the taxonomy of words regarding the natural world of animals and plants. Berlin (1972, 1992) and Berlin, Breedlove, and Raven (1973) reported that all human words have the hierarchical taxonomic structure of the natural world. In addition, layers of hierarchy were limited to the fifth or sixth. Limitations on the number of layers come from limitations of a human cognitive function.

The first layer, termed "unique beginner," corresponds to the words "plant" or "animal." The second layer, termed "life form category," corresponds to "tree, bush, vine, grass or herb" or "bird, fish or mammal." The third layer, termed "generic category," corresponds to "oak or maple" or "cat or sheep." This layer, which includes the maximum number of words, plays the role of a central layer in hierarchical taxonomy. The fourth layer is termed "specific category." The fifth layer is termed "varietals category."

Berlin (1972, 1992) reported the existence of an intermediate layer between the "life form category" and "generic category." He called this the "intermediate category." Rosch, Carolyn, Wayne, David, and Penny (1976) reported that the "generic category" functions as the basic level for the learning process of language.

Semantic Layer

The basic category is the first layer for children to learn words. The high taxonomic layer in the hierarchical structure is for children to learn words at the age of 4 years. Low-high taxonomic layer is for children to learn words at the age of 5 or 6 years. The three layers of the generic category (or basic level), plus the low-high taxonomic layer, are fundamental players in the construction of the semantic concept of words in a hierarchical structure in the developmental process.

The maximum number of taxonomic layers in folk biology was five (or low frequency six) from the limitation of human recognition. High-frequency layers include "life form category," "generic category," and "specific category." "Generic category" has been used for the first time in context, while the highest layer of "unique beginner" and lowest layer of "varietal category" have been used with low frequency. Many languages do not have a taxonomy layer of "unique beginner." The "varietal category has been used for objects of particular interest within a culture. Three semantic layers have played a role in the facilitation of categories in the recognition process.

The role of "basic level" was explained as a category with a maximum number of common factors and the same category members. This category has a minimal number of common factors with different category members. A high layer of category has some similar factors with other category members. A low layer of category has some similar factors with other category members. The "basic level" layer balances with the sameness of the belonging factor with internal and external category members.

This perspective is applicable for three layers. One or two layers are insufficient for the category group. When the category is able to include factors, it should be classified in another category. Four or five layers are too many. When the different categories include the same factor, they should be classified in another category. The three-layer structure was enhanced as an efficient categorization of the object.

Working Memory

Magical Number Seven

Miller (1956) discussed the limitations of short-term memory using the experimental results of the limits of one-dimensional absolute judgment. The experiment of a one-dimensional absolute judgment task was proposed to a subject having a corresponding response to a stimulus (previously learned). Stimuli were varied in one dimension (e.g., 10 different tones varying in pitch). The response of the subject was kept for accuracy of judgment, with the number up to five or six for different stimuli. The accuracy of judgment decreased with an increase in the number of stimuli over five or six.

Miller (1956) discussed limitations surrounding short-term memory based on differences in judgment accuracy with a number of stimuli (more or less than five or six). He termed the limitations of short-term memory using the results of the experiment as the magical number seven. According to Miller (1956), humans are limited to the number of objects they can hold in their short-term memory (7 ± 2). Memory limitation of the number or word is termed "chunk" (i.e., it does not exceed 7 ± 2). Chunk is different to objects (i.e., the number was 7, the letter was 6, and the word was 5, approximately). The short-term memory limitation of 7 ± 2 may play a role in the process of operability in memory. Various studies on short-term memory report a different number of limitations. It is clear that there is a limit to the number (nearly seven) related to the function of short-term memory.

Ericsson and Kintsch (1995) reported the taxonomic structure of chunks of the magical number seven.

Chase and Ericsson (1981) found the expert performance memorization of a list of 30 digits. They were able to assess a hierarchical retrieval organization that experts used to memorize lists of 30 digits. The function of long-term working memory was found in memorization of text. The memorization process of text uses retrieval cues to the hierarchical organization of encoded text to provide access to text information (Fletcher 1981) (Appendix 3)

Hypothesized Concepts: Semantic Layer and Magical Number Seven

Construction of the context memory threshold of 300–500 was hypothesized using the concepts of the semantic layer and magical number seven. The threshold number was explained with a three-layer structure of folk biology and limitation of the number seven for short-term memory. If the grouping role of short-term memory limitation seven was plated on all three layers, the number of manipulated objects could show 7 × 7 × 7 = 343 on the three layers.

The number 343 did not deviate from 300. The number of manipulated objects for variation was 8 × 8 × 8 = 512 and 9 × 9 × 9 = 729, which indicates that the magical number deviates according to the object. Numbers 512 and 729 exceeded 500. However, these numbers were clearly lower than 2,000. The number seven could facilitate the threshold for the manipulation of memory of the recall function in the process of grouping and layer forming.

Grouping and layer forming with the number seven may result in smooth memory recall without extreme effort. A three-layer structure and the number seven could play a factor in quantitative constraints for expression of the threshold of context memory on the background of the theories of folk biology, cognitive semantics, and short-term memory as the magical number seven.

The threshold in the range 300–500 was expressed with a grouping function in short-term memory on the magical number 7 ± 2 in three layers of the hierarchical structure of semantics. Three layers corresponded, as the first layer was the life form category and superordinate level, the second layer was generic category and basic level, and the third layer was a specific category and subordinate level, based on the theories of folk biology and cognitive semantics. Two concepts of the magical number seven and three major layers of the theories of folk biology and cognitive semantics allowed us to show the threshold of context memory of the numbers in the range 300–500. The concept of long-term working memory was possible to help develop the research threshold of 300–500 (Appendix 3).

Figure 7. Concept of expression of threshold contest memory constructed for the property of grouping and hierarchical arrangement by the theories of short-term memory as the magical number 7 ± 2, semantic layers of cognitive semantics, and folk biology

Establishment of memory and Anticipation and Wave form variation

Presentation Number of pun: 500 ⇔ 2000

Step (1):First presentation: First and Second phrase
Memorizing of pun: Established ⇔ Non Established

Step (2): Second Presentation:
Step (2-1): 1 st phrase presentation
First phrase of pun: Remember 2 nd pun⇔ Non Remember 2 nd pun
(Provocation of Anticipation)
Second phrase of pun: Anticipated 2 nd pun⇔ Non Anticipated 2 nd pun

Step (2-2): 2 nd phrase presentation
(Status of Anticipation)
Anticipation for Sound Arrival : Anticipated ⇔ Non Anticipated
Facilitation of informational processing: Occurs ⇔ Not Occurs
(Averaged Wave Form of AEF)
Peak Amplitude of AEF: Decreased ⇔ Not Decreased
Latency of AEF: Shortened ⇔ Not Shortened

DISCUSSION

Perspective of Limitation of the Chunk Layer

The significance of the context memory threshold was discussed from the perspective of long-term working memory. A study on long-term working memory presented the perspective of layer structure between chunk and memory key, and the component of chunk and memory key can also create a layer structure. The limitation of the chunk component layer expressed the limitation memory, which had a relationship with chunks. The concept of context memory threshold proposed the layer limitation of working memory as three layers from the perspective of cognitive semantics. The limitation of the layer brought the limitation of the number of memories as context. The concept of context memory threshold was possible to present the limitation of the chunk layer in the study of long-term working memory, as it was possible to play the limitation of memory in the number of components in the layer structure.

Overcoming the Limitation of Context Memory

The significance of context memory was discussed from the perspective of context generation with the AI system. It was difficult to generate an over-number context of threshold with human activity. In contrast, the context generation with AI system was almost infinite, as in the case of the "dajare machine." The AI system of infinite context generation led to the first investigation of the limitation and threshold of context memory. The AI system presents qualitative differences in the generated context of over or under the threshold.

Framework for Describing World

The significance of context memory threshold was discussed from the perspective of the utilization of AI for the external environment of humans.

The perspective of context memory provides a framework for describing the world. *Shippaigaku* indicated directions for understanding its background. An object number of 300 was described as a limitation number for recognition. This object number was a framework for others in describing the background for the world's failure. If a report of the outside world was produced with AI, a framework with limitation 300 was helpful in its understanding.

Stories generated about the outside world through narratology also utilized this framework. In *Shippaigaku*, a normal *rakugo* storyteller has approximately 100 stories. Several objects are possible within the framework, decreasing it from 300 to 100.

The context threshold is a tool in technology maintenance related to the novelty of context. It provides a framework for others when describing the world (Appendix 4).

Finiteness of Context and Music

Social significance and social function of context memory threshold were discussed from the perspective of the lack of construction of boring and monotonous feelings for the limitation of context and music.

The finiteness of music was discussed based on the law of conceptual design of music patents (Mizuno, 2017). Music (as melody), which was expressed in notes on musical score, was finite. The property of finite music was brought into Western music. Contemporary musicians have the assumption of finiteness of music, reuse of music, and a combination of musical components. They also attempted to produce new grooves, feelings, and ambience.

They have constructed their music with frequency variation over time; the number of patterns they have been able to express has been limited. Moreover, the chance to express new music patterns has been limited and has decreased. The depletion of the new music patterns was discussed in the phenomenon of the appearance of the same music on various irrelevant places of music composition. The reproduction environment of music has been greatly developed as a result of Internet technology, while the depletion of new music patterns has significantly progressed.

Conversely, music listeners have not had the pressure of feeling bored by music; rather, they have had fun listening to music and experienced mental well-being. It was possible to maintain the novelty of music through the function of the memory threshold.

In the modern world, highly developed technologies for the visual, auditory, and/or somatosensory presentation of information have led to a reduction in the novelty of information. The degree of loss of informational novelty has rapidly progressed through the establishment of photography technologies in media. Maintenance of the novelty of context is an important issue in the information management of various media. This perspective of context memory limitation may help technology in maintaining the novelty of context while controlling the number of presenting contexts.

Walter Benjamin

The quality of duplication technology has evolved. The development of technology has been ravaged with the presentation of context. The technology of visual, auditory, and somatosensory information has been greatly developed. The novelty of information in the inexperienced world has been seriously damaged. The degree of this loss of "aura," or informational novelty (Benjamin, 1936), has progressed from the establishment of phototechnology to an informational presentation of media. Benjamin (1936) discussed the loss of novelty of the context of aura presentation with duplication technology. He described this loss with the technology of photographs, which coincided with the publication of his book. According to Benjamin (1936),

Around 1900, technical reproduction had reached a standard that not only permitted it to reproduce all transmitted works of art and thus to cause the most profound change in their impact upon the public; it also had captured a place of its own among the artistic processes. (p. 12)

He also discussed the loss of the novelty element (Benjamin, 1936, p. 15) as follows:

Even the most perfect reproduction of a work of art is lacking in one element: Its presence in time and space, its unique existence at the place where it happens to be [...] One might subsume the eliminated element in the term "aura" and go on to say: That which withers in the age of mechanical reproduction is the aura of the work of art.

Adorno (1971) presented the risk of duplication technology to enhance the attitude of loss motivation without resistance. During the same period, there was an argument regarding the monotonous labor of the "Ford model." This provoked a sense of helplessness and violation of human mental health.

Actually, mental health of humans was not infringed upon with the loss of novelty of context presentation. Humans did not perceive the loss of novelty as boring. Quality mental health for the informational environment of human life was not a serious issue . People enjoyed music, videos, and movies.

Context memory with a threshold of 300–500 increases the sense of novelty for information. This remains true when information and novelty are damaged or provoke a sense of boredom.

CONCLUSION

The chapter discussed the threshold of context memory on numbers 300–500 from the perspective of the assumption experiment and concept for the threshold with semantic layer and working memory. The property of the AEF was able to express the difference in memory states over (2000) or under (500) the threshold of context memory. This chapter discussed the concept of threshold with a semantic layer and working memory. The perspective of the semantic layer presented from the investigation of the hierarchical structure of cognitive semantics, folk biology, and semantic layers of basic categories. The perspective of working memory presented the study on "magical number seven" as the function of short-term memory. The study on long-term working memory presented the bridge between the three semantic layers and magical number seven. The three semantic layer limitations expressed the limitation of the chunk layers of long-term working memory and the limitation of the number of memory, as the number of context memory thresholds.

REFERENCES

Abe, A. (2011). What is "Kotoba-kougaku" (language engineering) Intelligence and Information, (Journal of the Japan Society for Fuzzy Theory and Intelligent Informatics), 23(5), 666-675.

Adam, J. M. (1984). *Le récit*. Presses Universitaires de Paris.

Adorno, T. W. (1971). *Fukyo waon: Kanri syakai ni okeru ongaku* (N. Sanko & T. Takatsuji, Trans.). Ongakunotomosha. (Original work published 1938)

Amano, S. (1996). Lexical facilitation of phoneme perception in time-modified speech. *The Journal of the Acoustical Society of Japan (E)*, *17*(1), 9–14. doi:10.1250/ast.17.9

Amano, S., & Kondo, T. (1999). NTT database series, vol. 1, word familiarity. Tokyo, Japan: Sanseido.

Amano, S., Kondo, T., & Kakehi, K. (1995). Modality dependency of familiarity ratings of Japanese words. *Perception & Psychophysics*, *57*(5), 598–603. doi:10.3758/BF03213265 PMID:7644320

Aristotelēs. (1949). *Poetics*. Iwanami Shoten.

Barsalou, L. W. (1985). Ideals, central tendency, and frequency of instantiation as determinants of graded structure in categories. *Journal of Experimental Psychology. Learning, Memory, and Cognition*, *11*(4), 629–654. doi:10.1037/0278-7393.11.1-4.629 PMID:2932520

Benjamin, W. (1936). *Das Kunstwerk im Zeitalter seiner technischen Reproduzierbarkeit* [The work of art in the age of mechanical reproduction]. Shoubunsha.

Berlin, B. (1972). Speculations on the growth of ethno botanical nomenclature. *Language in Society*, *1*(1), 51–86. doi:10.1017/S0047404500006540

Berlin, B. (1992). *Ethno biological classification: Principle of categorization of plants and animals in traditional society*. Princeton University Press. doi:10.1515/9781400862597

Berlin, B., Breedlove, D. E., & Raven, P. H. (1973). General principles of classification and nomenclature in folk biology. *American Anthropologist*, *75*(1), 214–242. doi:10.1525/aa.1973.75.1.02a00140

Bernard, H. R. (2006, September 28). *Honoring Peter Killworth's contribution to social network theory*. Paper presented at the University of Southampton.

Chase, W. G., & Ericsson, K. A. (1981). Skilled memory. In J. R. Anderson (Ed.), *Cognitive skills and their acquisition* (pp. 141–189). Erlbaum.

Connine, C. M., Mullennix, J., Shernoff, E., & Yele, J. (1990). Word familiarity and frequency in visual and auditory word recognition. *Journal of Experimental Psychology*, *16*, 1084–1096. PMID:2148581

Cowan, N. (2001). The magical number 4 in short-term memory: A reconsideration of mental storage capacity. *Behavioral and Brain Sciences*, *24*(1), 87–185. doi:10.1017/S0140525X01003922 PMID:11515286

Engstrom, D. A., Kelso, J. A. S., & Holroyd, T. (1996). Reaction-anticipation transitions in human perception-action patterns. *Human Movement Science, 15*(6), 809–832. doi:10.1016/S0167-9457(96)00031-0

Ericsson, K. A., & Kintsch, W. (1995). Long-term working memory. *Psychological Review, 102*(2), 211–245. doi:10.1037/0033-295X.102.2.211 PMID:7740089

Eulitz, C., Diesch, E., Pantev, C., Hampson, S., & Elbert, T. (1995). Magnetic and electric brain activity evoked by the processing of tone and vowel stimuli. *The Journal of Neuroscience: The Official Journal of the Society for Neuroscience, 15*(4), 2748–2755. doi:10.1523/JNEUROSCI.15-04-02748.1995 PMID:7722626

Fletcher, C. R. (1981). Short-term memory processes in text comprehension. *Journal of Verbal Learning and Verbal Behavior, 20*(5), 264–274. doi:10.1016/S0022-5371(81)90183-3

Freedman, D. J., Riesenhuber, M., Poggio, T., & Miller, E. K. (2002). Visual categorization and the primate prefrontal cortex: Neurophysiology and behavior. *Journal of Neurophysiology, 88*(2), 929–941. doi:10.1152/jn.2002.88.2.929 PMID:12163542

Gladwell, M. (2000). *The tipping point: How little things can make a big difference.* Asuka Shin Sha.

Harada, N. (2019). Question of 300-500 context threshold and assumption experiment. *Proceedings of the Society for Language Engineering "Kotoba-Kougaku", 62*, 1-24.

Harada, N., Iwaki, S., Nakagawa, S., Amano, S., Yamaguchi, M., & Tonoike, M. (2005a). Effect of word familiarity on auditory evoked magnetic fields. In K. Zilles, J.-B. Poline, & C. Grady (Eds.), *Neuro Image: Proceeding of 11th Annual Meeting of the Organization for Human Brain Mapping (OHBM)* (26, Supplement (1), 1115). Amsterdam: Elsevier.

Harada, N., Iwaki, S., Nakagawa, S., Amano, S., Yamaguchi, M., & Tonoike, M. (2005b). Effect of word familiarity on auditory evoked magnetic fields. In *Proceeding of the 20th Annual Meeting of Japan Biomagnetism and Bioelectromagnetics Society* (pp.36-37). Academic Press.

Harada, N., Iwaki, S., & Tonoike, M. (2006a). Effect of the incongruity controlled by semantic distance on visually evoked magnetic fields. In K. Zilles, J.-B. Poline & C. Grady (Eds.), *Neuro Image: Proceeding of 12th Annual Meeting of the Organization for Human Brain Mapping (OHBM)* (31, 440). Amsterdam: Elsevier.

Harada, N., Iwaki, S., & Tonoike, M. (2006b). Effect of the incongruity controlled by semantic distance on visually evoked magnetic fields. In *Proceeding of the 21th Annual Meeting of Japan Biomagnetism and Bioelectromagnetics Society, 19*(1) (pp.56-57). Academic Press.

Harada, N., Iwaki, S., & Tonoike, M. (2006c). Effect of word familiarity for expression of incongruity. *Proceedings of the Society for Language Engineering "Kotoba-Kougaku", 24*, 36-54.

Harada, N., Masuda, T., Endo, H., Nakamura, Y., & Takeda, T. (2001). Evoked magnetic fields caused by auditory stimuli with 1/fn fluctuation in inter-stimulus intervals. *The Journal of Japan Biomagnetism and Bioelectromagnetics Society, 13*(2), 1–12.

Harada, N., Masuda, T., Endo, H., Yukihiro, N., Takeda, T., & Tonoike, M. (2005). Effect of 1/f fluctuation in inter-stimulus intervals on auditory evoked magnetic fields. *Neuroscience Letters*, *13*(379), 223–228. doi:10.1016/j.neulet.2005.01.012 PMID:15843068

Hari, R. (1991). Activation of the human auditory cortex by speech sounds. *Acta Oto-Laryngologica*, *111*(491), 132–137. doi:10.3109/00016489109136790 PMID:1814144

Hatanaka, T. (2000). *Recommendation of "Shippaigaku" (study of failure case analysis).* Koudansha.

Katz, J., & Fodor, J. (1963). Structure of semantic theory. *Language*, *39*(2), 170–210. doi:10.2307/411200

Katz, J., & Postal, P. (1964). *An integrated theory of linguistic description.* MIT Press.

Mäntysalo, S., & Näätänen, R. (1987). The duration of a neuronal trace of an auditory stimulus as indicated by event-related potentials. *Biological Psychology*, *24*(3), 183–195. doi:10.1016/0301-0511(87)90001-9 PMID:3663794

Martin, A. (2007). The representation of object concepts in the brain. *Annual Review of Psychology*, *58*(1), 25–45. doi:10.1146/annurev.psych.57.102904.190143 PMID:16968210

McCarty, C., Killworth, P. D., Bernard, H. R., Johnsen, E., & Shelley, G. (2000). Comparing two methods for estimating network size. *Human Organization*, *60*(1), 28–39. doi:10.17730/humo.60.1.efx5t9gjtgmga73y

McNab, F., Varrone, A., Farde, L., Jucaite, A., Bystritsky, P., Forssberg, H., & Klingberg, T. (2009). Changes in cortical dopamine D1 receptor binding associated with cognitive training. *Science*, *323*(5915), 800–802. doi:10.1126cience.1166102 PMID:19197069

Miller, G. A. (1956). The magical number seven, plus or minus two: Some limits on our capacity for processing information. *Psychological Review*, *63*(2), 81–97. doi:10.1037/h0043158 PMID:13310704

Mizuno, T. (2017). *Legal design, accelerating creativity and innovation through law.* Film Art Inc.

Näätänen, R. (1986). N2 wave of the evoked potential: Scalp reflection of neuronal mismach of orienting theory? In J. Strelau, F. H. Earley, & A. Gale (Eds.), *The biological bases of personality and behavior. Psychophysiology, performance, and applications. 2* (pp. 59–75). Hemisphere.

Näätänen, R., & Gaillard, A. W. K. (1983). The orienting refrex and the N2 deflection of the event-related potential (ERP). In A. W. K. Gaillard & W. Ritter (Eds.), *Tutorials in ERP: Endogenous components* (pp. 119–141). North-Holland. doi:10.1016/S0166-4115(08)62036-1

Näätänen, R., Sams, M., & Alho, K. (1986). The mismatch negativity: The ERP sign of cerebral mismatch process. *Electroencephalography and Clinical Neurophysiology. Supplement*, *38*, 172–178.

Näätänen, R., Simpson, M., & Loveless, N. E. (1982). Stimulus deviance and evoked potentials. *Biological Psychology*, *14*(1-2), 53–98. doi:10.1016/0301-0511(82)90017-5 PMID:7104425

O'Hanlon, W. H., & Hexum, A. L. (1990). *An uncommon casebook: The complete clinical work of Milton H. Erickson, M.D.* Norton.

Ogata, T. (2018). Monogatari-seisei no monogatari-ronaruiha posuto-naratoroji. *Cognitive Studies*, *25*(2), 200–217.

Purves, D. (2008). *Principles of cognitive neuroscience*. Sinauer Associates Inc.

Rosch, E. (1973). Natural categories. *Cognitive Psychology*, *4*(3), 328–350. doi:10.1016/0010-0285(73)90017-0

Rosch, E. (1977). Human categorization. In N. Warren (Ed.), *Advances in cross-cultural psychology* (pp. 1–72). Academic Press.

Rosch, E., Carolyn, M., Wayne, G., David, J., & Penny, B. B. (1976). Basic objects in natural categories. *Cognitive Psychology*, *8*(3), 382–439. doi:10.1016/0010-0285(76)90013-X

Sams, M., Hari, R., Rif, J., & Knuutila, J. (1993). The human auditory sensory memory trace persists about 10 sec: Neuromagnetic evidence. *Journal of Cognitive Neuroscience*, *5*(3), 363–370. doi:10.1162/jocn.1993.5.3.363 PMID:23972223

Squire, L. R., & Zola-Morgan, S. (1985). The neuropsychology of memory: New links between humans and experimental animals. *Annals of the New York Academy of Sciences*, *444*(1), 137–149. doi:10.1111/j.1749-6632.1985.tb37585.x PMID:3925849

Takakura, K., & Okubo, A. (Eds.). (1994). *MEG Noujizu no kiso to Rinshou*. Asakura Publishing Co. Ltd.

Tulving, E. (1972). Episodic and semantic memory. In E. Tulving & W. Donaldson (Eds.), *Organization of memory* (pp. 381–403). Academic Press.

KEY TERMS AND DEFINITIONS

Auditory Evoked Magnetic Field: The auditory evoked magnetic field was the component of the average magnetic response induced by the electrical current of the primary auditory nerve system. The largest and most widely studies component was the N100m component, reflecting the response activity of the primary auditory cortex on the right and left temporal lobes for tone presentation.

Class B Engine: The "class B engine" was an "almost infinite" Japanese pun-generating system, using a lexical engineering system of artificial intelligence. This system was developed to present the concept of the language engineering study group, *Kotoba Kougaku* (linking art and engineering), and research results.

Cognitive Semantics: Cognitive semantics was the research field of the semantics investigated with the perspective of human perception and recognition. This research field was developed by George Lakoff, Leonard Talmy, Ronald Langacker, and Gilles Fauconnier.

Context Memory: The context memory, which reflected the memory status of the narrative and pun, had a functional base of the long-term memory declarative memory. Long-term memory is divided into two types: semantic memory (constructed by repeated experience, situation of acquisition disappeared) and episodic memory (constructed by single trial experience, situation of acquisition was maintained with the memory of episode). Context memory, which is similar to the memory of the narrative plot and pun, is constructed by a single trial experience, while the memory of the situation of acquisition disappears. The context memory is a midway between semantic memory and episodic memory.

Human Brain Mapping: Human brain mapping is a noninvasive technology used to investigate human brain function for visual, auditory, and somatosensory stimulation. Representative technologies were electroencephalography, magnetoencephalography, functional magnetic resonance imaging, and near-infrared spectroscopy (NIRS).

Magical Number Seven: The magical number seven was the functional limitation of memorizing numbers and characters. The function reflected the activity of working memory, which is a system of short-term memory, and the functional activity of the frontal lobe.

Short-Term Memory: The human memory function was classified into short-term memory and long-term memory on the property of time course of the preservation of memory. This definition reflected the property of the time of status of memory preservation. Working memory is one component of the short-term memory.

APPENDIX 1

Human Brain Mapping of Context

The detection of the effect of narrative plot with provocation of emotional response was demanded for 90–100 times in the stimulus presentation of the plot. The response to the plot presentation with emotional reaction differed from that of the first time presentation to the second time or after trials. The response to the plot presentation significantly decreased from first to second or after. It was difficult to detect the response to the plot with emotional response with the repetition of the same plot presentation. The 90–100 numbers of different plots with emotional provocation were necessary for the detection of the effect of the narrative plot on the average waveform of magnetoencephalogram with auditory and visual stimulus presentation. The response for the pun had a similar property to the response to the presentation of the narrative plot. The response to pun presentation differed and decreased from first to second or after the presentation. The response decrement for the narrative plot and pun was a problem even in functional magnetic resonance imaging with repetitive stimulus presentation, which was in accordance with the experimental design.

The response of the sensory nervous system of the N100m component was gradually changed with the repetition of the stimulus presentation. The peak amplitude of N100m component was decreased exponentially with repetition of stimulus presentation, which showed the habituation of AEF (Harada, Masuda, Endo, Nakamura, & Takeda, 2001). In contrast, the response of the stimulus of the narrative plot and pun significantly decreased (as digitally) from first to second or after presentation.

APPENDIX 2

Anticipation and Decrement of Amplitude and Latency of Auditory Evoked Magnetic Field

The variation in decrement of amplitude and latency of average waveform reflected the function of the elicitation of the mismatch field (MMF) component with the anticipation for stimulus. The MMF component disappeared in the long ISI conditions. Instead, a reduced N100m component and increased MMF were observed with a shortened ISI, such as 1 or 2 s (Mäntysalo & Näätänen, 1987; Näätänen & Gaillard, 1983; Näätänen, Sams, & Alho, 1986; Näätänen, Simpson, & Loveless, 1982; Sams, Hari, Rif, & Knuutila, 1993). Mäntysalo and Näätänen (1897) reported that the inflexion point in the ISI range for change from elicitation to disappearance of the MMF was 2–4 s, while Sams et al. (1993) reported that it was 3–6 s. Variations in the length and randomness of the ISI produced opposite changes in the MMF and N100m component.

The range of ISI produced a Gaussian distribution that resulted in the range of times for the elicitation of the MMF overlapping with a short ISI and a sharp and narrow peak overlapping with the disappearance of the MMF with a long ISI. The disappearance of the MMF might have occurred because of the loss of anticipation of the stimulus with a long ISI.

The author suggests that the opposing changes in the MMF and N100m component are related to the temporal correlations of the input stimulus sequence and that this process is responsible for both the establishment and breakdown of anticipation behavior. The timing of the tone with a long ISI could not be anticipated by the subject; thus, a memory trace was not constructed for long ISIs, and the absence of a memory trace resulted in the disappearance of the MMF. Subjects could also not anticipate the timing of the stimulus with random changes in the $1/f^0$ ISI condition, which inhibited the formation of the memory trace, and the MMF was not elicited. Instead, the $1/f^\infty$ ISI condition resulted in anticipation and memory trace formation, and the MMF was elicited. The small change in ISI regularity from $1/f^0$ to $1/f^1$ enabled anticipation of the timing of the stimulus arrival. A low-level stochastic correlation in the stimulus train in $1/f^1$ changed the informational processing of the stimulus in the primary auditory area, which was reflected in the elicitation of the MMF. The temporal sensitivity of the auditory cortex has been determined from posttetanic potentiation in a rat's auditory cortex slice; posttetanic potentiation was maximally inhibited with an ISI time ranging from 1 to 2 s. The author considered that functional verification of the timing of stimulus arrival and internal monitoring clock for the stimulus timing may be performed in the primary auditory area.

APPENDIX 3

Function of the Frontal Lobe on Working Memory and Function of Taxonomical Expression

The categorization function of the human brain was discussed with a relationship between visual temporal lobe and frontal lobe in a functional magnetic resonance imaging study (Martin, 2007). The visual image was separated in the temporal lobe, and categorical judgment was executed in the frontal lobe. Short-term memory (e.g., working memory) was conducted in the frontal lobe. The frontal lobe was responsible for the function "magical number 7 ± 2" as the short-term memory.

The function "magical number 7 ± 2" was expressed with the function of the short-term memory, which was related to the "working memory" function located in the frontal lobe investigated in the functional brain mapping study (McNab et al., 2009).

The expression of the threshold function in the range of numbers 300–500 in context memory was described with the perspectives of the combination of theories of magical number 7 ± 2 and semantic layer structure.

Ericsson and Kintsch (1995) reported the taxonomic structure of chunks of the magical number seven, discussed long-term working memory, and discussed the over seven chunks used in human life.

The results of the study on the monkey's nerve cell and visual evoked magnetic field (VEF) of the human frontal and temporal lobe provided the semantic taxonomic classification.

The semantic discrimination of brain function was reported with the investigation of nerve cells of the frontal lobe and visual cortex of the monkey (Freedman, Riesenhuber, Poggio, & Miller, 2002). The semantic judgment of "dog" or "cat" was requested with the visual images constructed with the image mixing technology of "morphing" with the variation of mixing ratio from 0% to 100% on "dog" and "cat" images with semantic judgment. In the judgment of "dog," the activity of the nerve cell was gradually increased with the increment of mixing ratio of the "dog" image in the visual cortex. The activity of the nerve cell of the frontal lobe was digitally increased to > 50% mixing ratio of the "dog" image. The study described the results of the experiment, the increment of the activity of the nerve cell of the visual cortex reflected the form recognition, and the increment of the activity of the nerve cell of the frontal lobe reflected the semantic judgement.

The semantic provocation effect of animal body parts was investigated with the experiment of VEF (Harada, Iwaki, & Tonoike, 2006a, b). The visual image of the experiment was constructed with the "line" drawing of animals with changing the "body" and "head." The combination of parts of "body" and "head" was controlled with the semantic distance on the structure of the word constitution of Five Thesaurus of Japanese. The first condition was the combination of the original and normal "body" and "head" conditions. The second condition was the combination with changing in the group of taxonomy class relating to generic category on folk biology study. The third condition was the combination with changing intergroups of the taxonomy class. The semantic distance of the combination of the parts of body and head was increased in the condition from the first and second to third with the taxonomic joint number intergroups on the semantic layer structure of the thesaurus.

The VEF was recorded for the projection of the image of the three types of the "line" drawing animal images with the conditions of the combination of "body" and "head" on the semantic distance of the first (0 joint) and second (2 joints) to third (4 joints) condition. The average waveforms of the root mean square (RMS) value were calculated with the left frontal, right frontal, left occipital, and right occipital area channels. The average waveform of the VEF of the RMS in the left-occipital area was investigated with the peak amplitude of 170 ms and 240 ms. The peak amplitude of 170 ms of the third condition was significantly decreased with the amplitude of the first and second conditions. The peak amplitudes of 240 ms of the second and the third conditions were significantly decreased with the amplitude of the first condition. The peak amplitude of the average waveform of the left frontal area on 380 ms was significantly increased from the first, second, and third conditions.

The peak amplitude of 170 ms differed between the first, second, and third conditions and reflected the extraction process of the large difference of the visual image. The peak amplitude of 240 ms differed between the first, second, and third conditions and reflected the extraction process of the small differences in the visual image. The steps of informational processing of the visual image were indicated for two steps of processing of the large and small difference extractions. The two steps of extraction of difference of images allowed the classification of three levels of the taxonomic difference with the variation in semantic distance with the number of joints. The increment of semantic distance was used to enhance the activity of the left frontal area observed with a peak amplitude of 380 ms.

The variation of the peak amplitude of 170 ms and 240 ms of the visual cortex of the left occipital area showed the semantic classification of three layers of taxonomy. The activity of the left frontal lobe of 380 ms was enhanced with discrimination of the semantic definition of the visual cortex. This result showed that the visual cortex allowed the classification of the visual image with three layers of taxonomy.

The combination of the function of the short-term memory of the magical number 7 ± 2 on the frontal lobe, and the semantic classification function of three layers on the visual cortex enabled the expression of the function of the threshold of context memory of the range of numbers 300–500.

APPENDIX 4

Dunbar's Number

Dunbar reported the relationship (correlation) between the volume of the cortex (brain) and group size of the number of members in monkeys. The estimated value in humans was 150. This amount implied the hypothesis that 150 was the threshold of the cognitive function for the group size, which allowed to durably maintain the social relationship. A higher amount of group members would request a restriction rule to maintain stable social relationships. Bernard (2006) and McCarty, Killworth, Bernard, Johnsen, and Shelley (2000) reported the number of 290 (almost twice of 150) for the stable social group, which they estimated by field research on several social groups in America.

Gladwell (2000) analyzed the case study of the company of WL Goa, which had < 150 users of the office building and company parking lot. The limitation to < 150 users contributed to identifying and inhibiting social issues.

The number 290 was extremely close to the number 300 of the threshold of the context. The deviation of the magical number seven was reported plus and minus two. The number 7×7×7=343 could deviate to 6×6×6=216 and 5×5×5=125. The number 216 and 125 was closer to Dunbar's number between 150 and 290. Cowan (2001) reported the magical number was estimated as four with the strictly controlled experiment.

In the perspective of the case study of the story maker, Japanese novelist of Kenzo Kitakata made some comment of the relationship of memory between the person and the story in the FUJI TV program of in "Typewriters" at 2020/7/18 on 10.28-11.50. He made a comment that, "The first process story making was the provocation of person in the image field, the second process was the starting up the flue of story". This comment showed the linkage of memory function between person and story. The common cognitive base between personal identification and context recognition could be used. The system of personal identification can be adopted for the system of context recognition.

Section 2

Narrative Generation Systems and Narrative Content Generation

Chapter 6
Bridging the Gap Between Narrative Generation Systems and Narrative Contents With Kabuki-Oriented Narratology and Watakushi Monogatari

Takashi Ogata
https://orcid.org/0000-0001-9181-8876
Iwate Prefectural University, Japan

ABSTRACT

The author's narrative generation study is based on two types of systems: the integrated narrative generation system (INGS) as a single narrative generation and reception mechanism and the Geinō information system (GIS) as a multiple narrative production and consumption mechanism. The first theme of the chapter is to introduce an idea that deals with narrative phenomena as the integration of both systems. This theme is tied to the topic of narrative content creation by presenting kabuki narrative generation or kabuki-oriented narratology and Watakushi Monogatari as a collection of narrative content to be created by the author. Hence, the second objective is to describe kabuki-oriented narrative generation and the third is to explain the ideas, thoughts, and design underlying Watakushi Monogatari in the context of internal and external narrative generation to create and distribute narrative content. Through these three themes, this chapter bridges the gap between narrative generation systems and narrative content with kabuki-oriented narratology and Watakushi Monogatari.

INTRODUCTION

The author published a series of books (Ogata, 2020a, 2020e) that systematically Shōwan integrated approach to studying narrative generation. These books assert that research on content generation and narrative generation forms an important, challenging field that has a fundamentally interdisciplinary

DOI: 10.4018/978-1-7998-4864-6.ch006

characteristic. This characteristic is based on artificial intelligence (AI) (including cognitive science) and narratology (including literary theories), grounded in an integrated, interdisciplinary narrative generation system of study through a model of multiple narrative structures, as well as other philosophical considerations. In the first part of this chapter, the author describes the meanings of the concept of an integrated approach to narrative generation. "Integration," in this case, connotes a broad range of senses beyond the integration conveyed in the Integrated Narrative Generation System (INGS) that the author has been designing and developing. The term "integration" includes the following:

1. **Integration Among Modules in INGS:** This is the main meaning of "integration" in INGS. At the most macro-level of the module division, INGS has three modules: story generation, narrative discourse generation, and narrative representation. These modules have a mutual relationship based on the simple process of order. At the same time, they are associated with each other through various connections.
2. **Integration Between INGS and the *Geinō* Information System (GIS):** Although the author's narrative generation study can be seized as a grand whole that includes personal and individual narrative generation-reception process and social narrative production-consumption process, INGS functions under organic relationships within another system framework: GIS. In the author's research system, GIS corresponds to a system that deals with the abovementioned social narrative production-consumption aspect. Furthermore, when the author elaborates on the stage of *Watakushi Monogatari*, which will be described in this chapter, the mutual cooperation of INGS and GIS will be necessary to proceed in a comprehensive manner.
3. **Integration as Multiple Narrative Structures:** This is a manifold kind of integration in the sense that narrative generation phenomena are composed of three macro levels: the level of the brain (super personal level A), the symbolic/linguistic level, and the social/collective/collaborative/institutional/communal level (super personal level B). In terms of Takaaki Yoshimoto's (1924–2012) philosophical system or his main books that Ogata (2020e) frequently refers to, these three levels correspond to the levels of mental phenomena (Yoshimoto, 1971, 2008), linguistic phenomena (Yoshimoto 1965), and social phenomena or communal illusion (Yoshimoto, 1968).
4. **Integration Among Research, System Development, Social Development, and Creation:** The author's narrative generation study does not necessarily fall under academia alone; it covers the level of social development or distribution, and the private/personal levels of the production and creation of actual narrative works, via the level of system design and development (although basic, prototypical designs and development are academic, the level will be tied to a more pragmatic stage) from the angle of research. In a wide sense, the "research" is integrated because it is carried out in a broad, successive area.
5. **Integration Between Information and Literature (or Expanded Literary Theory):** The author's expanded literary theory (Ogata, 2002a, 2002b, 2003a, 2003b, 2014; Ogata, Imabuchi, & Akimoto, 2014) was the earliest approach to the theme of "integration" in the author's narrative generation study. It aims to build a new narrative study by synthesizing informational science, engineering (including AI and cognitive science), and literary theories (such as narratology and narrative theory).
6. **Integration Among Diverse Academic Areas:** Since developing the expanded literary theory described earlier, the author's integrated/synthetic narrative generation study has been extended to an integrated approach that covers a wide range of fields beyond the fusion of information technol-

ogy (centering on AI and cognitive science) and narratology (including literary theories). Chapter 3 (Ogata, 2020c) in Ogata (2020a) refers to the primary relevant areas.

7. **Integration Among Diverse Narrative Genres (or "Narrative Genre System"):** This study does not limit narratives to the field of literature, and understands the genre of narratives to exist in broader areas. Chapter 2 (Ogata, 2020b) in Ogata (2020a) outlines the narrative genre system, which is composed of five categories, and provides Japanese narrative genres for the central part, presenting concrete examples of the narrative genre system. This is partially related to the fact that the extreme goal of this integrated approach to narrative generation is to produce personal and private narrative content called *Watakushi Monogatari*. In particular, "I" (*watakushi*) is most closely linked to the production of *Watakushi Monogatari* and many parts of knowledge, forming the narrative content necessary for a narrative, and directly relating to the nation or history of Japan, which deeply connects with "I" (*watakushi*). Hence, the narrative genre system is constructed based on diverse Japanese narrative genres.

Although this chapter more or less pertains to all of the above topics, it has strong ties with both **2 (Integration between INGS and the *Geinō* Information System (GIS)**) and **4 (Integration among Research, System Development, Social Development, and Creation)**. Regarding the latter, this chapter aims to describe the fundamental strategy that bridges the research and development (R&D) of INGS and GIS for the production/creation and social development/distribution of R&D through GIS. The author intends to achieve this goal by considering the following three themes. As demonstrated by the title, this chapter attempts to bridge the gap between narrative generation systems and narrative content with *kabuki*-oriented narratology and *Watakushi Monogatari* in the author's research system.

- **The Study of Major Narrative Generation Systems, INGS and GIS:** In the following section, the study is called "**(A) INGS and GIS**."
- **The Study of Survey, Analysis, and the Systematization of *Kabuki*:** Similarly, this study is called "**(B) *Kabuki***."
- **The Study of External Narrative Generation and Internal Narrative Generation for *Watakushi Monogatari*:** This study is called "**(C) Internal and External**."

Ogata (2020e) includes these three themes, and Ogata (2020a), whose work corresponds to the first section in the sense of topics, examines these themes at a fundamental level. Through these two books, the author presents many (or almost all) concepts concerning the author's integrated approach to narrative generation and basic considerations of them. However, these books have not reached the scale of organic integration among systems (INGS and GIS), *kabuki* studies, or *Watakushi Monogatari* in the context of internal and external narrative generation. Stated in the previous part, the central strategy of this chapter is to aggregate the three kinds of themes to form a consistent research story. In the following section of this **INTRODUCTION**, concerning these three themes, the author summarizes the respective descriptions of future directions based on Ogata (2020a, 2020e) to establish the key topics and problems in this chapter.

First, concerning **(A) INGS and GIS**, Chapter 1 (Ogata, 2020f) in Ogata (2020e) remarks on INGS architecture, its philosophical ideas, and mechanisms, such as generation, as well as content mechanisms and application systems. INGS is currently being carried out at the experimental level. INGS develops the aspects of "expanded literary theory" by incorporating the literary theories of Propp (1968), Genette

(1972), and Jauss (1970) into the system and covers the implementation of a story, narrative discourse, and narrative representation. Through the developmental process of INGS, several components and mechanisms were also applied, including conceptual dictionaries, a state-event transformation knowledge base, and story generation mechanisms using these dictionaries and knowledge base The implementation enabled effective constructive research on the complex mechanisms in which an entire performance becomes possible through the interactions of diverse narrative techniques and expertise. INGS architecture has been put into effect as a modularized, flexible program for easy extension, transformation, and revision; this characteristic is tied to a kind of development that enables the gradual incorporation of various ideas and modules into the system. The author aims to address the issue of development based on a circulative, repetitive process by organically consolidating partial revisions into the entire architecture using its plastic feature. In addition to continuing the progressive development of INGS, in **FUTURE RESEARCH DIRECTIONS** in Chapter 1 (Ogata, 2020f) in Ogata (2020e), the author strives to bridge the system to the social and personal "practices" of narrative generation. The following topics are covered in the ongoing system development of INGS:

1. **Basic Generation Methods**
2. **Content Knowledge for Generation**
3. **Expanding Content Knowledge**
4. **Knowledge Acquisition and Learning Mechanisms**
5. **Natural Language Generation**
6. **GIS (the *Geinō* Information System) or the Social Narrative Production and Distribution System**
7. **Expanded Literary Theory**
8. **Japanese Literary Theories and Narratology**
9. **Philosophical Concepts Related to Expanded Theory**
10. **Multiple Narrative Structures**
11. **Social Narrative Distribution (in Plans for Social and Individual Development)**
12. **"My [the author's]" Narrative Creation (in Plans for Social and Individual Development)**
13. **Evaluating Narrative Generation Systems**

In these future works, **6**, **8**, **11**, and **12** are closely linked to this paper. In **6**, GIS is not only an application system of INGS, but is also an extension of INGS that allows it to be tied to a social/business framework for the production and distribution of generated narratives. In particular, GIS is designed to contain one or more INGS. In **8**, an especially important direction of the expanded literary theory is to analyze traditional or modern/contemporary narratological research and literary theories in Japan, and to connect them with INGS (and GIS). In **11**, this topic implies that narratives produced by a narrative generation system (which include descriptions created in collaboration with humans) and the narrative generation system itself are associated with society, namely the outside of the system. There are diverse possibilities, such as business. Forms like "virtual publishing" and "virtual entertainment production" by GIS may make clear images for introducing the aspect of business into narrative generation systems. In **12**, this concept refers to narratives pursuant to the author's private themes and materials. In particular, INGS introduces private knowledge content into the story content knowledge base and other narrative content knowledge bases.

Next, the author describes **(B) *Kabuki***. The following is the **FUTURE RESEARCH DIRECTIONS** in Chapter 4 (Ogata, 2020h) in Ogata (2020e). The main short-term plans concerning *kabuki*'s survey and analysis are presented:

1. **The Modeling and Design of GIS with INGS, Based on the Introduction of Elements and Knowledge of *Kabuki***
2. **Deepening the Survey and Analysis of *Kabuki* Itself as an Information System**
3. **An Experimental Implementation and the Verification of GIS with INGS**
4. **An Applicable System Design and Implementation, Including Social Distribution Experiments**

By the term "GIS with INGS," the author means the following. In the sense that GIS is a broad framework in which INGS (correctly, one or more INGS[s]) is included, the author has used the term GIS with INGS or "GIS including INGS." This chapter employs the terms, especially GIS with INGS. In the above list, **1**, **2**, and **3** are particularly relevant to this chapter. In **1**, the first objective is to design a more detailed GIS with the INGS model, extended through a survey of *kabuki*. This also draws on the next theme, specifically, deepening the investigation of *kabuki* itself. Regarding this direction, a literature survey by the author on *kabuki* describes, in depth, ***Kabuki* as Documents** in **BACKGROUND** in Chapter 2 (Ogata, 2020g) in Ogata (2020e). Furthermore, with a fresh perspective, the author delves into the concepts of "person," "story," *tsukushi* and *naimaze* in *kabuki* in the section **INTRODUCING *KABUKI* KNOWLEDGE INTO GIS WITH INGS: FROM ANALYSIS TO GENERATION**. A future study would involve producing deeper, broader examinations of numerous *kabuki* elements. Extending the study's range to other elements in *kabuki* is also regarded as a vital path for future research. In **2**, the *kabuki* analysis, along with the system design and implementation, forms a mutual process in that the exploration of *kabuki* shifts toward the systematization of GIS with INGS. In this vein, the system design and implementation require more detailed, precise, and systematic scrutiny of *kabuki*. Depending on the analysis, Ogata (2020e) describes experimental programs using the concepts of "person," "story," *naimaze* and *tsukushi*. One future possibility would be to conduct an experiment to incorporate the proposed program on persons and other elements into INGS and GIS. Although the topic of **3** accompanies the expansion and precision of INGS, it is possible to use the current, tentative version in the GIS experimental system, because it could act as an entire system. Another trajectory for future research could be to develop a meta-level mechanism to control the mutual relationship between the two mechanisms, in order to produce a sequential story of narrative work from the narrative content. In addition, multiple narrative structures could be seen in each constituent element in the narrative generation-reception and production-consumption mechanisms. INGS may be able to tackle this issue by enhancing knowledge bases for elements, including persons and other features.

Last, concerning **(C) Internal and External**, the following describes **FUTURE RESEARCH DIRECTIONS** in Chapter 4 (Ogata, 2020h) in Ogata (2020e):

1. **GIS**
2. **Developing *Watakushi Monogataris***
3. **Philosophical Considerations of Institutional Narratives in the Broadest Sense**

In the above list, **1** and **2** are closely related to this chapter. **1** covers designing and developing a system with INGS. An attempt to grasp its design and development is considered, incorporating, with more awareness, the experimental system based on the analysis of *kabuki* as a mechanism of GIS with INGS. Further, the author strives to design the virtual construction of a troupe or a *geinō* production company (entertainment agency)—an actual form of the GIS—and to prepare *geinō* characters in various components in GIS. Concerning **2**, the author plans to contemplate and concretize a myriad of information and generation methods in relation to the conceptual dictionaries in INGS and GIS.

Stated above, although Ogata (2020a, 2020e) describes the direction of future research for the three topics, this chapter covers challenges whereby the previous narratology cannot be extended to the pertinent approaches to solve them. In other words, the author tackles the gap between previous narratology and information modeling using AI and cognitive science. As concretely as possible, this chapter reveals the strategy needed to synthesize the above topics to finally actualize them as *Watakushi Monogatari*.

In regards to **(A) INGS and GIS**, based on the first chapter (Ogata, 2020f) in Ogata (2020e), a narrative generation system called INGS corresponds to a system that creates simple pieces of narrative content individually, while another system, called GIS, produces many pieces of narrative content in a circulative manner. By studying the narrative generation of *kabuki*, the author presents mechanisms and methods for how INGS and GIS can integrate and collaborate. In particular, this chapter aims to design, and partially develop, a multiple narrative generation mechanism in which INGS and GIS are organically and mutually executed.

In **(B) *Kabuki***, for the author, *kabuki* is an essential research theme that offers a comprehensive, interdisciplinary approach to narrative generation. The narrative multiplicity in *kabuki* is typical of a multiple narrative structures model, which serves as a critical analysis of the author's narrative generation strategy.

The author believes that the dramaturgy of *kabuki* provides interesting, crucial examples in which new narrative content is produced by combining and transforming previous narrative content, along with the fragments. Hence, the author scrutinizes the broad spectrum of analyses of *kabuki*'s elements, seen from the angle of producing a *kabuki* work on stage. After reviewing narratological studies of *kabuki* by the author, this section explores several narratological references or books, including *Sakusha Shikihō Kezairoku* [Authors' Ceremonial Methods Kezairoku] (1972) (a theoretical essay written by a *kabuki* writer during the Edo period) and *Sekai Kōmoku* (1916) (a database of *sekai*s [worlds] used to create *kabuki* works from the Edo period). Moreover, the author investigates two contemporary *kabuki* researchers and their books. The basic methods to approach the two references (i.e., books) include analyzing them from the angle of narrative generation, and utilizing the acquired results in the narrative generation system design and implementation. This chapter delves into the possibilities of their new usage and application through partial system design and program development. Specifically, this section on *kabuki* discusses the problems that previous *kabuki* narratology or narratological studies of *kabuki* could not handle, as well as new techniques to solve them. The author refers to the gap between previous narratology in literary theories and information modeling using AI and cognitive science. The author also systematically examines *kabuki* elements from the viewpoint of producing a *kabuki* work on stage.

Finally, in terms of **(C) Internal and External**, through narrative and literary histories, the author describes diverse methods to create and distribute narrative content. For instance, some narrative content in a specific genre is produced through collaborative efforts by many people who play different roles and distribute content at a physical site, such as the stage of a theater. In another genre such as novels, narrative content is written using papers and characters and is spread through printed books. Today, computational technology and narrative generation systems are regarded as tools or media that create new methods and possibilities for narrative creation and distribution.

The author presents two related concepts, namely "internal" and "external" narrative generation, which are associated with the issue of producing and circulating narratives. Internal narrative generation indicates that narrative content is created using a narrative generation system, especially GIS with INGS. Further, it relates back to the author himself, and includes a sub-genre called *Watakushi Monogatari*, which, in its current form, is a narrative vision to be written using the author's own materials. On the other hand, external narrative generation corresponds to the social distribution of narrative content, using a mechanism in GIS in particular. This section covers the basic designs of internal and external narrative generation, respectively. In his description of internal narrative generation, the author deals with the topic of narrative content creation supported by GIS with INGS, and narrative content that includes a concrete approach to *Watakushi Monogatari*. This chapter also presents the design of external narrative generation, which is the mechanism for organically allocating created narrative content. Moreover, both internal and external narrative generation are explained according to an organic relationship. This part discusses the problem whereby previous narratology should be bridged to informational and computational methods through the author's narrative generation study, regarding narrative creation and distribution. In other words, the author recognizes the gap between previous narratology and information modeling (which uses AI and cognitive science), and utilizes a research direction to connect them.

This chapter consists of the following sections. The **BACKGROUND** deals with **Post-Narratology and the Post-Narratology of Narrative Generation**, which examines an integrated narrative generation approach based on the multiple narrative structures model and other philosophical considerations by the author. Next, **Narrative Generation Mechanisms Based on Post-Narratology** focuses on the following: (1) INGS and (2) GIS as the most important elements in the above strategy and (3) previous attempts to combine INGS and GIS. The section on *kabuki*—**The Narrative Generation of *Kabuki*** and **Systematic Surveys and Analyses of *Kabuki* Elements, Seen from the Viewpoint of Producing a *Kabuki* Work on Stage**—summarizes preceding approaches to the narrative generation of *kabuki*. The former part reviews the author's narrative generation study of *kabuki* or *kabuki* as a narrative generation study. In the latter part, *kabuki*'s multiple elements (perceived based on the author's narrative generation system and multiple narrative structures model) form the core outcomes of the analysis of *kabuki*, or the starting point of the author's narrative generation study on *kabuki*. Finally, the subsequent section—**Internal and External Narrative Generation**—introduces former approaches to internal and external narrative generation.

The following three sections are located in the main parts of this chapter. The first is **A MULTIPLE NARRATIVE GENERATION MECHANISM THROUGH TWO TYPES OF NARRATIVE GENERATION SYSTEMS: GIS WITH INGS**, which includes **Basic Elements and the Process of GIS, Incorporating the *Watakushi Character***, and **The Overall Behavior of GIS, Including a *Watakushi* Character with INGS**. One of the three key parts of this section covers the conceptual design of an integrated mechanism for INGS and GIS or GIS with (i.e., including) INGS. The foremost section is **NARRATOLOGY OF *KABUKI*: ORIENTED TOWARD NARRATIVE GENERATION**. The author has used the term of "*kabuki* as language." By *kabuki* as language, the author means that *kabuki* has produced many types of linguistic documents (e.g., essays, reports, critiques) and existed alongside them, although an individual performance on stage is regarded as the most critical component of the art form. This section briefly reviews diverse types (or genres) of *kabuki* documents through examples. First, the paper centers on contemporary *kabuki* researchers and critics Yukio Hattori and Tamotsu Watanabe, as well as their books; perceived as theory-oriented writers, their thoughts are useful for the author's narrative generation approach to *kabuki*. Next, as older documents, the author examines *Sakusha Shikihō*

Kezairoku and *Sekai Kōmoku* in depth from the narrative generation system's perspective, which are to be applied to narrative generation mechanisms based on GIS with INGS. In the next section, the author explores ways to solve the challenges of the gap between former narratological studies related to *kabuki*, and narrative generation modeling based on AI and cognitive science; the author refers to several paths for future research. As for the final theme, **THE CONCEPT OF *WATAKUSHI MONOGATARI* AND DESIGNING INTERNAL AND EXTERNAL NARRATIVE GENERATION METHODS TO CREATE AND DISTRIBUTE NARRATIVE CONTENT**, the first part describes internal narrative generation—particularly its methods and practices supported by GIS with INGS—to produce narrative content such as *Watakushi Monogatari*. The next part details the basic procedures of external narrative generation, which organically allocate narrative content created by the proposed internal narrative generation mechanism. The third section outlines the integrated relationship between internal and external narrative generation to make this study count as an actual attempt. Finally, the chapter states how the author bridged the gap that was mentioned in the first section, and refers to several possibilities for future research.

The following portion, **SUPPLEMENTAL DISCUSSION AND DERECTIONS FOR FUTURE RESEARCH**, addresses four topics not dealt with previously, and additional themes related to GIS with INGS, *kabuki*, and *Watakushi Monogatari*. In addition, the author examines possibilities for future research, including problems faced when designing and executing GIS with INGS, systematizing *Sakusha Shikihō Kezairoku* and *Sekai Kōmoku*, and developing *Watakushi Monogatari*.

BACKGROUND

Based on the above, Ogata (2020a, 2020e) comprehensively and systematically describes his narrative generation study, and summarizes prior investigations. This background reviews the chief themes that emerge in these two books. The following topics are examined:

1. **Post-Narratology and the Post-Narratology of Narrative Generation**: This entails an integrated narrative generation approach based on the multiple narrative structures model and other philosophical considerations by the author.
2. **Narrative Generation Mechanisms Based on Post-Narratology**: This includes INGS and GIS as the most critical elements in the above approach, as well as previous attempts to combine INGS and GIS.
3. **The Narrative Generation of *Kabuki***: This is a new, interdisciplinary approach to *kabuki* from the angle of a constructive view of narrative generation.
4. **Systematic Surveys and Analyses of *Kabuki* Elements, Seen from the Viewpoint of Producing a *Kabuki* Work on Stage**: This part summarizes the systematic analyses of *kabuki* elements, seen from the perspective of producing a *kabuki* work on stage.
5. **Internal and External Narrative Generation**: The final part of the background gives examples that describe different methods to create and distribute narrative content through narrative and literary histories.

Post-Narratology and the Post-Narratology of Narrative Generation

The author summarizes an integrated narrative generation approach based on the multiple narrative structures model and other philosophical considerations. First, grounded in Ogata (2020a) (especially, chapters 3 (Ogata, 2020c) and 4 (Ogata, 2020d)), the author reviews the post-narratology and narrative generation or post-narratology of narrative generation and focuses on important issues. (Regarding the concept of post-narratology, the author has been discussing through Ogata (2019b) and other papers.) Narrative generation systems comprise a new narratology, and the evolution of their conceptual/technical construction is the process of building narratology itself. The author discusses narrative generation systems based on essential ideas of narratology.

First, in **Narrative as Multiple Communication**, Fujii (2004) observes the origin of narrative in the discourse on dialogue. He sees the opportunity for dialogue to be transformed into a story via the author's differentiation and multiplicity. Fujii presents the example of a conversation between a mother and her child, in which the mother takes on the child's role in speaking to the child using the first person. This real-life instance illustrates the narrator's multiplicity. In particular, a single narrator is differentiated into multiple, virtual narrators (in both role and function), revealing the narrator's manifold nature. In a narrative, the narrator and narratee are not substantive, realistic entities, but can rather be called "agents" in the sense that their roles/functions, and the information surrounding them, determine their existence. By referring to interactive discourse as communication, as well as to the functional differentiation of the narrator's/narratee's multiplicity/multiplexing, it is possible to seek a single property of a story in multiple communication. Multiple communication differs from the singular communication of daily conversation, in which there is a direct confrontation between sender and receiver, in that the actual narrator and narratee are functionally differentiated into multiple roles.

In **Narrative as Simulation**, the author considers several issues from the standpoint of narrative as a type of simulation, which occurs when a certain subject is expressed with a limited number of characteristics. Through the interactions with subjects, the object's behavior is artificially simulated, and we can take into account ways of simulating the human in the social state of a narrative (perhaps including nature as well). The temporal development and spatial expansion of the interaction between the characters and the environment (society and nature) is a means of carrying out a thick simulation. This thickness does not refer simply to detail but is linked to the issue of multiple communication, which surely permeates through the opportunity that the narrator offers. In a narrative, in addition to the story—or rather beyond it—the dimension of discourse is vital. The narrator harnesses the diverse functions of those who carry the discourse, or of the mediator between the story and the discourse. The narrative embodies the discourse, which explicitly or implicitly strongly involves the narrator. Accordingly, the narrative, as a simulation, juxtaposes the thickness of not only the content of events but also the mechanisms of language. There are various types of simulations, such as scenario/management simulations and simulations in the natural world. When the narrator's opportunity comes to the foreground (-should, -must be) in this way, the narrative that simulates the author is original and unique. Moreover, if we think about the matter in terms of multiple communication's special features, the concept of the narrator within the narrative is differentiated into the actual narrator (i.e., the author) and the fictional narrator that appears in the narrative (a kind of character). The discourse possesses a complex nature, which is realized through this convergence. In some cases, the narrator's unique qualities (e.g., uncertainty, ambiguity, multiplicity) can wholly obscure the concrete nature of the simulation of events.

In "**Transposition (1): Narratology as a Production and Narrative Generation System**, the author's goal is to actualize narratology as a narrative generation system, so it goes without saying that this project necessitates the transposition of traditional narratology. The first point entails a transposition from an analytical, hermeneutical direction to a production-oriented path. The achievements of narratology's great founders formed a very strong model for organization and orientation; it probably once contained the potential to develop the processes of generation and production, but in its successive phases, the practical application of textual analysis and interpretation became mainstream. Along with the trends of acceptance theory in literary investigations and the flourishing of cultural studies (as reading comprehension and the interpretation of texts became mainstream in literary research), narratological research came to coexist with other schools of thought and is now viewed as a tool for textual interpretation. Of course, logic and practice became established whereby interpretation and acceptance simultaneously became acts of projection. For example, reception theory implies that active interpretation on the receiver's side is somewhat a productive act (i.e., acceptance is brought into the category of production). According to this logic, in the end, in doing the opposite of what its name suggests, acceptance theory has contributed to the expansion of authorial nature. What the author wants to say here is different: that it is literally an act of production. Looking at the origin of this idea, narratology is tied to the genealogy of rhetoric, or was envisioned with an aim toward the present revival of rhetoric. For Genette (1972), the root concept of narratology explicitly involves reviving this kind of literary tradition. Propp (1968) holds that the field of dependent morphology is centered on the search for the morphogenesis of plants and animals, a generative way of thinking. From this perspective, early narratology focused on production, and hence became production oriented. In this sense, it is not necessary to deliberately stress transposition; later on, narratology came to be treated as a single method of literary comprehension and acceptance.

Concerning **Transposition (2): From Narratology as Informatics to Expanded Literary Theory**, the production-oriented transposition outlined above is realized more directly as a computer program that serves as a narrative generation system. For that reason, this narratology is based on informatics, information theory, and technology (especially AI and cognitive science), as well as the fusion of literary and narratology techniques. This section also denotes the transposition of narratology as information. From the author's perspective regarding narratology as a whole, however, he aims to bring a greater, more complex meaning to this field, including not only including information but also various other ideas and intentions; he has referred to this project as expanded literary theory. In speaking of "expanded," we should note that narratology originated as a dramatic expansion beyond the realm of text in a narrow sense as the object of literary research. Regarding methodology, narratology has been extended in terms of a deepening literary analysis by using methods of microscopic internal examination. Hence, narratology is already a broadened literary theory. Thus, the author's approach can be called "expanded literary theory."

In **Institutional Perspective**, thinking of literature and narrative in conjunction with the concept of the institution is more common. For narratives and literature, the author takes an "institutional perspective." The phrase has become a household word among members of the institutional population; it carries an academic meaning as well as an everyday sense, but the author does not necessarily (want to) use it in an academic way. The novelist Kenji Nakagami (1946–1992) frequently uses "institution" in his literary discourse. In particular, he uses "narrative institution." In general, this phrase refers to the structure of the spell of the narrative, which is unconsciously expressed by the writer and cannot be deliberately avoided. The author of this book was greatly influenced by Yukio Mishima (1925–1970) as a boy. Later on, the author read the latest works by Yukio Mishima (2000–2006) in their entirety, but given the concept of anti-psychologism and the pragmatic nature of words, the author realized once again that Mishima's literature had given birth to his affinity for the institutional perspective, as well as his attachment to the idea of familiarity. Mishima is an incredible writer with a direct interest in political and social institutions.

Regarding the section **Narrative Generation System**, for the author, the construction of narratology is synchronized with building a narrative generation system. To put it in extreme terms, in this research, the narrative generation system corresponds to the theory of narratology itself. A narrative is a cross-border concept; in response to it, narratology has evolved as movement to bring about a unified perspective. However, a narrative generation system goes further since it seeks to actualize narratology into a cohesive, convergent core that synthesizes analysis and production. A narrative generation system serves as a framework for incorporating the cross-border technique of narrative into this cohesive, convergent method. Chapter 1 (Ogata, 2020f) in Ogata (2020e) provides a concrete description of a narrative generation system (INGS), but, as mentioned above, the entire book is organized around a stance that expresses various aspects of the narrative generation system. The author touches upon the concepts of literature versus narrative. Like a narrative, literature is a cross-border notion. Within the framework of a narrative generation system, narrative and literature overlap, and contrast with each other at certain points like the points on a fan.

Narratology expands outward toward the scope of the target text (a literary text in a narrow sense); the author's research is based on this foundation. At the same time, the opposite tendency arises: to go against the current and return to the literary text. From this angle, the narrative generation system's specific output is literary text. Simultaneously, the narrative generation system is oriented toward diverse fields of study, actions, and practices that lie at the center of literary thought. "Narratology"—which actualizes the narrative generation system—has expanded since "literary theory" originates in the trend of the movement from narrative to literature.

Next, the author reviews the concept of circular narrative generation based on Chapter 4 (Ogata, 2020d) in Ogata, (2020a). This study deals with the phenomena of narrative generation, involving ever evolving and generated things at the social and personal levels. The three types of philosophical ideas in the same group are "circular narrative control," "norm-deviation," and "fluidity-fixation." At the macro level, they dynamically control the abovementioned, ongoing generating narrative generation.

"Circular narrative control" implies that a narrative generation procedure is performed through a sequential process without a beginning or end, and it continues across different media and genres. This concept can be used to control INGS, as described in Chapter 2 (Ogata, 2020g) in Ogata (2020e). Many narrative generation systems produce a specific type of output information from a determined type of input information. One basic idea is to be able to create diverse types of output information from various pieces of input information grounded in circular narrative control. The following description explains this notion using two examples.

This concept is related to developed mechanisms. An attempt is made in the narrative discourse phase of INGS (a more detailed explanation will be provided in a sub-section of the next section). The control mechanism—which applies the reception theory by Jauss (1970)—iteratively repeats the narrative discourse generation according to the interactions between the narrator and narratee involved in the system. The control mechanism embodies the idea of circular narrative control. Although this mechanism is meant for a particular component of generation, it was originally intended to perform a free, flexible narrative type of ordering, for which the sequence/order of the operating modules/phases is not preliminarily decided upon. INGS, in the current stage, partially encompasses the concept in the abovementioned sense by introducing a music generation and transformation mechanism, which the author has been developing in piecemeal fashion. The idea is not only that music is the accompanied expressive medium of semiotic narrative representation, but also that its input-output relationship can be quite flexible and diverse. This idea has been generalized to the connection between music and narrative as an essential notion for narrative generation, via the appearance of circular narrative control.

Circular narrative control entails the design of a narrative generation system envisioned by the author, in particular the current form of INGS. The first meaning is iterative narrative generation, as shown in the experimental, implemented mechanism for the narrative discourse phase, explained earlier. This concept of the first meaning will be tied to a method through which a narrative generation cycle is perpetuated to obtain numerous generated outcomes; it will be further synthesized for an edited, individual narrative work. In contrast, the second meaning refers to the free, flexible control of the applicable order of narrative generation phases. For example, INGS should not fix the generation order among all modules or generation units, including conceptual structure generation and surface expression generation. INGS should be designed to enable various generation courses and repletion. Furthermore, GIS is a medium for circular narrative control in that the narrative production and consumption mechanism causes the myriad levels of narratives to repeat (such as individual narrative works, an actor's authentic narratives, and the entire narrative of a sequence of each productive narrative).

Narrative Generation Mechanisms Based on Post-Narratology

This section introduces the following topics: overviews of (1) INGS and (2) GIS; and (3) previous attempts to combine INGS and GIS.

An Overview of the Integrated Narrative Generation System

This section deals with INGS as the most important element in the above approach, or rather, a concretization of the author's earlier described thoughts on narrative generation. The following is according to the flow of Chapter 1 (Ogata, 2020f) in Ogata (2020e).

THE ARCHITECTURE OF INTEGRATED NARRATIVE GENERATION SYSTEMS explains the architecture underlying INGS in terms of the mechanism (**The Generation Mechanism of INGS**) and content (**Content Knowledge in INGS**). The main generation mechanisms contain the entire process: the story generation, the narrative discourse, and the narrative representation mechanisms. The generation controls are also included in the explanation. The central components of content knowledge are conceptual dictionaries and narrative content knowledge bases. Two large sections, **The Generation Mechanism of INGS** and **Content Knowledge in INGS**, describe the above topics.

Figure 1 is shown. The primary generation modules are for the story, narrative discourse, and expression (including language, music, and picture images). The story and narrative discourse are embodied by their respective conceptual representations. All modules are controlled by each local control mechanism, which are collectively under global dominance. Story and discourse generation are performed using story and discourse techniques, respectively. Furthermore, INGS has dictionaries (e.g., for concepts and language) and various knowledge bases (e.g., for story content and state-event transformation). In addition, as stated in previously, several literary theories are incorporated into INGS for the practice of expanded literary theory. Although the figure below of INGS architecture has been used in prior papers, and several alterations have been made to the architecture, the author uses this figure in this version

Figure 1. The architecture of INGS

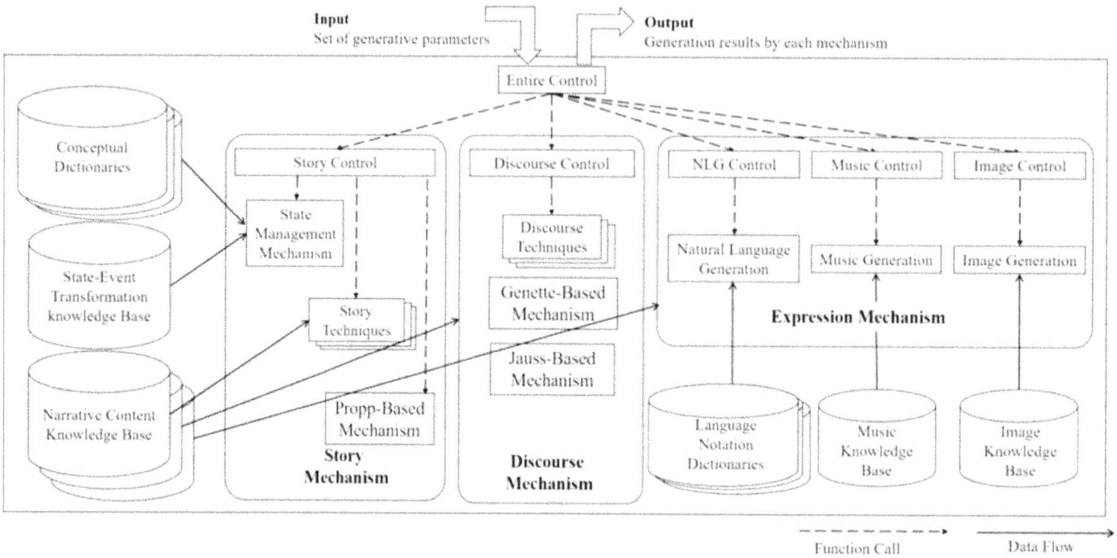

because the foundation will not change in this first draft (however, the next extended draft will use a developed version of the figure).

Regarding **The Generation Mechanism of INGS**, as mentioned earlier, the author considers aspects of INGS architecture through the division of content and mechanism (or form). Firstly, this section explores the formal attribute of mechanism.

For ***Entire Generation Process***, INGS proposes mechanisms to control the overall generation process, as well as several smaller points. The control mechanism's chief function at each stage of generation is to operate the characteristics of a narrative to direct the use of narrative techniques. Although many of these mechanisms are simple, the Jauss-based one (which, comparatively speaking, is precisely designed) is only employed in the narrative discourse mechanism.

One characteristic of narrative generation control in INGS is circulative control, which is grounded in the use of a fundamental concept of the author's narrative generation. INGS aims to implement a form of circular narrative control. The INGS mechanisms are partially developed through the diversity of generation routes. Ideally, this control mechanism means that all the generation mechanisms can receive any type of input information. The user can select the generation route at the starting time in a session in any of the following ways:

1. The user selects a possible route: The user can choose a free combination of "story, narrative discourse, sentence, visual image, original music, and variation music."
2. The system automatically determines a generation route.
3. The system automatically determines the next module at the ending time of a generation module.
4. If the user does not select any of the previous options, the system determines the order: "story → narrative discourse → sentence → visual image → variation music."

In general, a generation module harnesses the previous output data as the input data. However, for example, for the order "story → sentence → discourse," the discourse module takes the output data from the story module as the input data. Such details are explained in the respective sub-sections. The information in this table will be shown in the sections for each module.

For ***Story Generation Process***, INGS provides mechanisms that control both the overall narrative generation process and each point of the narrative generation. During the phase of story generation, the system produces a story using a tree structure with three basic elements: event, relation, and state, organized by time. Each event is constructed using an event concept and semantically related noun concepts. The conceptual representation is used as a frame structure to semantically categorize each noun concept, using eight basic cases ("agent," "counter-agent," "object," "location," "time," "instrument," "from," and "to") and nine special cases ("adverb," "possessive," "situation," "purpose," "experiencer," "source," "idiom," "information," and "as"). Each event accompanies the preposition state and the resulting state at the beginning and end of the event. At the same time, each relationship links an event or aspect of the tree structure (which includes events) to another event or aspect of the tree to create a new structure, whereby events and states have a top-level relationship. Many relationships (such as "cause-effect," "sequence," and "script") are created.

States and events correspond to static and dynamic informational elements, respectively. A state's central functions are to manage knowledge about the story's world and to handle the flow from event to event. Although a story is a temporary organization of events, its semantic consistency is governed by relationships and states.

For ***Narrative Discourse Mechanism***, like the story generation mechanism, the narrative discourse mechanism is composed of narrative discourse techniques applied to its developing structure. The similarity between the story generation and narrative discourse mechanisms is grounded in the following characteristic of INGS. Namely, both mechanisms play the role of conceptual processing in narrative generation in INGS. Here, conceptual processing refers to deep, semantic narrative generation in INGS, in contrast to surface narrative processing, which uses various representative media in the narrative expression mechanisms in INGS. Although the concrete processing of the narrative discourse mechanism differs from story generation processing, they perform similar procedures, formally speaking.

The narrative discourse module executes generation and circular that employ the narrative discourse theory by Genette (1972), and the reception theory by Jauss (1970), respectively. The use of these theories symbolize a concretization in the narrative discourse mechanism of expanded literary theory, along with the introduction of Propp's hypothesis on the story generation mechanism. The latter mechanism (grounded in Jauss's reception theory) reveals a possible implementation of narrative generation control based on numerous possibilities. A narrative discourse (generated or modified by the Genette-based mechanism) corresponds to the structural aspect of narrating a story tree structure, including a sequence or series of events. The discourse structure to be narrated and its required surface media (such as language and images) are points distinguished within the system, and are described in terms of conceptual representation, as stated above. The narrative discourse should correspond to a transformed story structure. From the angle of expanded literary theory (which simply uses narratological and literary theories), in the key part of the process, Genette's philosophy develops a set of discourse methods as the root of narrative techniques. The Genette-based and Jauss-based control mechanisms are linked to this phase. However, INGS, a comprehensive system, incorporates existing narratological and literary theories/approaches into the entire system framework. Hence, the author examines the generalization of theory-based mechanisms by looking at the design and development of INGS.

Regarding the concrete process of narrative discourse, a generated story structure is transformed into a narrative discourse (as the narration's structure) according to a temporal sequence of events and states. A narrative discourse structure can reorganize (or further ignore) the temporal order of events in the story. Additionally, in INGS, the narrative discourse structure to be narrated is differentiated from the narrative representation by the surface representation media (such as language, images, and music). Narrative discourse structures are described using corresponding conceptual representations, like in story structures. Thus, a narrative discourse coincides with a transformed story structure. Although it is a tree structure in which events are the main elements, the different relations from story structures are used. A narrative discourse structure is described by the tree form, transformed from a story tree; the events and states in the narrative discourse structure are same as in the story structure.

The ***Narrative Representation Mechanism*** engenders surface representations and contains the following sub-mechanisms:

1. The **natural language generation mechanism** creates a natural language representation or expression.
2. The **image generation mechanism** forms an image presentation or expression, namely visual images of moving and still pictures.
3. The **music generation mechanism** produces musical representations or expressions.

Previous studies on INGS have focused on creating conceptual representations, while the above mechanisms for producing surface representations (especially language and images) remain at the invention/framework stage. The following sub-section discusses natural language (1) and music generation (3).

In addition, an experimental function of the mechanism for generating narrative images (2) yields pictures that correspond to elements in INGS' conceptual dictionaries, using image management rules to synthesize them. In response to a story or narrative discourse structure, the mechanism generates a text command file, which delivers a series of processing commands to the image representation interface. In the future, the author aims to construct an image generation mechanism in INGS based on the similar concept of the music generation mechanism. In particular, the plan treats the generated tree structure of narrative conceptual representation as the tree structure of image representation; it inserts various rhetorical elements of imagery, corresponding to rhythm and tempo in the case of music, for example. The fundamental grammar of image representation—which corresponds to language-based case structures and sentence patterns—may effectively control concrete image representation through a camera work function or other means. Characters can be employed in the generated narrative visual representation. (The above part is the direct reference.)

An Overview of the Geinō Information System

GIS is the most vital element in the above approach, or a concretization of the philosophy underlying the author's narrative generation Chapter 3 (Ogata, 2020c) in Ogata (2020a). The following introduction is according to the reference.

In **The Sequential and Multiple Narrative Generation System and the *Geinō* Information System**, there are phenomena called narrative genres and mixed media. In the previous section, **The Development and Distribution Mechanism**, the author tried to capture such phenomena in terms of content, but they have occurred naturally since ancient times. For example, in the case of *kabuki*, in addition to texts that are performed, during the Edo period, there was also a type of critical (and even advertising) narrative, such as written commentary on each actor. Scripts have also been reading through books. At present, it is not uncommon for films, novels, comics, games, and other forms to be linked. In the case of music, various conversions are also performed, such as recording and editing live musical or theatrical performances at venues (e.g., records and CDs) and distributing them via TV or another kind of network. In many cases, the task of textual conversion between media is not done purely at the individual level but by an organization (such as a company). This category includes publishers, film and video production companies, TV and radio stations, advertising agencies and production businesses, game production enterprises, and entertainment production. These are not the only entities that engage in such profit-generating activities; some artistic groups promote a certain art form based on the artistic lineage or houses (the *iemoto* system), such as floral design (*ikebana*), the tea ceremony (*sadō*) (both popular rituals in Japan), and literature (novels, poetry, *tanka*, *haiku*, and other forms of literature).

Coterie associations focusing on *manga* and other interests, as well as (depending on one's point of view) religious groups, can be included too. It is normal for organizations, companies, and other groups to be involved in a "narrative generation system as genre production." However, up to the previous stage of simple narrative generation, in contrast to the organizational involvement in which the creation of a text remains internal, and at the stage of sequential or composite narrative generation, the center of gravity of organizational engagement shifts from a relationship with individual works to wielding control over them. In other words, the difference is that the production and control of multiple works is carried out at this time, whereas in the past, this stage entailed creating a single work. In this sense, there has been a qualitative change between the two. If we consider the problem in this way, the subject does not necessarily need to be a group or an organization. From this angle, we can also grasp the mechanism of the "author" as an individual, such as a novelist. This point is connected to the GIS described below. Yet broadly speaking, we discuss this issue in terms of the general trend of its superimposition on the organizational mechanism.

***Geinō* Information System: Overall Architecture and Thought** introduces GIS and addresses the basic concept of transforming INGS into GIS. The system architecture is currently fluid and will be expanded and revised in future concrete design/implementation. In addition, "why *geinō*" is an important point. For the author, narratives are not literature and arts but rather *geinō*. In the narrative genre system (Chapter 2 (Ogata, 2020b) in Ogata (2020a), narrative as literature and arts is only one part. Further, this is also related to the fact that the author's narrative generation study (or what the author ultimately desires to achieve with the narrative generation study) is not what is advanced and avant-garde but instead "commonplace" and "non-unique." According to the author, a narrative is the farthest from anything represented by words such as "advanced," "avant-garde," and "innovative."

Figure 2 portrays a conceptual composition of the GIS. The INGS mechanism is included as a component. The architecture will be changed in future versions. Regarding previous attempts, refer to papers such as Amino, Kawamura, and Ogata (2002). GIS corresponds to a narrative production-consumption mechanism that encompasses the multiple use of INGS as a narrative generation-reception mechanism. The system is, at the highest level, composed of a sender mechanism, a receiver mechanism, and a *geinō* history.

Simply speaking, GIS continuously produces *geinō* content through the interactions between the sender and receiver mechanisms. In the sender mechanism, a variety of methods coexist and, at the highest level, it branches out into *geinō* works and *geinōjins* (each a kind of sender mechanism). Thus, the GIS also creates content related to the subjects of *geinō* works, called *geinōjins*, in addition to engendering content

Figure 2. The architecture of GIS

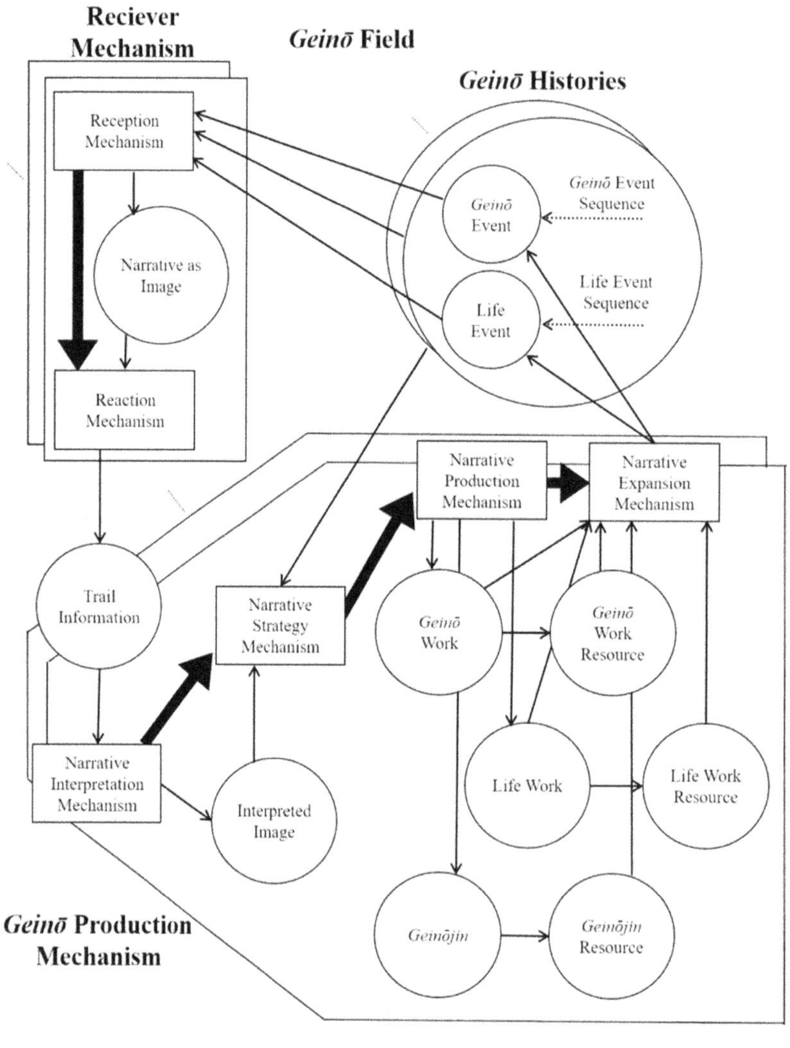

as *geinō* works. Further, as mentioned in the above review of informational narratology, receivers of a narrative frequently realize the level at which they themselves are the senders, beyond their positioning as pure reception subjects. In this sense, receivers can also be viewed as subjects, who create narratives. GIS produces diverse content through the above mutuality between the sender and receiver(s). However, the production-consumption process leads to a sequence of pieces of content, in addition to content as a simple object. A sequence of pieces of content is referred to as "*geinō* history." Namely, from a more macro angle, GIS produces *geinō* histories according to the mutuality between the sender and receiver(s).

Previous Approaches to Combining INGS and GIS

In **TOWARD INTRODUCING *KABUKI* KNOWLEDGE INTO GIS WITH INGS: FROM ANALYSIS TO GENERATION** in Chapter 2 (Ogata, 2020g) in Ogata (2020e), based on the above framework or components, the author surveyed several *kabuki* elements, including "person," "story," *tsukushi* or *zukushi*, and *naimaze*. Although the person and story aspects are included in the fifteen elements, *tsukushi* and *naimaze* are not. They belong to another group in the author's framework of the narrative generation study of *kabuki*. The latter elements are a group of universal narrative techniques that can be applied to various components. In particular, the fifteen *kabuki* elements are not concrete narrative techniques of *kabuki*, but instead are meant to construct a *kabuki* phenomenon. They can contain various narrative methods employed in *kabuki*. However, *tsukushi* and *naimaze* are respectively a narrative approach or set of narrative strategies that can be used in numerous elements of *kabuki*. For example, the narrative technique(s) of *naimaze* can be employed in terms of the person (character), story, and place. Although the examined narrative approaches for the person, story, *tsukushi*, and *naimaze* are not organically incorporated into INGS and GIS or GIS with INGS, they are positioned as important knowledge components of the unified system. Therefore, the following section reveals the macro-level framework of GIS with INGS.

Regarding **A Macro Framework of GIS with INGS**, a *kabuki* play and sequence are produced through a comprehensive narrative generation and production mechanism (the GIS with INGS in the author's model), using various types of *kabuki* knowledge, as stated in the previous section. These knowledge elements correspond to each knowledge base, storing historical, cultural, and other information on *kabuki*. The *kabuki*-related, domain specific knowledge bases are intended to be included in GIS, while more general, domain-dependent knowledge bases are intended to be developed in INGS. A single generation/reception of a *kabuki* play repeatedly extends to production and consumption at the social level.

The author considers a macro-level design plan for GIS with INGS. As described repeatedly in this chapter, INGS is a simple narrative generation system that corresponds to the first mechanism in modeling the author's comprehensive narrative generation research. However, the current version of INGS cannot create synthetic narratives using knowledge acquired through the analysis of *kabuki* from the angle of a narrative generation system. For the second mechanism, it is necessary to focus on the meta-level mechanism, which controls the use of one or more INGS(s), namely GIS. This is the meaning of the term "GIS with INGS." The above findings will contribute to any part of INGS and GIS. Yet it is too difficult to simply divide them, for instance, by simply thinking that the "character" and the "actor" in the "person" can, respectively, be associated with INGS and GIS. Notwithstanding, a character is partially a narrative element to be stored in a portion of GIS in its connection to the narrative world (*sekai*). Regarding the framework of the relationship between INGS and GIS, as shown in Figure 3, the INGS treats narrative generation as a simple subject. By contrast, the GIS produces sequential narratives on a broad level, which includes each piece of narrative work and narratives of *geinōjins* (entertainment performers) such as ordinary people, actors, and characters.

The Narrative Generation of *Kabuki*

This section covers the author's narrative generation study of *kabuki*, or *kabuki* as a narrative generation study. This paper examines *kabuki* according to the constructive (computational) method or the cognitive scientific approach. The author describes the meaning (although this has been the technique of cognitive science, many researchers do not know). This is also a pre-conditional concept of the author's expanded literary theory or post-narratology. The above is the theoretical direction. (As an engineering or constructive/productive method, this section uses the *kabuki* analysis based on the constructive method to produce *kabuki* narratives grounded in any direction. The following part refers to chapter 2 (Ogata, 2020g)

Figure 3. Expanded narrative generation by INGS and GIS

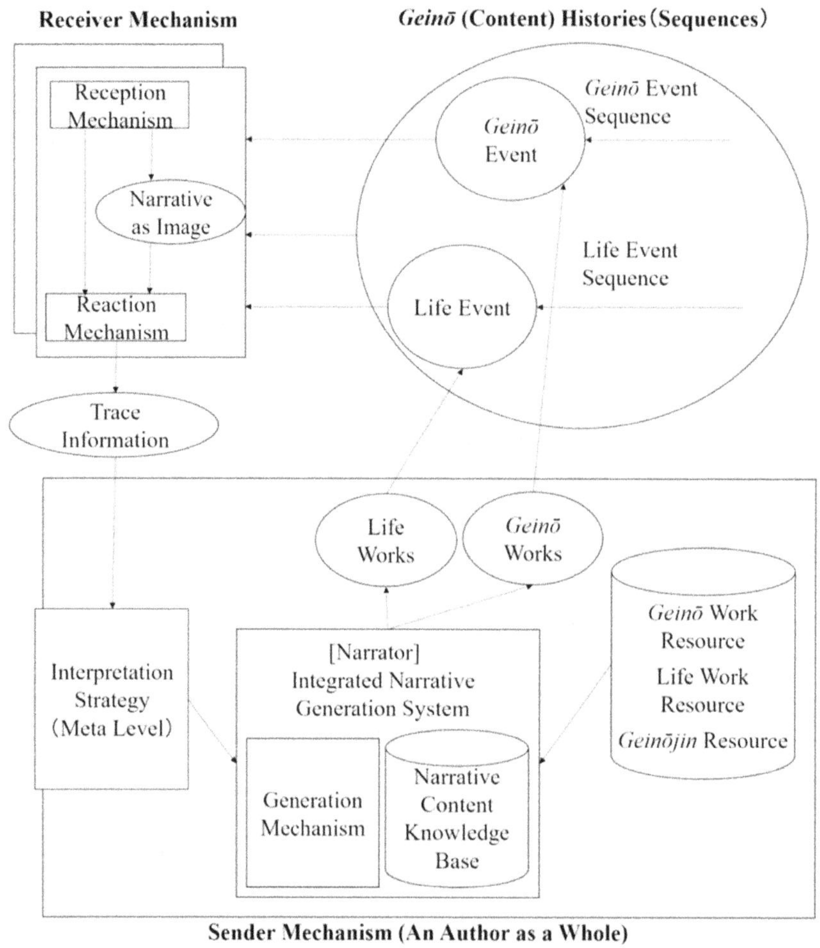

(TOWARD INTRODUCING *KABUKI* KNOWLEDGE INTO GIS WITH INGS: FROM ANALYSIS TO GENERATION) in Ogata (2020e), delving into previous concrete attempts or the current findings.

Based on the above framework or components, the author surveyed several *kabuki* elements including "person," "story," *tsukushi* or *zukushi*, and *naimaze*. Although the person and story are included in the fifteen elements, *tsukushi* and *naimaze* are not included. They belong to another group in the author's framework of the narrative generation study of *kabuki*. The latter elements are a group of universal narrative techniques that can be applied to various components. In particular, the fifteen *kabuki* elements are not firm narrative techniques of *kabuki*, but instead are intended to build a *kabuki* phenomenon. They contain diverse narrative strategies employed in *kabuki*. However, *tsukushi* and *naimaze* are, respectively, a narrative approach or set of narrative techniques that can be used in numerous aspects of *kabuki*. For example, the narrative strategy or methods of *naimaze* can be used for the person (character), story, or place. Although the investigated narrative approaches for the person, story, *tsukushi*, and *naimaze* are not organically incorporated into INGS and GIS, or GIS with INGS, they are positioned as critical knowledge components of a unified system. Hence, the following section reveals the macro-level framework of GIS with INGS.

Regarding **A Macro Framework of GIS with INGS**, a *kabuki* play and sequence are produced through a comprehensive narrative generation and production mechanism (the GIS with INGS in the author's model), using various types of *kabuki* knowledge, as stated in the previous section. These knowledge elements correspond to each knowledge base, storing historical, cultural, and other information on *kabuki*. The *kabuki*-related, domain-specific knowledge bases are intended to be included in GIS, while more general, domain-dependent knowledge bases are intended to be developed in INGS. A single generation/reception of a *kabuki* play repeatedly extends to production and consumption at the social level.

The author considers a macro-level design plan for GIS with INGS. As described repeatedly in this chapter, INGS is a simple narrative generation system that corresponds to the first mechanism in modeling the author's comprehensive narrative generation research. However, the current version of INGS cannot create synthetic narratives using knowledge acquired through the analysis of *kabuki* from the angle of a narrative generation system. For the second mechanism, it is necessary to focus on the meta-level mechanism, which controls the use of one or more INGS(s), namely GIS. This is the meaning of the term "GIS with INGS." The above findings will contribute to any part of INGS and GIS. Yet it is too difficult to simply divide them, for instance, by simply thinking that the "character" and the "actor" in the "person" can, respectively, be associated with INGS and GIS. Notwithstanding, a character is partially a narrative element to be stored in a portion of GIS in its connection to the narrative world (*sekai*). Regarding the framework of the relationship between INGS and GIS, as shown in Figure 3, the INGS treats narrative generation as a simple subject. By contrast, the GIS produces sequential narratives on a broad level, which includes each piece of narrative work and narratives of *geinōjins* (entertainment performers) such as ordinary people, actors, and characters.

As for Two Multiple Elements and Two Technical Concepts: Person, Story, *Naimaze*, and *Tsukushi*, based on prior studies (Ogata, 2016a, 2018e, 2019c, 2019d, 2020g; Ogata, Fukuda, Ono, & Ito, 2020), the author supplements two types of multiple elements—person and story—and two kinds of technical concepts—*naimaze* and *tsukushi* (*zukushi*). These investigations reveal the multiple structures of the persons or people of *kabuki*. In *kabuki*, there are numerous "persons," such as "actors" (performers), (dramatic) "characters" and "real persons." These personas temporally and spatially form (and possess) multiple existences and relationships.

In terms of Preparatory System Development Based on the Analysis of *Kabuki*, this section discusses preparatory system development using methods that the author acquired through prior studies on *kabuki*. One of the goals is to contemplate, in detail, narrative functions, mechanisms and techniques regarding *kabuki* in order to implement GIS and INGS in the future. Thus, the proposed systems are relatively simple, although they are grounded in a consistent approach: "generation based on editing." In particular, the systems employ actual *kabuki* texts and edit them using narrative techniques accumulated through the author's analysis of *kabuki*. As Ogata (2019c, 2019d, 2020g) outlines the fundamental technique, this section revises and expands the description based on the summary.

In A Sequential Generation Example, the author traces a narrative generation process through the connections of several generations using the methods of person, story, *naimaze*, and *tsukushi*.

For *Analysis and Generation of a Person's Information*, the author scrutinizeShōwacquired, described, and structuralized property information of *kabuki* actors is linked to the GIS and INGS. Figure 3 includes the conceptual figure of the GIS. *Kabuki* actors' *geinōjin* and life information is stored in the *geinō* work resource (*geinō* resource) and life work resource (life resource). Figure 3 also indicates that the GIS with INGS starts out as a narrative and becomes a sequence of narratives, receives them, and consumes them. Moreover, this figure displays the circulation of these processes.

In *Selecting and Editing Stories*, Ogata chooses two stories on *kabuki* works, *Sukeroku* and *Narukami*, that Ichikawa Ebizō XI frequently produced. *Kanjinchō* is added as another example. The three *kabuki* works have relatively simple, clear scenarios. The author shows these scenarios' synopses (from Wikipedia in Japanese), their property data, and generated sentences.

For *Using Naimaze*, the dramaturgy of *kabuki* is constructed based on *kata* and *shukō*. *Kata* refers to the typical patterns of diverse elements in *kabuki*, including characters, places/scenes, and stories/plots. The tradition of *kabuki* contains different *kata*. *Kata* is also related to the concept of *sekai* (world) in *kabuki*. The author provides explanations of *sekai* in a post-narratology chapter (Ogata, 2019b) and the INGS chapter (Ogata, 2019a) in a previous study (Ogata & Akimoto, 2019). Regarding the history of *kabuki*, the document *Sekai Kōmoku* (1916), from the Edo period, gives a systematic description (as a kind of database/knowledge base) of *sekai*. *Sekai Kōmoku* provides actual works in many narrative genres including *nō*, *ningyō jōruri*, novels, famous characters, and typical stories/plots. In the production of a *kabuki* work, one more *sekai* are chosen to determine the essential *kata* to carry out a work efficiently. However, early *kabuki* writers did not simply apply the *kata* to the work. The second concept, *shukō*, is equal to special, individual narrative techniques for transforming *kata* to make a work more interesting. *Naimaze*, which fuses different stories (as *kata*) to form a new story, is also a type of *shukō*. Applying a *naimaze* method to a story/plot is the most general usage. *Naimaze* is also related to the element of a "character" as a "person." In particular, although it is possible for *naimaze* to mix the "stories" or "plots" of two or more narratives, the use of the *naimaze* of "stories" or "plots" is tied to the *naimaze* of "characters" in different works. Similarly, introducing "characters" from different works into a new one engenders an associative relationship between new and old works. Ogata (2019c, 2019d, 2020g) presents an approach for designing the narrative techniques of *naimaze* for INGS (and GIS) by applying the method for combining "moves," as explained in Propp's (1968) narratological study. However, the following example is simpler. An example synthesizes the synopses of *Sukeroku*, *Narukami*, and *Kanjinchō* into a new scenario using *naimaze* techniques. The author employs the conceptual structures of *Sukeroku*, *Narukami*, and *Kanjinchō*. This form of a conceptual structure is the story representation that is used in INGS. Hence, the narrative data described here can be harnessed in the synthesis with INGS. First, the author transforms the sentences in each story into a collection of simple sentences using a semi-automatic script generation program developed by Arai, Ono, and Ogata (2016). Second, each modified sentence is altered manually using an input format. Using these input data, stories of *Sukeroku* and *Kanjinchō* are inserted into the story of *Narukami* as the foundation. *Sukeroku* and *Kanjinchō* are connected to *Narukami* via the points in which large, special changes appear. These connecting stories cut through the natural flow of *Narukami*'s story. Ogata and Kanai (2010) study this type of story-cutting rhetoric. Sukeroku, the hero in *Sukeroku*, reaches new places after the events of "Narukami is angry" or "Yoritomo is angry." Kumonotaemahime, the heroine in *Narukami*, is changed to Agemaki, who is the heroine in *Sukeroku*, based on the common trait of being female.

For *Using Tsukushi*, *tsukushi* is comprehended from the angle of techniques through which objects respectively relate, meaning that objects are inserted (or sprinkled) into a narrative work/part. Methods of *tsukushi* may correspond to lateral techniques for expanding objects. In the basic approaches of *tsukushi*, strategies for connecting objects to one another are included and divided into methods by semantic and non-semantic relations. The former, for instance, includes the geographical relationships of places and the conceptual relationships of objects. The latter contains relationships through *dajare* (puns) and linguistic sounds. Many spatial objects can be inserted into a narrative using geographical movement and temporal change. The typical structure is *michiyuki*.

For *A Final Example and the Positioning in GIS with INGS*, according to the story construction, the final story structures are produced. These include the generated explanation sentences of Ichikawa Ebizō XI, the generated explanations of the three *kabuki* works (i.e., *Sukeroku*, *Narukami*, and *Kanjinchō*), the works' generated synopses, and the synthesized (and expanded) *kabuki* synopses that apply the techniques of *naimaze* and *tsukushi*.

Systematic Surveys and Analyses of *Kabuki* Elements, Seen from the Viewpoint of Producing a *Kabuki* Work on Stage

The multiple elements of *kabuki*—perceived based on the author's narrative generation study and multiple narrative structures model (which the author addresses in this section)—are the central outcomes of the author's *kabuki* investigation. Alternatively, the consideration of these *kabuki* elements serves as a point of departure for a more full-fledged narrative generation study of *kabuki*. Regarding these multiple components, after uncovering initial findings, the author refines and extends the elements by providing abundant examples. However, these efforts are conducted using the same framework, which itself has not been altered or expanded. The narrative generation study of *kabuki* does not aim to merely perform an analysis. Since its goal is to be connected to the design and development of narrative generation systems based on the knowledge and techniques of *kabuki*, the author makes several attempts related to system modeling and system implementation, which depend on the same framework of multiple elements of *kabuki*. In contrast, currently, grounded in the previous accumulation of data, the author is rethinking the framework of the multiple elements of *kabuki* in order to change and expand the framework. The author will contemplate these issues in a future study.

The author analyzes the multiplicity of *kabuki* with regard to the fifteen elements, displayed in Figure 4. The term, "multiple," in the caption of this figure indicates that each element has multiplex aspects. The author regards the multiplicity as one of the important features in *kabuki*. For instance, the element of a "person" in *kabuki* can be divided into the following three aspects: the real "person" who lives in the actual world; the "person" as an actor who has a history of performing roles; and the "person" of the *kabuki* character that appears on stage. Each person has his/her own history or story. In this section, the author explains the main elements of *kabuki*, or narrates them as a story, from the perspective of an audience that is viewing *kabuki* plays in a *kabuki* theater. In the following, the author describes these multiple elements of *kabuki* according to a flow or story of an audience in a *kabuki* theater.

Let us start this story with a scene where an audience enters a *kabuki* theater; for example, Kabukiza in Tokyo. What is the audience of *kabuki* (**[a] Audience**)? *Kabuki* is primarily meant to be seen in a theater by an audience. Of course, this does not mean that viewing *kabuki* in the physical space of a theater is an absolute experience for the audience. However, *kabuki* is closely, essentially, and mutually bound to the theater's existence. The audience also plays a role in the multiple editing of *kabuki*. The difference between seeing a *kabuki* play as a movie and in the theater is greater than the difference between seeing a movie on TV and at the cinema. The perspective (camera work) in a movie has already been prepared based on the creator's intention and technique. The camera work is often regarded as the most creative narrative technique in a movie. Rather, when an audience views a movie at the cinema, the stance in the work does not change according to the place in which the audience witnesses it. For *kabuki*, there are

Figure 4. Multiple narrative structures of kabuki

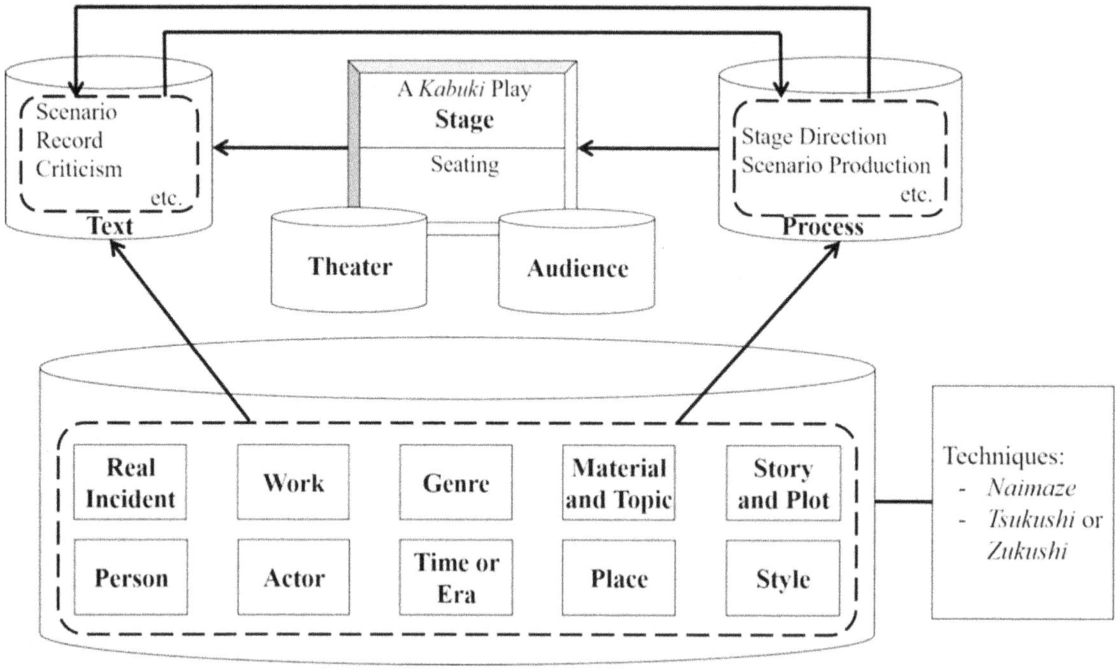

different outlooks according to the place in the theater. An audience's positioning in a theater is decided upon either intentionally or accidentally.

First, the audience experiences **[b] Theater, Stage, and Seating**. Traditional stage dramas in Japan incorporate theaters into necessary components in respective genres. In *kabuki* and other Japanese *geinō* genres, a theater is not a universal space where all (or diverse) theatrical genres are performed. In particular, *no*, *ningyō jōruri*, and *kabuki* take place in exclusive theaters and are performed on stages specifically suited for them. The *hanamichi* is a mechanism on the *kabuki* stage that has multiple uses as a unique device, and is effectively utilized in real *kabuki* theaters in Japan.

The audience members view a **[c] Stage Performance**. *Kabuki* during the Edo period was a contemporary art form of the time. *Kabuki* plays were performed in a style called *tōshi-jōen*, in which all parts are played continuously over one or two days. However, the actual form was not necessarily simple, as there were complex rules or performance systems. Two types of newly created works—*jidai-mono* and *sewa-mono*—were mutually performed over a series of *kabuki* performances. In contrast, after the Meiji era, the *midori-jōen*, a then-new performance style in which the classical works of Edo *kabuki* were repeatedly performed, gradually spread and became popular. Numerous new works (scenarios)—including the sub-genre of *shin-kabuki* (new/modern *kabuki*)—were published by many writers, including famous ones such as Kidō Okamoto (1872–1939) and Seika Mayama (1878-1948).

The performances and dances in the theater have respective **[d] Style(s) (Form(s) or Pattern(s)**. The term "style" here means "performance style." Different styles are adapted for the stage in *kabuki* plays such as dance, music, narration, and action forms or patterns by actors. A variety of multiplicities and possibilities exist based on the constraints of prepared elements and combinations. The music *uta-mono* (played as actors carry out actions) includes myriad genres such as *nagauta*, *kiyomoto*, *tokiwazu*, and *ō-zatsuma*. The narration by a *takemoto*, a type of *katari-mono* (a narrative genre)—corresponding to a *tayū* in *ningyō jōruri*—was transformed so that human actors could play the roles instead of artificial puppets. In *ningyō jōruri*, all parts of a work's scenario are narrated by several *tayū*; a *kabuki* work is divided into the *katari* (narration) by a *takemoto* and the *serifu* (speech), given by human actors. The artistic blend of *kabuki* and *ningyō jōruri* gave birth to another aesthetic technique called *ningyō-buri*, a special performance style in which a human actor imitates (or simulates) the action of a puppet in *ningyō jōruri*. An interesting, complex structure in *kabuki* is also represented in the example.

Although any method could be called a "direction" for a stage performance, **[e] Direction** in *kabuki* has unique features. The attribute of a *kabuki* play—whereby different historical periods and places in a narrative are mixed (rather than representing a realistic standard)—influences the production of each actual stage. Various styles of blending, or the overlapping of costumes, languages, speech, and music, sometimes exceed ordinary situations. The characters' make-up and costumes are designed according to the form or type of traditional norm to be used for each typical character. Such symbolic and formal qualities frequently ignore the realistic consistency and naturalness of the entire staging. The author considers that the above characteristics show the "expressionism" of *kabuki*. *Kabuki* frequently emphasizes the beauty and effectiveness of expression more than reality. One reason may be the disordered confusion of forms through historical accumulation. Some actors, such as Ichikawa Danjūrō IX (1838–1903) during the Meiji era, proposed innovative ways of restoring *kabuki* to a realistic style. However, since overly realistic and symbolic styles form a tense tradition and a unique beauty involving the audience, *kabuki* grew into a more heterogeneous, synthesized dramatic genre than one of pure realism.

In addition to the dramatic characters in dramas and dances on stage, players or actors appear. **[f] Person** in *kabuki* has multiple senses. A "person" can be an "actor" who performs on stage, or a (dramatic) "character" in the narrative of a work. These two kinds of individuals form numerous relationships on the temporal and spatial levels. First, a person plays a character in a *kabuki* play; at the same time, the existence accompanying the real name also has an essential meaning. The aspect that "an actor plays" in *kabuki* frequently has a more important significance than representing a universal drama or story. A *kabuki* scenario from the Edo era was not written with the names of the characters in the story but with the names of the actors who performed the drama. In particular, a scenario was ordinarily written according to the characteristics of the actors in classical *kabuki* works. Hence, the actors were not selected based on any artistic viewpoint grounded in a preliminary scenario. Although there are many critics of the weak points of such a star system and, after the Meiji era, scenario-centered drama production became a popular method, the tradition of a star system that emphasizes the actors has persisted in contemporary *kabuki*. This means that many audiences go to a *kabuki* theater to see their favorite actors.

Dramas and dances are called **[g] Work**s. Through the historical development of *kabuki*, a work was often adapted to other ones using various types of editing and revision methods. Although influential relationships among works are ordinary phenomena in the current artistic environment, very quick and active transformational phenomena, especially during the Edo period, would have partially relied on the lack of awareness of copywriting. A more practical reason for such an outstanding production style was that a revised work was easy to adapt to the unique traits of the author, theater, actor, or troupe. This is a very interesting problem to be considered in association with the theoretical and methodological narrative production techniques of *kabuki*, using the concepts of *sekai* ("world"), *naimaze* ("blending"), and *shukō* ("ingenuity" or "device" as ideas and contrivances).

In the previous section, the author uses words such as "drama" and "dance." These kinds of **[h] Genre**s exist in *kabuki*. Creating a new work based on one or more past ones in the same or another genre is a common approach in Japanese dramatic or *geinō* genres, such as *nō* and *ningyō jōruri*. This method of creation reached a unique, extreme level in *kabuki*. *Kabuki* is a synthetic collection of diverse *geinō* genres, including *nō*, *kyōgen*, and *ningyō jōruri* as dramatic *geinō* genres. Generally speaking, the transformation among diverse genres in narratives is a universal phenomenon in Japanese literature, *geinō*, and the arts. In this case, narratives are a type of versatile, general software. In contrast, genres are media for narratives in the form of software, and a narrative can be embodied in many different media.

Kabuki has a number of genres and works. They are narratives in the broad sense and commonly have a **[i] Story and Plot**. In *kabuki* narratives, we can see the multiplicity (overlapping features) of stories or plots. This facet is also tied to *sekai*, *naimaze*, and *shukō* mentioned in **[g] Work**. The history of *kabuki* resulted in the accumulation of stories and plots. Many narrative stories/plots are blended into fresh, more complex versions. Thus, audiences of *kabuki* can associate the fragments of stories/plots that they already know, and imagine another story/plot from the work performed on stage.

A *kabuki* work is usually performed by several actors, and takes place on the narrative stage in terms of place (**[j] Actor and Place**). According to Tamotsu Watanabe (1989), a limited number of actors in a theater troupe can play multiple repertories of *kabuki* by playing various dramatic characters. A theater troupe is needed to make preliminary preparations and the necessary arrangements to construct a myriad of places on stage: settings for fictional locales and costumes for the actors' roles. Since both characters and costumes are categorized in a patterned system, the combinations can be limited. This combinatorial mechanism enables actors to play different roles using a restricted number of costumes and closely related make-up. Likewise, places on the stage can be set up by using a certain number of settings and combinations. Such features of *kabuki* engender the artistic and narratological aspect in which the realism of performing is not respected.

Similarly, a *kabuki* narrative includes the unique temporal scope in which the narrative is developed; for instance, a specific **[k] Time (Era or Age)**. Different historical times are frequently mixed in a *kabuki* play. Such temporal multiplicity has been linked to the censorship system of the Edo government, in addition to artistic requirements. Direct representation of contemporary events was forbidden under the political circumstances of the time. A work based on a modern, real incident repeatedly changed the temporal setting to a past age, and was expressed in a different temporal setting, using the *sekai* of another era. Changes of time were simultaneously tied to changes in spatial elements, the components for space and place. We can see in *kabuki* an interesting phenomenon regarding temporal and spatial transformation in which the spaces and characters in different historical periods are overlapped.

Seeing a *kabuki* narrative from a different angle, each narrative has **[l] Material or Topic**. Materials or topics commonly used in *kabuki* works already exist and have been categorized under numerous groups. For instance, when a *kabuki* work is produced using certain existing material, and subsequent work is created based on editing a previous work, the latter work becomes a variation of the former while using the same material/topic as the original. This ongoing flow has resulted in multiple groups of *kabuki* works according to the same respective materials. Although this may be similar to the theme of a "real incident," the material or topic referred to here is not necessarily a genuine event. Fictional occurrences are also utilized. Further, "real" and "fictional" can be changed through a *kabuki* generation sequence or circulative evolution. A *kabuki* work produced from a true event can transform into a fictional narrative.

Kabuki works/narratives frequently use a **[m] Real Incident**, either from modern times or the past. Stories/plots in *kabuki* are often constructed based on actual incidents; for instance, historically famous events and facts for *jidai-mono*, namely "historical story type." In contrast, *sewa-mono* ("contemporary, realistic story type") frequently employs present-day events from the daily lives of ordinary people. Typically used historical sets of events have been repeatedly introduced and arranged in many *kabuki* works. The links between a real incident and its fictionalization are complex. The fictional parts of a true occurrence, such as a character's formation, sometimes have a large influence on people's imaginings. Such blending is similar to the cases of the above story/plot, time/era, and material/topic, which reveal *kabuki* dramaturgy based on a complicated mix (*naimaze* in the broad sense) of the original ingredients. Audiences who experience *kabuki* during a certain period may be aware of various *naimaze* techniques.

Audiences see a fixed number of works involving the above diverse elements played out on stage. Although audiences only view stage performances, we can imagine that these real plays have respective scenarios (which in the book of a scenario is called *daichō*, a kind of hidden existence before a scene is actually played on stage). This is the aspect of **[n] Production of Scenarios ("*Daichō*")**. The production of authentic performance scenarios had a very simple format in the early days of *kabuki*, which only contained a single structure with a few scenes. However, *kabuki* works gradually came to have large-scale, complicated structures. The production mechanisms also became more systematic, thanks to excellent authors, especially from *ningyō jōruri*, such as Chikamatsu Monzaemon (1653–1725) and Ki-no Kaion (1663–1742). As a result, a new type of scenario production style was invented, through which several authors created a scenario by working jointly. Moreover, as the above *naimaze* techniques came to be utilized, *kabuki daichōs* became long and intricate. Currently, one can buy scenario books for staging *kabuki* plays at the shop in Kokutitsu Gekijō (the National Theater in Tokyo). This audience in this story bought a pamphlet called *sujigaki* or *banzuke* at a shop. The book is not unlike the above scenario books, and includes explanations of things such as today's *kabuki* works, performances, actors, places, historical periods, and in-depth critical articles. Through the tradition of *kabuki*, **[o] Text** for stage performances became abundant. Many people buy *sujigaki* instead of pure scenario books in the theater. Unlike reading a novel, witnessing a drama on stage implies, in a primary sense, perceiving humans' physical actions in various settings and decors. A drama on stage is not a representation solely based on the enumeration of words, but is also a collection of bodily actions, and sometimes the stage itself accompanying these actions. At the same time, *kabuki* comes from a tradition of essays originating in the *yakusha hyōbanki* of the Edo period, a series of books that comment on *kabuki* actors. The style's existence has relied on the fact that an audience views a *kabuki* play on stage in a theater. This indicates that the texts of *kabuki* consist of individual stage performances as a collection of bodily actions by actors and stage settings. This custom has not yet been corrupted by the present-day *kabuki* scene. The *kabuki* tradition of text production has developed a different style of criticism, such as theoretical and methodological approaches that go beyond each actor toward *kabuki* narratives/stories.

Through scenarios/stories that audiences view at a theater as several *kabuki* works, the author reveals diverse elements for *kabuki* narratives. A vital concept in this story is "multiple narrative structure" in the sense that an element has relationships with invisible existences on stage. For example, a narrative character that the audience sees on stage synthesizes several elements in multiple methods such as actors, or a person in the real world. Moreover, a *kabuki* work that the audience is currently and truly observing has diverse relationships with the preceding works, as well as works of the future. In particular, a *kabuki* work appears, so to speak, as a bundle in the present time of myriad multiple relationships.

Internal and External Narrative Generation

The first topic for discussing internal and external narrative generation is the narrative generation-reception process and the narrative production-consumption process in a multiple narrative structures model. The narrative generation-reception process has individual humans as its units, whereby a novelist, using thought, narrows down a plan and expresses it in words and a reader interprets it. However, observing the process more realistically, in the case of a novel (a book), various tasks are involved and performed by individuals other than the author, such as the publishing company and specific people (e.g., editors) who structure the book; they insinuate themselves into the actual process. When a certain narrative goes into social circulation, diverse elements penetrate it so that it cannot be fully expressed by the generation-reception model at the individual level. Many people besides the author intervene and are engaged in this generation process; in totality, they make up what could be called the author/collective author.

The aim of the multiple narrative structures model is to develop a unified understanding of the narrative generation-reception process as an individual one and the narrative production-consumption process as a social one. The functions that connect the two, and the subject responsible, comprise the work content for narrative generation, aiming only for abstract effects, regardless of whether the objective of narrative generation is an individual thing, a group, or an organizational thing; the functions are techniques and technology for narrative generation.

The above multiple narrative structures model bridges narrative generation as a simple body to the problem of social development, specifically regarding how to develop and distribute socially generated narratives; the precondition is the element of a "generated narrative." In the author's framework of narrative generation study, the generated narrative simply means "a narrative generated by a narrative generation system" and the multiple narrative structures model is a synthesized framework for connecting it to the direction (problem) of how this narrative should be socially developed and spread. At the same time, the author extends the concept of the "generated narrative" to narrative content, which strengthens the characteristics of a work. This extended concept can include application systems associated with a narrative generation system, such as computer gaming systems. The author expands this flow to the direction of more literary/artistic narratives by the author himself. In the above discussion, the author calls the social development of narrative generation "external narrative generation" and the issue of the narrative as the work/content (by the author) "internal narrative generation." The framework of GIS with INGS corresponds to the concrete architecture for realizing external narrative generation, while internal narrative generation will be incarnated through *Watakushi Monogatari*. These are the author's future plans.

The following section describes the generation process of *Watakushi Monogagari* and GIS with INGS based on **TOWARD EXTERNAL OR SOCIAL PRACTICES OF NARRATIVE GENERATION** (Chapter 4 (Ogata, 2020h) in (Ogata, 2020e). This section covers the developmental plan for both the external (social) and internal (personal and individual) directions. These directions are related to the bridge, from research to content creation. This bridging may influence the growth of narrative generation systems, including INGS.

The entire narrative generation system (as technology) is a broad framework containing "the narrative generation system as a single system," which is equivalent to INGS. It refers to a comprehensive, narrative production-consumption model or system, in contrast to the narrative generation-reception model or system of INGS. Their overlapping is a concrete form of the multiple narrative structures model.

The external development (i.e., social expansion) of INGS corresponds to the narrative content produced by INGS, ranging from collaborative production forms with humans based on supporting methods, to societies as an external area of systems, which more firmly extend to the social distribution and consumption mechanisms. The area involved is positioned in the outermost zone of the multiple narrative structures model. A typical form is business; this can include new possibilities, such as a virtual *geinō* production company. In addition, trials of pure artistic and literary experiments also become possible.

Originally, the author's narrative generation system study was an extremely individual, personal attempt; hence, the development approaches to the internal, personal, and individual fields contrast with the above external directions. Simultaneously, internal development is conducted using INGS and the narrative production-consumption mechanism, including INGS. External and social development require more technological, mechanical, and formal discussion. In contrast, since internal and individual development require the author's "personal" knowledge as content knowledge, we need to discuss themes related to narrative contents.

In line with the above, the author introduces GIS, which contains INGS, to the larger system framework. It is possible to accomplish both the external (social) and internal (personal and individual) facets of INGS through GIS. Throughout the author's study, GIS was originally planned as a system for realizing the social development of narrative generation via the multiple, complex uses of the narrative generation system as a single system, namely INGS. Simultaneously, GIS aims to incorporate the author's individual and personal narrative content into the mechanism based on various *geinō* genres, especially *kabuki* and *ningyō jōruri*. Thus, the GIS system is a framework for expanding both externally and internally. Although it is different from INGS, GIS is still in the stage of fragmental implementation, and is explained according to the vision for execution.

In the GIS architecture, first, the sender mechanism receives the external information, namely traces of information output by the receiver mechanism. This occurs via the "interpretation strategy" mechanism inside the sender mechanism. Next, the receiver makes an interpretation according to the interpretation framework to produce a narrative. However, the narrative does not mean that something is being created according to the receiver's requirements and expectations. INGS is positioned in the center of the narrative generation by the sender mechanism in the GIS architecture.

The sequence and reception of the narrative generation by the sender mechanism form the next *geinō* history. For processing, more special knowledge bases not included in INGS (i.e. the "work resource," the "life resource," and the "*geinōjin* resource") are used. Although these resources are similar to the narrative content knowledge bases in INGS, these ones store GIS-specialized content knowledge. The *geinōjin* resource stores character or persona information on *geinōjins*. Through the life resource, the life histories of *geinōjins* are extracted from the *geinōjin* resource. The work resource refers to the processed knowledge of *geinō* works; compared to the content knowledge stored in the narrative content knowledge bases in INGS, this information is grounded in more specific narratives. In particular, the work resource aims to store narrative content knowledge that centers on *kabuki* and *ningyō jōruri*. Regarding the division between the narrative content knowledge bases in INGS and the work resource in GIS, *geinō*-related content knowledge is stored in the work resource.

The receiver mechanism includes any mechanical reception subjects, in addition to the human receiver. These output any reception results based on the reception experiences of an individual work, a sequence of works, and the life history of a *geinōjin*. A *geinō* history is sequentially formed through the circular and repetitive interactions between the sender and receiver mechanisms. The content of a *geinō* history is classified into both a *geinō* event and the sequence, as well as a life event and the sequence. The former means each generated work while the sequence and the latter correspond to the actor's flow of life in a *geinōjin*. Works here include scandals focused on an actor's life.

Regarding the division of the content knowledge of INGS and GIS, the knowledge of INGS is more general. Further, each dictionary corresponds to the most general knowledge group. GIS's knowledge, as more special expertise, is divided into large two groups along the axis of generality/specialty. The most unique group in GIS is the author's private, individual knowledge, and the various knowledge resources are prepared as more general knowledge.

Although the *kabuki* knowledge mechanisms are linked to INGS, in terms of relations with GIS, many elements are currently included in the work resource; these encompass "fact (event)," "work," "genre," "material," "story or plot," "place (stage)," "time (era)," "*geinō* style," "text," "production," "direction," and "staging." These were stated in the previous part of this chapter. However, works in GIS do not necessarily indicate *geinō* or artistic literary works in the narrow sense. They include scandals about the life stories of *geinōjins*. Instead, they are tied to the *geinōjin* and life resources. On the other hand, only a minor element is associated with the *geinō* history and reception mechanism. Precisely speaking, the *geinō* history is related to comparatively many elements, such as the historical features of works, characters, actors, and staging. In the future, analyzing *kabuki* by focusing on the dynamic aspects of narrative production and consumption will be required to extend the knowledge resources of GIS. Thus, GIS modeling will help to design the study of *kabuki* more organically.

The *Watakushi Monogatari* is created by using GIS or INGS in a special way. The systems consist of an independent, general organization, based on which it imports personal and private knowledge content. Potentially, the "I" (*watakushi*) that exists outside the systems can perceive the narratives created there as "fluid" and can express a narrative on another level: a *Watakushi Monogatari* that is fixed. Alternatively, the cycle or repetition of narrative generation by the GIS, equipped with INGS, may aggregate all fluid narratives formed at some point to produce a "*Watakushi Monogatari* that has been fixed." GIS is a mechanism for setting up the "external I" (*watakushi*) of INGS within the system. It is possible to imagine other possibilities. In the end, the aim to build an organization that brings together multiple concrete mechanisms.

Figure 5 shows, in abstract form, a narrative concretization mechanism based on the concepts of "fluid" and "fixed." Documents on the personal and private levels are correlated with narrative content knowledge bases grounded in diverse levels of INGS, as well as GIS locations. What the author wants to focus on in this chapter is not the aspect of narrative generation mechanisms, but rather the personal and private knowledge content itself, on the foundation of which the mechanism for creating the *Watakushi Monogatari* is built. This knowledge content makes up the most personal and private knowledge content and knowledge bases used by the GIS with INGS. The GIS as a whole, including the INGS, continuously and repetitively generates/gives birth to fluid narratives through its interface. The fluid narratives are cyclically, repetitively, and continuously generated using the GIS's troupe (including actors) and stage equipment. In this fluid narrative generation, the process can be identified as a work, where the receiver can observe myriad aspects of the process at any time using the interface. This expression can emerge in different forms. At present, this generation image is not especially distinguished from the general GIS narrative generation image. The work to reach a conclusion on this point remains ahead of us.

The descriptive text that forms the knowledge content needed for the *Watakushi Monogatari*, as discussed above, is also an independent unit by itself, or several partitioned units of writing. This is the knowledge used to create narrative content through the narrative generation system, yet at the same time, it makes up text that appears like narrative content in various senses. As such, it is not necessary for the narrative content to be used exclusively inside the system, but it is possible to try to publish it in books and magazines, on paper and online, as multiple writings appropriately partitioned and ordered in some form or content. We can imagine that a series has the words *The Narrative Generation of XXX* in all of its titles (in this case, "narrative generation" simultaneously includes the meaning "anti-/non-narrative generation" (Ogata & Kanai, 2010)). Although this is a connection to the "categorization of

Figure 5. Image of Watakushi Monogatari generation/production based on "fluid" and "fixed"
(Source: Figure 3, Chapter 4 (Ogata, 2020h) in Ogata (2020e))

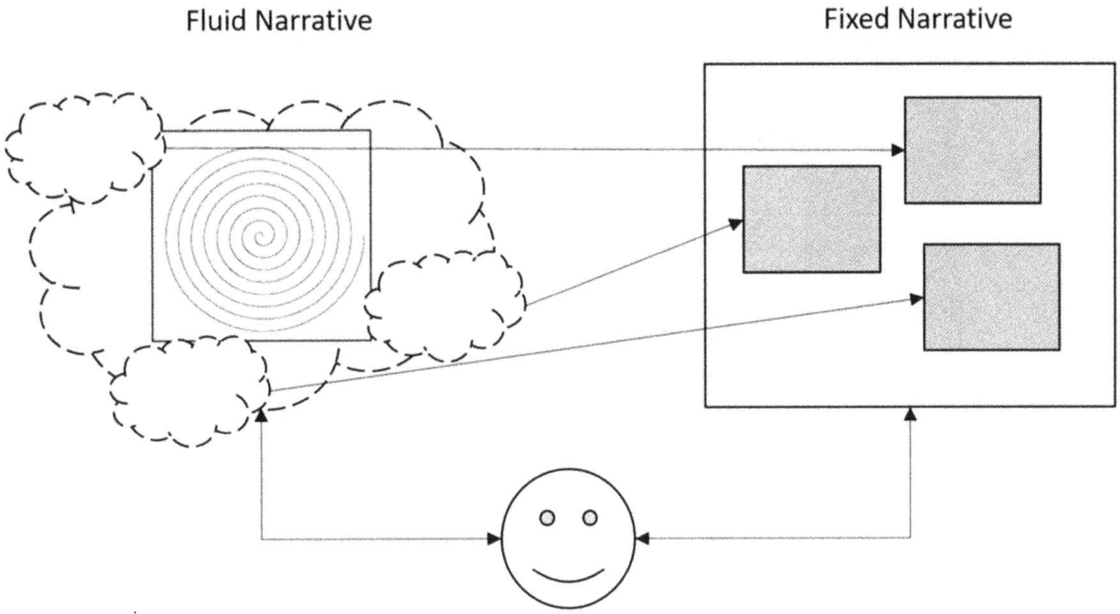

personal texts" discussed in the next section, its titles make up a unit belonging to the category of criticism (essays), contrasting with the academic papers that have been at the center of the author's writing so far. Moreover, the fluid form of the *Watakushi Monogatari* (operated in computers or online) and its fixed state (expressed in the forms of books, paper-based or electronic) make up another group called narrative works and content.

To prepare for producing or creating *Watakushi Monogatari* in the future, Figure 6 summarizes the knowledge content necessary for *Watakushi Monogatari*, in particular a kind of knowledge resource forming the main materials of *Watakushi Monogatari*. Figure 6 presents the overall private and personal knowledge content for producing *Watakushi Monogatari* by using GIS with INGS. The "temporal divisions" signify the time needed to cover the materials prepared, while the "text" implies the types of text to be concretely collected or created and edited within that timeframe.

The author considers different kinds of knowledge to create the *Watakushi Monogatari*. The first "temporal divisions" form the basic structure of private and personal knowledge content, i.e., texts correlated with four types of temporal division:

- **Present Continuous**: The first temporal division began March 11, 2011 and will undoubtedly continue for the time being.
- **Past**: The next temporal division comprises the memories that demarcate "my/myself," which the author defines as the period from November 25, 1970 to March 11, 2011. The first day is the day Yukio Mishima died; subjectively, that was when the author's conscious and self-aware days of life began.
- **Past Perfect**: This is the third temporal division.

Figure 6. Overview of knowledge content needed for Watakushi Monogatari
(Source: Figure 4, Chapter 4 (Ogata, 2020h) in Ogata (2020e))

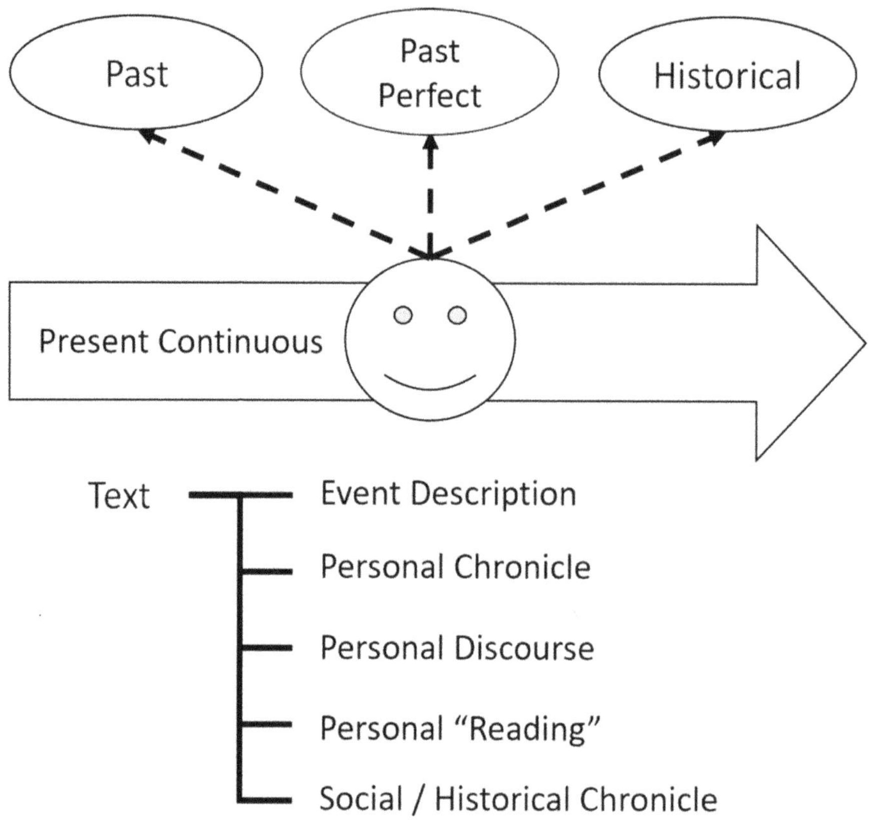

- **Historical**: This is the fourth temporal division. This refers to a temporal division where "I/myself" do not exist, but the existence of those connected to "me" is indeterminate. There is a vast historical space that transcends the private, where time intermingles without distinction.

Next, in accordance with the abovementioned temporal divisions, the author constructs a chronological account of events deemed facts, as represented by the following list. The content of "personal texts" and "personal readings" should not be written anew but rather edited and organized:

- **Event Descriptions:** A temporal, chronological list of past matters thought to have happened to the author (or which "I/myself" think have happened), as well as a somewhat detailed explanation thereof. Outside of that, there are descriptions of issues directly related to (but not directly experienced by) the author, forming the outer edge of the author's experiences.
- **Personal Chronologies:** Seen from a bottom-up perspective, this refers to the chronological information of the individual-level temporal divisions (past perfect, past, present continuous), made up of the abovementioned aggregate of event descriptions as "personal chronologies."
- **Social and Historical Chronologies:** These are general chronological descriptions that do not overlap with any account of "my own life" as the author, and can broadly be divided into three

areas: a chronology of Japan during and after the Meiji era, a chronology of Japan before that, and a chronology of other relevant regions that are not Japan, meaning the world.

There are additional two categories: "**personal texts**" and "**personal readings**." The "personal texts" is a chain of many statements written by the author in the past and present continuous. The list is a kind of discography but does not necessarily contain only presented or published statements. Instead, it is a more comprehensive, inclusive personal/private list that also encompasses unpublished content. Ideally, it is an aggregation of all texts written by the author to date, including texts that the author simply did not publish anywhere and texts that were not approved by the system (e.g., documents rejected, papers, research plans). It would be an enormous undertaking if the author wanted to include remnants of manuscript revisions, but it goes without saying that the author does not think to save them because they are excellent. Rather, the author saves them because they exist, doing nothing but saving them as a record of a man without qualities. In that sense, it might be more meaningful to save things that are not excellent. As such, these are private things, but they are also *other* things that are normally dissolved in statistics, and the *I/other narrative without qualities* includes such things as well. The author categorizes the author's personal texts from 1970 to 2011 into the following four types:

- **Published Academic Content**: These includes papers and essays in academic journals (paper-based and electronic), as well as books.
- **Unpublished Content Related to Academic Activities**: Unpublished texts related to academic papers and other research activities.
- **Materials from Conferences and Classes**: Projection materials and handouts (e.g., resumes) related to talks and lessons.
- **Content That Was Not Published**: Content stored privately, without it having been presented or published anywhere.
- **Lesson Records and Large Volumes of Notes from Self-study, Research Gatherings, and Other Occasions Written for Record or Contemplative Reasoning**: These encompass unarranged records and texts.

The "personal readings" cover all or main content types that the author has read up until this point. However, this is figurative, because it actually includes everything that the author has *heard* and *seen*, yet if the author were to summarize it all and call it *appreciated*, that would give it a strange nuance, so the author says *read* instead. This is likely an expression of the author's disposition to think of centering on reading. Specifically, this includes novels, music, pictures, movies, TV series, radio, newspapers, magazines, and plays. It also pertains to things the author is reading in present continuous. The author considers the following literary texts as the principal subjects of reading, investigation, analysis, and dissected use, as they are deeply tied to the content aspect of narrative generation systems, as well as the existential facet, since it is a list not formed by research necessity, but rather personal interest and curiosity. The categories mix; genres and authors are identified with other ones where that is not the case. Categories like *kabuki*, *ningyō jōruri*, and Yukio Mishima carry the meanings of subjects that should be treated or that the author wants to treat with exceptional care:

- **Japanese Literature from Antiquity to the Present:** Works: narratives, novels, poems, *waka*s (a genre of Japanese short songs), and *haiku*s.

- **World Literature from Antiquity to the Present**: Works in the world of narratives, novels, poems, and plays.
- **Japanese Literature from Antiquity to the Present:** Commentaries, critiques, and miscellaneous writings in Japan.
- **Music**: Western classical music and jazz were vital, yet, perhaps the truly important ones are the other genres.
- **Video Works:** Movies, TV dramas, and others.
- ***Kabuki* and *Ningyō Jōruri* and Texts**: Research and commentaries, in addition to the works.
- **Works by Yukio Mishima**: Related research and commentaries, in addition to his literary works.
- **Other Miscellaneous Narratives**: All narratives, such as *manga* (comics) and TV commercials.

In Chapter 4 (Ogata, 2010) in Ogata and Kanai (2010), the author presents a comprehensive list of books the author had read until a given point. Figure 7 is a conceptual image of a mechanism for generating *Watakushi Monogatari* that connects with other kinds of knowledge. A part of the above book list (1973–1974) is also described in this figure. In terms of private and personal stories for the author, the R&D of INGS and GIS is conducted to realize this kind of image.

Figure 7. Conceptual image of Watakushi Monogatari generation

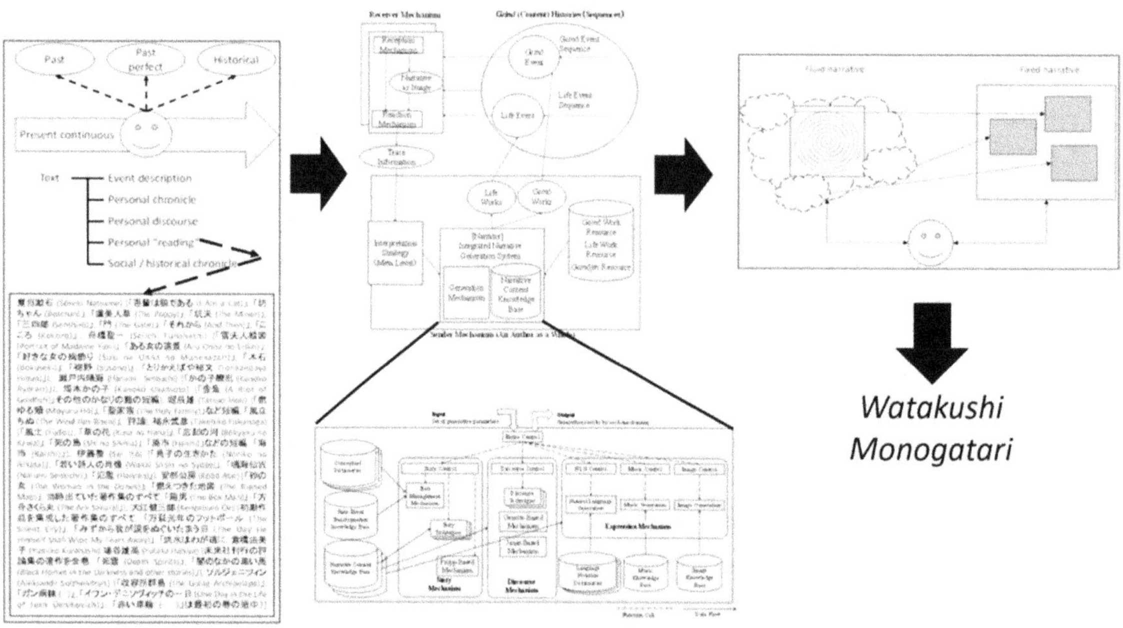

A MULTIPLE NARRATIVE GENERATION MECHANISM THROUGH TWO TYPES OF NARRATIVE GENERATION SYSTEMS: GIS WITH INGS

This section describes GIS with INGS based on the post-narratology of narrative generation or a macro design of an integrated mechanism of GIS and INGS. The author presents the conceptual design of an integrated mechanism of INGS and GIS, or GIS with (including) INGS. In particular, this section covers the macro-level design of the system, which the author has been calling GIS with INGS or GIS including INGS. Previously, although the consideration of the macro-level design has been addressed, the goal here is to link the GIS with INGS in order to connect to the level of design regarding actual program functions, structures, and data from simple conceptual designs.

Basic Elements and the Process of GIS

GIS functions as a whole and includes INGS. Ogata and Amino (2002) examine partial implementation via a computer program of GIS architecture that they conceptually provide. This attempt is an example of GIS's execution in the first section. The study sets a virtual *geinōjin* character, a woman called Kumiko, and simulates her *geinō* activities and life using Propp's literary theory (developed by Ogata and Amino), which defines the typical structure of folktales based on the relationships of 31 basic event concepts called "functions." These functions relate to the generated stories of *geinō* activities and life trajectories in the form of another level's narratives and narrative generation, beyond each narrative content and life event. Narratives/stories to be created by this system form a group of individual narratives/stories. In this sense, the proposed system is a collective system of narratives/stories. Hence, this study implements the concept of GIS, realizing it whereby a narrative/story at the collective level is produced through the sequence of much smaller narratives/stories. This study also presents the idea that each small narrative/story is generated continuously and sequentially under the generation strategy of grand narratives/stories in the above sense. Both the generation of each narrative/story and narrative/story as the sequence are equally included in narrative generation. In the author's current system framework, each small narrative/story is created using INGS, and large narratives/stories, as the sequence, are engendered through GIS. Further, GIS involves comprehensive positioning, and GIS encompasses INGS. Although Ogata and Amino focus on the generation of sequential and large narratives/stories, and do not deal with each narrative generation, in this chapter, GIS with INGS (which in parallel and in a duplicate way generates both aspects) corresponds to an extended model that complements the above study.

In this section, the author scrutinizes the development of the above study's primitive presentation, moving toward an efficient, thorough approach. However, the author reveals the macro design level's consideration before the system is implemented. Based on Figure 2, the main macro-level components in the GIS architecture are the *geinōjin* resource, the *geinō* work resource, and the life work resource. When a *geinōjin* character stored in the *geinōjin* resource is focused on, this *geinōjin* character produces individual *geinō* work, and the *geinō* work's line forms the sequence of each *geinō* work using the *geinō* work resource. At the same time, the *geinōjin* character produces each life work, and the life work's line functions according to each life work using the life work resource.

- **The *Geinōjin* Resource and the *Geinōjin* Character:** A GIS is like a *geinō* production agency or organization. Although the translation is close to "entertainment agency" in English, the author uses "*geinō* production" because the author's theory is founded on the Japanese form of *geinō*, literature/narrative, and is inspired by its classical or traditional forms such as *kabuki*, *ningyō-jōruri*, and *no*. GIS, as a *geinō* production agency, contains many *geinōjin* characters. GIS stores a *geinōjin* resource as an important component. As shown in the author's analysis of *kabuki*, each *geinōjin* character is a multiple existence and includes the following sub-components:

- **The Aspect of the *Geinōjin* Character as an Actress/Actor:** Each *geinōjin* character has unique property information. Each activity forms a *geinō* work, and the chain/sequence corresponds to the discography as a *geinō* work chain/sequence for the actress or actor. If the spatial information is temporal information, the sequence is temporal as well. The property information, as spatial information, also changes according to the temporal progression.
- **The Aspect of the *Geinōjin* Character as an Ordinary Person:** This aspect also has unique property information (spatial), and each life event and chronological (temporal) sequence is formed.
- **The *Geinōjin* Character as a Character in a *Geinō* Work:** Another multiple element in the *geinōjin* character is the facet as the dramatic character in a *geinō* work, which is a little different from the above two elements. Also, the property information, regarded as spatial information in the above elements, certainly gives a consistent feature whereby each *geinōjin* character has the potential to become a dramatic character. However, the *geinōjin* character, who plays diverse dramatic roles, needs to have new property information to play a different role each time. This type of property information can be interpreted as the element that forms part of the model of each dramatic character. Hence, we can divide the model of a dramatic character into such property information, and the part related to dramatic events is induced according to the property information.

- **The *Geinō* Work Resource:** A *geinōjin* character plays many *geinō* works/narratives through his/her discography, particularly a *geinō* history. The information related to the works that the *geinōjin* character appears in (or which is meant for production) is built into the *geinō* work resource. The *geinō* work resource also contains information relevant to the dramatic characters.
- **The Life Work Resource:** Likewise, the life work resource stores various types of information related to the lives of *geinōjin* characters, which helps to build their life narratives/stories.

In the following sub-sections, the author considers prospective concrete content.

The *Geinōjin* Resource and the *Geinōjin* Character

As described above, the most basic components of GIS architecture are the *geinōjin* character, the *geinō* work resource, and the life work resource. They are the internal elements, a sequential narrative production mechanism. At the same time, they are related to INGS, which is an individual narrative generation mechanism and is part of GIS. Although the basic role of a *geinōjin* character in the primary elements of GIS is to appear in works as an actress/actor in produced narratives, the same *geinōjin* characters perform in produced life narratives/stories as other existences or ordinary persons.

In the framework of GIS with INGS, "narrative authors" are simply positioned as INGS and GIS. As frequently stated, regarding the distinction, INGS is a mechanism for generating each narrative as a single body, and GIS is a mechanism for producing sequential narratives based on single narratives. In terms of a broader perspective, narratives in this case include narratives as life works, more than narratives as *geinō* works. At this point, although the Japanese term *geinō* is frequently translated as "entertainment" in English, there is a large gap between these two words. By *geinō*, in this chapter, the author does not mean mere content meant for entertainment. Instead, the author means the narrative, which encompasses a representation of the work through a one-time performance. INGS can create life events for each *geinōjin* character as a real person, and GIS incorporates the function to produce a corresponding life narrative/story as a line of life events. Yet although the "authors" in INGS and GIS may need to have corresponding and individual characters, at present, both INGS and GIS have not personified figures as narrative authors. Regarding this problem, the author believes that a *geinōjin* character can also create a work as an author, and he/she can perform in the work created. In early *kabuki*, the leader (*zagashira*) of a troupe (*ichiza*) wrote *kabuki* works to be performed by the *ichiza*. This was ordinary. In such a case, the *geinōjin* character, as a *kabuki* writer, overlaps with INGS and GIS. A question to be asked is, "Who is the author or writer of a life narrative work?" An adequate answer is to consider that a life narrative work is written by a *geinōjin* character, who proceeds along a lifeline. Hence, in creating a life narrative work, the corresponding *geinōjin* character overlaps with INGS and GIS.

In the above framework, each *geinōjin* character, in parallel, executes and repeats two types of practices and creations of *geinō* works and life works to make the prospective *geinō* history/story and life history/story. By focusing on a *geinōjin* character, he/she continues to participate in practices such as performances, and appears on stage; at the same time, he/she performs his/her life events. His/her life constitutes both a *geinō* history/story and a life history/story. These histories/stories are performed with a holistic viewpoint in mind. Furthermore, when we perceive the entire *geinō* production agency or a *geinō* group/troupe (which synthesizes and fuses the *geinō* history and life history of each *geinōjin* character included in these agencies and troupes), we also establish a grand narrative/story from a broader standpoint.

Now, what information is needed for the property information of a *geinōjin* character? As mentioned above, since a *geinōjin* character contains three different aspects (including characters as an actress/actor, as a real existence, and as a dramatic character), it is necessary to define the corresponding types of property information for the three elements. Because the property information of a dramatic character is the information to be incorporated into the next *geinō* work resource, the author will describe it in the next section.

As for the remaining two elements, as an actress/actor, the *geinōjin* character needs pertinent property information; this includes his/her name and the basic strategies/concepts of his/her career. The *geinōjin* character, as a real person, has a name, sexuality, birthplace, career, interests, hobbies, and a unique personality. In many cases, these two aspects of the *geinōjin* character cannot be seen as completely divided. Of course, there may be a type of *geinōjin* character who intentionally, consciously, and clearly differentiates the character as a real person from the character as a *geinōjin*. However, two kinds of property information for these characters are mutually mixed in many situations. In addition, characters, as real people, have a vagueness about them, and it is challenging to classify them under the above life work resource. One idea is to define characters as real persons in the life work resource. As described above, three types of property information (for the three categories of a *geinōjin* character) leave room for consideration to develop a clearer definition.

Figure 8 depicts the respective types of property information and problems that arise when categorizing the three multiple elements of a *geinōjin* character.

The *Geinō* Work Resource

By *geinō* work resource, the author means the resource necessary to create concrete *geinō* works/content in which *geinōjin* characters, as actresses/actors, perform using various styles. The *geinō* work resource is an essential information resource for GIS in the sense that it is most directly linked to GIS and INGS as a narrative generation system.

Figure 8. Categorization of a geinōjin character and related problems

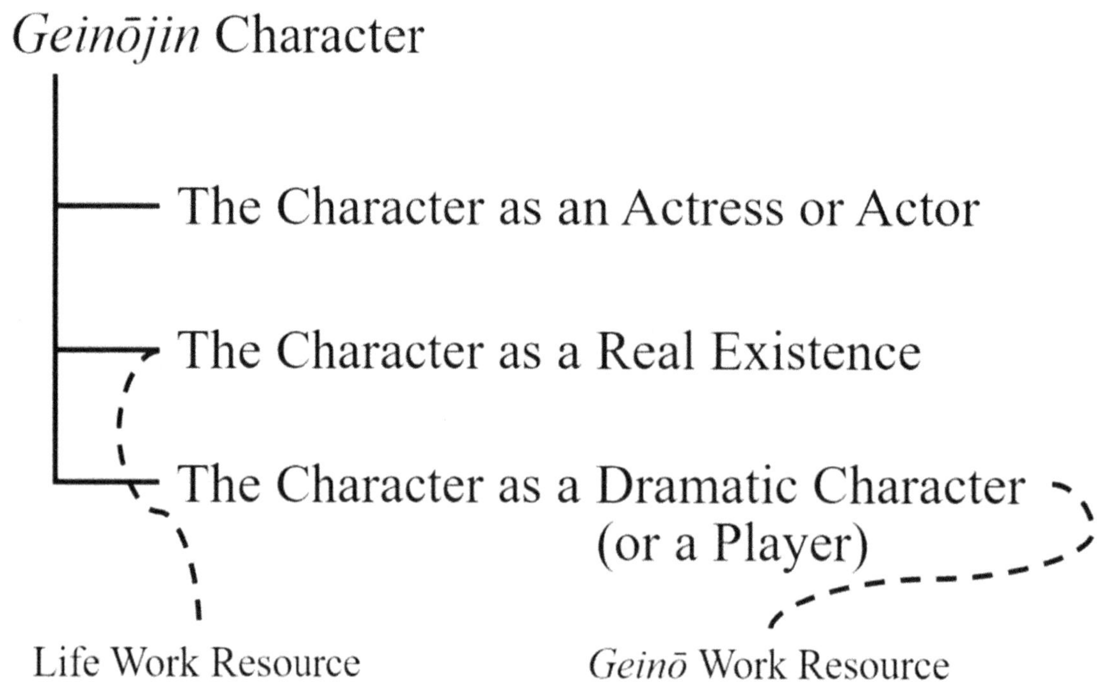

If we consider the *geinō* work resource in a broader sense, it is pertinent to the narrative content knowledge base in INGS. The narrative content knowledge base contains diverse techniques for utilizing the narrative generation of the narrative generation mechanism in INGS and mostly stores methods for story generation. The story content knowledge base covers know-how and strategies for the entire structure of a story for the partial structure of a story. Given the current version of the narrative content knowledge base (including the story content knowledge base), INGS is tentative systematization in progress. The author refers to different narratological studies and literary theories and analyzes narrative/story works, encompassing numerous genres (e.g., novels, folktales, theater, ads, *manga* [comics]). In this section, the author discusses the differences and relationship between the narrative content knowledge base in INGS and the *geinō* work resource in GIS.

The information stored in the *geinō* work resource is concrete, comprehensive information on narratives, compared to the information in the narrative content knowledge base. By "concrete" information, the author indicates the following: Narrative knowledge in the narrative content knowledge base is abstract, universal knowledge, effective for all narratives. In contrast, narrative knowledge for the *geinō* work resource is more specific, meant for particular narrative genres and works. By "comprehensive," the author implies a more delicate meaning. The narrative knowledge and techniques in the narrative content knowledge base can be used generally, regardless of specific narrative genres. In this sense, the narrative content knowledge base has a comprehensive characteristic. The comprehensiveness is the result. In contrast, the comprehensiveness of the *geinō* work resource is intended to apply to diverse genres of narratives. The narrative content knowledge base is comprehensive in a bottom-up way, while the *geinō* work resource is comprehensive in a top-down manner.

The concreteness and comprehensiveness in the above sense of the *geinō* work resource rely on the characteristic of intending to create discographies of *geinōjin* characters. They are related to the point of how a *geinōjin* character, as an actress/actor, performs in each *geinō* work in any style. For example, the discography of an actual *geinōjin* character includes different records regarding performances and other practices. He/she may perform as a dramatic character in a TV drama, as a talented performer on a TV variety Shōwas a singer in a popular song program, as a character in a TV ad, as a narrator in a documentary, as a radio performer in an animated TV program, as a dramatic character in a movie, and as a dramatic character in a play. Since it is difficult for the current GIS with INGS system(s) to concretize all of these narrative representations, they should be realized in simple and alternative forms. For instance, it is possible to substitute these narrative representations by generating narrative skeletons such as scenarios, plots, or stories of works in which *geinōjin* characters perform. In any case, the *geinō* work resource must be able to deal with myriad narrative genres. The following types of scenarios, plots, and stories are included:

- Folktale-like narrative scenarios, plots, or stories
- Ad-like narrative scenarios, plots, or stories
- Game-like narrative scenarios, plots, or stories
- Drama-like narrative scenarios, plots, or stories
- *Kabuki*-like narrative scenarios, plots, or stories
- *Manga*-like narrative scenarios, plots, or stories
- Movie-like narrative scenarios, plots, or stories
- TV drama-like narrative scenarios, plots, or stories
- Conte-like narrative scenarios, plots, or stories

These narratives can be applied to other genres beyond the ones mentioned here. In this case, names according to narrative content (instead of names of narrative genres) are also needed; for instance, narrative genres of family trouble and double suicide, which are popular and frequently appear in *kabuki* and *ningyō-jōoruri*. This kind of narrative is closely associated with narrative content, and is included in the narrative resource for "narrative types commonly used in many narrative genres."

Let us return to the discussion of narrative scenarios according to specific genres. If an ad-like narrative scenario is an original narrative work, its derivative narrative works may also exist. The derivative narrative works are distributed through their transformation into different media from the original medium. *Geinōjin* characters can also perform in derivative narratives.

- Deviated narrative works from folktale-like narrative scenarios: Ex.) A folktale is narrated on a radio or TV drama.
- Deviated narrative works from ad-like narrative scenarios: Ex.) A magazine ad is adapted to a novel or a book.
- Deviated narrative works from game-like narrative scenarios: Ex.) A game is adapted to a novel or a book.
- Deviated narrative works from *kabuki*-like narrative scenarios: Ex.) A *kabuki* work is adapted to a *ningyō jōruri* work.
- Deviated narrative works from *manga*-like narrative scenarios: Ex.) A *manga* work is adapted to a game and a *kabuki* work.

- Deviated narrative works from movie-like narrative scenarios: Ex.) A movie work is adapted to a TV drama and a game.
- Deviated narrative works from TV drama-like narrative scenarios: Ex.) A TV drama is adapted to a movie and *manga* works.
- Deviated narrative works from conte-like narrative scenarios: Ex.) A conte (short story) is adapted to a TV drama and film.

For example, a *geinōjin* character performs in TV dramas and movies with a "folktale-like scenario," TV and Internet ads with an "ad-like scenario," computer games and movies with a "game-like scenario," staged plays and their cinematized works with a "*kabuki*-like scenario," and adapted movies, TV dramas, and *kabuki*. An important term is "narrative media," which enables us to realize various narrative genres. These media need to be contained in the knowledge base of the *geinō* work resource.

At this point, let us organize the main types of narrative information to be encompassed in the *geinō* work resource. First, narrative genres exist. Chapter 2 (Ogata, 2020b) in Ogata (2020a) introduces the author's "narrative genre system," which divides narrative genres into five broad categories (listed below) and shows their sub-categories and examples, primarily based on Japanese literary genres:

1. Narratives as Works in the Narrow Sense or Works in the Narrow Sense in Which Narratives Appear (or Narratives Are Included)
2. Narratives as Works in the Broad Sense or Works in the Broad Sense in Which Narratives Appear (or Narratives Are Included)
3. Narratives as Social and Emergent Phenomena or Social and Emergent Phenomena in Which Narratives Appear (or Narratives Are Included)
4. Narratives That Invade Real Phenomena or Real Phenomena in Which Narratives Appear (or Narratives Are Included)
5. Narratives as Human Physiological and Psychological Natural Phenomena or Human Physiological and Psychological Natural Phenomena in Which Narratives Appear (or Narratives Are Included)

This narrative genre system is organized as a collection; it contains narrative genres in the broad sense, such as mental narratives (e.g., dreams) and social narratives (e.g., festivals, real events), in addition to ordinary narratives (e.g., literary and narrative works). Thus, for example, *geinōjin* characters can also perform the narration of a real event (news story).

Next, the *geinō* work resource covers the concept of "narrative media," which is divided into several levels: media as devices or machines (e.g., TV, radio); media as more direct representation (e.g., sound, characters), and media as social organizations or agencies (e.g., TV stations).

The *geinō* work resource includes the concept of "narrative type." Compared to the above genres and media, the narrative type is the most substantial kind of content information for narrative works. The information is most directly tied to narrative plots/stories. For example, there are events and their sequences (the author employs a temporal event sequence story and an event sequence for a narrative development plot), dramatic characters, objects, places (spaces), and times. In order to analyze this type of narrative information, the author begins each topic in the first phase to categorize the results into several groups, in which methods in a group can be used for a common goal from the angle of cross-genres or trans-genres. For instance, although a narrative type of *kabuki*, "family troubles" (*oie sodo mono*), emerged through the examination of *kabuki*, generalization beyond narrative genres enables us to apply the method to diverse narrative genres other than *kabuki*.

Regarding a concrete narrative work/content performed by a *geinōjin* character, first, there is a story/plot in which specific characters, objects, spaces or places, and times appear. The story/plot and its elements are defined using the narrative content knowledge bases and related narrative techniques. Furthermore, to realize a narrative in a specific *geinō* genre, the above abstract elements of narrative are concretized by combining certain media to be performed by *geinōjin* characters.

By blending diverse narrative elements (e.g., different genres' works and patterns, various dramatic characters in different works), myriad places and times (ages) become connected to present the concept of *naimaze*, which is a traditional method for creating a new work in *kabuki*. Strictly speaking, this concept means a specialized dramaturgy, in particular methods for creating new works by blending different *sekai* (worlds), produced by several excellent *kabuki* writers or creators in an age (from the middle to the latter half of the Edo period), especially Sakurada Jisuke I and Tsuruya Nanboku IV. This methodological concept of narrative generation will be generalized into the creative direction of a principle of narrative generation, which contains numerous concrete techniques for producing new works from the amalgamation of two or more works and their many components. The author aims to examine different problems of narrative generation, including narrative deviation and transformation among many genres and media, from the perspective of *naimaze* as a generalized methodological concept. The author introduces *naimaze* in **The Narrative Generation of *Kabuki*** in the **BACKGROUND** section of this chapter.

The Life Work Resource and *Geinōjin* Characters

Each *geinōjin* character lives in an actual world, and the continuous path of events shapes the *geinōjin* character's "life course" (also called "life story" or "life history"). Linking with INGS, GIS can produce the life stories of *geinōjin* characters, in addition to narratives as *geinō* works. Hence, life stories are part of the narratives that are generated regarding each *geinōjin* character. GIS stores the life work resource with the *geinō* work resource. The life work resource contains the material information for creating the life stories of *geinōjin* characters, mainly the information for constructing their life events. People commonly have typical life events, shown in the following list:

- Life events around birth: Father, mother, ancestor, place, time or era, etc.
- Life events around school: Entrance, graduation, registered absences, leaving school, etc.
- Life events around love: First love, lost love, marriage, immorality, etc.
- Life events around friendship: Friendship among members of the same sex, the opposite sex, etc.
- Life events around jobs: Part-time jobs, employment, job changes, social failures, social successes, etc.
- Life events around family: Relationships with/among mothers and fathers, brothers and sisters, husbands and wives (marriage, divorce), children, etc.
- Life events around hobbies: Sports, reading, music, etc.
- Life events around *geinō*: Debut, work, job, failures, successes, etc. (This is the boundary area with the *geinō* work resource.)

These categories and examples cover only some of the life events and are not sufficiently systematized, but this list provides tentative content according to particular themes such as birth, schooling, and love. Seen from a spatial angle, life events form a hierarchical structure that includes several high-level event categories, as well as low-level event categories belonging to the higher categories. The low-level categories can be subdivided into smaller units. Smaller categories encompass many events. At any hierarchical level, events move into the range of the narrative content knowledge base in INGS from the range of the life work resource, because events after the level of an elaboration can be interpreted as event development, to be used generally, and are not contained in life events. Since the information to be stored in the life work resource in GIS corresponds to the comparatively large categories for forming people's lives, knowledge for detailed, concrete development is stored in the narrative content knowledge base, especially the story content knowledge base. However, the above description is still a general direction, and the isolation of both areas will be considered in the future. Regarding the actual development of each *geinōjin* character's life events, a portion of the lower event categories in the life event structure is activated in each timing to construct a *geinō* sequence of events in an efficient form. Each event sequence is concretized through a specific temporal range, including eras and seasons, a certain space (place), participating people, and the relationships among them.

A *geinōjin* character's life events indicate the relationships among *geinō* events, which are performed according to his/her name as an actress/actor in order to produce his/her life history/story, which is a relatively individual event sequence. A *geinōjin* character's entire life story is a narrative, just as the entire *geinō* discography is a narrative/story. Additionally, since each narrative in a *geinō* discography is respectively a narrative, each event in a life story (e.g., birth, lost love, job success or failure) is also a narrative. As mentioned above, these events should be included in a narrative as a single body to be generated by INGS. Thus, INGS contains the mechanism for generating life events/narratives from the viewpoint of the linked model with GIS.

There may be various design methods and ideas about the implementation of the life narrative generation program. The typical directions are the top-down and bottom-up strategies. For the former, a pre-determined, basic, broad scenario of a life narrative is created at the first starting point of the simulation of an entire life narrative, or the large part, by a *geinōjin* character. Next, small events are produced according to the broad scenario. The bottom-up strategy did not previously have such a broad scenario. The strategy engenders the life events of a *geinōjin* character in his/her relationships with other *geinō* narratives' generation, and the entire life event sequence is continuously and sequentially created through the above repetition. The most useful approach is the hybrid method of the top-down and bottom-up strategies. For example, in the first stage, a basic and broad scenario for the extension of a life story, many of the parts remain in a non-detailed state. In the next step, the system completes in detail through the simulation of concretization and elaboration in engagement with context and situations. In this process, a planned or predicted broad life story sometimes shifts to a different life story. Both the top-down and bottom-up strategies for generating life narratives can be applied to create *geinō* work sequences using the *geinō* work resource.

Lastly, *Watakushi Monogatari*, described in **THE CONCEPT OF *WATAKUSHI MONOGATARI* AND DESIGNING INTERNAL AND EXTERNAL NARRATIVE GENERATION METHODS TO CREATE AND DISTRIBUTE NARRATIVE CONTENT**, is a vision of the narrative(s) to be created based on the *Watakushi Character*, which was discussed in **Incorporating the *Watakushi Character***. Regarding the material information for creating *Watakushi Monogatari*, the author partly systematizes the life information based on "I" as the author. This was stated in **Internal and External Narrative Generation** in the **BACKGROUND** of this chapter. The life work resource in this section is an information unit for storing the common knowledge resource for *geinōjin* characters. The systematization of life narratives, as the material information for *Watakushi Monogatari*, is the information for more a concrete and individual "I" (*Watakushi Character*). In the future, the systematization of life narratives for *Watakushi Monogatari* and the *Watakushi Character* will indicate how to construct life narratives as more general knowledge.

GIS and INGS in the Integrated Mechanism

First, the author summarizes the above description. The construction of GIS requires establishing *geinōjin* characters and providing the corresponding property information for each *geinōjin* character. The property information is divided into data for an actress/actor and for an ordinary human. The former and latter pieces of information are respectively related to the discography of the *geinō* work performance (*geinō* history) and the life narrative. Moreover, the above discographies (a discography is a type of narrative) are tied to the *geinō* work resource, and the life narratives are linked to the life (work) resource. These narratives are collective in the sense that many individual narratives are gathered. GIS can produce an entire narrative, which is a collection of many individual narratives, while INGS generates each narrative included in the narrative as a collection. At the same time, the division of roles is relevant to the characteristics of the knowledge bases and resources for producing narratives. In particular, the *geinō* work and life work resources of GIS deal with a more macro-level narrative, such as entire narrative structures in a specific narrative genre, and the framework for the development of important events in lives. The narrative content knowledge bases (especially the story content knowledge base) deal with knowledge for developing more detailed, general events.

The author discusses the division of roles of GIS and INGS in the productions of *geinō* and life narratives. As described above, although the knowledge contained in INGS is only general knowledge for narrative generation, GIS's knowledge is more concrete, individual knowledge based on each narrative work. For example, the story content knowledge base in INGS only contains fragmented narratives or story knowledge. The author asserts that the image of knowledge resources to be contained in GIS can become clear by applying *Sekai Kōmoku* in *kabuki* to it. This means that *Sekai Kōmoku*, or the knowledge structures seen in *Sekai Kōmoku*, can be beneficially applied to the *geinō* work resource in GIS. A detailed description of *Sekai Kōmoku* is provided in the following section, ***The Survey and Analysis of* Sekai Kōmoku**; it is a professional knowledge system that was used for writing *kabuki* scenarios during part of the Edo period. *Sekai Kōmoku* categorizes the typical patterns of *kabuki* narratives, and, for all *sekai* (worlds), describes or lists main dramatic characters; the original books, including novels and histories, of the *sekai*; and *ninghyo-joruri* works written based on the *sekai*. *Kabuki* writers from that time period referred to diverse *sekai* mentioned in *Sekai Kōmoku*, and added new *shukō* to a *sekai* or several *sekai*(s) to create a new *kabuki* scenario. Although *Sekai Kōmoku* does not directly address narrative plots and places, as the *kabuki* writers in that age completely knew the details, they selected one or more *sekai*(s) and determined basic characters, plots, spaces, and times to add *shukō* or twists to the *sekai*(s) for differentiating them from previous and existing narratives. Thus, they produced fresh narrative scenarios. The author believes that GIS, especially the *geinō* work resource, can be defined and stored in a similar form with the narrative knowledge structures of *Sekai Kōmoku*. For instance, regarding a narrative *sekai* as a kind of retrieval item, related narrative works, characters, and others are organically represented. In the real *Sekai Kōmoku*, the readers arrange and edit the content as the names of titles and characters. Given the possibility of a new version of *Sekai Kōmoku* in GIS, as the related information stored in the *Sekai Kōmoku* is computationally reorganized, the users decide on narrative macro-level structures, forms, and content linked to GIS and INGS, complete with narrative scenarios using more detailed information.

As stated above, in an actual narrative generation process based on the macro-micro incorporation of INGS and GIS, the *geinō* work and life work resources are provided by GIS to produce relative macro narratives. Therefore, GIS can create comparatively macro narrative structures and content, in addition to sequential narrative generation, including each narrative generation.

Incorporating the *Watakushi* Character

Earlier, the author deals with the general structure of GIS with INGS. This section states the GIS with INGS from the viewpoint of relationships with the author's *Watakushi Monogatari*. *Watakushi Monogatari* is also created through GIS with INGS. As mentioned above, *geinōjin* characters in GIS basically perform in produced narratives. For example, they perform various *geinō* narrative works. Furthermore, they can create narrative works as authors or writers. Although the concept of "author" is not clearly articulated in GIS with INGS, the *geinōjin* characters can be narrative authors or writers who create narrative works by themselves. Moreover, GIS with INGS, which focuses on *Watakushi Monogatari*, is necessary to prepare one or more *geinōjin* character(s) who can be author(s) or writer(s).

By *Watakushi Monogatari*, the author intends to produce a unique narrative work. In the plan of *Watakushi Monogatari*, the author designs the form in which this real "I" (*watakushi*) exists with one or more *geinōjin* character(s) as virtual characters, or enters into the *geinōjin* characters. Although GIS is basically composed of a collection of *geinōjin* characters as virtual characters, this real "I" (*watakushi*) invades the *geinōjin* characters. In particular, the *geinōjin* character, who has half a virtual existence and half a real "I" (*watakushi*) existence, produces narrative works tentatively called *Watakushi Monogatari*, and socially distributes the narrative works through GIS. Hence, the process of the *geinōjin* character, including "I" (*watakushi*), encompasses the act of creating *Watakushi Monogatari*, in addition to the act of distribution.

Simultaneously, regarding the production, creation, and distribution of *Watakushi Monogatari*, the production and distribution of other narratives by GIS are continuously facilitated through other authors, narrators, and actresses/actors. The acts of creation and distribution of *Watakushi Monogatari* by "I" (*watakushi*), as a part of *geinōjin* characters, are embedded in diverse narrative acts in GIS. In addition, this "I" (*watakushi*) and *Watakushi Monogatari* are not completely independent things separated from narrative generation acts by other subjects. They are existences that are influenced from the external elements and situations and can influence external things.

Figure 9 shows an overview of the creation process of GIS with INGS and *Watakushi Monogatari*. It is vital to discuss the concrete, detailed form of the concept in this figure.

For example, when the *geinōjin* character mixed with "I" (*watakushi*) creates *Watakushi Monogatari*, the *geinōjin* character creates the *Watakushi Monogatari* by using or editing various types of knowledge, which are analyzed in **Toward Internal Narrative Generation: The Creation of *Watakushi Monogatari*** in this chapter. At the same time, the knowledge and resources stored in GIS and INGS are also used in the creation. They can be employed from the angle of automatic narrative generation. Furthermore, *Watakushi Monogatari* and the part created once can be restructured using the knowledge and resources in GIS and INGS. The following tentative list contains several methods for the creation (production) of *Watakushi Monogatari* based on "I" (*watakushi*), a *geinōjin* character including an "I" (*watakushi*) or a *geinōjin* character as an "I" (*watakushi*), and GIS with INGS:

Figure 9. The concept of the creation process of GIS with INGS and Watakushi Monogatari

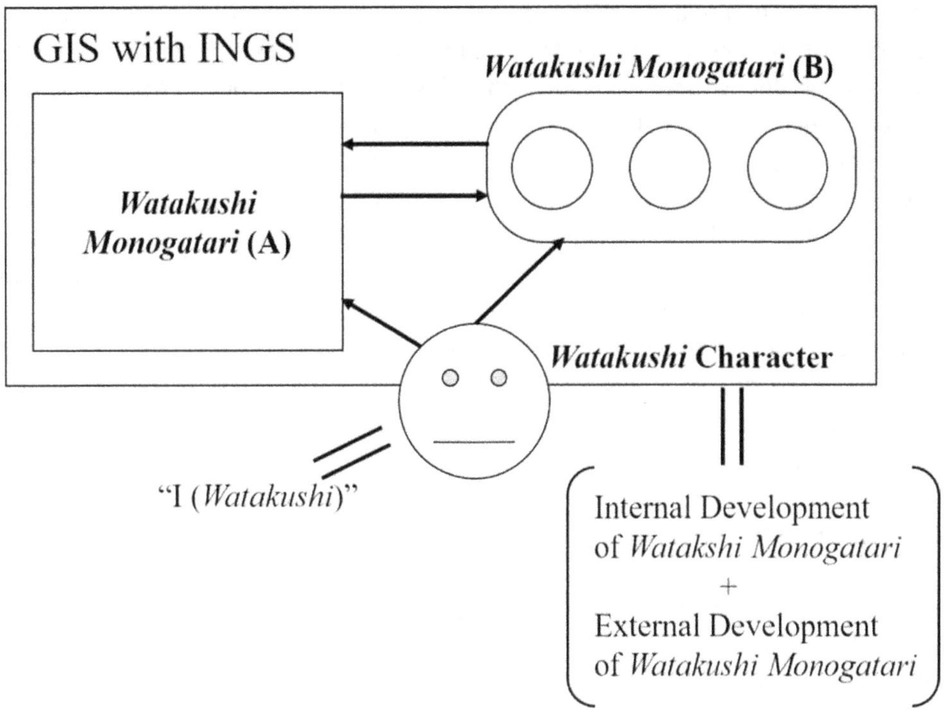

- The *geinōjin* character as an "I" (*watakushi*) thinks up the plots and stories of a *Watakushi Monogatari* and the parts through GIS and INGS, in addition to only the living "I" (*watakushi*).
- "I" (*watakushi*) can develop the plots or parts for *Watakushi Monogatari*. Simultaneously, this "I" (*watakushi*) refers to the plots or parts that the GIS with INGS system generates. Further, "I" (*watakushi*) can expand a provisionally made plot through GIS with INGS and translate a one-time narrative created by the system. Like these, "I" (*watakushi*) can create the parts of *Watakushi Monogatari* through various interactions between the system of GIS with INGS.
- More micro parts for narrative development and expansion, such as event development, explanations, and descriptions of constituent elements of a story and event, and narrative discourse techniques at the level of the transformation of event development, can use GIS with INGS. This generates the possibilities of narrative micro development, other than the possibilities thought by "I" (*watakushi*) or beyond them, in addition to the elements of macro narrative expansion.

"I" (*watakushi*) can create *Watakushi Monogatari* considering the above possibilities. The above description covers formal methods regarding the creation of *Watakushi Monogatari* by "I" (*watakushi*). Regarding formal methods of the next distribution process of *Watakushi Monogatari*, the author considers the following possibilities:

- "I" (*watakushi*) makes it possible to send *Watakushi Monogatari* to the distribution process via the created parts. These parts can be like initial "seeds" for the final forms, which do not comprise a fixed narrative, and can be transformed in continuous phases.

- Narratives, as initial "seeds," can be modified after the distribution process. In this case, the distribution process is part of a production process. (This type of creation is compared to the popular creation and distribution of novels and stories in modern Japanese literature, though it is doubtful whether many writers clearly understand that this kind of creation-distribution is a unified creation process.) In any case, first, a part of a novel is published in a literary magazine and, continuously, the novel is published in the form of a book. During transformation, a novel, which is serially published in the medium of a magazine, is altered through corrections, revisions, and further reorganization. Using this comparison, part of *Watakushi Monogatari*—which was first published in a "magazine"—changes the structures and content of an entire *Watakushi Monogatari* and its parts before the *Watakushi Monogatari* is finished. A difference is whether the objects to be modified comprise the entirety or parts of a novel/narrative. However, if we regard a part in the case of a whole, the same structure is recursively repeated. The difference in phenomenological form (in the case of *Watakushi Monogatari*) strengthens the characteristic whereby the narrative creation process enters the narrative distribution process.

The Overall Behavior of GIS, Including a *Watakushi* Character with INGS

Since the *Watakushi Character* here is a *geinōjin* character in GIS, it functions like other *geinōjin* characters or as one of the many *geinōjin* characters. The operation means that a *Watakushi Character* cyclically creates and performs narrative works. However, the genres or narrative kinds it creates are limited. In particular, only *Watakushi Monogatari* itself or the content for preparing *Watakushi Monogatari* are the narrative genres/types produced by the *Watakushi Character*. Furthermore, the *Watakushi Character* is not a pure or completed virtual AI character; half of it is occupied by the "I" (*watakushi*) in the real world. Hence, *Watakushi Monogatari* is completed through the collaborative act of the *Watakushi Character* and the "I" (*watakushi*) in reality.

Although an overview of the concrete repetition and cyclical process is mentioned in the following list, a detailed discussion will be given in **THE CONCEPT OF *WATAKUSHI MONOGATARI* AND DESIGNING INTERNAL AND EXTERNAL NARRATIVE GENERATION METHODS TO CREATE AND DISTRIBUTE NARRATIVE CONTENT**:

- First, there is a situation in which many *geinōjin* characters cyclically continue to produce and perform diverse narratives. In other words, it is a fluid narrative situation.
- In this fluid situation of narrative generation, the *Watakushi Character* intermittently creates fixed, comparatively short narrative works through concerted action with the real "I" (*watakushi*) included in the *Watakushi Character*. The author tentatively calls the narrative collection "fragmented narratives."
- Moreover, based on the "fragmental narratives," the *Watakushi Character* and the real "I" (*watakushi*), who is involved in the *Watakushi Character*, continue the creative act toward *Watakushi Monogatari*, which is a more systematic, fixed narrative. The author calls the systematic, fixed narrative *Watakushi Monogatari*. Alternatively, it could be called "the narrative of fluidity and fixation." However, the *Watakushi Monogatari* in the above meaning may be the *Watakushi Monogatari* in the narrow sense. In this case, the total narrative, combined the "fragmented narratives" and "the narrative of fluidity and fixation," may be called *Watakushi Monogatari*.

Abstractly speaking, the fixed narratives, which are sometimes produced in the forever fluid situation, evolve into fragmented narratives. Further, the narrative of fluidity and fixation—as the fixed narrative that integrates, edits, and adapts fragmented narratives—is produced. At the theoretical level, the narrative of fluidity and fixation is also fluidized toward other circulative, repetitive narrative situations.

NARRATOLOGY OF *KABUKI*: ORIENTED TOWARD NARRATIVE GENERATION

The author has frequently argued that in general, *kabuki* is discussed with a focus on the stage, and moreover, on so-called "patterns," while *kabuki* retains an extremely large volume of "language" since the Edo period, and this diversity of language in *kabuki* forms one factor that has supported the resilience and continuity of *kabuki*. Therefore, *kabuki* should be studied through the stage and also through language. By *kabuki* as language, the author means that *kabuki* has produced many types of linguistic documents, including essays, reports, and critiques, and existed with them, though an individual performance on the stage is regarded as the most important element in *kabuki*. Chapter 2 (Ogata, 2020g) in Ogata (2020e) organizes and classifies the types of "language" found in *kabuki* as follows:

- Research
    - General and introductory studies
    - Studies focusing on actors
    - Studies focusing on authors
    - Studies focusing on works or plays
    - Studies focusing on music
    - Studies focusing on theaters and environments
    - Folkloric and *geinō* studies
    - Historical studies
    - Sociological studies
    - Literary, artistic, aesthetic, and critical studies
    - Fringe studies
    - Informational and cognitive studies
    - Dictionaries and encyclopedias
- **Dramatic Critics**
    - Dramatic critics (from early modern to modern) [synchronic]
    - Dramatic critics (contemporary) [diachronic]
- **Reports**
    - Reports (a) (pamphlets in theaters)
    - Reports (b) (other documents including picture collections)
    - Magazines
    - Actor's talks
- **Works**
    - Scenarios (*daichō*) in *kabuki* and related genres including *no* and *ningyō jōruri*

This section introduces and discusses topics relating to "Literary, artistic, aesthetic, and critical studies" from previous **Research**, and "Dramatic critics (from early modern to modern) [synchronic]" and "Dramatic critics (contemporary) [diachronic]" from **Dramatic Critics**. In the aforementioned classification, "Literary, artistic, aesthetic, and critical studies" from **Research** and **Dramatic Critics** ("Dramatic critics (from early modern to modern) [synchronic]" and "Dramatic critics (contemporary) [diachronic]") are similar; however, this indicates that in the case of *kabuki*, some critiques may be more like comparative studies, while others may be similar to criticism. First, the author mentions Yukio Hattori and Tamotsu Watanabe, two present-day *kabuki* researchers who have discussed numerous topics of thought and philosophy concerning *kabuki*, and present their statements from perspectives of thought and philosophy. Then, the author focuses on a work of *kabuki* theory (*Sakusha Shikihō Kezairoku*, which means the rules and principles that *kabuki* writers and directors create and produce *kabuki*'s scenarios and stage performances) and a type of instructional work or knowledge base (the aforementioned *Sekai Kōmoku*, which means the outline and details of the *sekai*s (worlds) that are used for writing *kabuki* scnerios) written in the Edo period, and considers how each is linked to the author's studies of narrative generation in *kabuki* in an essential sense. It should be noted that *Sekai Kōmoku* is neither a research work nor a critical work, but a type of instructional text or guide for creating *kabuki*. The aforementioned classification does not have an item suited to this type of document and therefore requires expansion in the future.

The Thought and Philosophy of Two Present-Day *Kabuki* Researchers

While a large volume of *kabuki*-related "language" accumulated through the Edo period, academic study of *kabuki* flourished particularly in the Meiji era. *Kabuki* has had a rich tradition of critical and commentary texts since the Edo period, and thus, the modern *kabuki* world has many critical researchers or research-based critics who practice texts of a type that combine scholarly methods and critical methods. Yukio Hattori and Tamotsu Watanabe, discussed here, are a new type of scholars we will call researchers-cum-critics. If pressed, the author would say that Hattori is a writer who leans towards being a researcher and Watanabe is a writer who leans towards being a critic, and that both authors are exceptionally skilled researchers and critics. The author of this chapter is not prepared to systematically discuss an overview of their research/criticism, so the author will merely consider one book by each and present their outlines.

Yukio Hattori's The Structure of *Kabuki*

Yukio Hattori is a noted *kabuki* researcher who worked at Kokuritsu Gekijō (the National Theater of Japan) and produced many works. His early work, *The Structure of Kabuki* (Hattori, 1970), is an important text in the literature on *kabuki* research. It was published on November 25, 1970—the very date on which author Yukio Mishima died by suicide at the Self-Defense Force Base in Ichigaya, Tokyo after inciting force members to rise to action. Around this period, the climate concerning speech was heavily influenced by the wave of structuralism sweeping across the world of Japanese thought and philosophy. Hattori, who was still comparatively young, probably wrote this work while conscious of riding this tide of modern thought and modern philosophy. As previously described, structuralism also formed the broad tides of thought and philosophy in Japan, and many thinkers and researchers in the humanities and social fields were influenced by it to an extent. Examples include Masao Yamaguchi (1931–2013) and Chizuko Ueno (1948–). However, few researchers seriously pursued formal or structural analyses of narratives, as observed in structuralist narratology. Kazuhiko Komatsu (1947–), a folklorist and folktale researcher, was a representative of this kind of researchers. Moreover, very few researchers went beyond this and assimilated their attempts and outcomes to incorporate them into systems of thought and philosophy. In this sense, structuralism and narratology were generally inclined to remain a passing trend in thought and philosophy and did not go beyond the level of scholarship for the sake of scholarship in Japan.

To supplement the above, Masao Yamaguchi's structuralism swept through the journalistic world of thought and philosophy in Japan at one time (around the 1970s and 1980s), directly influencing the novel-writing techniques of authors such as Kenzaburō Ōe (1935–) (for example, *Dōjidai Game* [The Game of Contemporaneity] (Ōe, 1979)). Although his structuralist theory of cultural anthropology, scattered with concepts such as "center vs. periphery" and "the trickster," was of interest to the author of this chapter, it was not attractive enough to make the author comprehensively explore it. Around that time, Japan seemed to have developed a formula closely linking structuralism with Yamaguchi, however, this structuralism differs from what the author regards as structuralism. From the author's perspective of structuralism, merely in the sense symbolized by formal, structured narratology (the author believed that Yamaguchi and others lacked this perspective of detailed formal, structural object analysis), Japanese structuralism as represented by Yamaguchi seemed extremely journalistic, particularly emphasizing only the obvious parts or the remarkable parts in the structure of the object. In other words, in the aurhor's opinion, their analysis and examination seemed to be from an extremely "elevated" vantage point. They had seemed to suddenly leap to semantics before conducting a formal and structural analysis of the object at a "lower" level. However, this is merely the author's impression, and a careful reexamination may produce countervailing evidence. This trend also extended to economics and other fields, leading to the appearance of "the trickster theory of business administration" and the like, which applied folklorist and narratological concepts mechanically or arbitrarily to business and organizational phenomena (for example, Teramoto and Nakanishi (1993)). If such a thing were called dynamic structuralism, what the author sought would be—to put it ironically and in extreme terms—static structuralism, and could even be deemed as something desirable or something that had already ended by the 1980s and 1990s.

Hattori's structuralist approach to *kabuki* can probably be considered to function on the foundation of Hattori's own approach to his later works and research. However, it was possibly linked to the aforementioned trends concerning structuralism in general, and approaches such as this were rarely developed further within *kabuki* research. (The author has not yet investigated works and papers on structuralist *kabuki* theory, but may find related works.) For example, Toshio Kawatake's (1924–2013) *kabuki* studies (Kawatake, 1989, 1992, 2003) are widely read, and while they are individually highly attractive, they focus on the "style" of *kabuki* (such as the beautiful and peculiar style); although they are organized, they are not systematic in an essential or structural sense). However, the author intends to focus on Hattori's structuralist *kabuki* theory and draws attention to it again.

Hattori's structural approach to *kabuki* focuses on structure rather than style. In extreme terms, it is a study centered on extracting the elements of *kabuki* and identifying the relationships between them. In practicing this attempt, Hattori first noted the people involved in *kabuki* and then examined them beginning with those towards the outer circles—the financiers of the troupes—and moving in order through the actors to the creators.

Hattori pays particular attention to the existence of "actors" ("*yakusha*s" or "*haiyū*s" in Japanese) in *kabuki*. In other words, he places actors at the core of the structure of *kabuki* and clarifies that the concept is more comprehensive, overreaching, and fundamental than the concept of actors as we usually imagine them. In this way, Hattori considers actors to be a central element in the structure of *kabuki*. In fact, passages concerning actors account for the largest portion of this book. To wit, Hattori discusses the creation and development of *kabuki* actors in historical terms, explains the structure of actors' "art" and its making by invoking many concepts, such as "styles," "roles," "forms," and "spirit," and finally explains that the authorial function known as "*kyōgen tsukuri*" (lit. play-maker) was originally performed by the actors.

This attention to actors does not mean that Hattori simply regarded actors as more important among the many elements; rather, it means that the actors were the most important in actual *kabuki*. However, Hattori also notes the "creators," who separated and became independent from actors over time, and specifically investigates and analyzes structural characteristics of *kabuki* "works." At the same time, the reason creators were traditionally positioned below actors in *kabuki* is revealed with statements about the roles of actors. Specifically, in early *kabuki*, the duties of creators could not be considered to occupy a clear position among the various structural elements of *kabuki*. It is possible that creators existed only as a function, and there were no actual creators in a pure sense. In fact, in early *kabuki*, a certain work was written by the stronger actors of a troupe (company) called "*ichiza*" in Japanese, who then performed it themselves. The task of creating and performing works was the responsibility of the actors. At last, creators came to exist separately from actors, but this characteristic in early *kabuki* also influenced the later development of *kabuki* and formed the tradition that still continues, wherein actors occupy a central, core position in *kabuki*.

A mysterious phenomenon that exists even today is that *kabuki* has a designation called "playwright" (*kyōgen sakusha*). Inferring from this title, most people would normally think it refers to a person who creates *kabuki* plays. However, their actual job is not to write plays, but to perform general duties around the stage. Naturally, this does not mean that jobs like this are not important, and it is a highly important role in the process and setting of constructing an actual stage performance. However, this is not typically the task we would associate with a creator, nor is it the job of a producer. In modern terms, it could provisionally be considered similar to the role of an assistant director in film or television. However, in the world of *kabuki* in the Edo period, the aforementioned designation in which actors also functioned as creators and the duties of a creator were subsumed within those of actors was gradually followed by the formation and independence of the creator as a separate position. Moreover, this position came to be borne by a group of playwrights or *kyōgen sakusha*s belonging within a so-called department known as the "play office" (*kyōgen-beya*) that was included in a *kabuki* troupe or *ichiza*. In these circumstances, it is true that the creator arose as an occupation that was independent of the actor in *kabuki*. Nevertheless, they were still positioned below the actor in charge of the troupe (*ichiza*), especially below one or more key actors in charge of other actors in the *ichiza*, produced individual plays, and oversaw all the work needed in the *ichiza*, such as formulating strategies for the flow of several plays. Later, after many *kabuki* works came to be written by external creators not belonging to a specific troupe in the Meiji era, playwrights (*kyōgen tsukuri*s) and play offices (*kyōgen-beya*s) lost power and eventually became mere shadows of their former selves, as discussed.

This work by Hattori views "actors" as the most important people involved in *kabuki*, but in the initial section, places the "financiers," who have the capital capacity to actually produce *kabuki*, further outside them, thus revealing the true state of *kabuki* against the background of the economic infrastructure within society. Based on this, statements concerning actors account for the largest portion of this book, and "the structure of [their] 'art' and its making" is clarified. However, this book does not simply discuss actors. It also focuses on the play-maker (*kyōgen tsukuri*), which was originally positioned as an actor's ability, and discusses the internal structure of *kabuki* works they created. In this regard, references are also made to the *kabuki* theoretical work *Sakusha Shikihō Kezairoku* and the instructional text *Sekai*

Kōmoku—both written in the Edo period (discussed in detail in the section **Surveys and Analyses of Two Documents from the Narrative Generation System's Point of View** in this chapter), and theories of related narratives and literature such as *nō* and *haiku* as theories on the "*shigumi*" of *kabuki*. *Shigumi* approximately means methods for creating the structures and developments of *kabuki* works and plays.

Hattori's structural theory of *kabuki* can be considered in comparison with the multipl narrative structures model the author has proposed. Alternatively, as it closely resembles the author's multipl narrative structures model, it can be viewed by incorporating the structural model of *kabuki* into the multipl narrative structures model. Another possibility is to understand the model of the structure of *kabuki* as an example of the multipl narrative structures model. However, the multipl narrative structures model is a model for narrative generation, and naturally, it places the concept of the "creator" at its center. Nevertheless, as described, Hattori convincingly states in this book that *kabuki* has traditionally (and presently) regarded the "actor" as more important than the "creator." How should this point be considered?—In this regard, however it would be useful to consider Hattori's concept of the "actor" to be different from—or greatly surpass or expand upon—the concept of actors as we imagine them today. According to Hattori, the "actor" in *kabuki* was not someone originally defined as a person who only acts; it was nearly the only position that made up a troupe. In other words, *kabuki* was born from troupes (*ichiza*s) as a gathering of actors. Watanabe (1989) also emphasizes in his writing that *kabuki* is a form of *geinō* born out of troups or *ichiza*s. Actors, in this case, therefore did not merely perform plays on a stage, but comprehensively undertook a range of jobs that accompanied this. We can then say that these duties also included a job that was equivalent to that of a "creator." In other words, *kabuki* did not originally have a "creator" as an independent person, and the so-called "actor" was not an actor who (merely) performed in the narrow sense in which actors are presently understood, but in an expanded sense, who also performed the roles of the creator, the director, and the producer in a troupe (*ichiza*). Accordingly, although they have the designation of being an "actor," they are not actors in the narrow sense that we imagine today.

This situation is not particularly mysterious from the perspective of the aurhor's multipl narrative structures model. In other words, in the multipl narrative structures model, the title of an agent as a person or organization involved in narrative generation is no more than a symbol used for convenience, and what is important is the actuality. Therefore, in this case, the agent known as the "actor" contains the "actor," the "creator," the "director," the "producer," or a swathe of other functions in the narrative. The entirety of this is fulfilled by several people known as actors. In this sense, the actor is not merely an agent operating just as an "actor." In more definite terms, the author has said that some actors are key actors while others are not, and so on and so forth, but will omit that here. In the preceding section, the author examined the structure of GIS with INGS, and the author noted there that GIS with INGS does not explicitly deconstruct the function of "creator," and also that in the narrow sense, a *geinōjin* character (entertainer, performer, or player character) equivalent to the "actor" may also simultaneously function as a creator. The reason this structure arose is probably that GIS with INGS is not a narrative generation system in the narrow sense, but no more than an "*geinō* information system." If "*geinō*" is placed at the center, the existence and concept of the "creator" in a narrow sense shifts to the background or is rendered dormant.

It should be noted that another reason the concept of the creator is not brought forward in *geinō* is that *geinō* works are often not new works that exhibit creativity in the way that we think of it. Many *geinō* works are repeated performances or revisions of existing works. In *nō*, a narrative *geinō* predating *kabuki*, the job of the creator existed more explicitly broken down than in *kabuki*, although there is considerable overlap with the actor, and many works that have been passed down clearly show the name of the creator. However, Zeami (c. 1363–c. 1443), the most important *nō* creator and its greatest theoretician, states that *nō* works should always use characters appearing in existing works or well known historical or fictional characters and add new innovations to them (Zeami, 1983). *Ningyō jōruri* and *kabuki* also follow this kind of tradition (although at the same time, *ningyō jōruri* and *kabuki* create narratives that use topics contemporary to the period more often than *nō*).

However, notwithstanding these circumstances, Hattori also writes about more specialized creators as those opposed to creators in the sense of actors who are also creators. This is due to the process observed as we pass through the early developmental stages of *kabuki*, in which the substance of the job the actor must do to also be a creator gradually becomes more complicated, and more importantly, *kabuki* narratives gradually become more complicated, thereby requiring the position of the creator in a narrow sense, the creator as a specialist. Amid these statements, Hattori refers to *Sakusha Shikihō Kezairoku* and *Sekai Kōmoku*, a *kabuki* theoretical work (the former) and an instructional text and knowledge base (the latter), both written in the Edo period, and also mentions the creator's mechanism and the mechanism within the creator's process for creating a narrative.

The final chapter of *The Structure of Kabuki* refers to something called "*kabuki* works." Like many *kabuki* texts, Hattori does not provide style-based commentary or explanations of "*kata*s," which mean a kind of typical and settled patterns or forms for various aspects in *kabuki*, on individual works performed on stage or individual genres, but rather pursues the core topic of what the fundamental principles of *kabuki* narratives are. Here, Hattori primarily considers "*jo-ha-kyū*", which has underlined Japanese *geinō* from ancient times and was positioned by Zeami as the most fundamental principle of *nō*. (Although it is diffilcut to explain the concept of *jo-ha-kyū* because it has deiverse meanings, oen of the imterpretations is "slow introduction, normal-speed development, and rapid finale." However, it is not simpley an idea according to speed. For example, "*ha*" means a kind of breakdown or interruption of the first state, i.e. "*jo*.") However, more interestingly, Hattori locates the fundamental principles of *kabuki* narratives in *haikai* as a literary genre of short poems including *haiku*, and moreover in the Danrin-ha (Danrin school) of *haikai*. Particular attention is paid to *renga*-like and *rengu*-like methods in which several individual *waka* and *haiku* are connected, rather than the structure and form of the individual *haiku*, and Hattori considers this method to display phenomenon coupling principles in narratives unique to *kabuki*. In addition, *renga* and *rengu* are two types of linked-verse forms, and *waka* is a genre of Japanese classical short songs.

Incidentally, Hattori describes, as a type of generative grammar, the process by which early *kabuki*'s structure as a Shōwattained more complexity than a simple structure containing only a single phenomenon and then one part became fixed as a *kata*. Descriptions like this have probably been influenced by the structuralism and narratology from when this book was written, but this is a characteristic no longer seen in later *kabuki* study, or in humanities study that includes it. However, such an idea is useful in identifying phases of complication in *kabuki*, in addition to narrative and *geinō* works in general, and merits reexamination from the perspective of the author's narrative generation system research in particular.

Tamotsu Wanatabe's Philosophical or Narratological Thought of *Kabuki*

Tamotsu Watanabe is a present-day *kabuki* researcher and *kabuki* critic, but unlike other purely intellectual researchers, critics, and commentators who provide impressionistic criticism, he is supported by a foundation of thought and philosophy, and moreover, is an author of timely *kabuki* commentary, making him a rarity. Watanabe has written sporadically on thought and philosophy (Watanabe, 2004,

2012, 2018), but here, the author presents and discusses the unique, particular thought and philosophy written in *Musume Dōjōji* (Watanabe, 1986), a research text and critique concerning the *kabuki* known as "*Dōjōji-mono*", which the author and other researchers are currently studying, and in particular, the work *Kyōganoko Musume Dōjōji,* which is considered a masterpiece of *kabuki* dance. (*Kabuki* script: *Kyōganoko Musume Dōjōji*, 1970; Video: Bandō, 2003)

The *kabuki* dance *Kyōganoko Musume Dōjōji* was not an original *kabuki* work in the beginning. At the same time as the ancient legendary tale called *Dōjōji Densetsu* ("The Legend of Dōjōji") passed through Buddhist fable, *nō* and other writing and illustrated scrolls came to be incorporated into *kabuki* and made into various works as *Dōjōji-mono,* and were crystallized as *Kyōganoko Musume Dōjōji* in *kabuki* dance. In this way, narratives concerning Dōjōji have a long history of transition stretching over more than a thousand years. In addition, Dōjōji is a temple in Kishū or Kii province in the Edo period (modern-day Wakayama prefecture) and *Dōjōji-mono* includes many narrative works in diverse genres based on *Dōjōji Densetsu.*

Specifically, *Kyōganoko Musume Dōjōji* forms a type of sequel, originating in the Legend of Dōjōji, an ancient legend relating to the temple of Dōjōji. Once the Legend of Dōjōji was remade as a *kabuki* dance, it developed into a work with its own contents beyond the original legend, even though many parts were omitted or simplified in the plot of the narrative. According to Yasuda (1989), the Legend of Dōjōji underwent the following transitions to become *Kyōganoko Musume Dōjōji* in *kabuki*: The narrative of Dōjōji has long historical transitions over one thousand years.

The oldest records of the Legend of Dōjōji can be seen in the scene 129, *Ki no Kuni Muro no Kōri no Ashiki On'na* [The Evil Woman from Muro District, Kii Province] (No. 129) (1974) in *Dainipponkoku Hokkekyō Genki (or Honchō Hokkekyō Genki)* (1974). The first half is a tale of female passion, in which a woman who fell in love with a young monk but was betrayed burned the young monk together with a bell, while the second half is a tale of salvation, in which the woman and young monk were saved by the power of Buddhist law. Next, act 14, scene 3 of *Konjaku Monogatari Shū* (1993, 1994, 1996, 1999a, 1999b) also describes this as *Ki no Kuni no Dōjōji no Sō Hōke wo Utsushi Hemi wo Sukueru Koto* [The Tale of a Monk of Dōjōji in Kii Province Who Transcribes the Lotus Sutra and Saves a Snake] (Vol. 14, No. 3) (1993). The story is mostly the same as in *Dainipponkoku Hokkekyō Genki,* although there is more interest in the young man's difficulties in the first half. *Genkyō Shakusho* (1965) describes the man as "Anchin" from Kuramadera (Kuramadera temple), but refers to the woman only as "a widow" and emphasizes the man's perspective. In the illustrated scroll *Dōjōji Engi* (1982), held by Dōjōji in Wakayama and said to date to the Ōei era (1394–1428) in the Muromachi period, the woman transforms while chasing the man and takes the form of a snake to cross a river and leaps into the river. This woman is also described in specific detail as "the wife of the master, a person known as Kiyotsugi Shōji," of the inn in "Masago, Muro District, Kii Province," and the perspective from the woman's side is growing. The name "Kiyohime" first appears in *Dōjōji Genzai Uroko* (Asada & Namiki, 1995), a *ningyō jōruri* dated around the 18 th century. The first volume of *Dōjōji Monogatari* (1982), which is a story composed of three volumes that was published in Manji 3 (1661), depicts a travel scene in which the monk makes a pilgrimage to Kumano Province (modern-day Wakayama Prefecture). The second volume reveals the story of the women transforming into a serpent out of passion and burning the monk to death, and the third volume presents the story of the woman who comes to perform a memorial service for the bell, transforms into a serpent and hides in the bell, and a tale of salvation in which the monk and the woman perform a memorial service by transcribing the Lotus Sutra. The scene in the tale of salvation where the bell is restored and the memorial service is performed was incorporated into the *nō Kanemaki* by

Kanze Kojirō Nobumitsu (?1435 or 1450–1516) (at present, this work is rarely played), and the later *nō Dōjōji* (1998) which is of unknown provenance, and simplifies the dialog and narration in *Kanemaki* and emphasizes the dance section. In *Dōjōji*, the serpent transforms into a courtesan and appears where the memorial service for the bell is being performed, then causes the bell to fall from hatred and hides within it. This serpent does not pray to enter Nirvana, but vents her hatred on the bell. *Kanemaki* focused on the male perspective, which views the woman's acts as evil, but *Dōjōji* gives rise to a perspective that examines the woman's sadness. To this point, the Legend of Dōjōji has transmuted from a tale in which the man has a spiritual awakening and attains Buddhahood to an expression of the woman's sadness and hatred, and later, the Legend of Dōjōji comes to be reread as a story from the woman's perspective.

In the early modern period, the Legend of Dōjōji comes to be incorporated into *kabuki* and *ningyō jōruri*. The *kabuki* dance plays with the bell of Dōjōji begin with *Katari Dōjōji* or *Ko-Dōjōji* (also known as *Furu-Dōjōji*). "Dōjōji Kane no Dan (Act of Dōjōji's Bell)" in *Kokon Hauta Taizen* (Kurihara, 1886) provides the major verse and prose for the *nō* plays *Dōjōji* and *Miidera* [Miidera Temple] (1998) and is considered a pioneering work in Dōjōji dance. The first *Dōjōji-mono* in *kabuki* is *Wakanoura Kataonami* (premiered in Genroku 2 (1689) in the Edo period). *Kabuki* and *ningyō jōruri Dōjōji-mono* developed under each other's influence. The first half of the tale of the Dōjōji has been reconstituted in performances to better integrate it with early-modern realistic developments, and today, only the part where Kiyohime turns into a snake and crosses the Hidaka-gawa (Hidake River) remains in dance plays in both *kabuki* and *ningyō jōruri*. However, the main interest in the tale of Dōjōji was instead focused on the sequel. In other words, interest in the Legend of Dōjōji moved away from the story as a whole and towards connections between the bell, the serpent, and the woman.

In the third month of Hōreki 3 (1753), *Musume Dōjōji* saw one type of completion as a *kabuki* dance in *Kyōganoko Musume Dōjōji*, danced by Nakamura Tomijūrō I (1719–1786), in the third scene of the *kabuki*, *Otokodate Hatsugai Soga* (1985) at the Edo Nakamura-za. This work inherited the form of the *nō Dōjōji* without change, but the verse and prose did not concretely show the Legend of Dōjōji itself and did not tell the tale of the past of the woman, the main character. This work focused, in the end, on showing the dance itself with its many changes, rather than the development of the story. In other words, the Legend of Dōjōji is not expressed in the verse and prose and has been relegated to information that provides the background for the work. In this way, *Kyōganoko Musume Dōjōji* wandered far from the original legend. Moreover, the tone of the dance is still bright and flashy on the surface. However, a terrible sense of fear leaks from it, and ultimately, the climax of the story explodes. This extreme contrast between the flashy gaudiness and terrible darkness forms the attraction of this work.

This important work that showcases *kabuki* dance is known as a work played by the representative on'nagata (male performers who play female roles) of the time, such as Nakamura Utaemon VI (1917–2001) in the Shōwa era (1926–1989) and Bandō Tamasaburō V (1950–) in the Shōwa era, Heisei era (1989–2019), and current Reiwa era (2019–) in recent years, and is a masterpiece that exhibits to the greatest extent the properties of *kabuki* dance in its delicateness, fineness, dramatic characteristics, and subtle changes, among other features.

Tamotsu Watanabe's *Musume Dōjōji* is a significant work that comprehensively describes the historical origin of *Kyōganoko Musume Dōjōji* and its development, combined with the paths of genius on'nagata actors such as Segawa Kikunojō I (1693–1749) and Nakamura Tomijūrō I in the Edo period. The statements in *Musume Dōjōji* mostly follow the process of the origin and development of *Kyōganoko Musume Dōjōji*, closely relating these to actors (on'nagata). The main on'nagata who appear here include Segawa Kikunojō I, who first developed *Musume Dōjōji* (*Momochidori Musume Dōjōji* (1960; Nagauta Genpon

Shūsei Kankōkai, 1927)), Nakamura Tomijūrō I, who completed *Kyōganoko Musume Dōjōji*, Segawa Kikunojō II (1741–1773) and III (1751–1810), Arashi Hinasuke I (1741–1796), Bandō Mitsugorō I (1745–1782), Nakamura Nakazō I (1736–1790), Ichikawa Uzaemon IX (1791–1819), Onoe Matsusuke I (1744–1815), Iwai Hanshirō IV (1747–1800) and V (1776–1847), Yoshizawa Iroha I (1755–1810), Sawamura Tanosuke II (1845–1878), and Ichikawa Dan'nosuke I (1684–1740).

The final chapter is the first to provide the reader with statements concerning the structure of the narrative of *Kyōganoko Musume Dōjōji* itself. Specifically, Watanabe deconstructs the acting (dancing) of the female main character of *Kyōganoko Musume Dōjōji* into the three elements of mind, gestures, and lyrics, and has created a table containing the concrete contents (values) for 11 parts of the modern performance form consisting of 14 parts in total. Here, "mind" is a term that expresses the true nature of the character (at the time), such as "girl" or "ghost," "gestures" means the type of physical movement, such as the dances *shirabyōshi no odori*, *hanagasa odori*, or *nō-gakari*, and "lyrics" refers to the song or music with lyrics, such as *hauta*, *taue-uta*, or *wakite-bushi*. As seen in Table 1 (equivalent in the structure to *Musume Dōjōji*, p. 479, table, upper 4 rows), the contents of the mind, gestures, and lyrics continually change for each part.

Table 1. Mind, gestures, and lyrics of the female main character in Kyōganoko Musume Dōjōji and the phases of changes therein

Scenes	心 (*Kokoro*: Mind)	振り (*Furi*: Gestures)	歌詞 (*Kashi*: Lyrics)
1. 道行 [*Michiyuki*]	亡霊(嫉妬—芝居) [*Bōrei* (*Shitto—Shibai*): Departed spirit (Jealousy—Play)]	生娘(流行唄) [*Kimusume* (*Ryūkō-uta*): Innocent girl (Popular song]	白拍子(能の道行) [*Shirabyōshi* (The *michiyuki* of *Nō*]
2. 乱拍子 [*Ran-byōshi*]	娘 [*Musume*: Girl]	白拍子 [*Shirabyōshi*]	能 [*Nō*]
3. 中啓の舞 [*Chūkei no Mai*]	娘	白拍子	能(鐘づくし) [*Nō* (*Kanezukushi*)]
4. 手踊り [*Te-odori*]	真奈児の娘の嫉妬 [*Manago no musume no shitto*: The jealousy of Manago's daughter]	生娘	端唄(娼婦のエロティシズム) [*Ha-uta* (*Shōfu no eroticism*)]
5. 鞠唄 [*Kemari-uta*]	娘	鞠つき(踊り) [*Maritsuki* (*Odori*)]	廓づくし(遊女) [*Kuruwa-zukushi* (*Yūjo*)]
6. 花笠 [*Hanagasa*]	娘	花笠踊 [*Hanagasa-odori*]	わきて節 [*Wakite-bushi*]
7. くどき [*Kudoki*]	恋する女(裏切られた女) [*Koisuru on'na* (*Uragirareta on'na*): Woman in love (Betrayed woman)]	娘	恋と嫉妬の小唄 [*Koi to shitto no kouta* (A song of love and jealousy)]
8. 山づくし [*Yama-zukushi*]	娘	羯鼓の踊り [*Kakko no Odori*]	山づくし [*Yamazukushi*]
9. 手踊り [*Te-odori*]	真奈児の娘の嫉妬	生娘	端唄(娼婦のエロティシズム)
10. 鈴太鼓 [*Suzu-daiko*: Bell and drum]	娘	鈴太鼓 [*Suzu-daiko*]	田植唄(早乙女踊) [*Taue-uta* (*Saotome-odori*)]
11. 鐘入り [*Kaneiri*]	本性 [*Honshō*: True character]	能がかり [*Nō-gakari*: *Nō*-like gestures]	能
12. 祈り [*Inori*]	——	——	——
13. 蛇体 [*Jatai*]	亡霊	能がかり	能
14. 押戻し [*Oshimodoshi*]	亡霊	能がかり	芝居とのつながり [*Shibai to no tsunagari* (The connection with the Dōjōji story)]

Watanabe proposes an extended table that adds the elements "costume," "props," and "physical points" to the structure of the dance in *Kyōganoko Musume Dōjōji* based on these three elements, and completes it by inserting values showing the concrete contents for these elements of "mind," "gestures," "lyrics," "costume," "props," and "physical points" for each of the 14 parts of *Kyōganoko Musume Dōjōji*.

Watanabe's work contains many keen observations, and the author has frequently cited it in his previous research (Ogata, 2018b, 2019c, 2020g; Kawai, & Ogata, 2019, 2020). With regard to *Musume Dōjōji*, much of the final chapter in particular is spent on considerations of thought and philosophy. Watanabe's analysis of *Kyōganoko Musume Dōjōji* is based on his thoughts described below, which the author deeply identifys with. The major aspects of his thought are demonstrated through the following quotations.

First, the combination of the elements of mind, gestures, and lyrics in the development of *Kyōganoko Musume Dōjōji* does not have a consistent method (Watanabe's "law"). Watanabe states the following in regard to this: "The six elements, including not only mind, gestures, and lyrics, but also the color of the costumes, props, and physical points, are in opposition in each scene, and moreover, they are independent between scenes. The contrast between what appears to be a ghost (mind) but is a girl (gestures), or between what appears to be a girl but is a courtesan (lyrics) becomes in other scenes what appears to be a ghost (mind) but is a girl (gestures), what appears to be a girl but is a courtesan (gestures), and what appears to be a theatrical travel scene depicting travel in *nō* then becomes a *ranbyōshi* dance while still remaining *nō*. This completely lacks logic, consistency, and unity, and has no law. If there were a law, it would be the law that there is no law. The law of "absence." Completely contradictory things and unrelated things are combined together to establish a relationship." **[p. 473]**

However, it is pointed out that the law of no laws has the following type of method: "Each of the elements in this table are like the colored glass in a kaleidoscope. The elements themselves, like the pieces of colored glass, have no relation to each other. They are scraped together and form a pattern, and if you give them a shake (after each scene), they create a different pattern. They create mostly accidental relationships from moment to moment. These relationships are not created by each part linking to the others organically, but merely by differences in shape. One scene is formed by these relationships. This kind of thing is repeated infinitely, and can be inserted or substituted freely anywhere." **[pp. 473–476]**

The lack of law here really means the lack of the following: "Here, things like an overall, coherent structure, topic, meaning, which modern artworks are certain to have, are completely absent. Usually, each of the elements operates together to express the topic or meaning of a work, and each of the elements are a part that forms the whole. No matter how many parts are gathered, they cannot make the whole on their own. There is a center that gathers the parts, and the whole takes form only through this center. This center is the topic and the meaning of the work." **[p. 476]**

However, Watanabe holds that *Kyōganoko Musume Dōjōji* lacks a center or topic or meaning in the sense set out above. ("However, this table lacks what is known as a center." **[p. 476]**) In addition, "What is here are independent parts, and just the arbitrary relationships created by the parts, and these meaningless relationships alone continue infinitely. Odd relationships, in which the parts are themselves the whole, and the whole is in other words the parts, continues." **[p. 476]**

This so-called methodless method, however, is not completely without target, but actually forms a strategy constituted with a view to a target. To explain, this is because Nakamura Tomijūrō I, who first created this work, was able to implement the aforementioned methodless method and thereby "absorb all manner of entertainment genres. *Nō*, *kabuki*, dance (*mai* and *odori*), folk entertainment (*minzoku geinō*), party songs (*zashiki-uta*), popular songs. It extends from the *nō* favored by the upper classes to the popular songs of the cities, and from the performances of *geinin* (entertainers) in the pleasure quarters to the local entertainment (*geinō*) in the tall-grassed countryside. These things were all absorbed." **[p. 476]** What is more, "What is important here is that there is one center and these things are not things that have been inserted towards the center. They have not been inserted, but synthesized. In other words, there was not a single center initially, and the thing itself was formed by synthesis. If there had been one center, the things that are all independent from one another would certainly not have formed a harmonious whole to this degree. They formed a harmonious whole because in the first place, they were nothing but synthesis." **[pp. 476–477]**

According to Watanabe, this is also a narrative structure he calls "absence—the structure of the void." "Everything has a structure that it can be placed into, and once it enters this structure, each of its parts become independent and assert their own existence. For example, the spoken passage in the seventh scene (*dan* or act) seems like a confession of love by a betrayed woman, but the woman seems like a girl and also like a courtesan, and each of these personalities assert themselves, and from this discord arises a multifaceted, multiplex vision of the woman. This vision is precisely what is composed of the relationships, and the individual relationships themselves form a single whole (the vision). The whole exists only within these relations. This is a property of the structure of the void." **[p. 477]**

Thus, this structure of the void is "a mishmash, a patchwork," **[p. 477]**, but "the fact that this mishmash, patchwork is neither merely a mishmash nor a patchwork is due to the presence of the relationships that make the whole by themselves, and what establishes these relationships is the structure of the void, which rejects any topic or meaning. The mishmash, patchwork itself is the source of life that makes *Kyōganoko Musume Dōjōji*." **[p. 477]**

The aforementioned may be summarized thus: *Kyōganoko Musume Dōjōji* is not a narrative of a type where central elements (important semantic elements) and other elements exist in orderly organization within a story with a climax, such as in Western narratives. It is simply a narrative in which diverse narrative elements bridging various genres are just synthesized, and that rejects its own structure in which these converge into the meaning or topic as a whole. This creates an organized, semantic whole, or rather a synthetic whole. This explains the structure of *Kyōganoko Musume Dōjōji* as a work in which various parts covered by "all manner of entertainment genres" are arranged haphazardly within the broader narrative framework (which largely consists of the undepicted scenes of the narrative).

The aforementioned is a discussion of the uniqueness of so-called narrative structure (in a broad sense) in *Kyōganoko Musume Dōjōji*, but Watanabe further continues to attempt a discussion of the characteristics of the dances in this work. Specifically, he examines the dance in *Musume Dōjōji* under the Shigayama school (a school of dance), and states that the dance is not a symbol that points to something outside of itself. ("The choreography in the descriptive explanation is overwhelmingly small, and almost all of it consists of gestures unrelated to the meaning of the lyrics or the dramatic setting." **[p. 480]**) This also underlies the aforementioned fact that narratives are merely a synthesis of various elements. Narratives and dances are not organized as tributes to something, but exist as independent entities, and neither exists as some sort of meaning or something that can be reconstituted as a "narrative" in the form of a certain aggregation of meaning. Speaking boldly, *Kyōganoko Musume Dōjōji* as discussed by Watanabe is a pure narrative that does not serve as a tribute to anything other than a narrative being a narrative, and is entirely different from narratives like advertisements, for instance, whose existence is evidenced by serving something other than the narrative. In the case of advertising, we are forced to be conscious of the effect in correlation with the purpose. However, pure narratives do not have a certain purpose, and "assessment" itself is not possible. For example, "The idea here is that dance is the language of the body itself, nothing more, nothing less. This language is uttered only from within the very movements of the body, and is not controlled by anything else. Of course, it cannot be translated into everyday language at all. Moreover, it cannot be concealed or diluted by psychological descriptions or emotions. If it rejects the 'meaning' of the lyrics on one hand and excludes 'psychological descriptions' and 'emotions' on the other hand, all that remains is just the very movements of the body." **[p. 481]**

In this chapter's author's experience of examining *kabuki*, he has frequently felt uneasy upon seeing modern *kabuki* from the Meiji era and later (which is often called *Shin Kabuki* (New *Kabuki*)). For example, in Kidō Okamoto's (1969) *Banchō Sarayashiki* [*The Dish Mansion at Banchō*], the master, who would not have punished the chambermaid he secretly loved for merely breaking one of the family's heirloom plates, could not forgive her for doubting him for even a moment, and in the end, brutally murdered her. This work retains the start and end of the original legend of the dish mansion, but replaces the history linking the two with a mentalist narrative. This psychological story could be construed as a simple explanation to draw out the morals of the brutal murder in accordance with *kabuki* aesthetics, but in present-day *kabuki* performances, it is usual for the psychological drama section to occupy a central position and the final brutal murder to be treated as a mere addition. This direction is the "progress and development" in modern *kabuki* or *Shin Kabuki* and was regarded as an achievement in *Kyōganoko Musume Dōjōji* and stands far from the meaninglessness or "nature as amusement" of *kabuki*; however, in truth, the progress and development is no more than a change biased towards the meaning of individual psychology and things. The author constantly has this impression. (Yet, if *Banchō Sarayashiki*, which the author has presented here as an example, is seen as a rare work that grotesquely materializes complicated psychological reasons, its narrative could be interpreted anew as something with a range exceeding a mere psychological drama. In the author's experience, he had a similar feeling when he saw Yūzō Yamamoto's (1887–1974) *Shin Kabuki, Sakazaki Dewanokami* (1977). Since the Meiji era, Japanese literature and narratives incorporated and applied "mentalist realism" that rushed in from the West, but this climate notwithstanding, *kabuki* plays have their own biases—perhaps we should think that the so-called "expressionist" bias exceeding realism from the Edo period and later still remained, without going into decline.)

So, Tamotsu Watanabe's views on *kabuki*, as presented above, offer an important suggestion regarding the author's narrative generation research for *kabuki*, or his narrative generation research in general, in terms of proposing narratives as a broad framework capable of arbitrarily synthesizing narratives from various genres.

Surveys and Analyses of Two Documents from the Narrative Generation System's Point of View

The author surveys and analyzes in detail *Sakusha Shikihō Kezairoku* and *Sekai Komōku* from the narrative generation system's point of view to be actually applied to narrative generation mechanisms based on INGS and GIS.

The Survey and Analysis of *Sakusha Shikihō Kezairoku*

In this section, based on the Ogata (2019e) and Hattoris's (1970), the author shows the structure and content of *Kezairoku*. Namiki Shōza II (?–1807), an Edo-period *kabuki* writer (and not Namiki Shōza I (1730–1773), who was famous for *Yadonashi Danshichi* (*Yadonashi Danshichi Shigure no Karakasa*) [The first performance 1768] and other kabuki works), penned the renowned work *Sakusha Shikihō*

Kezairoku (1972) under the name Nyūgatei Ganyū in Kyōwa 1 (1801). Namiki Shōza II is not regarded as an exceptional *kabuki* writer, but his theoretical abilities and writing abilities were regarded extremely highly (Iura, 1990; Kominami, 1998), and although this particular work is not especially long, it is known as a systematic, theoretical description on the creation of *kabuki*, which was rare in the Edo period and even more so today.

This book exists in several printed versions, each differing from the others (Iura, 1990, 1991), but here, the author mainly refers to the revised edition by Masakatsu Gunji (1913–1998) (Sakusha Shikihō Kezairoku, 1972), which is easier to obtain today, and its revisions, annotations, and commentary. However, Iura points out several issues with Gunji's revisions and annotations, so at times, the author also adds Iura's opinions on amendments. Moreover, the aforementioned Yukio Hattori's *The Structure of Kabuki* also mentions *Sakusha Shikihō Kezairoku*, so the author has also referred to it. The author (Ogata, & Ono, 2019) has evaluated this book and the following statements are also an expansion thereof.

The overall composition of *Sakusha Shikihō Kezairoku* is:

- [Introductory part]
- 作者差別之事 [*Sakusha Sabetsu no Koto*]
- 浄瑠璃作者連名 [*Jōruri Sakusha Renmei*]: Nyūgatei Ganyū regards Chikamatsu Monzaemon, Ki no Kaion, Bunkōdō (Matsuda Wakichi) (the mid Edo period), and Namiki Senryū (Namiki Sōsuke) (1695–1751) as *Kyōgen Sakusha Shiten'no* ("The Greatest Four Jōruri Writers") and shows other 56 jōruri writers.
- 歌舞伎作者之部 [*Kabuki Sakusha no Koto*]: This section shows 82 *kabuki* writers are listed and also includes the figures of *Goka Jūyō no Den* and *Nidō Shibetsu no Den* (which are introduced in the following part by the author of this chapter).
- 看板之心得 [*Kanban no Kokoroe*]
- 三都狂言替り有事 [*Santo Kyōgen Kawari Aru Koto*]
- 竪筋横筋之事 [*Tatesuji Yokosuji no Koto*]
- 四季人情差別之事 [*Shiki Ninjō Sabetsu no Koto*]
- 狂言之場行工合之事 [*Kyōgen no Ba Yuki Guai no Koto*]
- 作者金言之事 [*Sakusha Kingen no Koto*]
- 作者役場心得之事 [*Sakusha Yakuba Kokoroe no Koto*]
- 役者役場之事 [*Yakusha Yakuba no Koto*]
- 役場甲乙之事 [*Yakuba Kōotsu no Koto*]
- 役割番附之事 [*Yakuwari Banzuke no Koto*]: 36 *kabuki* actors are treated.
- 表八枚看板の事 [*Omote Hachimai Kanban no Koto*]
- 作者支配之事 [*Sakusha Shihai no Koto*]
- 江戸紋番附之事 [*Edo-mon Banzuke no Koto*]
- 大阪絵看板之事 [*Osaka Ekanban no Koto*]
- 一夜附の事 [*Ichiyazuke no Koto*]/役納る之事 [*Yaku Osamuru no Koto*]/作者出勤之事 [*Sakusha Shukkin no Koto*]
- 幕明の景様之事 [*Makuake no Keiyō no Koto*]/引上て下に置ぬ [*Hikiagete Shita ni Okanu*]/一件男女にて引しめる事 [*Ikken Danjo nite Hikishimeru Koto*]

The aforementioned includes several interesting items, but the one that has attracted attention for many years is the figure titled *Sakusha Kokoroe no Koto: Goka Jūyō no Den* (The Writer's Knowledge:

The Commentary of Five Flowers and Ten Leaves)," listed in "*Kabuki* Sakusha no Bu (*Kabuki* writer's section)." The figure is reproduced as Figure 10 by adding the English alphabet. The main text does not directly explain this figure, but Iura (1990) argues that the text following the figure offers clues to its meaning. Furthermore, the *Sakusha Shikihō Kezairoku* book was not originally published for and distributed to the public, but a secret book passed down through generations of *kabuki* writers and the associated people. However, given the central position occupied by this figure and the book's secret char-

Figure 10. Sakusha kokoroe no koto: Goka jūyō no den

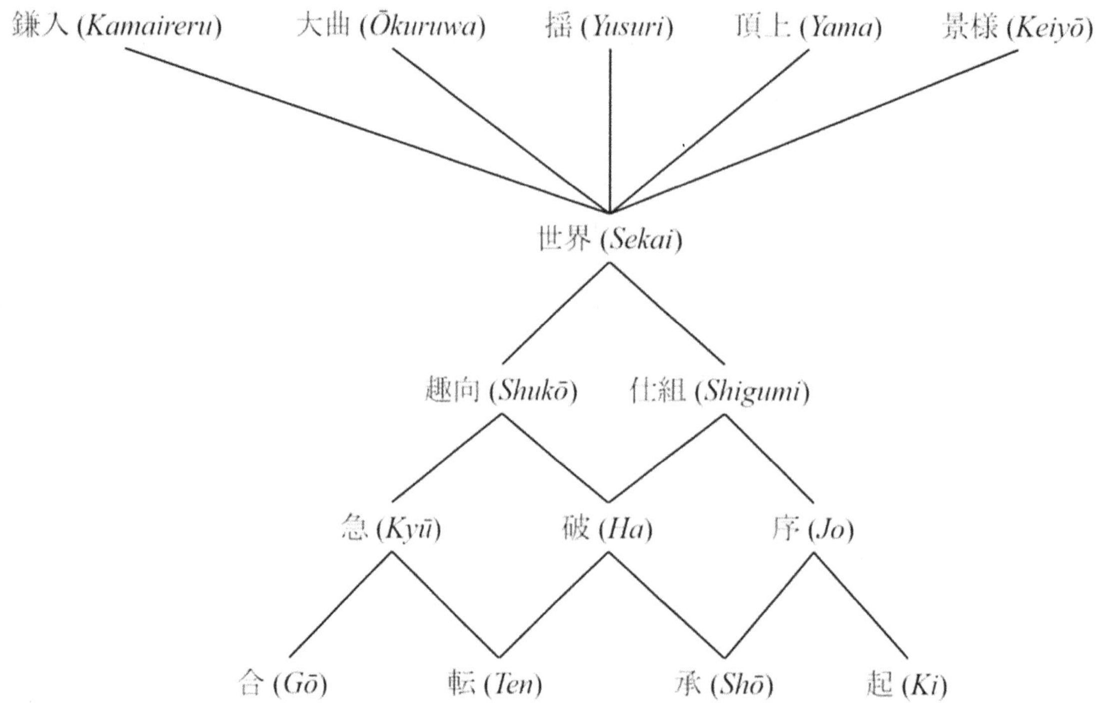

acter, it can be imagined that verbal explanation may have been deliberately omitted (Iura, 1990, 1991).

The *Goka Jūyō no Den* forms a type of method for creating *kabuki* plays. It is explained below with reference to Gunji's revisions and annotations and the aforementioned Iura paper. From this figure, the narrative world develops within the following narrative structure. This is equivalent to the five-stage composition for plays that is popular in *kabuki* and *ningyō jōruri.*

- **景様** (*Keiyō*): Gunji writes this using different characters that mean "appearance" and holds that it means showing the form of things in the opening act. Conversely, Iura believes that it rather means visual appeal, as is suggested by the similar terms *keigoto* (dance plays in *ningyō jōruri*) and *keibutsu* (seasonal features), among others. In other words, it is not the internal features, but the beauty of form, and mainly a good outward appearance.
- **頂上** (*Yama*): Gunji considers this to correspond to "gradation." Conversely, Iura argues that it means "the peak in the latter half within the first half of the entire drama." This is equivalent to the

peak in the "world" in *kabuki* works considered period pieces. Two major classifications of *kabuki* narratives are *jidaimono* (period pieces) and *sewamono* (pieces about everyday life). *Jidaimono* are narratives developed with a focus on the samurai class and nobles, rather than commoners, in periods earlier than when they were created, while *sewamono* are narratives developed based on the actual lives of commoners in the current period when they were created. Many *kabuki* works had a structure that develops as *jidaimono*-like narratives in the first half and continues with development as *sewamono*-like narratives in the second half. The *yama* referred to here, according to Iura, refers to the peak in the *jidaimono* section in the first half of the work.

- 揺 (*Yusuri*): According to Gunji, this means the thing that fluctuates or is the most active. Conversely, Iura states that it refers to a dispute that again overturns the situation that had settled down in the yama, and that this portion occupies a mini-scene-like position that is not equal to the other four parts and has a range of functions and contents that draw out the next scene.
- 大曲 (*Ōkuruwa*): According to Gunji, this is the scene where the narrative makes a single large development. Conversely, Iura argues that it contrasts with the yama, and means a major narrative development in the contemporary world.
- 鎌入 (*Kamaireru*): This means reaping, or in other words, the denouement (Gunji), or the scene with the resolution of all the complications and changes (Iura).

Kominami (1998) compares the narrative structure depicted in this figure with Freytag's structural diagram for Western classical plays. As a comparative theory, it has not been described in detail, but can be thought of as an awareness of a problem that views the five-stage structural diagram in *Sakusha Shikihō Kezairoku* in a more generalized form. In broad terms, this direction aligns with the perspectives and targets of my research, which understands *Sakusha Shikihō Kezairoku* as a narrative generation model to be a form of "narratology," or positions it as a collection of groups of techniques for a narrative generation system. In this sense, comparing the diagram in *Goka Jūyō no Den* with the *jo–ha–kyū* model, the *ki–shō–ten–gō* ("introduction–development–turn–conclusion") model, Propp's morphology model, Prince's narrative grammar model, and the cognitive science-based story grammar model, among others, and assessing it are both important tasks.

Returning to the contents of *Sakusha Shikihō Kezairoku,* what is considered important after the aforementioned portions in relation to the narrative structure and techniques in *kabuki* is the statement portion in *Tatesuji Yokosuji no Koto* (On Vertical and Horizontal Stripes). It states that *kabuki* narratives develop through a certain *shigumi* (plot) and *shukō*. Here, *shigumi* means an ordinary method of development along the lines of narrative pattern, and *shukō* means the creator's own unique innovations and ideas for that. *Goka Jūyō no Den* also uses the terms *shigumi* and *shukō*, and below that, describes the two representative patterns for methods of narrative development, *jo–ha–kyū* and below that, *ki–shō–ten–gō*. The *shigumi* means to basically follow these, while the *shukō* means inserting unique mutations and differences into it.

Note that Zeami viewed *jo–ha–kyū* systematically as a comprehensive basic principle that stretched from single *nō* narratives and their performance compositions that would last from one day to several days, to finely detailed compositions of a single sound or word. *Sakusha Shikihō Kezairoku* invokes *jo–ha–kyū* as a concept that shows (one face of) the basic structure of a single work. By contrast, *ki–shō–ten–gō* is a concept that indicates a typical (normal) composition of a four-line Chinese poem (a poem of four lines with five characters in each). The relationship between *jo–ha–kyū* and *ki–shō–ten–gō* in *Sakusha Shikihō Kezairoku* is such that *jo* is linked to *ki* and *shō*, *ha* is linked to *shō* and *ten*, and *kyū* is linked to *ten* and *gō*. The interpretation that the transitions between elements in *jo–ha–kyū* uses and is supplemented by *ki–shō–ten–gō* is also possible.

The *Tatesuji Yokosuji no Koto* section of *Sakusha Shikihō Kezairoku* also contains the following statement that appears to relate to the *sekai* ("world") seen in the *Goka Jūyō no Den* section of the same work, and the *naimaze* (blending) (Miura, 2000) relating to that: "As the world also serves to establish the main stripes, an ordinary world does not work on a play. Because the stripes are assembled, they are called vertical and horizontal stripes. Let us insert the horizontal stripes of Ishikawa Gozaemon into the vertical stripes of *Taikōki* [*The Life of Toyotomi Hideyoshi*] (1996). In addition, Kashiwade, Kinnari, Sakurako, Katsurako, Keyamura Rokusuke, etc., are all horizontal stripes. The vertical stripes are the *sekai* (world), and the horizontal stripes are the *shukō*. The vertical is *jo*. Even if the stripes are mixed until the end, it does not work. If the horizontal stripes are brought out towards the middle, they begin to work and make the play look new. This is an important aspect."

However, according to Urayama (1974), the method described here cannot necessarily be considered the blending of *sekai* in the original sense. The method called the *shukō* here means inserting into the *sekai* specific people in the narrative before the *sekai* is formed. In contrast, "blending" means a method of fusing two or more *sekai* in one narrative. (However, Iura notes that if the *shukō* part is expanded or systematized by the *sekai*, *Tatesuji Yokosuji no Koto* will approach blending in its original sense.)

Conversely, the above-cited Kominami mentions the ideas of *Tatesuji Yokosuji no Koto*. First, he adopts a more generalized view of plot as a method of structuring the world of the narrative with multiple layers. In addition, he understands this as an idea broadly seen in renga, the use of legendary backgrounds in painting, *waka*, *haiku*, and other genres of Japanese poetry and art. On the other hand, *shukō* is understood as a comprehensive, method-based concept relating to all elements involved in stage production, including production, acting, techniques, and costumes, which include a variety of techniques to create changes in the narrative.

Next, another figure *Nidō Shibetsu no Den* (The Commentary of Two Paths and Four Differences) which Gunji annotated with "Precepts for living on the creator's path," is depicted as Figure 11. Even here, the figure is not explained in the main text and Gunji's commentary is only brief, and it is therefore difficult to interpret accurately. The author would like to think that it can be reinterpreted as a type classification for characters, slightly modifying the understanding that it is a description concerning creators. A more detailed explanation than Gunji's commentary is provided by Iura (1991), but that also ultimately expands this *Nidō Shibetsu no Den* beyond its narrow meaning of the path of the creator and believes, "The creator must write these diverse people and the world they form," and "The world of people is a huge stage complicated and constructed by the fame and wealth, and the living, as well as the *jōkon* (ability to follow the Buddhist path), *jitsujō* (true feelings), *gekon* (lack of ability to follow the Buddhist path), and *hakujō* (unfeelingness), as shown in the 'two paths and four differences,' and various things."

In addition, the *Santo Kyōgen Kawari Aru Koto* (That Plays Change in Three Cities) section raises the *kawa* ("skin"), "*niku* ("flesh"), and *hone* ("bones") theory," which represents the character of *kabuki* performed in the three cities where *kabuki* was held with state approval in the Edo period as Kyō (*kawa*), Edo (*hone*), and Ōsaka (*niku*), but *Sakusha Shikihō Kezairoku* develops a generalized theory related to the characterization of *kabuki* works. It argues, "The stripes are the bones, the plot (*shigumi*) is the flesh, and the script (*serifu*) is the skin," and "If the bones, flesh, and skin are all continuous, they create an excellent work." According to Iura, *hone* refers to a strong, filled out character, *niku* refers to a richly filled out character, and *kawa* refers to a smooth, beautiful character. In addition, the points that should be considered important in *jidaimono*, *oiemono* (plays about upheavals in feudal clans),

Figure 11. Nidō shibetsu no den

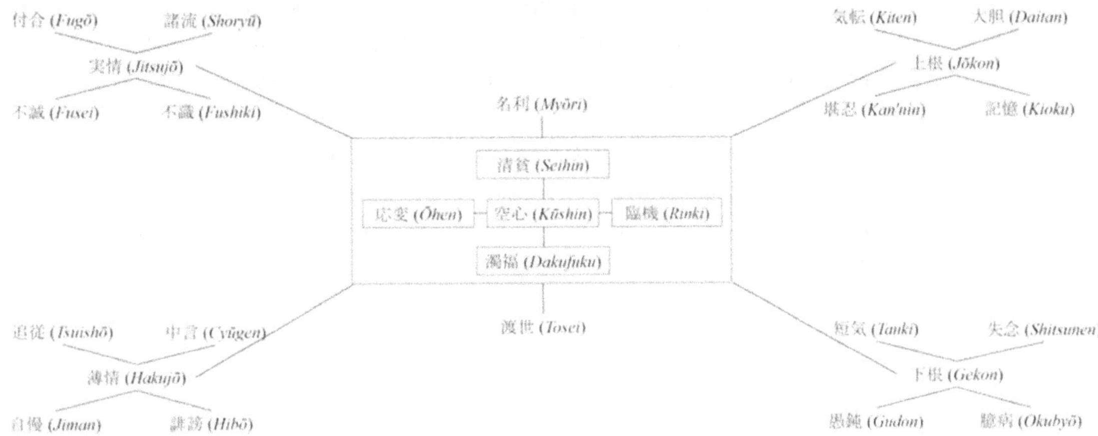

and *sewamono* (which according to Iura includes the everyday life parts in the entirety of a narrative, in which case they become everyday life scenes) are represented through the metaphor of bones, flesh, and skin, as "*Jidaimono* should have solid bones, *oiemono* should have abundant flesh, and *sewamono* should have fine skin." (Note that this theory of *kawa*, *niku*, and *hone* is thought to be originally based on discussion by Zeami (1983).)

The Survey and Analysis of *Sekai Kōmoku*

The author shows and discusses the content and structure of *Sekai Kōmoku* based on Ogata (2019e), Hattori's *The Structure of Kabuki* (1972), and other references.

The *sekai* (world) in *Sekai Komōku* means the typical or normal systematized knowledge for each individual narrative that has been used in creating *kabuki*. One *sekai* is a system consisting of the plot of the narrative, the main characters, the main settings (stage or spaces), the main temporal scope (period, season, etc.), and other elements. A set of *sekai*s can be thought of as a type of knowledge base shared between people involved in creating *kabuki* and the recipients. *Sekai Kōmoku* (1916) has been regarded as an Edo-period document concerning "*sekai*s" It shows in concrete detail how *sekai*s were understood *kabuki* at the time. *Sekai Kōmoku* was not a document providing abstract explanations of what *sekai*s are in *kabuki*, but a type of practical handbook or manual—albeit a secret one permitted for use only by top-class creators accepted by their fellows, and not for use by the general public. *Sekai Kōmoku* lists many practical examples of works that have been or could be used as raw material for *kabuki* at the time, using a consistent description method.

Katō (1997a, 1997b) states that *Sekai Kōmoku* began when the famous *kabuki* writer Kanai Sanshō (1732–1798)—viewed as a rival of the equally famous writer Sakurada Jisuke I (1734–1806)—gathered and documented the *sekai*s of many renowned *kabuki* of the time. *Sekai Kōmoku* is a type of practical secret text for *sekai sadame* (world-fixing)—a regular function held at the start of autumn every year in the mid-Edo period to determine the basic direction for *kabuki* plays that would be performed in each theater in the coming year—and was handed down mainly among *kabuki* creators who followed Sakurada Jisuke's line, and once the custom of *sekai sadame* faded towards the end of the Edo period, it came to be distributed more widely. In any case, it is not the kind of book that was in general distribution. Furthermore, *Sekai Kōmoku* exists in three editions—the National Diet Library (Kokuritsu Kokkai Toshokan) edition, the University of Tokyo College of Arts and Sciences edition, and the Tsukuba University edition—and although there are significant differences between two of these editions in particular, they are all considered to spring from the same root. The reader is referred to Katō's aforementioned paper. Incidentally, the University of Tokyo College of Arts and Sciences edition is a transcription of the National Diet Library edition by *kabuki* and *ningyō jōruri* researcher Kuroki Kanzō (1882–1930).

For reference, the aforementioned *sekai sadame* is discussed briefly in terms of its relation to the present research. Kominami (1998) describes *kabuki* performances particularly in the period when the *sekai sadame* function was still actually performed as a magnificent dramatic world over a span of one year. This idea connects with my GIS narrative generation = narrative production–consumption. GIS is positioned as a mechanism to cyclically create a narrative at a broader level than as a chain of individual narrative works, but the formulation of an annual performance strategy based on specific world under *sekai sadame* in the Edo period can be viewed as a concrete, exceptional example of a manner of operation for the GIS narrative production–consumption system. In Japan, this kind of idea of identifying a set of individual narratives as a narrative on another level has existed for some time. For example, Zeami viewed the concept of *jo–ha–kyū* in terms of the structure of a single *nō* work, or in terms of the structure of a fragment of acting by an actor, zooming in to the closest level, and conversely in terms of the overall structure of one or several days' performance, zooming out to the furthest level. For Zeami, *jo–ha–kyū* was not a theory or method fixed or tied to a specific level, but occupied a position as a universal formal and structural theory, and a method that went beyond identification with or affixing to any level. The author's concept of narrative generation is similar, and narrative generation is equally valid on various levels in the multipl narrative structures, from the micro to the macro. GIS with INGS is a term that indicates narrative generation in general, from the micro to the macro, in which INGS bears the generation of narratives as single units and GIS bears the production of compound or linked narratives. The aforementioned, *sekai sadame* to formulate a strategy for a chain of performances in a year-long time framework means, in the author's words, a strategy to ascertain a representative segment for compound or linked narrative production in the *kabuki* GIS.

Thus, according to the National Diet Library edition, *Sekai Kōmoku* is divided into the following four sections, each of which contains many or few world descriptions:

1. **歌舞伎時代狂言世界之部 (*Kabuki Jidai Kyōgen Sekai no Bu*) [58 *sekai*]**: (1) 日本武尊 (*Yamato Takeru*), (2) 神功皇后 (*Jingū Kōgō*), (3) 仁徳天皇 (*Nintoku Ten'nō*), (4) 衣通姫 (*Sotoorihime*), (5) 浦島 (*Urashima*), (6) 松浦佐用姫 (*Matsurasayohime*), (7) 聖徳太子 (*Shōtokutaishi*), (8) 大職冠 (*Taishōkkan*), (9) 天智天皇 (*Tenchi Ten'nō*), (10) 大友皇子 (*Ōtomo no Ōji*), (11) 大友眞鳥 (*Ōtomo no Matori*), (12) 百合若 (*Yuriwaka*), (13) 安部仲麿 (*Abe no Nakamaro*), (14) 弓削道鏡 (*Yuge Dōkyō*), (15) 中将姫 (*Chuujōhime*), (16) 田村丸 (*Tamuramaru*), (17) 融大臣 (*Tōru Daijin*), (18) 小野篁 (*Ono no Takamura*), (19) 小町 (*Komachi*), (20) 松風村雨 (*MatsukazeMurasame*), (21) 業平 (Narihira), (22) 道風 (*Tōfū*), (23) 北野御記 (*Kitanogyoki*), (24) 蝉丸 (*Semimaru*), (25) 将門純友 (*Shōmon Sumitomo*), (26) 惟茂 (*Koremochi*), (27) 源氏六十帖 (*Genji Rokujujō*), (28) 四天王 (*Shiten'nō*), (29) 奥州攻 (*Ōshūzeme*), (30) 殺生石 (*Sesshōseki*), (31) 保元物語 (*Hōgen Monogatari*), (32) 平治物語 (*Heiji Monogatari*), (33) 平家物語 (*Heike Monogatari*), (34) 頼政 (*Yorimasa*), (35) 伊豆日記 (*Izu Nikki*), (36) 木曽 (*Kiso*), (37) 義経記 (*Gikeiki*), (38) 源平軍並に頼朝治世 (*Geipei Ikusa Narabini Yoritomo Chisei*), (39) 曾我 (*Soga*), (40) 頼家治世 (*Yoriie Chisei*),

(41) 實朝治世 (*Sanetomo Chisei*), (42) 和田合戦 (*Wada Kassen*), (43) 鉢木 (*Hachinoki*), (44) 太平記 (*Taiheiki*), (45) 東山 (*Higashiyama*), (46) 甲陽軍 (*Kōyōgun*), (47) 出世奴 (*Shusseyakko*), (48) 小栗 (*Woguri*), (49) 苅萱 (*Karukaya*), (50) 三庄太夫 (*Sanshō Dayū*), (51) 愛護 (*Aigo*), (52) 角田川 (*Sumidagawa*), (53) 信田 (*Shinoda*), (54) 俊徳丸 (*Shuntokumaru*), (55) 望月 (*Mochizuki*), (56) 放下僧 (*Hōgesō*), (57) 大内之介 (*Ōuchinosuke*), (58) 甲賀三郎 (*Kōga Saburō*)

2. 御家狂言之内敵討之部ならびに類 (***Oie Kyōgen no Uchi Tekiuchi no Bu Narabini Rui***) **[7 *sekai*]**: (1) 義士傳 (*Gishiden*), (2) 伊賀上野 (*Iga Ueno*), (3) 浄瑠璃坂 (*Jōrurizaka*), (4) 亀山 (*Kameyama*), (5) 御堂前 (*Midōmae*), (6) 巌柳島 (*Ganryūjima*), (7) 非人敵討 (*Hinintekiuchi*)
3. 歌舞妓世話狂言世界之部 (***Kabuki Sewa Kyōgen Sekai no Bu***) **[75 *sekai*]**: (1) 一代男 (*Ichidai Otoko*), (2) 與之助 (*Yonosuke*), (3) お七 (*Oshichi*), (4) 曽根崎 (*Sonezaki*), (5) 重井筒(*Shigei no Tsutsu*), (6) お千代半兵衛 (*Ochiyo Hanbei*), (7) お花半七 (*Ohana Hanshichi*), (8) 三勝半七 (*Sankatsu Hanshichi*), (9) お染久松 (*Osome Hisamatsu*), (10) お菊幸助 (*Okiku Kōsuke*), (11) お梅粂之助 (*Oume Kumenosuke*), (12) 丹波與作 (*Tanba Yosaku*), (13) 山崎與治兵衛 (*Yamazaki Yojibei*), (14) お夏清十郎 (*Onatsu Seijūrō*), (15) 夕霧伊左衛門 (*Yūgiri Izaemon*), (16) 玉屋新兵衛 (*Tamaya Shinbei*), (17) 梅川忠兵衛 (*Umagawa Chūbei*), (18) 小金彦惣 (*Okane Hikosō*), (19) お三茂兵衛 (*Osan Mohei*), (20) 小春治兵衛 (*Koharu Jihei*), (21) 金村屋 (*Kanamuraya*), (22) 助六総角 (*Sukeroku Agemaki*), (23) 椀久松山 (*Wankyū Matsuyama*), (24) 淀屋 (*Yodoya*), (25) 本町二丁目 (*Honchō Nichōme*), (26) お亀與兵衛 (*Okame Yohei*), (27) お萬源兵衛 (*Oman Genbei*), (28) おきさ治郎兵衛 (*Okisa Jirobei*), (29) おしゅん傳兵衛 (*Oshun Denbei*), (30) 梅川新七 (*Umegawa Shinshichi*), (31) 小いな半兵衛 (*Koina Hanbei*), (32) お勘半九郎 (*Okan Hankurō*), (33) 樽屋おせん (*Taruya Osen*), (34) おさい道徳 (*Osai Dōtoku*), (35) 神田與吉萬字屋高崎 (*Kanda Yokichi Manjiya Takasaki*), (36) 金屋金五郎額の小さん (*Kanaya Kingorō Hitai no Osan*), (37) おはん長右衛門 (*Ohan Chōemon*), (38) 信田妻 (*Shinodazuma*), (39) 道成寺 (*Dōjōji*), (40) 富士淺間 (*Fuji Asama*), (41) 薄雪 (*Usuyuki*), (42) 清玄 (*Seigen*), (43) 鳴神上人 (*Narukami Shōnin*), (44) 七草四郎 (*Shichikusa Shirō*), (45) 高尾 (*Takao*), (46) 粂豊勝 (*Kume Toyokatsu*), (47) 天竺徳兵衛 (*Tenjiku Tokubei*), (48) あこぎ平治 (*Akogi Heiji*), (49) 累 (*Kasane*), (50) 東金茂衛門 (*Tōgane Moemon*), (51) 鑓の権三 (*Yari no Gonza*), (52) 五人男 (*Gonin Otoko*), (53) 團七 (*Danshichi*), (54) 黒船 (*Kurofune*), (55) 濡髪 (*Nuregami*), (56) 梅の由兵衛 (*Ume no Yoshibei*), (57) 源五兵衛 (*Gengobei*), (58) 湯の勘右衛門 (*Yu no Kan'emon*), (59) 帯刀男達, (60) 神祇組 (*Jingigumi*), (61) 銘職人 (*Mei Shokunin*), (62) 佛教大師 (*Bukkyō Daishi*), (63) 役行者 (*Eki no Gyōja*), (64) 弘法大師 (*Kōbō Daishi*), (65) 柿本記僧正 (*Kakinomoto no Kisōjō*), (66) 空也上人 (*Kūya Shōnin*), (67) 法然上人 (*Hōnen Shōnin*), (68) 新泉馬上人, (69) 日蓮上人 (*Nichiren Shōnin*), (70) 遊行上人 (*Yugyō Shōnin*), (71) 西行 (*Saigyō*), (72) 兼好 (*Kenkō*), (73) 一休 (*Ikkyū*), (74) 自然居士 (*Jinenkoji*), (75) 釋氏名目 (*Shakushi Meimoku*)
4. 神祇之部 (***Jingi no Bu***) **[3 *sekai*]**: 淡島 (*Awashima*),常陸帯 (*Hitachiobi*),高砂 (*Takasago*)

Each world consists of the following items:

① 役名 **[*Yakumei*: Role names]**: A list of major characters who appear in the *sekai*.
② 引書 **[*Hikigaki*: References]**: A list of ancient (authoritative) literature in which the *sekai* appears.
③ 義太夫 **[*Gidayū*: Gidayū jōruri]**: Lists the names of *gidayū jōruri* (*jōruri* works with dramatic recitation accompanied on the shamisen) works that use the *sekai*.

For example, the *Woguri* world, which has comparatively few items, in the *Kabuki* Jidai Kyōgen Sekai no Bu (*Kabuki* Period Play World Section), includes the following contents:

- **Role names**: 小栗判官兼氏實は孫五郎 (*Woguri Hangan Kaneuji, Jitsu ha Magogorō*), 横山治郎 (*Yokoyama Jirō*), 栗橋太郎 (*Kurihashi Tarō*), 戌亥局 (*Inui no Tsubone*), 青墓の長 (*Aohaka no Osa*), 鬼次新左衛門 (*Oniji Shinzaemon*), 常陸小萩 (*Hitachi no Kohagi*), 遊行上人 (*Yugyō Shōnin*), 池庄司時門 (*Ike no Shōji Tokikado*), 照手の姫 (*Terute no Hime*), 鬼王源太 (*Oniō Genta*), 後藤新左衛門 (*Gotō Shinzaemon*), 後藤三郎 (*Gotō Saburō*), 横山郡司(式部太郎) (*Yokoyama Gunji* (*Shikibu Tarō*)).
- **References**: 『小栗實記』 (W*oguri Jikki*), 『新編鎌倉志』 (*Shinpen Kamakurashi*).
- ***Gidayū jōruri***: 『小栗判官』 (*Weguri Hangan*), 『小栗判官軍街道』 (W*oguri Hangan Ikusa Kaidō*), 『今様小栗判官』 (*Imayō Woguri Hangan*), 『忠臣金短冊』 (*Chūshingura Kin no Tanzaku*), 『鬼鹿毛武蔵鐙』 (*Onikage Musashi Abumi*).

To explain, the aforementioned example means that the Woguri, which consists of 14 characters, was originally recognized as an independent world based on *Wguri Jikki* (*The Authentic Account of Woguri*) and *Shinpen Kamakura Shi* (*New Topography of Kamakura*), and W*oguri Hangan* (*Judge Woguri*) and the following works were formed based on this world. Many of us presently do not have detailed knowledge about *Woguri Jikki* or *Woguri Hangan*, but the writers or creators and associated people at the time when they were used in *sekai sadame* would have been entirely familiar with their contents. In fact, the contents of various narratives listed here combine to provide concrete form to the attributes and characteristics of the characters, the period (time), and place or stage (space) where the narrative is developed, and outlines of the narrative's story and plot in the world of *Woguri*.

According to Urayama (1974), *sekai*s did not exist as a continually fixed thing, but changed and interchanged over time. Using the discussion of vertical and horizontal stripes, that is, the *shigumi* and *shukō*, in the previously mentioned *Sakusha Shikihō Kezairoku* as an example, much of what was considered a *shukō* in this case were elements such as characters in *sewamono*. However, in *Sekai Kōmoku*, *sewamono* also form many of the worlds as previously described. From this, the very method of combining *shigumi* and *shukō* that was mentioned in *Sakusha Shikihō Kezairoku* may not be able to be termed blending as discussed, but when a *sewamono*-like narrative that forms the foundation for a *shukō*, as seen in *Sekai Kōmoku*, is established as a *sekai*, the aforementioned discussion of horizontal stripes–vertical stripes or *shigumi–shikō* approaches the discussion of blending methods or for strategy.

As a discussion relating to *sekai* in *Kabuki*, the author wishes to present research by Shigeo Moriyama (1914–2000), who analyzed *sekai*s and blending based on many actual works and subsequently presented a more generalized theory of *sekai*. Moriyama's (1993) *Tsuruya Nanboku: Naimaze no Sekai* [Tsuruya Nanboku: The World of Blending] gives hardly any conceptual explanation in the body, and relates in extremely concrete detail (or "in a surge") across more than 370 pages the state of blending in the various works of Tsuruya Nanboku IV (1755–1829), a *kabuki* writer from the Edo period, and to fully comprehend this, the reader is forced to read the group of works set as source materials and then make careful notes while reading through, but in the two pages of the "Afterword," a general explanation of blending finally appears. According to this explanation:

- *Naimaze* (blending) is destroying existing *sekai*s and reorganizing them as different worlds. Doing so not only demolishes the consistent order, patterns, and narratives of the traditional worldview, but creates a new, larger-scale, compound *sekai.*
- *Naimaze* historically began with mixed performances of the highlights of a few works, and gradually built up to the doubling or compounding of different *sekai*s from different works. However, this historical development or complication was influenced by the actors' desire for a diversity of performance, especially their desire to unveil transformation performances, in addition to the growing complication of audiences' wishes for dramatic performances.
- The evolution of creators' dramaturgic skills was linked to the discovery of a new drama in the conflict between the orders of multiple *sekai*s. Works that had been the norm were parodied, and works that appeared to be parties came to have meaning as new norms.
- Tsuruya Nanboku IV freely used *naimaze* as a conscious dramaturgic method. Nanboku used *sekai*s from various narratives and inherited, invaded, or transmuted them to give birth to one new *sekai* after another. Thus, the strategic operation of *sekai* leads to the birth of novel *sekai*s.

Moriyama's description of *naimaze* is concrete, using Nanboku as its raw material, but it is treated as a highly comprehensive narrative technique or strategy. The aurhot also thinks that *naimaze* is not a mere means for creating *kabuki* narratives, but the most important composition principle. Here, *naimaze* may be thought of as the narrative creation principle, by which the "*sekai*s"—characters, events, historical background, plot-lines, etc.—of existing narratives—myths, legends, folk tales, stories, history, etc.—are used as the base for creating a new narrative work through the synthesis and correlation of other narrative *sekai*s by a range of methods. This is an overarching term that contains diverse methods and techniques.

It should also be noted that Uchida, Kondō, and Funato (1987) propose the creation of a computer system from *Sekai Kōmoku*. The proposed system has functions to attach furigana (reading information) to all items that appear in *Sekai Kōmoku* and can search for the names of relevant works from the name of a character, among others. Although it is rather simple, the author presents it as a pioneering achievement.

Inspired by *Sekai Kōmoku* in the Edo period, the author is drawn to the idea that it could be possible to create a modern-day *Sekai Kōmoku*. In this chapter, the author discusses the scheme for *Watakushi Monogatari*, but from the perspective of using various knowledge sources as raw materials for it. They include a variety of events from ancient times to the early modern period and the present day, and certain types among them could be considered to already exist as worlds. In particular, early modern and modern history has several event units that can be recognized as worlds. Thus, as seen in the Edo-period *Sekai Kōmoku*, *sekai*s are often given concrete form through the writing as works of literature and narratives, rather than history texts. Based on this idea, the construction and establishment of *sekai*s through works of literature and narratives is needed for the early modern and modern periods. In any case, the author aims to compose a certain something among the knowledge sources for *Watakushi Monogatari*, either by borrowing the concept of a *sekai* as discovered by *kabuki*, or by referring to the composition of *Sekai Kōmoku.*

In the previous sections, the author attempted to consider *Sakusha Shikihō Kezairoku* and *Sekai Kōmoku* from the perspective of narrative generation system research, and as demonstrated in several places, these works have parts that overlap. In particular, the *Goka Jūyō no Den* and the theory of horizontal and vertical stripes (*shigumi–shukō*), which form the heart of *Sakusha Shikihō Kezairoku*, are intimately related to the concept of worlds. In the future, I aim to use my considerations mixing the two as a bridge to reach the construction of one or more methods within a narrative generation organization that utilizes *kabuki*'s narrative methods or GIS with INGS, incorporating sekais, naimaze, *shigumi–shukō*, horizontal and vertical stripes, and *jo–ha–kyū*.

THE CONCEPT OF *WATAKUSHI MONOGATARI* AND DESIGING INTERNAL AND EXTERNAL NARRATIVE GENERATION METHODS TO CREATE AND DISTRIBUTE NARRATIVE CONTENT

In the two sections that have thus far formed the bulk of this chapter, **A MULTIPLE NARRATIVE GENERATION MECHANISM THROUGH TWO TYPES OF NARRATIVE GENERATION SYSTEMS: GIS WITH INGS** and **NARRATOLOGY OF KABUKI: ORIENTED TOWARD NARRATIVE GENERATION**, the author first outlined the structure of GIS with INGS and in so doing established the *Watakushi Character* as a type of *geinōjin* resource with the aim and purpose of creating *Watakushi Monogatari*. In the subsequent section on *kabuki*, the author introduced several theoretical and practical studies and documents surrounding *kabuki* in the modern and Edo periods. On the basis of this, in this section the author will discuss *Watakushi Monogatari* as concretely as possible. First, the

Figure 12. The structure of this section

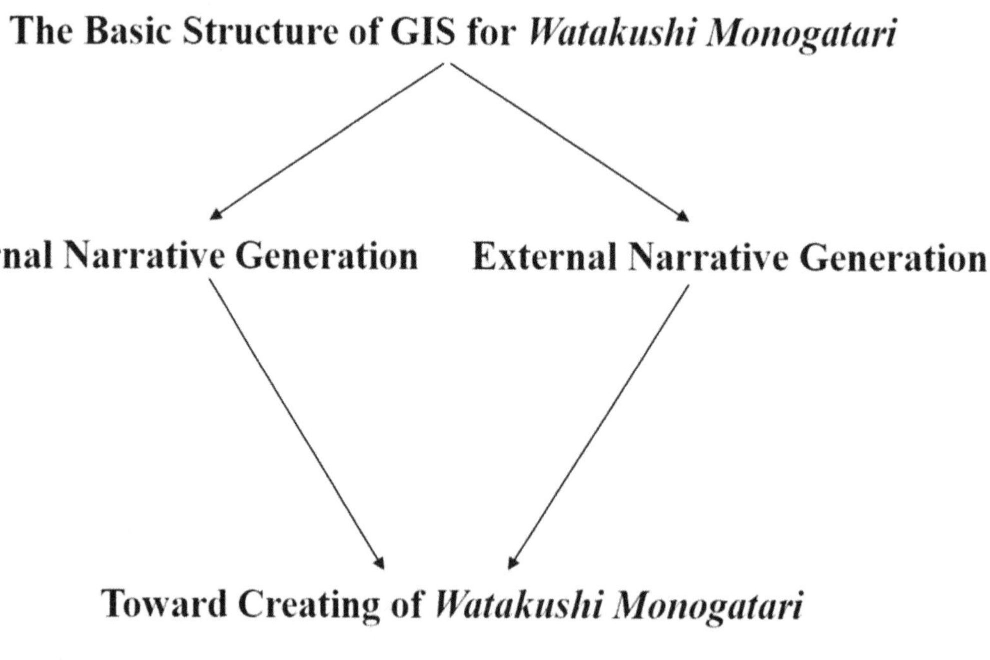

author will discuss how private knowledgea for *Watakushi Monogatari* fundamentally constitutes the content of the *Watakushi Character*, and then the author will describe the internal and external plans for *Watakushi Monogatari*. Here, the internal plan is a plan or direction in relation to aspects concerning the contents of narrative content (which is further separated into content and formal aspects). At the same time, the external plan directly signifies how the contents of the narrative content unfold externally and socially; at the same time, this also has a meaning in relation to the method by which it creates the narrative contents of *Watakushi Monogatari*. In other words, in *Watakushi Monogatari* the process of narrative creation and the process of societal distribution are not cut and dried, but rather they are more closely related in an essential sense, and the internal and external plans or directions of *Watakushi Monogatari* here are discussed as an intimately intertwined and integrative form. Based on the above, the structure of this section can be illustrated as shown in Figure 12.

The Basic Structure of GIS for *Watakushi Monogatari*

The GIS was originally envisioned as neutral and generic. In other words, it was a type of tool that was positioned as an applied system for narrative generation research. Thus, the traditional design excluded consideration of the specific "author" or "*geinōjin* character" elements that create and distribute narrative content through the GIS. During the design and development of a system that is intended to function as a tool or an application, the system designer or developer is usually the provider of that tool or application and, as such, he/she is sharply distinguished from those who use the tool to produce content. However, one of the most important goals of this study on narrative generation systems is to examine the author's production of personal and private works in a format that is more than a simple autobiography or a first-person novel. Moreover, in order to go beyond the level on which there is a tool, and that tool has a use, the author is also aiming for a tool form that is simultaneously a work of art. In this section's title, the term "internal narrative generation" refers to an author's generation of narratives on the personal and private levels, and signifies a narrative work or narrative content that is produced by processing a private narrative and *Watakushi Monogatari*.

The author considers this narrative and its generation on the personal and private levels from the perspective of *Watakushi Monogatari*'s conceptual goal. Having had many opportunities to discuss *Watakushi Monogatari*, the author has long had the idea for a personal study as well as the intention to bring that idea to fruition (Ogata, 2018c, 2018d, 2020d, 2020h). As set out earlier in this chapter, *Watakushi Monogatari* refers to a narrative based on the author's personal subject matter, objects, and topics, and is further generated using GIS with INGS.

To confirm *Watakushi Monogatari*'s basic concepts, the author would first like to describe it in a somewhat abstract sense, based on the various abovementioned papers. *Watakushi Monogatari* is created using GIS with INGS in a special way. The "I" (the *Watakushi Character*) that exists half within the GIS and half outside of it is able to perceive the fluid narrative created and expressed through GIS with INGS. Furthermore, the character is able to generate a "narrative as a fixed thing" on a separate level. Through these processes, the *Watakushi Character*, in conjunction with GIS with INGS, can collect all the fluid narratives generated at a specific point in a cycle or iteration of narrative generation using GIS with INGS to generate a "fixed *Watakushi Monogatari*" on a higher level. In short, GIS with INGS and the *Watakushi Character*, half of which is positioned internally and half of which is positioned externally, ultimately arrive at a fixed *Watakushi Monogatari* through a cycle of fluid and fixed narrative generation on a number of levels (although this provisionally final fixed narrative, when viewed from a higher level still, can be one phase of a fluid narrative).

Both the concept of narrative generation in *Watakushi Monogatari* through the flow and fixation described above and the personal and private knowledge that should be used to create *Watakushi Monogatari* are presented in the section titled **Concepts of Internal and External Narrative Generation** in the **BACKGROUND** in this chapter, and the author will explore them in more detail in the next section. Chapter 4 (Ogata, 2020h) in Ogata (2020e) includes a detailed explanation of the relevant framework.

Moreover, although the conceptual image of *Watakushi Monogatari* generation is shown in Figure 9, the author presents a more specific and newly extended view below. The *geinōjin* resource within the GIS includes a *Watakushi Character* as one of the *geinōjin* characters. This *Watakushi Character* is half artificial, by virtue of belonging to the GIS, while the other half corresponds to an "I" (*watakushi*) of reality, signifying that there are aspects of the *Watakushi Character* that automatically generate narrative as well as aspects of narrative that are directly generated by this "I" (*watakushi*) of reality. The *Watakushi Character* contains both private and personal knowledge related to this "I" (*watakushi*), and a discussion of its specific contents, namely what kind of content comprises each respective category, will follow in the upcoming section titled **Toward Internal Narrative Generation: The Creation of *Watakushi Monogatari***. As described above, the actual generation of narrative takes place through GIS with INGS and the *Watakushi Character*, and "fluidity and fixation" are concepts that are involved in this generation. Chapter 4 (Ogata, 2020d) in Ogata (2020a) offers a foundational view of this concept. At present, the author considers real narrative generation through fluidity and fixation to be represented by the following image. First, to the author, *Watakushi Monogatari* as the final narrative content is a fixed narrative. Indeed, this emerges from the interaction between fluid and fixed narrative generation. The specific generation process can be understood as follows (details of which may be subject to change in the future). That is to say, the *Watakushi Character* is constantly monitoring (as an artificial entity, this is primarily the *Watakushi Character*'s performative role) the narrative (automatically) being constantly generated by GIS with INGS (this has so far been viewed as fluid narrative). Based on these narratives (though, of course, not only these), the *Watakushi Character*, as a real entity, creates a fixed narrative. Even during this generation process, it may be possible to use an automatic generation mechanism in some form. In this sense, fixed narratives are assumed to be fragmentary, and none are very long. Their short duration allows those fragmentary fixed narratives to be subsumed again into the GIS with INGS and used for fluid narrative generation. Through this recursive repetition, the *Watakushi Character* creates the final *Watakushi Monogatari* as a larger fixed narrative.

Toward Internal Narrative Generation: The Creation of *Watakushi Monogatari*

In the sense of narrative content, the term "content" refers, in short, to what is written in that narrative. By contrast, the narrative form—for example, where a narrative is written in text, as in a novel—refers to narrative structure and the type of composition used (i.e., sentence length, word choice, whether action is emphasized over psychological description or vice versa, etc.). It is easy to imagine that every narrative has different narrative content. It is also true that not all narratives are written in the same form. However, there are a great many narratives that take very similar forms but have different narrative content. Perhaps changing or even innovating narrative forms is a task that requires a more conscious strategy than changes to or innovation in narrative content. Furthermore, while the formal aspect is a crucial element for the narrative, many authors are not explicitly aware of the narrative's form. Although various arguments can be made, narrative consists of both content and formal aspects. In the following pages, the author will discuss the narrative created by *Watakushi Monogatari* by separating it into two parts, namely the formal aspect and the content aspect, to clarify what the author means by internal narrative generation.

The Conception of Content in Internal Narrative Generation

The author will now expand on Ogata's (2020e) description of narrative generation research's development toward the internal, specifically the development and creation of *Watakushi Monogatari*. This section describes according to the following order, **Fundamental Knowledge Sources for the Narrative Content of *Watakushi Monogatari*, Two Directions of the Narrative Content of *Watakushi Monogatari* and Their Fusion: From Fluidity/Fixation To In Motion/Static**.

(1) *Fundamental Knowledge Sources for the Narrative Content of* Watakushi Monogatari

Although this does not constitute the narrative content of *Watakushi Monogatari* itself, the author will first consider the knowledge group that should be the source of its development materials. This knowledge group was first organized by Ogata (2018c, 2018d), and Chapter 4 (Ogata, 2020h) in Ogata's (2020e) provides an extended description. These are summarized in the section titled **Internal and External Narrative Generation** in the **BACKGROUND** section in this chapter. Here, as an internal concept of narrative, the author will provide what can be considered "material" for the as-is realization of *Watakushi Monogatari*'s narrative content; however, a worthwhile task for the future will be to provide such actual, specific material information in a more organized form and structure the narrative content material using fuller descriptions.

The following list shows the knowledge that is needed for *Watakushi Monogatari*:

【A. Temporal Division】
1. The temporal division of "Present Continuous"
2. The temporal division of "Past"
3. The temporal division of "Past Perfect"
4. The temporal division of "Historical"

【B. Text Categorization】
1. Event descriptions
2. Personal chronologies
3. Social and historical chronologies
4. Personal texts
    a. Published academic contents
    b. Unpublished contents related to academic activities
    c. Materials from conferences and classes
    d. Contents that was not published (contents stored privately without it having been presented or published anywhere)
    e. Lesson records and large volumes of notes from self-study, research gatherings, and other occasions written for record or contemplative reasons

【C. Personal readings】
1. Japanese literature from antiquity to the present (works: narratives, novels, poems, *waka*, *haiku*, etc.)
2. World literature from antiquity to the present (works: narratives, novels, poems, plays, etc.)
3. Japanese literature from antiquity to the present (commentaries, critiques, miscellaneous writings, etc.)

4. World literature from antiquity to the present (commentaries, critiques, miscellaneous writings, etc.)
5. Music (especially Western classical music and jazz, but perhaps the truly important are the other genres)
6. Video works like movies and TV
7. *Kabuki* and *ningyō jōruri* and texts (research and commentaries)
8. Works by Yukio Mishima (and related research and commentaries)
9. Other miscellaneous narratives

Of the above, **B-4** "Personal discourse" consists of the existing personal documents themselves, but the question exists as to how specifically these should be linked to *Watakushi Monogatari*'s narrative content. The same holds for "personal 'reading'" in **C**. Closely examining these is a task for the future; however, broadly speaking, there is both a direct as well as a more indirect method for addressing the issue of bridging this gap. The direct method makes direct use of individual texts, while the indirect method processes and uses the text in some way. A close examination of the latter, i.e., "personal 'reading,'" is also a task for the future, but it may be possible to use this in various forms, including based on direct quotation and processing. Under "personal 'reading,'" the author has listed only the novels the author himself has read from junior high school up to university (the majority of which are modern classics, although there are also a number of Japanese classics) with the intention of later using them as narrative content for what the author calls *Watakushi Monogatari*. In addition to this, in the future, specific consideration must be given to the question of how information should be added, processed, and used, and in Appendix 1 to this chapter, the author provides an expanded list containing detailed information. However, this list is a tentative version to aim at future completion for introducing it as narrative networking knowledge into GIS and the creation of *Watakushi Monogatari*.

Next, the knowledge that is to be examined, considered, and, more specifically, created as the substance of *Watakushi Monogatari*'s narrative content corresponds to items **1-3** in **B**, namely "Event description," "Personal chronicle," and "Social and historical chronicles" for each of the "time categories" that are recorded in A above. Here, **B-1** "event description" refers to the description or explanation of a relevant past event that the author has directly or indirectly experienced. The event description groups are arranged chronologically, corresponding to "Personal chronicle" and "Social and historical chronicle." Here, "Personal chronicle" refers to a chronicle that includes, at its core, a group of events that are directly related to the author, while also including a group of events that are not directly related to the author, but are closely, albeit indirectly, linked to the former set of events. On the other hand, the phrase "Social and historical chronicle" refers to a general chronicle that does not overlap with the author's life and can be very roughly separated into the following three fields: the first is the social and historical chronicle of Japan since the Meiji period; the second is the social and historical chronicle of an historically earlier Japan; and the third is the social and historical chronicle of relevant regions outside of Japan, that is, the world. These "Event descriptions," "Personal chronicles," and "Social and historical chronicles" have the following relationship. When one clicks, so to speak, on a particular description under a "Personal chronicle" or a "Social and historical chronicle," a relatively detailed description of the corresponding event description appears. There are two types of event descriptions, namely personal/private ("Personal/private events") and social ("Social/historical events"), which, respectively, correspond to **1**. "Present continuous time segments," **2**. "Past time segments," **3**. "Things spoken of in the past perfect tense," and **4**. "Historical periods and epochs" in A, namely "Time segments."

Unlike **B-4**, writing these anew is time consuming. It is likely that these descriptions alone would be sufficient to produce one or more narratives. However, these are not the final *Watakushi Monogatari*, but rather information and texts that form the materials for it. In the section titled **The Overall Behavior of GIS, Including a *Watakushi Character* with INGS**, the author describes how the fluid narratives constantly created by GIS with INGS constitute a collection of fragmentary narratives from which one or more fixed narratives are created through the *Watakushi Character*, but conversely, these event descriptions and personal as well as social and historical chronicles may be fragmentary narrative groups created by the *Watakushi Character*'s "I (*watakushi*)" and then forming a bridge to *Watakushi Monogatari*. A deeper, more substantial description remains a task for the future, but below the author sets a general policy for each of these items with respect to the information that should be described in the future. Thus, as indicated above, the author would like to stress that the specific events in each of the time categories are only described tentatively for the purpose of formulating a general policy; thus, not only may they be expanded and detail or greater specificity added, they may also be changed, removed, or restructured.

[The temporal division of "Present Continuous"] This is the present form that has been in existence since March 11, 2011 (Heisei 23 in Japanese era name), and it must continue until a certain point in time in the future. It is demarcated as a time segment, containing personal and private events as well as social and historical events. However, while the author uses the term "historical" as a shared classification term, social events, in this case, are either prior to or in the process of being historicized, and so they are an intermediate step. In addition, private and personal events, and social and historical events have overlapping boundaries that cannot be clearly separated from one another. This situation is common to other time segments as well; however, especially in the present, which is an ongoing time segment, personal and private events are strongly shaped by social and historical events, and so the area of ambiguity is also broader. The lists below provide examples of each event:

A. Private and personal events
   1. Personal and private situations during and immediately following the Great East Japan Earthquake.
   2. Personal and private situations at universities.
   3. Personal and private situations in the "countryside" in Japan (rural life).
   4. Private and personal "pair illusion" situations (Yoshimoto, 1968; Ogata, 2020h).
   5. Private and personal life situations amidst the novel coronavirus pandemic that began around the end of 2019 and spread globally (i.e., a further present ongoing situation in the present ongoing time segment).

B. Social and historical events
   1. Social (as well as political, economic, and cultural) conditions in Japan during and immediately following the Great East Japan Earthquake.
   2. Social (as well as political, economic, and cultural) conditions in Japan since the Great East Japan Earthquake and continuing up to the present.
   3. Global social (as well as political, economic, and cultural) conditions during the same period.
   4. The situation in Japan and around the world surrounding the novel coronavirus pandemic that began globally around the end of 2019 (i.e., a further present ongoing situation in the present ongoing time segment).

[The temporal division of "Past"] This is a time segment that was demarcated based on the author's memories, which span the period beginning on November 25, 1970 (Shōwa 45), and ending on March 11, 2011 (Heisei 23). As the author has mentioned several times, November 25, 1970, is the date on which the writer Yukio Mishima died; to the author, this is not only a date of social and historical significance, it also holds significance as the boundary beyond which the author gained a real awareness of his interest in literature and narrative, ergo, this is a date that marks an artistic period for the author on both a personal and a private level. Many important social and historical events occurred between that day and the day of the Great East Japan Earthquake; however, the precision or sharpness with which the author recalls them differs depending upon the strength of their connection to his personal and private events. As such, the author shall only cite those social and historical events with which he has a strong connection. The following are examples of personal and private as well as social and historical events that can be categorized as such:

A. Private and personal events
    1. Memories of personal and private events on the date of Yukio Mishima's suicide at the Self-Defense Force Base in Ichigaya, Tokyo, i.e., the Mishima Incident.
    2. Private and personal situations in "school" society.
    3. Private and personal situations in "family" society.
    4. Private and personal situations as individual illusions.
    5. Private and personal situations during the experience of "work" at a "company."
    6. Private and personal situations during the experience of "work" (i.e., research, teaching, etc.) at universities.
B. Social and historical events
    1. Yukio Mishima commits suicide at the Self-Defense Force Ichigaya Camp.
    2. Serial lynchings and murders are carried out by the Coalition Red Army and the Asama-Sanso Incident.
    3. The Japan Airlines Osutaka-yama (Mt. Osutaka) crash.
    4. The chain of collapse of socialist states and regimes that began with the overthrow of the Soviet Union.
    5. The Gulf War.
    6. Chaos and the sudden rise of emerging Islamic states.
    7. China's move from recovery to establishing economic hegemony.
    8. The collapse of Japan's "bubble" society.
    9. Emperor Shōwa's (Hirohito's) death.
    10. The Great Hanshin earthquake.
    11. The Aum Shinrikyo incidents.
    12. The acceleration of the general weakening of Japanese society under the Jun'ichirō Koizumi (1942–) and Heizo Takenaka (1951–) regimes.
    13. Economic and social stagnation and decline in Japan.
    14. The Lehman Brothers' fall.
    15. Emperor Heisei's (Akihito's) retirement and the Crown Prince's (Hironomiya's) ascension as Emperor Reiwa (Naruhito).
    16. The continuation of Shinzō Abe's (1954–) cabinet.

[The temporal division of "Past Perfect"] This is the period of time in which the author's "I (*watakushi*)" becomes indeterminate or where his "myself" objectively did not exist; however, people directly connected to the author, such as his parents, did exist, covering the period from around 1925 (Shōwa 1) to November 25, 1970 (Shōwa 45). Therefore, for many periods of time during which the author was not yet in the world, his personal and private events also did not exist. Even after the author came into the world, he, for a while, had no memory of private or personal events (although there are some such events that precede his own memory, like the date of his birth); memory-based descriptions of the author's personal or private events became possible from around the early 1960s, and they become clearer after 1964 (Shōwa 39).

A. **Private and personal events**
   1. Memories originate mostly from hearsay, the author's parent's event groups, and those of the people around them.
   2. Dim, vague memories of event groups (several episodes) from the author's early childhood (and the people surrounding them).
   3. A group of events from the years between kindergarten and elementary school, when the author's memory becomes clearer, and the events group (of people) related to it (numerous episodes).

B. **Social and historical events**
   1 The eve of World War II and the Pacific War.
   2 World War II and the Pacific War.
   3 The period of chaos and confusion following Japan's defeat.
   4 A period of strong economic growth.

[The temporal division of "Historical"] This denotes a temporal segment during which the "I (*watakushi*)" in the author's mind does not exist within it, the existences of the various people who are connected with the "I (*watakushi*)" itself become indeterminate, and moreover, time indistinctly mixes with the vast historical space that goes beyond the private. A criterion linked to the author's "I (*watakushi*)" may make it possible to jump to a certain place in history. This criterion is basically considered to be private and personal interest as well as curiosity, since there is no direct personal relationship of any kind between the events that are categorized here and the "I (*watakushi*)." However, if this relationship is not always direct, and spiritual elements can also be involved, it can be said that personal and private interest and curiosity may, for some reason, in the context of the author's own "I (*watakushi*)," also become private and personal necessities. For example, the fact that the land where the author's ancestors lived for a long time was (indirectly) on the side of the losers in the Battle of Sekigahara (1600), which is of great historical significance, and also (directly) on the losing side in the Boshin War (1868–1869), which was of vital significance over 150 years ago, both serve as strong psychological impetuses to remember these conflicts and their accompanying social conditions. Of course, the author neither directly experienced the battle or the war nor did the author has any kind of exchange with anyone who observed them; however, such a statement can be made to the extent that it is not impossible for events of that kind to stand out as leaving a somewhat prominent impression among the many groups of events that are, for example, arranged chronologically in textbooks.

There are many other issues that must be examined and compiled with respect to these social and historical events (and the same can also be said for all other types of events). A systematic discussion on this topic is a task for the future; however, it is important to mention the difference in the character of the so-called universality of social and historical events. In other words, this is the extent to which a social or historical event is global or, conversely, the extent to which it is local. Supposing there was an incident or event that had a fairly equal impact on all the regions of the world, this event would have a highly global character. The novel coronavirus that is currently sweeping the globe is certainly likely to be such a highly global phenomenon. Conversely, there are social and historical events that have only a localized impact on a particular region. For example, many of the events that occurred in Edo-period Japan were very localized events when viewed on a global scale. However, the social and historical events here are to be taken up and organized neither from the perspective of universality nor from the perspective of global to local ranking; rather, they are things to be extracted and organized from among a large number of events, from the perspective of this performative "I (*watakushi*)" character that exists within a specific, limited time and space. Thus, if we say they have a hierarchy of importance (even if this is no more than a simple reflection of the amount of description), it would mean that these events are organized from the personal and private perspective, which is distinct from the global–local or universal–specific perspective. In the author's case, these social and historical events tend to be organized around narrative and literature, and below is a mixture of general historical nomenclature and period-specific characteristics of narrative, literature, and theater and the performing arts (it is particularly odd that the descriptions of each period include both parts, incorporating narrative and literature as well as theater and performing arts and their historical attributes). This may need to be reexamined and improved in the future. Furthermore, since there is a no definitive picture of a method for organizing social and historical events in the world, the author records only topics that are of interest to him below, in no particular order.

A. Social and historical events (Japan)
    1. The society and history of the Japanese islands and their surroundings prior to the genesis of written language—The Jōmon and Yayoi periods.
    2. The birth of Chinese civilization and culture, particularly the period during which the Japanese language was born and explored as a system of written language.
    3. The age of ancient ballads and myths.
    4. The Heian period, which witnessed the creation and explosive development of Japanese poetry and "*monogatari* (narratives)."
    5. Society in the age of the samurai, from the Kamakura period to the Muromachi period, which witnessed a multistage transition from *monogatari* to "*setsuwa* (legends)" and then to "*katarimono* (narrating arts)."
    6. From the Warring States period to the era of the three revolutionary figures, namely Oda Nobunaga (1534–1582), Toyotomi Hideyoshi (1537–1598), and Tokugawa Ieyasu (1543–1616).
    7. The Edo period, in which there was a sphere of stability and circulation.
    8. The independent world of Japanese puppet theater (*ningyō jōruri*) and *kabuki* in the Edo period.
    9. [Overview of history] The history and status of theater and the performing arts in Japan.
    10. [Overview of history] The history and status of literature and narratives in Japan.

B. **Social and historical events (world)**
    1. The East Asian sphere centered on ancient China.
    2. The South Asian sphere centered on ancient India.
    3. The ancient Middle East.

4. The European world from ancient Greece to ancient Rome.
5. The golden age of China (i.e., the Tang and the Song dynasties).
6. The Western world and the Mongolian Empire.
7. India before colonization, and Southeast Asia and China.
8. Islamic society.
9. Modern Europe and colonialism.
10. The existence of the United States of America.
11. The existence of Africa.
12. Social ideologies and their conflicts, i.e., democracy and socialism, capitalism and socialism, and liberalism and imperialism.

(2) Two Directions of the Narrative Content of *Watakushi Monogatari* and Their Fusion: From Fluidity and Fixation to In Motion/Static

The concepts of fluidity and fixation are not related to narrative content; instead, they comprise an idea that is related to a narrative generation strategy. However, the author also applies these contrasting concepts to narrative content aspects as they relate to *Watakushi Monogatari*.

For that reason, the author will modify the terms a bit, terming the narrative content concept corresponding to fluidity "in motion" and the term corresponding to fixation "static." Motion and static are two polar directions relating to narrative content in *Watakushi Monogatari*, with the text's final narrative content being created through their fusion. In this way, the terms "in motion" and "static" are both concepts that are involved in creating *Watakushi Monogatari* as well as polar concepts for controlling *Watakushi Monogatari*, so to speak. However, unlike fluidity/fixation, motion and static are involved in ascribing direction on a level that is close to the narrative's content.

What then are the respective characteristics of "in motion" and "static" in narrative content? In what ways are "movement" and "stillness" respectively realized in relation to *Watakushi Monogatari*? This is not the sort of question for which there is an objective answer, and so the author's answer must be woven together in a form that is either in line with or constrained by his preferences. In this way, the closer one gets to a narrative's content aspect, the stronger the need to weave together or create solutions based on one's own personal and private interests and preferences. Although this is no more than an initial consideration, the author proposes the (personal and private) idea of linking "in motion" and "static" to "a picaresque novel-type narrative" and "a speculative novel-type narrative," respectively.

First, the author will explain how in motion = a picaresque novel narrative. When the author graduated from the various stories "aimed at children" (or arranged for children), such as fairy tales, folk tales, and mystery and science fiction novels, among others, the works that held the most appeal around the time the author began reading "adult" novels were those by Ernest Hemingway (1899–1961), André Malraux (1901–1976), and Takeo Arishima (1878–1923). It seems to the author that a kind of "adventure" (including in the symbolic sense) resides in each of their novels. The author feels that their novels depict characters who take an active role in terms of seizing control of their own life directions and taking action that reflects their own ideas, concepts, and intuitions. For example, in Arishima's (1964) relatively short novel *Umareizuru Nayami* [Troubled Birth/The Agony of Coming into the World], we witness the emergence of a type of character who, rather than assuming other people's affairs through, for instance, psychological confrontation and personal relations, resignedly tries to take life actions according to his fate, although his life changes fatalistically (in *Oedipus* (1972) by Sophocles, which Aristotle (384–322 BC) references in *Poetics* (Aristotle, 1997), the protagonist's adversary is the same "fate") as a result of conflicts arising from his ideas, concepts, and intentions. This novel does not belong to the adventure genre, but the author considers it to be an extremely adventurous novel that tells an adventurous story, both at the time of its publication and still today. *Aru On'na* [A Certain Woman] (Arishima, 1995), which is by the same author, is a similarly adventurous novel. The author also perceived *For Whom the Bell Tolls* (1994), *The Human Condition* (Malraux, 1990), *Gone with the Wind* (Mitchell, 2014), and *Wuthering Heights* (Bronte, 2002) as adventurous novels with adventurous narratives.

When adventure or adventurous novels and stories are considered in this broad sense, just about any novel or story can be viewed as one or the other. There is a particularly strong trend of saying that novels and stories that evince great strength, so to speak, were so from the outset. Although the author thinks that this idea is partially correct, the characteristics of the author's concept of this type of adventurous novel or story can ultimately also be seen in *kabuki* and puppet theater (particularly in the Edo period) as well as in various Yukio Mishima's novels (e.g., *Ao no Jidai* [*The Age of Blue*] (Mishima, 1950), *Kinjiki* [*Forbidden Colors*] (Mishima, 1951, 1953), *Kinkakuji* [*The Temple of the Golden Pavilion*] (Mishima, 1956), *Honba* [*Runaway Horses*] (1969b), etc.). From the rather broad view that all novels and stories are adventurous ones, the author would like to briefly consider a slightly more limited view, that is, just what (in his estimation) are adventure novels or adventure stories?

When considering this, the term "picaresque novel" comes to my mind. In a narrow sense, this refers to a type of novel that was popular in Spain in the 16th and 17th centuries, and the author shall expand on the concept a little here, treating it as a "picaresque novel or story." Under this broad definition, there are many picaresque novels and stories among the literature from other periods and regions. For example, as stated above, Mishima's *Ao no Jidai*, *Kinjiki*, *Kinkakuji*, and *Honba* all seem to the author to be picaresque novels and stories. The protagonists in these novels include some who are not so-called "bad guys" and who all try to act and go about their lives according to their own beliefs, ideas, and outlooks. These beliefs, ideas, and outlooks are, in many cases, not the kind that other people and society at large admire; on the contrary, they are the kind that are reviled and hated. Many of Seichō Matsumoto's (1909–1992) novels (he wrote several Japanese mystery novels, in the broad sense of the genre, that many people still read, such as *Yakō no Kaidan* [Stairs That Shine at Night] (Matsumoto, 1961), *Linkage* [Renkan] (Matsumoto, 1962), *Warui Yatsura* [Bad Sorts] (Matsumoto, 1980), *Kurokawa no Techō* [Pocketbook of Black Leather] (Matsumoto, 1981), etc.) are more directly picaresque. These Matsumoto novels' protagonists are criminals, but they each have a strong will to act and live their lives on the basis of their beliefs and ideas. Moreover, in many cases, the detective role, which the typical mystery novel's protagonist would fill, is relegated to minor characters in these novels, with the reader's strongest emotions being reserved for the criminal protagonist instead.

When the author considers the characteristics of those novels and stories that he deems picaresque, the author devises character categories such as "the unrepentant character" and the "character who rushes in." These traits make for appealing characters. The unrepentant character, in particular, is heavily influenced by *kabuki*. *Kabuki* characters are significantly stylized; heroes are heroes and bad guys (villains) are bad guys (villains). Moreover, in *kabuki*, the bad guy (villain) is central to the play, with the best actors often playing those roles. Indeed, in *kabuki*, unrepentance is a villainous characteristic; hence, the bad guys' actions and lives are villainous to the end. In many cases in *kabuki*, the bad guys (villains) are ultimately punished with tragic deaths, yet audiences cheer and applaud or show enthusiasm for them for their consistent behavior as bad guys (villains). As character traits, unrepentance and rushing in are more important to the author than classifying other character traits, such as whether a character is active and socially appealing or, conversely, whether a character has no social appeal whatsoever and is repellent. Since the protagonists in the original picaresque novels are so-called villains, naturally, they are shunned by society, and they bear a black mark, but the author is desirous of seeing unrepentance and the headstrong character trait paired with the so-called "B-grade" or "averageness" and "(though contradictory) the characteristics of a character that does not stand out in one's memory." A protagonist who is a wholehearted "bad guy" is a reversal of the typical protagonist characterization as a perfectly sensible, respectable person; that is, the two are alike after all. The author would like to see a protagonist character who is simultaneously a "bad guy" and an ordinary, average person. The picaresque novel (or story) the author is seeking is, as described above, one that balances the "unrepentant, headstrong" characteristics with those of the "ordinary, average" person.

The author will now discuss another characteristic, namely "static," that the author is seeking in *Watakushi Monogatari*'s narrative content. Static is symbolized by speculative novel narratives. As used here, the term "speculative novel" is the author's own; that is, it is distinct from the science fiction genre, referring instead to novels and stories that contain a substantial amount of or focus on philosophical thought or critical discourse. Static = speculative novel narratives. Broadly speaking, by "speculative fiction," the author means novels and stories that feature a character who is opposite to those in the picaresque novels that were discussed above. That is, if in a picaresque novel, the protagonist's fundamental strategy involves taking action, rushing in, making breakthroughs, and maintaining a non-reflective psychological state, the speculative novel narrative is without action and is therefore reflective and static. Speculative fiction stories' protagonists are motionless, static, and stagnant, constantly reflecting and indulging in internal thoughts. If picaresque novels' protagonists "act without thinking," then those in speculative fiction stories "think without acting." However, if we consider the behaviors that are depicted in picaresque novels to be external, while the actions in speculative fiction are internal, the difference between the two can be attributed to a difference in behavioral direction; taking the above picaresque and speculative fiction narratives to be "external picaresque novel narratives" and "internal picaresque novel narratives," respectively, these distinctions may constitute a broad internal classification of picaresque novel storylines. Of course, this is the author's idea. We can conclude here that speculative fiction stories, which the author has called external picaresque novel narratives, are narratives in which the protagonists' internal actions, that is, their thoughts, are extremely active and unreflective.

Which novels or stories, then, does the author feel correspond to speculative fiction or internal picaresque novel stories? Japanese author Yutaka Haniya (1909–1997) wrote only a few novels; his most voluminous works were essays, in which the most important topics pertained to the relationship between politics and humanity. *Shirei* [Death Spirits] (Haniya, 1998) is among Haniya's novels; he wrote it intermittently, starting in his youth and continuing into his later years, but he never finished it. *Shirei* is set in pre-Pacific War Japan, and several characters are developed through ideological dialogue and philosophical considerations about humanity, the universe, and politics. In his novels and stories, Haniya proposes philosophical concepts such as *jidōritsu no fukai* (the discomfort of the law of identity) and *kyotai* (the void). *Shirei* is contemporary Japan's most important work of speculative fiction. Other Japanese novels, especially those from the contemporary era, including Hiroshi Noma's (1915–1991) (1983-1984) *Seinen no Wa* [The Ring of Young People] and Yukio Mishima's *Kinkakuji* and *Utsukushii Hoshi* [A Beautiful Star] (Mishima, 1962), seem to fall into the speculative fiction or internal picaresque novel stories category. Furthermore, Yukio Mishima's *Taiyō to Tetsu* [Sun and Steel] (Mishima, 1968) represents a new speculative narrative style, seemingly fusing the novel, essay, and poem forms. *Seinen no Wa* is an extremely long novel that is set in Osaka, Japan, before the Pacific War. The plot develops extremely slowly and somewhat redundantly, with multiple protagonists engaging in a great many philosophical discussions about humanity and political issues. In Mishima's *Kinkakuji*, the introverted, unsociable protagonist creates his own unique, dogmatic notion, and a philosophy of "beauty" based on his own complexes and sense of inferiority; on the basis of this, he then sets fire to the real-life *Kinkaku* (Golden Pavilion), which is a symbol of corrupted beauty, and in the final scene, he feels a sense of bliss watching the Golden Pavilion burn. Mishima's *Utsukushii Hoshi* has not been translated into English, but a magnificent cosmology of UFOs unfolds through several characters' speech, damaging the novelic structure. Among modern and contemporary literary works outside of Japan, Victor Hugo's (1802–1885) (1982) *Les Misérables*, Herman Melville's (1819–1991) (2009) *Moby Dick*, Fyodor Dostoevsky's (1821–1881) *Crime and Punishment* (Dostoevsky, 2014), *The Brothers Karamazov* (Dostoevsky, 2003), etc., Thomas Mann's (1875–1955) (2005) *The Magic Mountain*, and Hermann Hesse's (1877–1962) *Demian* (2013), *Steppenwolf* (2015), etc., all seem to fall into the category of stories the author calls speculative fiction or internal picaresque novel narratives. In *Les Misérables*, the narrator pauses the plot (for quite some time) to observe the development of external events, often providing lengthy commentaries on particular topics (especially the monastery and the sewers). This novel is closer to "explanatory" than "speculative" (since it offers an overwhelming amount of knowledge or is concerned with presenting details or special knowledge about specific topics), and the author would like to cite this as a special form of speculative fiction or internal picaresque narrative. Hesse's *Demian* discusses the possibility of a dualistic ideology and philosophy similar to ancient Persia's Zoroastrianism through dialogue between the young boy Demian and the narrator, Sinclair, which helps Sinclair overcome the spiritual crisis of his youth.

To summarize the above, what the author seeks from the narrative content of *Watakushi Monogatari*, as well as the policy or strategy at the macro-level of its production, is, on the one hand, a picaresque novel narrative or an external picaresque novel narrative, and, on the other, a speculative fiction story or an internal picaresque novel narrative. Moreover, the author seeks a story in which both these characteristics are present. When considering such a coexistence, the author is reminded, for example, of Yukio Mishima's *Kinkakuji*, which was mentioned above. This is a type of picaresque novel in which the protagonist sets fire to the Golden Pavilion in a grand, *kabuki*-like event. More broadly speaking, as a whole, it is also a kind of adventure novel (in which as a player in an adventure game the protagonist achieves the goal of setting fire to the Golden Pavilion in the final scene after a series of trials and errors). At the same time, the protagonist is a brooding, introverted boy/young man who is tormented by various complexes, including one relating to his appearance and a sense of inferiority. He never receives others' approval, and since he is the protagonist, his thoughts and ideas, which could be called philosophical, are interspersed in various episodes throughout the story. When read, this novel assumes the form of memories emanating from the protagonist's "I," and, as readers, we wonder how this gloomy, introverted, uneducated young man is able to conceptualize and express such fine philosophical thought in writing. To the author of this chapter, *Kinkakuji* is among a small number of novels (narratives) that fuse the characteristics of a picaresque or external picaresque novel narrative and a speculative fiction or an internal picaresque novel narrative. We can also point to *Seinen no Wa* as another example of this fusion in Japanese novels. Besides this novel's protagonist, Taguchi, who is a young man with a crippled leg and an unbalanced personality, plays a major role. In his booklet about the *zentai shōsetsu* (total novel), Inoue (2015) discusses Mishima's *Hōjō no Umi* [The Sea of Fertility] (1969c-1971) (*Haru no Yuki* [Spring Snow] (1969a), *Honba* [Rusnaway Hoses] (1969), *Akatsuki no Tera* [The Temple of Dawn] (1970), and *Ten'nin Gosui* [The Decay of the Angel] (1971)) and Hiroshi Noma's *Seinen no Wa* as two representative examples of contemporary Japanese novels that are total novels or that have a strong orientation toward being complete novels; however, here, the author has unintentionally employed these two novelists' works as representative examples of "picaresque novel stories or external picaresque novel narratives and speculative fiction stories or internal picaresque novel narratives" in contemporary Japan. While it can be said that the two authors contrast starkly in terms of their respective novels' styles and ideas, as Inoue pointed out, they can be considered to share an unexpected commonality with respect to novelic methodology.

Of course, what the author has described above is ultimately only a policy or a strategy for creating the narrative content in *Watakushi Monogatari*, and it does not mean that the narrative content uses or borrows from the stories that are told in *Kinkakuji*, *Hōjō no Umi*, or *Seinen no Wa*. In that sense, the kind of policy or strategy for narrative content that the author describes here should be classified as a relatively formal one among the elements relating to *Watakushi Monogatari*'s narrative content. In practice, *Watakushi Monogatari*'s specific narrative content is to be invented independently, by the author, using the abovementioned materials, subject matter, and automatic narrative generation function.

The Conception of Form in Internal Narrative Generation

Form in stories such as novels is a highly intriguing subject. When the author first read the Japanese translation of James Joyce's (1882–1941) (2010) *Ulysses,* what struck the author most was the form in which it presented adventure and the variety of that form. One of the translators was the novelist Saiichi Maruya (1925–2012). Maruya was able to translate one chapter, in which Joyce delivers narrative content while tracing the history of the English language from ancient English up to contemporary colloquial English, into Japanese by tracing the history of Japanese expressions from ancient Japanese to medieval and early modern (Edo-period) Japanese, the Meiji period, and finally, contemporary colloquial Japanese. What lingered in the author's memory was not so much the content of the story as the story's expression and the intensity of the awareness of form that made such an adventure possible. The final chapter is written using a formal technique known as stream of consciousness. The stream of consciousness technique was attempted prior to *Ulysses* in Natsume Sōseki's (1867–1916) (2014) *Kōfu* [The Miner]. Writers whose work the author read around that time, including William Faulkner (1897–1962), Yasunari Kawabata (1899–1972), Kenji Nakagami (1946–1992), etc., and Riichi Yokomitsu (1898–1947), whom the author read later, were all highly aware of novelic and story form. At the same time, Yukio Mishima's classical form is also a kind of formal adventure, going in the opposite direction. For example, Mishima's novel *Shiosai* [The Sound of Waves] (Mishima, 1954a) deliberately imitates the Greek novel *Daphnis and Chloe* (Longus, 2019), and several of his *kabuki* works also imitated the classical *kabuki* style of the Edo period (the current custom in *kabuki* performances is to use *jōshiki-maku* (drawing curtain of three colors) for so-called classical Edo-period *kabuki* and to use *donchō* (drop curtain) for dance and modern or New *kabuki* since the Meiji era; however, performances of Mishima's *Iwashi Uri Koi no Hikiami* [Sardine Vender Love's Trawl] (1954b) written in a pseudo-classical style, use the same *jōshiki-maku* as in classical *kabuki*). Mishima also wrote many other works, incorporating formal techniques such as those used in French psychologism novels and classical drama. Narratology was originally influenced by Russian Formalism, which is one of the sources from which it arose. It emphasized aspects of form in stories and narratives, and argued that innovation in form is equivalent to the innovation of a narrative. Natsume Sōseki's (2007) *Bungakuron* [Literary Theory] deals with narrative content in literature (mainly novels, poems, and plays), but its handling is very formal. That is to say, Sōseki does not directly criticize literary content (i.e., characters' personalities and ideas, temporal and spatial characteristics, etc.) as is usually the case with literary criticism; rather, he formally categorizes the literary content, so to speak. In that sense, he takes a very formal approach. Takaaki Yoshimoto's (1965) *Gengo ni totte Bi to ha Nani ka* [What is Beauty in Language?] also engages in a historical discussion of the question of the evolution of literary works based on the formal characteristics of the language in which they are written. Various other examples can also be mentioned, but it is clear that form is an extremely important topic in literature.

However, when the author speaks of formal concepts here, he does not suggest that "form" only includes those literary works that are oriented toward the form of adventure. It goes without saying that formality exists equally in works that are avant-garde and those that are not. Otherwise, neither novels nor stories could possess substance. In the broadest sense, form in narratives encompasses all the methods by which narrative content is actually expressed and can be seen equally in any masterpiece as in any slipshod work, and in any artistic as in any popular one. In literature, that is oriented toward the venturous form or the diversity of form mentioned above, special attention is paid to the question of mediating, so to speak, the value assessment of narratives or literary forms. Preceding this type of value assessment however, there is the question of form as a more universal subject. Naturally, the form the author is referring to here, and in a more substantive sense toward *Watakushi Monogatari*, is form in the sense of being in line with the specific realities of the work. Form in the sense of what form the narrative work to be generated takes is form at the level of the work itself.

In light of the above, the author will separate the discussion into "higher-level" and "lower-level" form, based on this issue or awareness of it in narratives and literature. Here, lower-level form is form in the sense of encompassing all forms that must be present in all literature. Higher-level form, by contrast, encompasses issues of the conscientious use and application of and experimentation with form, such as superior, adventurous, avant-garde, and (intentionally used) classical forms in narrative and literature. Note that the terms higher and lower do not imply value or significance. Lower-level suggests the sense of the foundational formal foundation, so to speak, while higher-level suggests the sense of conscientious use and application based on it.

As preparation for full-scale systematization in the future, the author will now discuss lower-level form, meaning all the methods and techniques that are used in narratives and literature to actually express and convey narrative content. The (lower-level) form in narratives can be further separated into two levels. The author will use the example of a novel, since narratives are expressed entirely in language (note that various issues and difficulties arise when we make such a consideration with regard to other types of narratives that, unlike the novel, are not expressed entirely using language, such as theater and the performing arts).

One is form on the level preceding individual sentences, and another is form on the level of the set of sentences. Using the INGS that the author has developed, the generative process is separated into three parts: the story generation process, the narrative discourse (generation) process, and the narrative representation (generation) process (particularly the linguistic representation (generation) process in relation to this discussion). The narrative discourse generation and the narrative representation generation processes are relevant to form as discussed here, and it follows that "form at the level preceding individual sentences" above corresponds to form for narrative representation ("form as narrative representation"), while "form at the set of sentences level" corresponds to form for narrative discourse ("form as narrative discourse").

The first, form as narrative representation, deals primarily with form in relation to sentences, words, letters, and sounds. What rules and methods exist within these and, moreover, what manner of information is present depends, of course, on the type of language. The main language the author uses in *Watakushi*

Monogatari is Japanese. In Japanese, words and sounds are, of course, different from other languages such as English, but there are many other differences. For example, it is safe to say that Japanese lacks English's strict syntax rules. In Japanese sentences, it is the "particles" or "*te-ni-o (wo)-ha*" that often play a major role in establishing meaning, while the order in which the words in a sentence are arranged is irrelevant to establishing the sentence's core meaning. Furthermore, while in English a vast number of words can be created from a small number of characters, there is an extremely large number of characters that are used in Japanese, particularly *kanji*, as well as other types of characters, such as the *hiragana* and *katakana* alphabets, and Roman characters, which are also present. Japanese sentences are mixtures of these different character types. In considering the form that is used to express narrative content in *Watakushi Monogatari*, we must examine these characteristic features of the Japanese language and devise a formal system based on it. However, English and many other languages can be considered to share the use of sentences, words, letters, and sounds as the main elements of narrative expression.

The author would now like to consider two vectors of form, "the form of narrative expression unrelated to the semantic content of the narrative" and, conversely, "the form of narrative expression related to the semantic content of the narrative." For example, "sentence length" is a form of narrative expression that is extremely important, with some writers, such as Hemingway and Yasunari Kawabata, frequently using short sentences, while others, such as Marcel Proust (1871–1922) and Yutaka Haniya, often compose long ones. Riichi Yokomitsu's short novel *Kikai* [Machine] (Yokomitsu, 1931) is also written primarily in extremely long sentences, although Yokomitsu is famous for changing his style of narrative expression forms in each new work. This demonstrates that sentence length is an essential element that creates stylistic differences between writers. On the premise that stories' styles are included in their meanings, the form individual sentence length is also involved in meaning; however, this differs from narrative expression forms in the sense of, for example, "what words one uses to describe roughly the same things at the story or event level." In the latter, every word retains its own particular meaning, and word choice corresponds to the form of narrative content on a more semantic, content-related level. In relation to this, the different forms relating to character types in Japanese, such as whether a word is expressed in *kanji*, *hiragana*, *katakana*, or using the Roman alphabet, create a large, essential characteristic in written narratives, but this kind of character selection is a form of narrative expression on a level that is further removed from semantic content than word choice. Thus, "involved or uninvolved in the semantic content of the narrative" is perhaps a comparatively binary concept, and while a strict dichotomy is impossible, this kind of distinction can certainly be made.

The next category, that is, the narrative discourse form, is, in the simplest terms, a level of form that focuses on the composition as a set of sentences, as described above. However, it is extremely difficult to clearly and comprehensively consider and organize its true state. Nevertheless, attempts to accomplish this, including the task of integrating with the form at the level of narrative expression above, are essential in aiming for a true system of narrative discourse theory (in the sense of being integrated with narrative expression) that extends well beyond the scope of Genette's (1972) narrative discourse theory. This can be demonstrated with an example. Suppose that, in a given story, we are handling "A scene from the day of the Great East Japan Earthquake" and "Events on the day of Mishima's ritual suicide" (both of which are episodes). Form on the narrative discourse level discussed here can be seen in the following examples: "The events of the day of the Great East Japan Earthquake are narrated through the narrator's evocation of memory in the present time and space, through the order in which the narrator recounts their memories," and "The events of the day of Mishima's ritual suicide are narrated based on a stance of objective coverage of the event following the passage of time and with the aid of newspaper articles and other articles from the time." It becomes clear that there are differences in the form of narrative discourse, such as certain events in the story being narrated as organized by memory, and narration using a style of incident reporting that follows a chronological order. Moreover, it is possible to go on to list numerous examples of this form of narrative discourse one after the other, such as the number of times the incidents or events in the abovementioned materials are expressed in the narrative overall, whether through brief summary descriptions or using lengthy descriptions, etc. A more definitive level of the form of narrative discourse appears next. For example, these include themes relating to the expressive style of narrative discourse, such as description, conversations, mental representations, etc., and themes relating to depicting actions (description), explaining external and internal character attributes, and providing exposition through descriptions, topics, and objects. In determining the various elements for the form of narrative discourse, each of its parts will be superficially realized on the narrative expression form level. For example, characters' actions in descriptive passages are expressed in short sentences using complex words, and conversations are expressed in long sentences using only *hiragana*. Conversely, in a more macroscopic direction, there is also form in the sense of the "composition" of each relatively large block that constitutes the whole narrative or a part thereof, as in a novel. *Ulysses* is, on the one hand, a novel that takes an avant-garde form, but in many respects it is a novel that adopts a classical form that was created based on the structure of Homer's (2006) epic *Odysseus*; that is, it was constructed using a form that fuses the two. Moreover, this novel was also created using a variety of formal structural principles that vary almost from chapter to chapter. However, if we view the individual chapters that constitute it as a whole, each is carried through by a consistent formal principle. In this sense, there are innumerable possibilities for composition when viewed from a macro perspective, but the extent of the temporal scope in novels of equal length, for instance, will be reflected in the narrative's characteristics. *Ulysses*, *The Brothers Karamazov*, *Shirei* (*Death Spirits*), and *Seinen no Wa* are novels that unfold within a relatively short span of time and are composed of a condensed set of tightly interconnected episodes. On the other hand, *Genji Monogatari* [The Tale of Genji] (1993, 1994, 1995, 1996, 1997), *War and Peace* (Tolstoy, 2010), *Jean-Christophe* (Rolland, 2011, 2018, 2007), and *Hōjō no Umi* (*The Sea of Fertility*) are narratives that unfold over an extremely long span of time, depicting a person's life, or at least half of their lifetime, and even generations of people, through a loosely connected series of incidents or episodes.

Find below a list of the major categories above as well as the main formal elements that might be included within each. It should be noted, however, that this list is only tentative, containing a few examples that attempt to encompass the narrative form:

A. Forms of narrative representation
   - Sentence length
   - The amount of conversation and description in a sentence, and whether these are many or few sentences as well as how they are controlled
   - Conversational response duration (long or short)
   - Paragraph length
   - The frequency at which *kanji* are used in the text
   - Similarly, the frequency of *hiragana*'s use
   - Similarly, whether (or not) a lot of *katakana* is used
   - Similarly, the frequency at which Roman alphabet characters are used
   - The distribution of character types, i.e., *kanji*, *hiragana*, *katakana*, and Roman alphabet characters, in the text

- ◦ Whether complex words are used (frequently or infrequently) in the text
- ◦ How heavily (or lightly) a sentence is punctuated
- ◦ The types of word endings used in a sentence
- ◦ Is the text written in literary or spoken language?

B. Forms of narrative discourse

- ◦ How is the timeframe of the story handled as the narrative unfolds? Does the narrative develop through the story or is the narrative structured in an order that is different from the story's?
- ◦ To what extent is the "narrator" actualized in the narrative and to what extent do they have power? Is the narrator visible or invisible? Is the narrator strong or weak?
- ◦ Is the development of the narrative scene- or summary-centric?
- ◦ How long (or short) are the scenes in the context of narrative development?
- ◦ As a method for segmenting the entire narrative, divide it into several "sections" or, alternatively, "chapters," divide it using numbers, etc., divide it without using symbols (a new paragraph, etc.), undivided, etc.
- ◦ How many parts of speech, such as nouns, verbs, adjectives, adjectival nouns, etc., are included, and how are they balanced?
- ◦ Is there a lot of (or a little) description?
- ◦ Is there a lot of (or a little) explanation?
- ◦ How detailed is each description? Is it relatively detailed or rough?
- ◦ What is the length of each description? Is it relatively long or short?
- ◦ How detailed is each explanation? Is it relatively detailed or rough?
- ◦ How long is each explanation? Is it comparatively long or short?
- ◦ How does personhood, i.e., first-, second-, and third-person, and others (e.g., Riichi Yokomitsu introduced the concept of the fourth-person in his novel *Kikai* (*Machine*)).
- ◦ How many characters appear, and how many of them are major characters?
- ◦ Is there a protagonist?
- ◦ Is there a single protagonist or are there several?
- ◦ Does the character (the protagonist) have a broad (or a narrow) scope of action?
- ◦ For how long does a character (e.g., the protagonist) appear in the narrative?
- ◦ What personality does the character (the protagonist) have?
- ◦ To what extent are material elements featured?
- ◦ To what extent are spatial elements (places or stages) featured?
- ◦ Is there much movement between locations?
- ◦ To what extent are abstract elements featured?
- ◦ To what extent are emotional components featured?
- ◦ From what point of view, i.e., the single (internal), omniscient, or dynamic (non-fixed) perspective, etc., does the narrative develop?
- ◦ Does the narrative have an overall comedic or tragic effect?
- ◦ Is the narrative as a whole mundane or extraordinary (fantastical)?
- ◦ Are animals or inanimate objects anthropomorphized in the narrative?
- ◦ Are humans animalized or objectified (reverse anthropomorphism) in the narrative?

In contrast to the lower-rank narrative forms described above, there are high-order narrative forms. This is related to the so-called "adventurous form" in *Watakushi Monogatari*, and it can be said that this adventurous form was one of the major objectives of *Watakushi Monogatari*'s creation; although higher-level narrative form is an important topic, this is something the author wishes to discuss in a separate paper.

Toward External Narrative Generation: The Sociological Distribution of *Watakushi Monogatari*

In the previous section titled "Internal narrative generation," the author addressed the issue of the *Watakushi Monogatari* content that should be written in the future or, more specifically, the materials that should be used to prepare for this content's creation. In this sense, the discussion has yet to address the full narrative content, but in *Watakushi Monogatari*, the generation and distribution processes are separate topics. The generation process is naturally performed through GIS with INGS (as described

Figure 13. The fluidity and fixation in Watakushi Monogatari or the narrative of possibility and the narrative of fixation

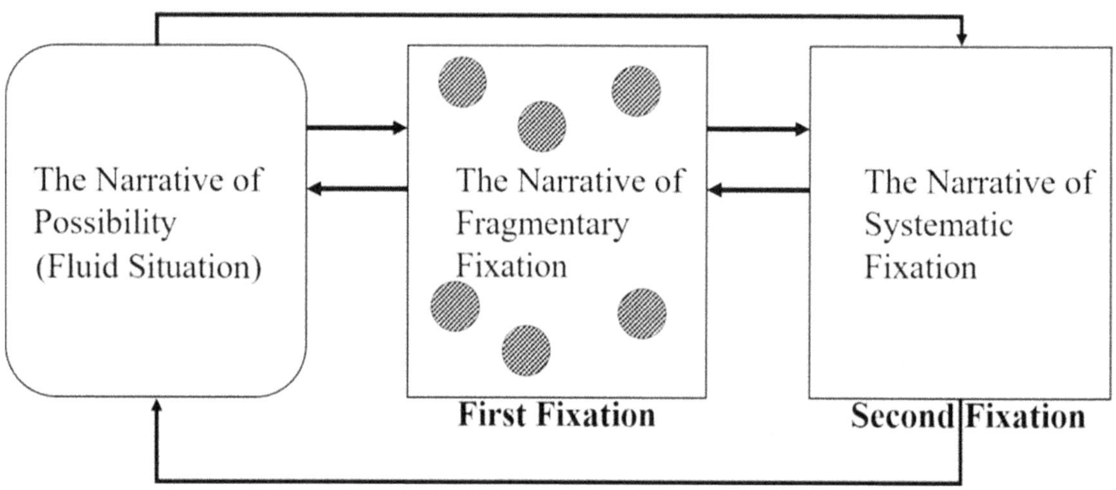

above) and the *Watakushi Character* that overlaps this "I (*watakushi*)." Meanwhile, distribution is accomplished through GIS and other distributive media. However, in *Watakushi Monogatari*, the generation and distribution processes are not completely distinct; rather, the two are interrelated. The author now withes to explain this interrelatedness in *Watakushi Monogatari*. As a precondition for clarifying the interrelated generation and distribution processes in *Watakushi Monogatari*, Figure 13 demonstrates fixation and fluidity.

In Figure 13, the "narrative possibility (fluid state)" is in constant motion. From there, the "fragmentary fixation of narrative" occurs, at times, in various forms. This process is called "primary fixation." Moreover, from there, or possibly directly from the fluid narrative, a "systemic fixation of narrative" arises. This process is called "secondary fixation," but this can be further separated into "secondary fixation A" and "secondary fixation B." Furthermore, fragmentary and systemic fixed narratives forward

Figure 14. The creation and distribution of Watakushi Monogatari and their cyclical process

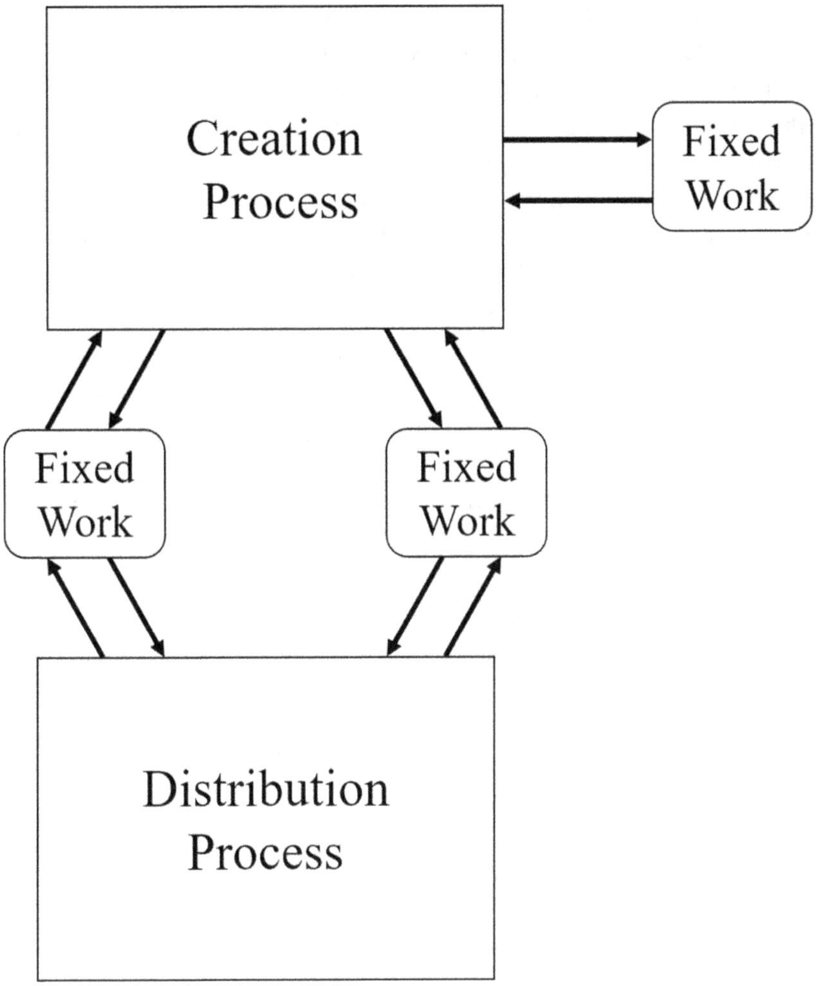

information back to "possible narratives (fluid state)" on various levels. In other words, this process gives rise to a great cycle.

Here, part of the fixation of narrative, in any form, escapes the generation process and enters the distribution process. We can try tweaking this slightly to accommodate a diagram that incorporates the three elements: the generation and distribution processes, and the abovementioned circulation (Figure 14).

In Figure14, the process of narrative generation (fluid state) produces a fixed work. Some never enter the actual distribution process, while others will. However, these processes are not unidirectional. In other words, once a narrative is fixed or once a narrative has entered the distribution process, it can return to the generation process as a type of material. Fixed narratives that return play the same role in the generation process as other materials and resources that the process utilizes. In this way, the generation process and fixed narratives and the generation and distribution processes form a reciprocal, cyclical process. The final stage of this process is the complete, systematic *Watakushi Monogatari*.

The external directions of narrative generation or external narrative generation can be simply interpreted the distribution from the production or creation through an internal process. However, as shown in the two figures above, Figure 13 and Figure 14, the *Watakushi Monogatari* method includes a cyclical process comprised of generation and distribution, and the narrative generation process can also be included in external narrative generation. Alternatively, external narrative generation should be understood in terms of the cyclical relationship between the narrative generation and the narrative distribution processes.

Incidentally, the author mentioned above that "The final stage of this [cyclical] process is the complete, systematic *Watakushi Monogatari*." Since Figure 13 focuses only on the cyclical process of narrative production and circulation, the distinction between the work as a work-in-progress and the final work is not made explicit. In Figure 14, however, a distinction is made between a "fragmentary fixed narrative" and a "systematic fixed narrative," and a distinction can be made between the former as a narrative-in-progress and the latter as a final narrative. This shows that, on the basis of the two reciprocal foundational concepts of fluidity and fixation, the distinction between a work-in-progress and the final work becomes less apparent; however, at the same time, the two must, of course, be distinguished from each other. There is, however, likely no essential difference between the two with regard to the internal logic of the narrative. At a certain point in the *Watakushi Monogatari* generation process—this is probably the point at which the idea of "completion" carries a great deal of weight in the author's mind—the final form of *Watakushi Monogatari* is transformed into a book (in this case, "book" is not limited to printed works but is more symbolic in nature). In this sense, in the final "book" form, at the point of conservative fixation, the narrative of the purely fluid process (perhaps *Watakushi Monogatari* in "the fluid process as a generative process") is ultimately fixed to form *Watakushi Monogatari*. An important consideration in relation to personal strategy is the degree to which we bring that to a point in time or, to put it another way, the place to which we bring the "end of the narrative."

The question also arises as to how the narrative, as a final, fixed work, should be distinguished from the narrative as a fixed work in the course of narrative generation. In this regard, the author proposes that there are two types of fixation: fixation that is performed through the author own intent and fixation that is done for other reasons. For example, there is a form in which narrative is fixed, where the recipient takes some manner of action within the vortex of fluid narrative generation and, for *Watakushi Monogatari,* fixation is performed deliberately by a part of the *Watakushi Character* other than the real "I (*watakushi*)." Alternatively, we can also consider that fixation during the process and fixation in the final work are realized on different scales. When considering this issue, while there is the idea of calling the final fixed narrative *Watakushi Monogatari* and referring to the in-progress narrative by a different title, there is also the idea of considering the scope that encompasses both to be *Watakushi Monogatari* and the in-progress narrative to be a special part of the same *Watakushi Monogatari*. Adopting the latter view, the final fixation refers to one large work in its entirety, the *Watakushi Monogatari*, whereas the fixation of the work-in-progress refers to pieces, fragments, and parts of the narrative related to *Watakushi Monogatari* on a smaller scale. In other words, in the course of generating the *Watakushi Monogatari*, independent, fixed narratives that are smaller than the final *Watakushi Monogatari* arise, and these are differentiated from *Watakushi Monogatari*'s main body in the form of, for example, "short narratives" or "narratives of medium-length (novellas)." Of course, this or part of the source of a type of narrative from which that fixed narrative originated can in some ways influence parts that should again be part of the *Watakushi Monogatari* as a whole.

Incidentally, fixed narratives realized through the medium of (paper) books where they go beyond the fluid process are an important form to the author, and the possibility of the circulation and development of narratives related to *Watakushi Monogatari* in the form of "paper books" cannot be excluded. In this case, the fixed narrative that enters the distribution process will be published as various books. At the same time, the author believes there is also a need to extend this concept of the book to include books in a sense that goes beyond the antiquated form of paper books to include "superbooks." A superbook is the form that a fluid narrative takes when it becomes a fixed narrative, but additional fluid elements are incorporated in its fixation, and this form is the idea the author means to describe with the term "superbook." A fixed narrative can be fixed not only in the traditional form of a paper book, but it can also be realized as digital books, which are fluid in themselves. In such cases, the boundary between fixed and fluid narratives becomes extremely blurred or (possibly) extremely creative, fluid, and ambiguous.

The author would like to further consider this notion of a superbook. In its simplest form this would be a "hypertext schema" in such stories where the text of the final work is linked to other parts of the work, annotations, explanatory text, etc., through the use of hyperlinks. The simplest form of this hypertext narrative scheme—with links to annotations and explanations—is also achieved in paper books in the form of annotations (and annotations of annotations). Some novels are heavily annotated. For example, the Japanese novelist and politician Yasuo Tanaka (1956–) included numerous annotations on various things that were popular at the time at the end of his book *Nantonaku Kurisutaru* [Somehow Crystal] (Tanaka, 1981). A more complex "hypertext schema" can also be considered. For example, this could be a form in which one part of a narrative is connected with another part of the same narrative through links. For example, there may be intra-narrative relationships, such as a particular word being connected to the same word or an associated word in another section or a certain event may be connected to the same or an associated (single) event in another part, or a particular structural unit may be connected to the same or an associated structural unit in another section, etc. There may also be various other relationships. If we can say that various readings of a single narrative become possible through these relationships, we will have a complex version of a novel, like the type represented by Cortázar's (1914–1984) (1991) *Hopscotch*. The above examples evince methods based on internal textual relationships within a single fixed narrative work. By contrast, we can envisage a situation in which behind a fixed narrative work another type of information is associated with it. This is possible when there is a mass of connected information behind a fixed narrative work. For example, if the final output of the narrative generated from INGS is (in the case of novel-type narrative generation) language, where a conceptual information group is present behind the narrative's superficial language, conceptual information that is, in some sense, related to that superficial language can be extracted. In other words, this means a multi-layered narrative. However, this is not multiplicity in a narrative that is directly expressed in language (as is the case for multiplicity in traditional narratives); rather, this is a form of narrative multiplicity in which there is another layer or multiple layers of narrative hidden behind the narrative that is expressed via language which, while, of course, related, exists on a different level. For hypertext mechanisms in ordinary narratives, links point to another part of the narrative that is expressed directly using language; however, in the sense of the superbook that is described here, links point to specific locations behind the narrative, so to speak, which exist on a different level than the point at which the narrative begins. Links that point to a level other than the narrative's include, among their types, annotations in and of themselves, but in the sense that we are discussing here with respect to a different level of narrative, these are not the narratives behind the narrative. This means that as a superbook *Watakushi Monogatari* incorporates multiple relationships between the multiple background narratives that may be present. (Taking the example of a narrative and the annotation of its individual parts and supposing that the individual annotations are based on another large narrative chunk that organically systematizes them, we come close to the idea of a superbook, as the term is used here.) The entire work, including background narratives, is *Watakushi Monogatari* as a whole, but the background narratives cannot be directly viewed from the surface. However, it is also true that one or more underlying narratives can be produced as similar to and in conjunction with the narrative that is visible from the surface. Perhaps these underlying narratives could be a collection of fragments or a mechanical chronicle that lacks a normal narrative's structure, and thus cannot be called individual works in their own right. As a superbook, then, *Watakushi Monogatari* not only allows for a hypertext scheme between superficial linguistic expressions, it also enables a type of relationship between those superficial linguistic expressions and a deeper narrative structure, i.e., underlying narratives. The latter signifies one aspect of the multiple narrative structure in *Watakushi Monogatari*. This also hints at "fixed narrative = the issue of a fluid narrative appearing in the final work." In this sense, in relation to *Watakushi Monogatari*, there is the opportunity for "fluidity and fixation" in all stages of narrative generation (including the final stage). It is also present at the final stage. Since such a form is unlikely to be fully expressed in the conventional "book" form, a superbook is needed, and it is essential to explore the possibilities of such a form. (With regard to multiplicity in *Watakushi Monogatari*, one can imagine that many superbooks are constantly being written in a fluid manner emanating from a single, extremely condensed superbook built into GIS with INGS.)

An Integrated Method Between Internal and External Narrative Generations

The next section designs a basic mechanism of the external narrative generation that organically distributes narrative contents created by the proposed internal narrative generation mechanism. In particular, the author focuses on the integrated relationships between the internal narrative generation and external narrative generation to progress this research to an actual attempt.

As is clear from the above, internal narrative generation, that is, the act of generating (producing) a real narrative work, and the act of external narrative generation or the external presentation (the distribution and circulation) of that generated (produced) narrative are not two separate, distinct processes within the *Watakushi Monogatari* framework of narrative generation research; rather, the two processes are very closely related to one another. Moreover, it is important to note that the two are quite intrinsically—that is, in a way that would not be possible without the use of such a framework—linked.

In the course of *Watakushi Monogatari*'s creation, there can be various forms of fixed narrative prior to the final fixed state. However, while remaining fixed, this state is simultaneously involved in various ways in the creation of *Watakushi Monogatari* and the final work. For example, where, in the course of *Watakushi Monogatari*'s creation, a part or the current form is fixed in the form of a "book," although this certainly represents the fixation of a narrative at that point in time, at the same time *Watakushi Monogatari* and that "book" may have some relationship as something that continues to change. For example, the fixed narrative (as pieces or fragments of a larger whole) can rejoin the *Watakushi Monogatari* generation process as a fluid subprocess, changing its fixed form, and influencing various parts of the fluid narrative.

Considered in this way, the concept of fixation here is not necessarily associated with the concept of an "end," but is rather established in very close connection with the concept of a fluid process. In other words, the narrative generation process surrounding *Watakushi Monogatari* is not simply a fixed process in the sense that the overall *Watakushi Monogatari* process ends when fluidity becomes fixed. If we consider this more pragmatically, that is, if we consider the question of the specific process by which the author's "I (*watakushi*)" becomes the *Watakushi Character* and creates a narrative, we can publish fixed fragmentary or fragmented narratives from time to time during this process as fixed media, for example, as books. However, this is not simply a case of combining many such books to construct or

Figure 15. From GIS with INGS and kabuki methods to Watakushi Monogatari

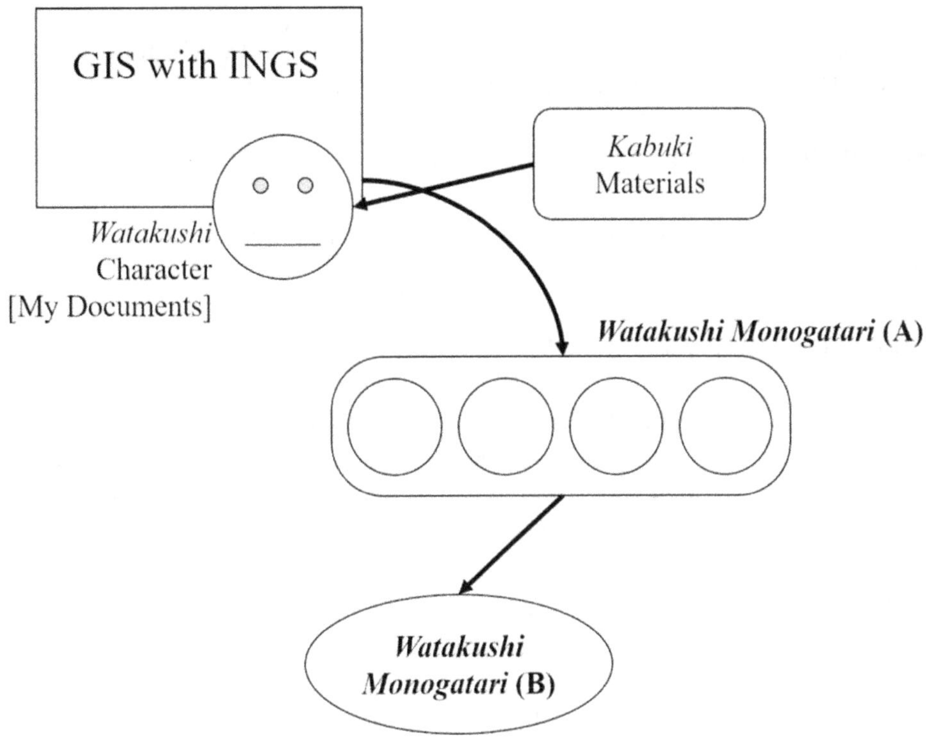

restore the whole. Those pieces, fragments, and fragmentary narratives are not later incorporated as-is into the overall *Watakushi Monogatari*; fixation promptly moves to a fluid phase, and the fixed narrative enters a phase in which it is again changed and used in the fluid process. The phenomena of internal narrative generation (creation and production) and external narrative generation (production–consumption) in *Watakushi Monogatari* can be understood as this representation of the narrative generation process. To simplify, from a highly pragmatic perspective, and as shown in Figure 15, the *Watakushi Character* and "I (*watakushi*)," using GIS with INGS, generate a number of intermediate *Watakushi Monogatari* as a group of fragmentary narratives, utilizing knowledge and methods related to *kabuki* as the narrative technique that this paper introduced and connecting them in the final *Watakushi Monogatari* as a kind of compilation.

The fragmentary group of *Watakushi Monogatari* in this case (*Watakushi Monogatari* in a broad sense) and the final *Watakushi Monogatari* (*Watakushi Monogatari* in a narrow sense) are all fixed narratives; however, the process that generates *Watakushi Monogatari* in the broad sense and the one that produces the final *Watakushi Monogatari* in the narrow sense are not solely fluid states of narrative generation and, particularly in the final *Watakushi Monogatari*, as described above, they are understood as fluid narratives through multiplicitous structures of underlying narratives or through a superbook's special function.

SUPPLEMENTAL DISCUSSION AND DIRECTIONS FOR FUTURE RESEARCH

In this section, the author provides a supplemental consideration based on the following four topics, as well as directions for future research regarding the entirety of this chapter.

Supplemental Discussion of Several Topics

This section addresses four themes that the author does not refer to in previous parts or additional topics for this chapter.

Utilizing and Extending *Sekai Kōmoku*

The first topic is tied to both GIS and *kabuki*. One of the new ideas in this chapter is that the author aims to construct the basic structure of the *geinō* work resource in the resources in GIS based on *Sekai Kōmoku* in *kabuki*. This part attempts to provide a supplemental discussion. The author continues to analyze *Sekai Kōmoku* in detail in the context of the narrative generation study of *kabuki*. The essential goal of the investigation is not necessarily to uncover new possibilities for the study of *kabuki*, but to utilize these documents for the author's development of narrative generation systems. From a broader viewpoint, dealing with prior narratives and literature from the perspective of *Sekai Kōmoku*, and expanding and changing the range of their uses, are themes relevant to the expanded literary theory and post-narratology that the author considers. The author's expanded literary theory and post-narratology (which is the recent term) are substantially developing, introducing analyses of *kabuki* studies, as well as further "narratologies in Japanese literature" (which is not the narratological interpretation of Japanese literature) beyond the previous usage of the narratology of Propp, Junette, and Jause.

The author affirms that *Sekai Kōmoku* can be utilized as a general, universal method/framework for addressing narratives both systematically and structurally. The author's idea, in a more pragmatic sense, is to develop a universal, general knowledge source based on *Sekai Kōmoku* or its essential ideas. *Sekai Kōmoku*, collected in a document or organized as text, is simply the enumeration of many narrative titles, characters, and so on. Readers or users during the Edo era employed personal and organizational knowledge obtained by professional people; they expanded, embodied, and concretized such knowledge to produce *kabuki* works. Consequently, to introduce *Sekai Kōmoku* and its essential ideas into the author's narrative generation study, it is necessary to analyze the invisible knowledge (which is frequently "tacit knowledge") to formalize and give structure to the underlying principles. In turn, the representation can be transformed into a conceptual, structured description for use in narrative generation systems, especially GIS with INGS. Actually positioning and utilizing *Sekai Kōmoku* as a major method in GIS requires the difficult restructuring of *Sekai Kōmoku*, including the above tasks.

The author introduces a previous attempt related to *Sekai Kōmoku* or its future use and expansion. The author describes the systematic collection of Japanese folktales as the conceptual representation structure by Common Lisp. he collection can be used as a narrative example in the GIS with INGS version of *Sekai Kōmoku*. In the next stage, the author considers the intertextual synthesis of several narrative works using the technique of *naimaze* in *kabuki*. In the processing order by GIS with INGS, GIS is in charge of defining a macro-level narrative structure, and INGS elaborates and transforms its structure to generate the final narrative scenario.

Life Narratives in GIS

GIS with INGS treats life narratives as a narrative type. This chapter concretely discusses life narratives, focusing on positioning and usage in GIS. Events in the lives of *geinōjin* characters need to be stored in the system. For example, scripts presented by Schank and Abelson (1977), which are a typical knowledge representation in semiotic AI, are related to life narratives. Existing scripts are organized based on each event's unit in people's daily lives. In contrast, the subject in the context of the author's life narrative is how the life events, as broader units, are organically defined. The specific consideration of life narratives requires collecting the author's life and other people's lives as materials, and organizing them for use in GIS with INGS. The information to be contained in GIS is the broad (hierarchical) framework of lives. INGS elaborates the framework according to more detailed information. As with life narratives, this detailed information can be commonly used for ordinary narratives. These kinds of fragmented forms in the narrative knowledge of INGS can be generally used for both ordinary and life narratives.

As shown in **Internal and External Narrative Generation**, in the **BACKGROUND**, the consideration about the author's "life" occurs in association with *Watakushi Monogatari*. This analysis classifies the "time" regarding the author's life into several types, and categorizes related and other information. The primitive idea for the respective content is described in **The Conception of Content in Internal Narrative Generation**. This is studied as an approach bound in the author's "I" (*watakushi*). The content is unique in the author's "I" (*watakushi*). It is possible to divert and adapt the framework for the design and development of life narratives in GIS.

Developing *Watakushi Monogatari*

Watakushi Monogatari provides the idea for a narrative (or narratives) to be created by the author. Generally, studies of computer systems assume use by external people other than the systems' designers or developers. Many computer systems are examined in terms of the development of tools, no matter whether they are support systems or alternative systems. Different computational content studies develop AI computers and narrative generation systems that select existing or past artists as materials and learn and imitate their methods of content generation and representation to produce similar content.

In contrast, *Watakushi Monogatari*, presented in this chapter, is the conception of narrative content creation to be written by the author. Therefore, it is very private and personal. As a similar example, Hori (2007) writes his essays using the creative activity aid program that he himself implemented. Although he originally designed his creative activity aid system as a general one for supporting users' creative activities, he added himself as a user and supported his writing of papers beyond simply verifying the usage. He employs the system that he researched and developed for his own creative activities.

The author intends to use a developed system for the narrative(s) he will write, *Watakushi Monogatari*, instead of practical documents, such as presentations and papers. *Watakushi Monogatari* does not simply support (through imitation) objects written based on existing content methods without altering the framework of previous content methods. This study is different from similar research, and attempts to provide a new approach to narrative creation by developing the method of producing *Watakushi Monogatari* using the author's narrative generation system study, integrated in GIS with INGS. One of the most unique points of *Watakushi Monogatari*, under the framework of the narrative generation study, is that it presents a narrative generation method or model in which both the narrative generation process (a kind of fluid situation) and its representation process (similarly, a fixed situation) are essentially liked and interact with each other.

The author considers the above topic from the angle of generality and concreteness. For example, several studies (including the author's) have been conducted that formalize Propp's narrative theory ("the morphology of the folktale") and utilize it in narrative generation systems in cognitive science and AI, but they have been frequently criticized. One opinion (Tanaka, 2010) asserts that although Propp's narrative research (in addition to similar narratological studies) can be used as a theory of narrative generation (as a general theory), when we concretize research to apply it to actual narrative generation, it is difficult to assign sufficient narrative generation abilities to the theoretical framework, owing to problems related to the data's details and common sense narrative processing. As for criticisms of Propp's theory from the perspective of AI and cognitive science, like Gervás (2013) in Spain, the author believes that Propp's narrative research includes very rich, concrete narratological knowledge and insights. The surface analysis of Propp's theory by researches in AI and cognitive science has ignored the rich characteristics that Propp's theory has. The author has sometimes mentioned this fact (for instance, Ogata and Kanai (2010)). In the first place, research that comprehensively systematizes Propp's fertile narratology does not exist. Scholars like Takana maintain that sufficient systematization, as a narrative generation mechanism of Propp's theory, is difficult, despite the theoretical possibility, perhaps because these scholars do not comprehend the material content. In contrast, the above R&D—based on the imitation of specific artists and content—forms an approach in which general narrative generation, as well as other genres' content generation, shift in the direction of concrete content production according to each objective. Regarding this point, the focus moves toward the concreteness of content, based on generality as a tool. In many cases, such research has a general tendency that positions a specific artist's imitation as a means for aiming toward performance improvement.

The author's conception of *Watakushi Monogatari* aims for the creation of the author's narrative(s), though they are, at first, certainly experimental and trial work(s). In this sense, the author strives for very concrete, special, and personal results. The final evaluation criteria include pragmatic, actual, and concrete achievements of purposes; for example, with narrative generation systems, whether: good works are created (though the evaluation of good works is not single), works that the author himself can satisfy are made, and narratives can be socially distributed. Of course, at several stages of the development process, the feedback from concrete works and content is addressed in the system as a general tool. This cycle is extremely turned around, contending for actual works and content, and finishes with the completion of *Watakushi Monogatari*. Although this book's title is *Bridging the Gap between Narratology, Artificial Intelligence, and Cognitive Science Through Narrative Generation*, from the above description, the trial of *Watakushi Monogatari* in this chapter goes beyond narratology and literary theories. Rather, the author endeavors to bridge the gap between the narrative itself, and AI and cognitive science, through narrative generation.

The Creation and Distribution of Novels and *Kabuki* in Japan

In this chapter, the author asserts that the creation and distribution of *Watakushi Monogatari* are not two divided things, but rather a mutually related process. In other words, the internal narrative generation in the author's proposed concept (i.e., the process of creating narrative content) and the external narrative generation (i.e., the social distribution and development of created narrative content) are mutually complex. In line with this explanation, the author believes that the internal creation and the external development of previous and actual narrative production in various genres may have similar mutual relationships; hence, the author will pursue such analysis in future studies. In this section, the author only reveals a tentative consideration of the pertinence between creation and social development (e.g., distribution), especially in modern novels and *kabuki*.

First, the author examines a typical method that has frequently been adopted in modern Japanese novels. In the following phenomena, creation and distribution occur based through a mutual (i.e., an obvious awareness regarding narrative creation and distribution), blended, whole process. A novel is partially published via the medium of a magazine. Next, the novel of the magazine is changed to the form of a published book. In the process, the novel, which was continuously published in the magazine, receives small primary corrections and revisions. In comparing *Watakushi Monogatari* with the above process, the sections for *Watakushi Monogatari* may correspond to the "fragmented narratives" shown in Figure 13. At this point, the stated process and the case of *Watakushi Monogatari* are similar. However, in the continuous change from magazine to book in the previous example, most parts of a novel published in a magazine (or several magazines) are synthesized into a novel as a book, notwithstanding the number of corrections and revisions. In contrast, in *Watakushi Monogatari*, only the collection and synthesis of fragmented narratives do not directly construct a larger *Watakushi Monogatari* to a substantial extent. An essential change takes place in the case of *Watakushi Monogatari*, beyond the number of corrections and revisions in the narrative transformation, from the fragmented narratives as the *Watakushi Monogatari* in a broad sense, to the *Watakushi Monogatari* in a narrow sense. *Watakushi Monogatari* forms more closed, fundamentally inseparable mutual relationships between narrative creation (i.e., internal narrative generation) and narrative distribution and social development (i.e., external narrative generation).

The author deals with *kabuki* as another topic. The author scrutinizes the creation method of a scenario (the scenario book was traditionally called *daichō* in the world of *kabuki*) in *kabuki* narratives. Watanabe (1989) respectively calls the levels of an actual performance and the scenario (*daichō*) "first text" and "second text." This indicates that scenarios do not dominate performances on stage in *kabuki*. In the history of *kabuki*, the text's level of scenarios (*daichō*s) existed, except during the period of early *kabuki*. Notwithstanding, *kabuki* stresses the role of the first text, and many *kabuki* critiques have traditionally been carried out based on materials of the first text instead of the second one. This fact is connected to the tradition of the strong power of actors. Hattori (1970) states that, when we compare actors to writers, the existence of actors is primary, and the existence of writers is secondary.

The author is interested in discovering what kinds of processes and method scenarios (*daichō*s) in *kabuki*, at the level of the second text, have been created. In other words, what type of mutual relationship has emerged regarding scenario (*daichō* in the second text) creation (i.e., internal narrative generation in this study) and actors' stage performances (i.e., external narrative generation)? The author divides the scenario (*daichō*) creation process into three types:

- A situation where there is no scenario (*daichō*). Actors produce a drama through their mutual practical acts. The scenario (*daichō*) is sometimes written as the play's report after the actual performance.
- Actors begin to perform a play according to the narrative's outline. Through this process, various additions and alterations take place, and the scenario (*daichō*) is changed to reflect them.
- A *kabuki* playwright composes a steady scenario (*daichō*), and the actors perform the drama based on it. However, during the performance, including the rehearsal, diverse corrections are

incorporated into the scenario, such as detailed and structural revisions. A final first text, a stage performance, is differentiated and transformed from the original second text, a scenario.

Kabuki scenarios (*daichō*s), as the second texts, are created in accordance with numerous ways. Actors, as subjects in a stage performance, play an essential role in scenario creation. This tradition has been passed on to contemporary *kabuki*, in which both narrative social distribution and delivery as a performance on stage, and narrative creation, as a scenario (*daichō*), are more or less merged.

The author may be straying from the central topic here, but an influential method of narrative generation system studies in AI is "narrative generation as the simulation of actions and the mentality of characters." Actors' role is great in *kabuki* and is often confused with the AI method for narrative generation. However, these two things are clearly not the same. The narrative creation method in *kabuki* is not "narrative generation by character(s) in a story," but "narrative generation by actor(s)" or "narrative generation by both actor(s) and writer(s)." Of course, when one or more actors create the narrative of a stage performance in *kabuki*, an actor can simulate a specific character's action and mentality. However, an actor in *kabuki* does not completely become one with a character. In many *kabuki* plays, an actor can play several roles. *Kabuki* actors do not completely emphasize specific characters. Rather, actors in a *kabuki* performance play many parts overlapping with writers. During the first period of *kabuki* history, the positions of "writer" and "actor" in *kabuki* were not separate. A certain type of actor played characters on the *kabuki* stage and was also a writer. Many *kabuki* works in the early period were created by actors. This means that stage performances by actors included the function of narrative works' creation. Both of the above two things are inseparable. The mutual connectiveness between the internal and external narrative generation in *kabuki* will influence the creation of *Watakushi Monogatari*.

Future Research Directions Regarding This Whole Chapter

There are diverse paths for future research regarding this chapter. A main theme is the system development and content creation in terms of various aspects, such as the following:

- **Design and implementation of GIS with INGS:** A key future research topic is the program development and implementation based on the policy and macro design described in this chapter. This program tentatively requires the following modules: (1) A program module for generating and changing the *geinō* information (discography) and life information of a *geinōjin* character at a certain moment. (2) A program module for a narrative genre and work that the *geinōjin* character appears and plays. (3) A program module for creating a narrative scenario using INGS based on the *geinō* work resource, which is similar to the knowledge system presented in *Sekai Kōmoku* in GIS. The *geinōjin* character is also formed as a dramatic character, in addition to an actress/actor. (4) A program module connected with the above program module (3) to generate a life story of the *geinōjin* character at a certain point based on the life work resource in GIS. (5) A *geinō* narrative work and life narrative work, at a certain point for the *geinōjin* character, are reflected in the extension of the *geinōjin* character's discography and life story. Narratives produced using GIS with INGS, which roughly has the above program modules, are approximately divided into the following parts: (1) The entire discography of a *geinōjin* character; (2) The entire life story of a *geinōjin* character; (3) *Geinō* narratives (scenarios) created at a certain point in time; (4) The part of the life story created at a certain point in time; and (5) The *geinōjin* character expanded at a certain point of time. The basic criteria for the investigation of the first stage system are the following: The first criterion is whether various features of the narrative, generated based on the thought of the macro design of GIS with INGS, stated through this chapter, are realized according to the thoughts of macro design. The second is whether the program, developed in accordance with the macro design's thoughts, can produce knowledge and insight that transcend the first prediction in some way. The results of the first verification will be linked with the next programming actions, such as the revision, expansion, and correction of the program's configuration, algorithms, and data structures. The second verification results will cause more radical actions toward reconsidering the program's basic macro design thought itself.

- **Systematization of *Sakusha Shikihō Kezairoku* and *Sekai Kōmoku*:** Further analysis and modeling of *Sekai Kōmoku* from the angle of narrative generation will be continued according to the policy described in ***The Survey and Analysis of Sekai Kōmoku***. In future studies of *Sakusha Shikihō Kezairoku,* the author will determine whether the schema of a multiple grammatical structure, shown in Figure 10, can be formed as a set of computer programing descriptions. Moreover, the author will advance other elements' systematization. The author predicts that the first *Sekai Kōmoku* will be able to be positioned in GIS with INGS and, further, that *Watakushi Monogatari,* as a useful document, will provide a generalized method beyond an object for the analysis of *kabuki*. The latter *Sakusha Shikihō Kezairoku* will be reconsidered from a similar standpoint. The next sub-section complements these topics.
- **Creating and Developing *Watakushi Monogatari*:** One of the most important paths for future research is to proceed with the actual creation of *Watakushi Monogatari* to move toward the pragmatic phase from the theoretical and conceptual stages. In the future, detailed, individual topics will be discovered based on a comprehensive theme. As made clear through the discussion in this chapter, the fixed narrative(s) in *Watakushi Monogatari* not only indicate the final fixed narrative. Through all phases of the generation, creation, and production of *Watakushi Monogatari*, fixed narratives are made in flexible, free, and diverse forms. The degree to which fixed narratives were limited and regulated in terms of influencing later narrative generation is small; their narratives can also be used effectively, freely, and flexibly for later processes of fluid narrative generation. The creation of *Watakushi Monogatari* will be experimentally moved ahead by tentatively making fixed narratives from fluid narrative generation, namely performing narrative creation as a kind of trial and error. As stated above, the author will enter the pragmatic phase for the creation of *Watakushi Monogatari* in the future.

Kezairoku and *Sekai Kōmoku* as Narrative Generation

What do we observe in Japanese narratologies and literary theories when viewing them from the angle of narrative generation study? This means that the knowledge acquired through the analysis of narratives is introduced into narrative generation models and systems as narrative generation techniques. Even if we could survey many narratological and literary studies and criticisms in Japan, we would probably see diverse research and criticisms with different characteristics. Even if we could model and further systematize research or critiques partially and formally, how can they be stored in an integrated manner into one or more narrative generation systems?

Simply speaking, there are methods or directions between two extremes. The first approach is the direction that individually examines numerous narratological and literary theoretical studies and critiques in Japan to develop one or more broad narratological and literary theoretical systems, based on the idea of a higher-level, more synthesized narratological system. Another path stresses individually using surveyed methods without an entire systematization like the above example. The author provisionally calls the former direction a systematic method and the latter direction an arbitrary one. They both have pros and cons. For example, although the former is clear and valuable theoretically or from an academic viewpoint, it requires enormous efforts and is inconvenient when we directly try to apply it to narrative generation systems in a pragmatic sense. In contrast, the pros and cons of the arbitrary approach are the inverse of the above case. It is easy to directly and practically use a narrative technique and the technical groups in narrative generation systems. However, such an approach may be interpreted as too individual and arbitrary in theoretical and academic senses. In any case, the author affirms that an eclectic method, including both directions or approaches, is realistically appropriate for the practical goal of the author's narrative generation study. The aim is, based on the process considering each object of Japanese narratological and literary theoretical research and critiques, to model and systematize them from the perspective of narrative generation research, and to build an entire theoretical system. Like this, a progression that blends both directions is in demand.

Meanwhile, narratologies and literary theories in Japan, like those addressed in this chapter, can be arranged assuming reflection on the narrative content knowledge bases in INGS and the narrative work resources in GIS. Specifically, a kind of editing method of narrative content seen in *Sekai Kōmoku* will give an indication as to the systematic construction method in GIS. Figures of *Goka Jūyō no Den*" and *Nidō Shibetsu no Den* in *Sakusha Shikihō Kezairoku*, as well as several theories related to *sekai*s (worlds) and *naimaze*, such as *shigumi–shuko* and *yokosuji–tatesuji*, are parts that need a detailed description and conceptualization. In addition, algorithms reference contemporary narratology and narrative theories. *Goka Jūyō no Den*, which includes the concepts of *jo–ha–kyū* by Zeami and *ki–shō–ten–gō* by Chinese poets, indicates a framework of synthesized narrative grammar. Other narratologies and literary theories in Japan also have diverse types of knowledge and insights that may appear through productive features in the connection with narrative generation. For instance, the idea of *kyojitsu hiniku* by Chikamatsu Monzaemon, is linked to the discussion about the adjustments of fiction and fact in narrative. There are books that can contribute to shedding light on the techniques of narrative generation, including *Bungakuron* by Natsume Sōseki, *Gengo ni totte Bi toha Nani ka* by Takaaki Yoshimoto, and *Nihon Shōsetsu Gijutsushi* by Naomi Watanabe (2012). The books by Yoshimoto and Watanabe are especially related to the aspect of narrative discourse. The literary theory by Sōseki covers both narrative content/stories and narrative discourse. For example, *Nihon Mukashibanashi Taisei* (Seki, 1978a, 1978b, 1978c, 1978d, 1978e, 1979a, 1979b, 1979c, 1979d, 1980; Seki, Nomura, & Ōshima, 1979, 1980) collects Japanese folktales and categorizes them into many kinds. Such studies will contribute to the facet of narrative content/stories. At the same time, this study accounts for the topic of narrative variation and transformation. It may be useful to examine the concrete methods of narrative discourse. As stated above, *kabuki* and other approaches to narratives and literature will make a substantial contribution to the author's narrative generation study in various senses. The following works will be scrutinized in parallel in the future:

- Analysis of narratologies and literary theories in Japan
- The organization from the angle of narrative generation:
    - Modeling and systematization toward/as INGS
    - Modeling and systematization toward/as GIS

This chapter covers two books by two contemporary researchers or critics and two classical documents written in the Edo period for the sake of introducing the author's narrative generation study in any form. In this chapter, the author could not present the computer programs for these materials; notwithstanding, the author describes the elementary plan for modeling and systematizing the above two classical results as mechanisms of narrative generation.

In recent *kabuki*, imaging techniques such as CG are often used on stage. In examples the author recently observed, *Konjaku Kyōen Senbonzakura* by Nakamura Shidō II (1972–), Sawamura Kuniya I(1978–), and Hatsune Miku (as a virtual character), etc. (Minamiza, Kyoto, August, 2019) and *Oguri* by Ichikawa En'nosuke IV (1975–) and Nakamura Hayato I (1993–), etc. (Shinbashi Enbujō, Ttokyo, Oct., 2019) are used frequently. *Yukinojō Henge* by Bandō Tamasaburō V, Ichikawa Chūsha IX (1965–), Nakamura Shichinosuke II (1983–), etc. (Kabukiza, Tokyo, August, 2019) are often used as visual images as a movie. However, there are no studies or practices that generate narratives such as scenarios, stories, and plots. As Ogata (2018a, 2018b) describes in detail, *kabuki* means a theatrical art that is performed on stage and witnessed by an audience. At the same time, *kabuki* involves enormous documents through language and images. They include documents called narratology or literary theories. *Sakusha Shikihō Kezairoku* (1972) and *Sekai Kōmoku* (1916) are documents of that type. Regarding this point, narratives such as scenarios, stories, and plots are sufficiently addressed in the case of *kabuki*. In the author's examination of *kabuki* (Ogata, 2016b, 2018a, 2018b, 2019c, 2020g; Ogata & Ono, 2019), the author analyzes and categorizes the multiple structural elements composing a *kabuki* work to be performed on stage, as well as related technical and rhetorical components. Next, the author proceeds to inspect each element and its relationships, and studies several topics. Furthermore, the study is extended to the level of sequential, repetitive, and circulative work generation. These surveys are done to introduce the results into GIS with INGS. Since a goal of GIS with INGS is develop the narrative generation mechanisms of *kabuki* in various senses, they include generating scenarios like *kabuki* narratives, and producing stage performance structures like *kabuki* from a story or plot, etc. Surveying *Sakusha Shikihō Kezairoku* and *Sekai Kōmoku* will also be a future meaningful topic in terms of modeling and systematization as concrete mechanisms in GIS with INGS.

Sakusha Shikihō Kezairoku as Narrative Generation

In various theories written in *Sakusha Shikihō Kezairoku*, the figure of *Goka Jūyō no Den* (Figure 10) provides a kind of formalized grammatical description. However, this book's author, Nyūgatei Ganyū, does not explain the figure. Other researchers do not state detailed explanations. The author would like to interpret this grammatical figure as a narrative development method based on a multiple structure. In particular, the author interprets this figure based on the division of "sequence (or chain)," "concept," and "method."

- [Sequence a]
    - 景様 (*Keiyō*)-頂上 (*Yama*)-揺 (*Yusuri*)-大曲 (*Ōkuruwa*)-鎌入 (*Kamaireru*)
- [Concept a]
    - *Sekai* (World)
- [Concept b]
    - *Shigumi*
    - *Shukō*

- [Sequence b]
    - *Jo–Ha–Kyū*
- [Sequence c]
    - *Ki–Shō–Ten–Gō*
- [Complicated Method]
    - *Jo* is connected with *Ki–Shō*, *Ha* is connected with *Shō–Ten*, and *Kyū* is connected with *Ten–Gō*. The moving among elements in *Jo–Ha–Kyū* are expanded using *Ki–Shō–Ten–Gō*.

The author interprets this hierarchical, multiple story grammar of *kabuki*. First, the above "Sequence a" is a global structure of a *kabuki* scenario to be finally composed. "Sequence b" is the most abstract structure and "Sequence c" is the intermediate structure. In contrast to these, "*Sekai*" in "Concept a" corresponds to the most primary principle of *kabuki* scenarios, and both "*Shigumi*" and "*Shukō*" (in "Concept b") guide the *sekai* toward a specific scenario structure. Moreover, the concept of *Tate-suji* (vertical line) and *Yoko-suji* (lateral line) may be related to the above narrative development. Based on the above ideas, the author advances more concrete consideration of multiple grammar for *kabuki* scenarios.

Sekai Kōmoku as Narrative Generation

Next, the author examines how *Sekai Kōmoku* can be described as a computer program. First, the author explores a method that faithfully follows the description shown in *Sekai Kōmoku*. For the preparation, elements in each unit in *Sekai Kōmoku* need to be conceptually structured, represented, and described in some manner. The key elements include names of characters (including fictional characters and real ones), prior works (including various narrative genres), and *ningyō jōruri* works. As the first "names of characters" also appear in prior narrative works, the decomposition and anatomy of related narrative works in some way are essential tasks for program design and implementation. The decomposition and anatomy can be executed for the names of characters, plots and stories, dramatic places, (important) objects, etc. The first subject for programming is decomposing and dissecting necessary and possible elements for narrative generation until a necessary and possible level is reached. If the knowledge structure of *Sekai Kōmoku* corresponds to the data in the computer program, the use and operation should be performed using other mechanisms. For example, the narrative grammar and narrative techniques written in *Sakusha Shikihō Kezairoku* may be able to be implemented for use and operation. Utilizing examples of characters and narrative works stored in the description of *Sekai Kōmoku*, the mechanism can be composed of relatively global narrative frameworks, which are included in the narrative production part of GIS in the author's study framework. Thereafter, a type of narrative skeleton, as a result of the above process, is passed on to the module of INGS to place various concreteness and elaboration within it.

How the Narrative Generation Study of *Kabuki* is Tied to GIS with INGS and *Watakushi Monogatari*

In this section, the author considers a way to introduce theories and methods in *Sakusha Shikihō Kezairoku* and *Sekai Kōmoku* into GIS with INGS. Generally speaking, the computer program that models and implements the theories of *Sakusha Shikihō Kezairoku* and the knowledge structure of *Sekai Kōmoku* is positioned as parts of GIS. It, namely GIS with these elements, produces the most global structure of a narrative scenario, and INGS generates a detailed description, namely conceptual representation structures of a narrative. Although this process reveals narrative generation as a single substance, its consecutive processing forms a narrative sequence for *geinō*. Narratives produced through these processes correspond to narratives as works, except for life narratives. If we, so to speak, overestimate the value in the theories and methods of *kabuki* discussed here, the *geinō* in GIS means that *kabuki* and GIS form a system to be concretely developed, reflecting and introducing the results of the author's narrative generation study of *kabuki*. This is an important problem for this study.

Sekai Kōmoku indicates the systematic composition method of the *geinō* work resource. One future topic based on the extension of *Sekai Kōmoku* is to construct a new type of narrative knowledge base in which personally defined and collected narrative knowledge is stored. Works of *Watakushi Monogatari*, through GIS with INGS, are created by both the systems themselves and a *Watakushi Character* as the author's "I" (*wataskushi*), who utilizes the knowledge base systematized like *Sekai Kōmoku*, and the development methods.

CONCLUSION

This chapter dealt with the following themes in three central sections except for the **INTRODUCTION, BACKGROUND, SUPPLEMENTAL DISCUSSION AND DIRECTIONS FOR FUTURE RESEARCH**, and this **CONCLUSION**:

1. **A MULTIPLE NARRATIVE GENERATION MECHANISM THROUGH TWO TYPES OF NARRATIVE GENERATION SYSTEMS: GIS WITH INGS** addressed topics regarding the design of GIS with INGS.
2. **NARRATOLOGY OF *KABUKI*: ORIENTED TOWARD NARRATIVE GENERATION** focused on two contemporary *kabuki* researchers and two classical documents of *kabuki*.
3. **THE CONCEPT OF *WATAKUSHI MONOGATARI* AND DESIGNING INTERNAL AND EXTERNAL NARRATIVE GENERATION METHODS TO CREATE AND DISTRIBUTE NARRATIVE CONTENT** discussed the concepts of internal and external narrative generation, as well as their mutual relationships.

In **1**, the author described major elements and processes of GIS. The elements include the *geinōjin* resource, the *geinō* work resource, the life work resource, and *geinōjin* characters. Furthermore, this broader section discusses how the function of *Watakushi Monogatari*'s creation is incorporated into GIS with INGS, and explains the overall behavior of GIS with INGS in the context of associations with *Watakushi Monogatari*.

In **2**, the author surveyed the philosophies of two contemporary *kabuki* researchers, namely Yukio Hattori's *The Structure of Kabuki* and Tamotsu Wanatabe's philosophical and narratological thought of *kabuki* through a typical *kabuki*-dance, *Kyōganoko Musume Dōjōji*. Next, as classical theories of *kabuki*, the author scrutinized two documents, *Sakusha Shikihō Kezairoku* and *Sekai Kōmoku* from the narrative generation system's point of view. Furthermore, the following sections considered the design in terms of the narrative generation of these documents or related parts.

In **3**, after this section shows the basic configuration of GIS with INGS for *Watakushi Monogatari*, the author discussed pertinent themes around internal and external narrative generation. Internal narrative generation means the creation of *Watakushi Monogatari*; part of the policy of content by internal narrative generation includes fundamental knowledge sources for the narrative content of *Watakushi Monogatari*, and two directions of narrative content for *Watakushi Monogatari* based on the concepts of fluidity and fixation or movement and stillness. Although the external narrative generation is regarded as the sociological distribution of *Watakushi Monogatari*, it is essentially associated with the aspect of internal narrative generation. Hence, the final part addressed an integrated method between internal and external narrative generation.

Finally, the macro-level goal of this chapter is for the author to concretely examine how to bridge the gap between narrative generation systems and narrative content with *kabuki*-oriented narratology and *Watakushi Monogatari*. This goal is also partially related to the title, a theme of the entirety of this book. Through this chapter, the author achieves this goal using the following methods. First, the narrative generation systems correspond to GIS with INGS, which the author has been proposing and developing, and the target of narrative content is set, including *Watakushi Monogatari* and related narrative content. Moreover, the gap between GIS with INGS and *Watakushi Monogatari* is bridged based on narrative knowledge, including *kabuki*-oriented narratological knowledge and diverse knowledge resources, such as knowledge related to the author's "I" (*watakushi*). Based on the above established framework, which depends on GIS with INGS, *Watakushi Monogatari*, and other knowledge resources, the author would like to develop necessary systems and knowledge, and experimentally create fragmented content for *Watakushi Monogatari*.

ACKNOWLEDGMENT

The research for this chapter was supported by the Japan Society for the Promotion of Science (JSPS KAKENHI), Grant No. 18K18509.

REFERENCES

Amino, T., Kawamura, Y., & Ogata, T. (2002). Hierarchical generation of geinō-idol stories—Toward geinō information system and narrative marketing. In *Proceedings of 17th Congress of the International Association of Empirical Aesthetics* (pp. 549–552). Rome, Italy: University of Rome Tre.

Arai, T., Ono, J., & Ogata, T. (2016). Semi-automatic generation of events sequence knowledge for narrative generation: The use in an integrated narrative generation system. In *Proceedings of the 30th Annual Conference of the Japanese Society for Artificial Intelligence* (3P1-1in2). Tokyo, Japan: Japanese Society for Artificial Intelligence.

Arishima, T. (1964). Umareizuru nayami. In Arishima Takeo shū (pp. 350–393). Tokyo, Japan: Chikuma Shobō. (Original work published 1918)

Arishima, T. (1995). Aru on'na. Tokyo, Japan: Shinchōsha. (Original work published 1919)

Aristotle. (1997). Poetics (M. Heath, Trans.). London, UK: Penguin Classics. (Original work published 4th century BC)

Asada, I., & Namiki, S. (1995). Dōjōji genzai uroko. In Toyotakeza jōruri syū, 3 (pp. 7-100). Tokyo, Japan: Kokusho Kankōkai. (First performed in 1742)

Bandō Tamasaburō, V. (2003). *Bandō Tamasaburō buyōsyū* [Motion Picture DVD]. Syōchiku Home Video.

Bronte, E. (2002). Wuthering heights. London, UK: Penguin Books. (Original work publishued 1847)

Cortázar, J. (1991). *Hopscotch: A novel* (M. Heath, Trans.). Pantheon. (Original work published 1963)

Dainipponkoku hokkekyō genki. (1974). In *Ōjōden Hokkegenki* (pp. 43-219). Tokyo, Japan: Iwanami Shoten. (Original work published 1040–1043)

Dōjōji. (1998). In Yōkyoku hyakuban, Shin nihon koten bungaku taikei, 57 [Nō 100 songs, New Japanese classic literature collection, 57] (pp. 52–57). Tokyo, Japan: Iwanami Shoten. (Original work published the Muromachi period)

Dōjōji engi. (1982). In *Muromachi jidai monogatari taisei, daiju, ten-hamo* [The Muromachi period narrative collection, No. 10, ten-hamo] (pp. 97–103). Tokyo, Japan: Kadokawa Shoten. (Original work published the Muromachi period)

Dōjōji monogatari. (1982). In *Muromachi jidai monogatari taisei, daiju, ten-hamo* [The Muromachi period narrative collection, No. 10, ten-hamo] (pp. 104–124). Tokyo, Japan: Kadokawa Shoten. (Original work published 1660, the Edo period)

Dostoevsky, F. (2003). *The brothers Karamazov: A novel in four parts and an epilogue* (D. McDuff, Trans.). Penguin Classics. (Original work published 1880)

Dostoevsky, F. (2014). *Crime and punishment* (O. Ready, Trans.). Penguin Classics. (Original work published 1866)

Fujii, S. (2004). *Monogatari riron kōgi* [Lectures on the theory of narrative]. Tokyo Daigaku Shuppankai.

Genette, G. (1972). *Discours du récit, Essai de méthode, Figures III.* Paris: Seuil.

Genji monogatari. (1993). *Shin nihon koten bungaku taikei, 19* [New Japanese classic literature collection, 19]. Tokyo, Japan: Iwanami Shoten. (Original work published early 11th century, the Heian period)

Genji monogatari. (1994). *Shin nihon koten bungaku taikei, 20* [New Japanese classic literature collection, 20]. Tokyo, Japan: Iwanami Shoten. (Original work published early 11th century, the Heian period)

Genji monogatari. (1995). *Shin nihon koten bungaku taikei, 21* [New Japanese classic literature collection, 21]. Tokyo, Japan: Iwanami Shoten. (Original work published early 11th century, the Heian period)

Genji monogatari. (1996). *Shin nihon koten bungaku taikei, 22* [New Japanese classic literature collection, 22]. Tokyo, Japan: Iwanami Shoten. (Original work published early 11th century, the Heian period)

Genji monogatari. (1997). *Shin nihon koten bungaku taikei, 23* [New Japanese classic literature collection, 23]. Tokyo, Japan: Iwanami Shoten. (Original work published early 11th century, the Heian period)

Genkyō shakusho. (1965). In *Shintei zōho kokushi taikei, 31* [New revision, supplement, national histories collection, 31]. Tokyo, Japan: Toshikawa Kobunkan. (Original work published 1322 (the Kamakura period)

Gervás, P. (2013). Propp's morphology of the folk tale as a grammar for generation. *OpenAccess Series in Informatics., 32*, 106–122. doi:10.4230/OASIcs.CMN.2013.106

Haniya, Y. (1998). Shirei [Death Spirits]. Tokyo, Japan: Kōdansha (Original work publishued 1948-1995)

Hattori, Y. (1970). Kabuki no kōzō [The structure of kabuki]. Tokyo, Japan: Iwanami Shoten.

Hemingway, E. (1994). *For whom the bell tolls*. Arrow. (Original work published 1940)

Hesse, H. (2013). *Demian: The Story of Emil Sinclair's Youth* (D. Searls, Trans.). Penguin Random House. (Original work published 1919)

Hesse, H. (2015). *Steppenwolf* (D. Horrocks, Trans.). Picador USA. (Original work published 1927)

Homer. (2006). The Iliad & The Odyssey (B. Williams, Trans.). TX: El Paso Norte Press. (Original work published the late eighth or early seventh century BC)

Hori, K. (2007). Sōzō katsudō shien system no riron to ōyō [Theories and applications of creativive activity aid]. Tokyo, Japan: Ohmsha.

Hugo, V. (1982). *Les miserable* (N. Denny, Trans.). Penguin Classics. (Original work published 1862)

Inoue, T. (2015). *Mishima Yukio "Hōjō no umi" vs Noma Hiroshi "Seinen no wa"—Sengo bungaku to zentai shōsetsu* [Mishima Yukio *The sea of fertility* vs. Noma Hirosi *The ring of young people*: Post-war literature and total novels]. Shintensha.

Iura, Y. (1990). "Sakusha shikihō kezairoku" kō (zen)—Mondaiten no shiteki to kaimei [Considering "Sakusha shikihō kezairoku" (the first part): Indication and elucidation of problems]. *Gakuen, 610*, 2–13.

Iura, Y. (1991). "Sakusha shikihō kezairoku" kō (go)—Mondaiten no shiteki to kaimei [Considerating "Sakusha shikihō kezairoku" (the second part): Indication and elucidation of problems). *Gakuen, 615*, 1–13.

Jauss, H. R. (1970). Literaturgeschichte als provokation. Frankfurt am Main: Suhrkamp Verlag.

Joyce, J. (2010). *Ulysses*. Wordsworth Editions. (Original work published 1922)

Katō, T. (1997a). "Sekai kōmoku" no shohon—Sono ichizuke to tensha no keii [Various books of "sekai kōmoku": Their positioning and the process of transcription]. *Gakushūin Daigaku Kokugo Kokubungaku Kaishi, 40*, 79–94.

Katō, T. (1997b). "Sekai kōmoku" no seiritsu [The establishment of "Sekai kōmoku"]. *Gakshūuin Diagaku Jinbun Kagaku Ronshū, 6*, 77–100.

Kawai, M., & Ogata, T. (2019). An analysis of "*Kyōganoko musume Dōjōji*" for the generation of the staging structures of *kabuki* from narrative plots. In *Proceedings of the 63th Special Interest Group on Language Sense Processing Engineering* (pp. 25–61). Tokyo, Japan: The Japanese Society for Artificial Intelligence.

Kawai, M., Ono, J., & Ogata, T. (2020). Analyzing the stage performance structure of *kabuki*-dance "*Kyoganoko Musume Dojoji*." In *Proceedings of the 34th Annual Conference of the Japanese Society of Artificial Intelligence* (3D1-OS-22a-04). Tokyo, Japan: Japanese Society of Artificial Intelligence.

Kawatake, T. (1989). *Kabuki-bi ron* [On *kabuki*'s beauty]. Tokyo Daigaku Shuppankai.

Kawatake, T. (1992). *Kabuki: Its beauty and tradition* (C. Holmes, Trans.). Japan Arts Council.

Kawatake, T. (2003). Kabuki: Baroque fusion of the arts (Connel Hoff, F., & J., Trans.). Tokyo, Japan: The International House of Japan.

Ki no kuni muro no kōri no ashiki on'na. (1974). In *Ōjōden Hokkegenki* (pp. 217–219). Tokyo, Japan: Iwanami Shoten. (Original work published 1040–1043)

Ki no kuni no Dōjōji no sō hōke wo utsushi hemi wo sukueru koto. (1993). In *Konjaku monogatari shū, 3* (pp. 286–290). Tokyo, Japan: Iwanami Shoten. (Original work published the end of the Heian period)

Kominami, T. (1998). "Kezairoku" no kōsatsu—Engeki ni okeru kankyaku ninshiki no shomondai [Consideration of "Kezairoku": Problems of audience recognition in a play]. *Otaru Joshi Tanki Daigaku Kenkyū Kiyō, 27*, 1–42.

Konjaku monogatari shū, 1. (1999a). In *Shin nihon koten bungaku taikei, 33* [New Japanese classic literature collection, 33]. Tokyo, Japan: Iwanami Shoten. (Original work published the end of the Heian period)

Konjaku monogatari shū, 2. (1999b). In *Shin nihon koten bungaku taikei, 34* [New Japanese classic literature collection, 34]. Tokyo, Japan: Iwanami Shoten. (Original work published the end of the Heian period)

Konjaku monogatari shū, 3. (1993). In *Shin nihon koten bungaku taikei, 35* [New Japanese classic literature collection, 35]. Tokyo, Japan: Iwanami Shoten. (Original work published the end of the Heian period)

Konjaku monogatari shū, 4. (1994). In *Shin nihon koten bungaku taikei, 36* [New Japanese classic literature collection, 36]. Tokyo, Japan: Iwanami Shoten. (Original work published the end of the Heian period)

Konjaku monogatari shū, 5. (1996). In *Shin nihon koten bungaku taikei, 37* [New Japanese classic literature collection, 37]. Tokyo, Japan: Iwanami Shoten. (Original work published the end of the Heian period)

Kurihara, K. (Ed.). (1886). *Kokon hauta taizen (Kaichū keikobon)*. Eirakudō.

Kyōganoko musume Dōjōji. (1970). In *Meisaku kabuki zenshū, 19, Buyōgeki shū* [Masterpeace *kabuki* collection, 19, *Kabuki*-dance works] (pp. 15–30). Tokyo, Japan: Tokyo Sōgen Shinsha. (First performed in 1753. Written by Fujimoto Tobun and composed by Kineya Yasaburō)

Longus. (2019). *Daphnis and Chloe* (E. Bowie, Trans.). Cambridge, UK: Cambridge University Press. (Original work published the second century)

Malraux, A. (1990). *Man's fate* (H. M. Chevalier, Trans.). Vintage. (Original work published 1933)

Mann, T. (2005). *The magic mountain* (J. E. Woods, Trans.). Everyman. (Original work published 1924)

Matsumoto, S. (1961). Warui yatsura [Bad Sorts]. Tokyo, Japan: Shinchōsha.

Matsumoto, S. (1962). Renkan [Linkage]. Tokyo, Japan: Kōdansha.

Matsumoto, S. (1980). Kurokawa no techō [Pocketbook of black leather]. Tokyo, Japan: Shinchōsha.

Matsumoto, S. (1981). Yakō no Kaidan [Stairs That Shine at Night]. Tokyo, Japan: Shinchōsha.

Melville, H. 's (2009). Moby-Dick: Or, the whale. London, UK: Penguin Classics. (Original work published 1851)

Miidera (1998). In *Yōkyoku hyakuban, Shin nihon koten bungaku taikei, 57* [Nō 100 songs, New Japanese classic literature collection, 57] (pp. 52–57). Tokyo, Japan: Iwanami Shoten. (Original work published the Muromachi period)

Mishima, Y. (1950). Ao no Jidai. Tokyo, Japan: Shinchōsha.

Mishima, Y. (1954a). Shiosai. Tokyo, Japan: Shinchōsha.

Mishima, Y. (1954b). Iwashiuri koi no hikiami [The sardine seller's net of love]. *Engekikai*, 11.

Mishima, Y. (1956). Kinkakuji. Tokyo, Japan: Shinchōsha.

Mishima, Y. (1962). Utsukushii hoshi. Tokyo, Japan: Shinchōsha.

Mishima, Y. (1968). Taiyō to tetsu. Tokyo, Japan: Kōdansha.

Mishima, Y. (1969a). Haru no yuki. Tokyo, Japan: Shinchōsha.

Mishima, Y. (1969b). Honba. Tokyo, Japan: Shinchōsha.

Mishima, Y. (1970). Akatsuki no tera. Tokyo, Japan: Shinchōsha.

Mishima, Y. (1971). Ten'nin gosui. Tokyo, Japan: Shinchōsha.

Mishima, Y. (1951, 1953). Kinjiki. Tokyo, Japan: Shinchōsha.

Mishima, Y. (1969c-1971). *Hōjō no umi* (4 books). Tokyo, Japan: Shinchōsha.

Mishima, Y. (2000–2006). *Ketteiban Mishima Yukio zenshū, 1–42* [Complete works of Mishima Yukio (definitive edition), Vols. 1–42]. Tokyo, Japan: Shinchōsha.

Mitchell, M. (2014). *Gone with the wind.* Pan Books. (Original work published 1936)

Miura, H. (2000). Naimaze. In Y. Hattori, T. Tomita, & T. Hirosue, (Eds.). Shinpan kabuki jiten [New edition kabuki dictionry] (p. 310). Tokyo, Japan: Heibonsha.

Momochidori musume Dōjōji. (1960). In T. Takano (Ed.), Nihon kayō shūsei, 9, Kinseihen (p. 86). Tokyo, Japan: Tokyodō Shuppan.

Moriyama, S. (1993). Tsuruya Nanboku—Naimaze no sekai [Tsuruya Nanboku: Naimaze's world]. Tokyo, Japan: San'ichi Shobō.

Nagauta genpon shūsei kankōkai. (Ed.). (1927). *Momochidori musume Dōjōji (Nagauta genpon shūsei, Vol. 1).* Tokyo, Japan: Nagauta Genpon Shūsei Kankōkai.

Natsume, S. (2007). Bungaku ron (I, II) [Literary theory (I, II)]. Tokyo, Japan: Iwanami Shoten. (Original work published 1907)

Natsume, S. (2014). Kōfu. Tokyo, Japan: Iwanami Shoten. (Original work published 1908)

Noma, H. (1983-84). *Seinen no wa, 1–5*. Tokyo, Japan: Iwanami Shoten. (Original work published 1949–1968)

Ōe, K. (1979). Dōjidai game. Tokyo, Japan: Shinchōsha.

Ogata, T. (2002a). Expanded literary theory: Cognitive/computational expansion of literary theories and narratology. In *Proceedings of the 17th Congress of the International Association of Empirical Aesthetics* (pp. 163–166). Rome, Italy: University of Rome Tre.

Ogata, T. (2002b). The concept of system narratology: From the viewpoint of expanded literary theory. In *Proceedings of the 3rd International Workshop of Literature in Cognition and Computer in held in PRICAI2002* (W7-1, pp. 1–10). Tokyo, Japan: Japanese Cognitive Science Society.

Ogata, T. (2003a). Monogatari no tajūsei to kakuchō bungakuriron no gainen—System narratology ni mukete, I [Narrative multiplicity and the concept of expanded literary theory: Toward a system narratology, I]. In M. Yoshida (Ed.), Fukuzatsu-kei shakai riron no shin chihei [New paradigm of complex social system theory] (pp. 127–181). Tokyo, Japan: Senshu Daigaku Shuppankyoku.

Ogata, T. (2003b). Kakuchō bungakuriron no kokoromi—System narratology ni mukete, II [Attempts of expanded literary theory: Toward a system narratology II]. In M. Yoshida (Ed.), Fukuzatsu-kei shakai riron no shin chihei [New paradigm of complex social system theory] (pp. 309–356). Tokyo, Japan: Senshu Daigaku Shuppankyoku.

Ogata, T. (2010). Shōsetsu—Ryūdō to kotei, Sakuhin no hō he [Novels: Fluidity and fixation, toward works]. In T. Ogata & A. Kanai (Eds.), Monogatariron no jōhōgaku josetsu—Monogatari seisei no shisō to gijutsu wo megutte [An introduction to informatics of narratology: Around the thoughts and technologies of narrative generation] (pp. 130–169). Tokyo, Japan: Gakubunsha.

Ogata, T. (2014). Expanded literary theory for automatic narrative generation. In *Proceedings of Joint 7th International Conference on Soft Computing and Intelligent Systems and 15th International Symposium on Advanced Intelligent Systems* (pp. 1558–1563). Tokyo, Japan: Japan Society of Fuzzy Theory and Intelligent Informatics.

Ogata, T. (2016a). Narrative generation of *zukushi*—Towards its systematization through surveying *kabuki* and developing a proper noun conceptual dictionary in INGS. In *Proceedings of the 30th Annual Conference of the Japanese Society of Artificial Intelligence* (1K5-OS-06b-2). Tokyo, Japan: The Japanese Society for Artificial Intelligence.

Ogata, T. (2016b). Kabuki as multiple narrative structures. In T. Ogata & T. Akimoto (Eds.), Computational and cognitive approaches to narratology (pp. 391–422). Hershey, PA: Information Science Reference (IGI Global).

Ogata, T. (2018a). Kabuki ni mukete (1)—Shiisei to henshūsei no monogatari kara tajūsei to jitsuzon no monogatari he [Toward kabuki (1): From narratives of arbitrariness and editing to narratives of multiplicity and real existence]. In T. Ogata, Y. Kawamura, & A. Kanai (Eds.), Jōhō monogatariron—Jinkōchinō, ninchi, shakai katei to monogatari seisei [Informational narratology: Artificial intelligence/cognition/social process and narrative generation] (pp. 187–208). Tokyo, Japan: Hakutō Shobō.

Ogata, T. (2018b). Kabuki ni mukete (2)—Tajū monogatari kōzō no shosō [Toward kabuki (2): Aspects of multiple narrative structures]. In T. Ogata, Y. Kawamura, & A. Kanai (Eds.), Jōhō monogatariron—Jinkōchinō, ninchi, shakai katei to monogatari seisei [Informational narratology: Artificial intelligence/cognition/social process and narrative generation] (pp. 209–244). Tokyo, Japan: Hakutō Shobō.

Ogata, T. (2018c). Gaibu he no monogatari seisei mata ha geinō jōhō system ni mukete [Narrative generation to the outside or toward a geinō information system]. In T. Ogata, Y. Kawamura, & A. Kanai (Eds.), Jōhō monogatariron—Jinkōchinō, ninchi, shakai katei to monogatari seisei [Informational narratology: Artificial intelligence/cognition/social process and narrative generation] (pp. 327–353). Tokyo, Japan: Hakutō Shobō.

Ogata, T. (2018d). Naibu he no monogatari seisei mata ha watakushi monogatari ni mukete [Narrative generation to the inside or toward I narratives]. In T. Ogata, Y. Kawamura, & A. Kanai (Eds.), Jōhō monogatariron—Jinkōchinō, ninchi, shakai katei to monogatari seisei [Informational narratology: Artificial intelligence/cognition/social process and narrative generation] (pp. 355–370). Tokyo, Japan: Hakutō Shobō.

Ogata, T. (2018e). An integrated approach to narrative generation: From Mishima and kabuki to narrative generation systems. In T. Ogata & S. Asakawa (Eds.), Content generation through narrative communication and simulation (pp. 49–147). Hershey, PA: Information Science Reference (IGI Global)

Ogata, T. (2019a). A computational, cognitive, and narratological approach to narrative generation. In T. Ogata, & T. Akimoto (Eds.), Post-narratology through computational and cognitive approaches (pp.1-84). Hershey, PA: Information Science Reference (IGI Global). doi:10.4018/978-1-5225-7979-3.ch001

Ogata, T. (2019b). Toward a post-narratology or the narratology of narrative Generation. In T. Ogata, & T. Akimoto (Eds.), Post-narratology through computational and cognitive approaches (pp. 85–142). Hershey, PA: Information Science Reference (IGI Global). doi:10.4018/978-1-5225-7979-3.ch002

Ogata, T. (2019c). Kabuki as Multiple Narrative Structures and Narrative Generation. In T. Ogata & T. Akimoto (Eds.), Post-narratology through computational and cognitive approaches (pp. 192–275). Hershey, PA: Information Science Reference (IGI Global). doi:10.4018/978-1-5225-7979-3.ch005

Ogata, T. (2019d). A method of *naimaze* of narratives based on *kabuki* analyses and Propp's move techniques for an automated narrative generation system. *Journal of Robotics. Networking and Artificial Life*, *6*(2), 71–78. doi:10.2991/jrnal.k.190829.001

Ogata, T. (2019e). Toward the incorporation of Japan's narratologies and literary theories into the narrative generation systems by the author. In *Proceedings of the 36th Annual Conference of the Japan Cognitive Science Society* (pp. 442–451). Kanagawa, Japan: Cognitive Science Society.

Ogata, T. (2020a). Toward an integrated approach to narrative generation: Emerging eesearch and opportunities. Hershey, PA: Information Science Reference (IGI Global). doi:10.4018/978-1-5225-9693-6

Ogata, T. (2020b). Areas of narratives or narrative genres. In Toward an integrated approach to narrative generation: Emerging research and opportunities (pp. 59–161). Hershey, PA: Information Science Reference (IGI Global). doi:10.4018/978-1-5225-9693-6.ch002

Ogata, T. (2020c). Narratology and post-narratology. In Toward an integrated approach to narrative generation: Emerging research and opportunities (pp. 162–314). Hershey, PA: Information Science Reference (IGI Global). doi:10.4018/978-1-5225-9693-6.ch003

Ogata, T. (2020d). Theoretical or philosophical considerations for an integrated narrative generation approach. In Toward an integrated approach to narrative generation: Emerging research and opportunities (pp. 315–403). Hershey, PA: Information Science Reference (IGI Global). doi:10.4018/978-1-5225-9693-6.ch004

Ogata, T. (2020e). Internal and external narrative generation based on post-narratology: Emerging research and opportunities. Hershey, PA: Information Science Reference (IGI Global). doi:10.4018/978-1-5225-9943-2

Ogata, T. (2020f). An integrated narrative generation system: Synthesis and expansion. In Internal and external narrative generation based on post-narratology: Emerging research and opportunities (pp. 1–108). Hershey, PA: Information Science Reference (IGI Global).

Ogata, T. (2020g). Kabuki as a synthetic narrative: Synthesis and expansion. In Internal and external narrative generation based on post-narratology: Emerging research and opportunities (pp. 109–254). Hershey, PA: Information Science Reference (IGI Global).

Ogata, T. (2020h). Toward external and internal practices of narrative generation. In Internal and external narrative generation based on post-narratology: Emerging research and opportunities (pp. 328–427). Hershey, PA: Information Science Reference (IGI Global). doi:10.4018/978-1-5225-9943-2.ch004

Ogata, T., & Akimoto, T. (Eds.). (2019). Post-narratology through computational and cognitive approaches. Hershey, PA: Information Science Reference (IGI Global). doi:10.4018/978-1-5225-7979-3

Ogata, T., & Amino, T. (2002). Tajū monogatari kōzō to *geinō* character no monogatari seisei system [A narrative generation system of multiple narrative structures and a *geinō* character]. In *Proceedings of the 19th Annual Meeting of the Japanese Cognitive Science Society* (pp. 130–131). Tokyo, Japan: Japanese Cognitive Science Society.

Ogata, T., Fukuda, K., Ono, J., & Ito, K. (2020). Considering *naimaze* in *kabuki* narratives and a method. In *Proceedings of the 34th Annual Conference of the Japanese Society of Artificial Intelligence* (3P5-OS-16b-02). Tokyo, Japan: Japanese Society of Artificial Intelligence.

Ogata, T., Imabuchi, S., & Akimoto, T. (2014). Narratology and narrative generation: Expanded literary theory and the Integration as a narrative generation system (2). In *Proceedings of the 5th Augmented Human International Conference* (P-10). ACM Digital Library. 10.1145/2582051.2582077

Ogata, T., & Kanai, A. (2010). Monogatariron no jōhōgaku josetsu—Monogatari seisei no shisō to gijutsu wo megutte [An introduction to informatics of narratology: Around the thoughts and technologies of narrative generation]. Tokyo, Japan: Gakubunsha.

Ogata, T., & Ono, J. (2019). Two elements and two techniques for the narrative generation of *kabuki*. In *Proceedings of the 33th Annual Conference of the Japanese Society of Artificial Intelligence* (1F2-NFC-1-03). Tokyo, Japan: Japanese Society of Artificial Intelligence.

Okamoto, K. (1969). Banchō sarayashiki. In Meisaku kabuki zenshū, 20, Shinkabuki shū [Masterpeace kabuki collection, 20, New kabuki works] (pp. 189–244). Tokyo, Japan: Tokyo Sōgen Shinsha. (First performed in 1916)

Otokodate hatsugai soga. (1985). In *Kabuki daichō shūsei, 8* [*Kabuki* script collection, 8] (pp. 75–240). Tokyo, Japan: Benseisha. (First performed in 1759. Written by Fujimoto Tobun)

Propp, V. Y. (1968). *Morphology of the folktale* (L. Scott, Trans.). University of Texas Press. (Original work published 1928)

Rolland, R. (2007). *Jean-Christophe: Journey's end* (G. Cannan, Trans.). Echo Library.

Rolland, R. (2011). *Jean-Christophe* (Vol. I; G. Cannan, Trans.). Aeterna.

Rolland, R. (2018). *Jean-Christophe in Paris: The market-place, Antoinette, the house* (G. Cannan, Trans.). Sagwan Press.

Sakusha shikihō kezairoku. (1972). In M. Gunji (Ed.), *Kinsei geidōron* (*Nihon shisō taikei, 61*) [Theories of early modern times' arts (Japanese classic phirosophical thoughts collection, Vol. 61)] (pp. 493–532). Tokyo, Japan: Iwanami Shoten. (Original work published 1801)

Schank, R. C., & Abelson, R. P. (1977). *Scripts, plans, goals, and understanding: An inquiry into human knowledge structures*. Lawrence Erlbaum.

Sekai kōmoku. (1916). In *Sekai kōmoku, shibai nenjū gyōji, gekikai chinwa* (*Chinsho kankōkai sōsyo, 9*) [Compendium of narrative worlds, annual events related to kabuki plays, rare tales in theatrical circles (Rare books publication society series, Vol. 9)] (pp. 6–46). Tokyo, Japan: Chinsyo kankōkai (Written in the late Edo period (before 1791))

Seki, K. (1978a). Nihon mukashibanashi taisei, 2—Honkaku mukashibanashi, 1 [Compilation of Japanese folktales, Vol. 2: Ordinary folktales, No. 1]. Tokyo, Japan: Kadokawa Shoten.

Seki, K. (1978b). Nihon mukashibanashi taisei, 3—Honkaku mukashibanashi, 2 [Compilation of Japanese folktales, Vol. 3: Ordinary folktales, No. 2]. Tokyo, Japan: Kadokawa Shoten.

Seki, K. (1978c). Nihon mukashibanashi taisei, 4—Honkaku mukashibanashi, 3 [Compilation of Japanese folktales, Vol. 4: Ordinary folktales, No. 3]. Tokyo, Japan: Kadokawa Shoten.

Seki, K. (1978d). Nihon mukashibanashi taisei, 5—Honkaku mukashibanashi, 4 [Compilation of Japanese folktales, Vol. 5: Ordinary folktales, No. 4]. Tokyo, Japan: Kadokawa Shoten.

Seki, K. (1978e). Nihon mukashibanashi taisei, 6—Honkaku mukashibanashi, 5 [Compilation of Japanese folktales, Vol. 6: Ordinary folktales, No. 5]. Tokyo, Japan: Kadokawa Shoten.

Seki, K. (1979a). Nihon mukashibanashi taisei, 1—Dōbutsu monogatari [Compilation of Japanese folktales, Vol. 1: Animal tales]. Tokyo, Japan: Kadokawa Shoten.

Seki, K. (1979b). Nihon mukashibanashi taisei, 7—Honkaku mukashibanashi, 6 [Compilation of Japanese folktales, Vol. 7: Ordinary folktales, No. 6]. Tokyo, Japan: Kadokawa Shoten.

Seki, K. (1979c). Nihon mukashibanashi taisei, 8—Warai banashi, 1 [Compilation of Japanese folktales, Vol. 8: Jokes and anecdotes, No. 1]. Tokyo, Japan: Kadokawa Shoten.

Seki, K. (1979d). Nihon mukashibanashi taisei, 9—Warai-banashi (Shōwa), 2 [Compilation of Japanese folktales, Vol. 9: Jokes and anecdotes, No. 2]. Tokyo, Japan: Kadokawa Shoten.

Seki, K. (1980). Nihon mukashibanashi taisei, 10—Warai-banashi (Shōwa), 3 [Compilation of Japanese folktales, Vol. 10: Jokes and anecdotes, No. 3]. Tokyo, Japan: Kadokawa Shoten.

Seki, K., Nomura, J., & Ōshima, H. (Eds.). (1979). Nihon mukashibanashi taisei, 12—Kenkyū hen [Compilation of Japanese folktales, 12: Researches]. Tokyo, Japan: Kadokawa Shoten.

Seki, K., Nomura, J., & Ōshima, H. (Eds.). (1980). Nihon mukashibanashi taisei, 11—Shiryō hen [Compilation of Japanese folktales, 11: Data]. Tokyo, Japan: Kadokawa Shoten.

Sophocles. (1972). Oedipus the king (A. Burgess, Trans.). Minneapolis, MN: University of Minnesota Press. (Original work first performed around 429 BC)

Taikōki. (1996). In Shin nihon koten bungaku taikei, 60 [New Japanese classic literature collection, Vol. 60]. Tokyo, Japan: Iwanami Shoten (Original work published 1626)

Tanaka, K. (2010). Kigō to saiki—Kigōron no keishiki, Program no hitsuzen [Symbol and recursion: The forms of semiotics, the inevitability of programs]. Tokyo, Japan: Tokyo Daigaku Shuppankai (The University of Tokyo Press)

Tanaka, Y. (1981). Nantonaku kurisutaru. Tokyo, Japan: Chikuma Shobō Shinsha.

Teramoto, Y., & Nakanishi, A. (1993). Chi no tenkansha tachi—Business trickstar [Transformers of knowledge: Business tricksters]. Tokyo, Japan: NTT Shuppan.

Tolstoy, L. (2010). *War and peace* (A. Mandelker, Ed., MaudeL.MaudeA., Trans.). Oxford University Press. (Original work published 1969)

Uchida, Y., Kondō, H., & Funato, M. (1987). *"Sekai kōmoku"* sakuin—1985 nendo kyōdō kenkyu hōkoku [*"Sekai kōmoku"* index: 1985 joint study report]. *Kyōritsu Joshi Daigaku Bungei Gakubu Kiyō, 33*, 173–286.

Urayama, M. (1974). Naimaze to sekai. *Geinō no kagaku, 5*, 103–120.

Watanabe, N. (2012). Nihon shōsetsu gijutsushi [A history of technologies of Japanese novels]. Tokyo, Japan: Shinchōsha.

Watanabe, T. (1986). Musume Dōjōji. Tokyo, Japan: Shinshindo Shuppan.

Watanabe, T. (1989). Kabuki—Kajō naru kigō no mori [Kabuki: The forest of excessive signs]. Tokyo, Japan: Shin'yōsha.

Watanabe, T. (2004). Kabuki—Kata no miryoku [Kabuki: Charm of forms]. Tokyo, Japan: Kadokawa Shoten.

Watanabe, T. (2012). Watashi no kabuki henreki—Aru gekihyō-ka no kokuhaku [My kabuki wandering: Confession of a drama critic]. Tokyo, Japan: Engeki Shuppansha.

Watanabe, T. (2018). Kudaime Danjūrō [Danjūrō IX]. Tokyo, Japan: Engeki Shuppansha.

Yamamoto, Y. (1977). Sakazaki Dewanokami. In Yamamoto Yūzō zenshū, 1 [The complete works of Yamamoto Yūzō, 1]. (pp. 287–351). Tokyo, Japan: (Original work published 1921)

Yasuda, N. (1989). Dōjōji densetssu kara "Kyōganoko Musume Dōjōji" he [From Dōjōji legnds to *Kyōganoko Musume Dōjōji*]. *Shōtoku Gakuen Gifu Kyōiku Daigaku Kokugo Kokubungaku, 8*, 1–12.

Yokomitsu, R. (1931). Kikai. Tokyo, Japan: Hakusuisha.

Yoshimoto, T. (1965). Gengo ni totte bi toha nanika [What is beauty for language?]. Tokyo, Japan: Keisō Shobō.

Yoshimoto, T. (1968). Kyōdō gensō ron [The theory of communal Illusion]. Tokyo, Japan: Kawade Shobō Shinsha.

Yoshimoto, T. (1971). Shinteki genshōron josetsu [A theory of mental phenomena: Introduction]. Tokyo, Japan: Hokuyōsha.

Yoshimoto, T. (2008). Shinteki genshōron honron [A theory of mental phenomena: Main issues]. Tokyo, Japan: Bunka Kagaku Kōtō Kenkyūin Shuppankyoku.

Zeami, M. (1983). Sandō. In M. Yamazaki (Ed.), Zeami (pp. 253–268). Tokyo, Japan: Chūōkōronsha. (Original work published 1423)

KEY TERMS AND DEFINITIONS

External Narrative Generation: External narrative generation refers to the social development and distribution of the narrative generation systems, including the Integrated Narrative Generation System (INGS) and the *Geinō* Information System (GIS), as well as content produced by the narrative generation systems, such as computer games and automatic generation stories and novels. The first goal of external narrative generation is to develop a computational environment for using the above systems.

***Geinō* Information System (GIS):** This indicates a system model of *geinō* production and consumption. GIS has been considered a framework in which various levels of narrative production processes are driven by writers, receivers, characters, and actors. Furthermore, GIS is a model for circulative narrative generation, which repeats the narrative generation process. In contrast, in the author's narrative generation study, the INGS corresponds to a single-level narrative generation system.

GIS With INGS: Although GIS and INGS are relatively different narrative generation systems, they function as a linked mechanism. INGS plays a role in creating individual narratives, and GIS produces each sequential narrative by connecting the narratives produced by INGS into a broader story line. GIS is a more comprehensive framework of narrative generation that contains INGS.

Integrated Narrative Generation System (INGS): INGS is a synthetic narrative generation system architecture that integrates previous studies by the author. From the broadest angle, INGS is divided into knowledge and procedures. INGS aims to be a narrative synthesizer that integrates a variety of narrative techniques, methods, rhetoric, and knowledge into an organic, dynamic generation framework.

Internal Narrative Generation: This term forms the counterpart to the above external narrative generation. The internal narrative generation corresponds to the aspect of creating narrative works by the author, using developed narrative generation systems, mainly INGS and GIS. The author integrates such narrative works according to the concept of *Watakushi Monogatari*.

***Kabuki*-Oriented Narratology:** The author's goal, based on *kabuki*-oriented narratology, is to scrutinize narratological research and critiques in the world of *kabuki* and apply the results to the modeling, design, and development of the narrative generation mechanisms of *kabuki*. In this chapter, the author focuses on two contemporary *kabuki* scholars, Yukio Hattori (1932–2007) and Tamotsu Watanabe (1936–), and two classical *kabuki* documents, *Sakusha Shikihō Kezairoku* (1972) and *Sekai Kōmoku* (1916).

Narrative Generation of *Kabuki*: The narrative generation study of *kabuki* is a type of *kabuki* research according to constructive, productive, engineering, computational, and cognitive scientific methods. It aims to analyze diverse components of *kabuki* based on one or more narrative generation systems. Further, the narrative generation of *kabuki* attempts to utilize the acquired results for various narrative generation mechanisms as narrative techniques and knowledge.

Post-Narratology: A concept for a field in which narratology and literary theories are organically introduced into computational and cognitive approaches. The author calls this type of narratology post-narratology, in contrast to classical narratology, which only focuses on only human and social scientific fields. However, the author's concept of post-narratology deals with classical narrative knowledge, such as *kabuki*, and positively introduces the Japanese cultural worlds to which the author himself belongs. In various senses, the conception of post-narratology exceeds an interdisciplinary narratology between literary and informational areas.

***Sakusha Shikihō Kezairoku*:** This is a comprehensive, theoretical document written during the Edo era (1801) by a real *kabuki* writer, Namiki Shōza II. The author pays attention to the parts in which unique theories and ways of how *kabuki* scenarios should be written and created are described, including a type of grammatical statement showing a narrative development structure and a technique for inserting narrative contrivances (*shukō* in Japanese) into a determined or stereotyped flow (*shigumi*) according to a *sekai* (*world*) as a typical type of narrative.

***Sekai Kōmoku*:** *Sekai* means a narrative system or collection (a kind of knowledge base) that has been used to produce *kabuki* scenarios. *Sekai Kōmoku* is a document written in the middle of the Edo era that systematically collects many *sekai*s as a knowledge base. In this chapter, the author analyzes the book's entire structure and the mechanism of each *sekai*'s unit. In the future, an interesting and practical theme would be to pursue the author's unique, new *Sekai Kōmoku* that perhaps treats modern and contemporary periods and topics.

***Watakushi Monogatari*:** *Watakushi Monogatari* ("My Story" or "My Narrative," or "I-Story" or "I-Narrative") means a new narrative genre that should be practiced using narrative generation systems. Although it is a type of story or narrative genre in which narrative worlds and narrations are pursued and represented through "I" (*watakushi*), this does not necessarily mean that the narrative world to be generated should be narrow and the field related only to "I." Rather, it means that the first-person world gradually opens to the second-person and third-person worlds to construct an entire narrative dependent on their organic relationships.

APPENDIX

Table 1.

Author	Title	Publication	Period(s) within the work	Location(s) within the work
有島武郎[Arishima Takeo]	生れ出る悩み [*Umareizuru Nayami*: The Agony of Coming into the World]	1918	The Meiji era	Iwanai, Sapporo (Hokkaido, Japan)
	カインの末裔 [*Cain no Matsuei*: The Descendants of Cain]	1918	The Meiji era	Matsukawa farm (Hokkaido, Japan)
	小さき者へ [*Chiisaki Mono He*]	1918	Taishō 6 (1917)	Tokyo
	或る女 [*Aru On'na*: A Certain Woman]	1919	The Meiji era	Yokohama, Japan, A ship bound for USA, Seattle, USA
Ernest Hemingway	A Farewell to Arms	1929	During World War I	Italy, Switzerland
	The Sun Also Rises	1926	During World War I	USA, Paris, Pamplona (Spain)
	For Whom the Bell Tolls	1940	During Spanish Civil War	Spain
	The Old Man and the Sea	1952	20th century	Cuba, Gulf of Mexico, Africa
	The Killers	1927	20th century	New Jersey, Brentwood
	The Snows of Kilimanjaro	1936	20th century	Africa,Mt. Kilimanjaro, Europe
Franz Kafka	The Metamorphosis	1915		Gregor's house
	The Judgment	1916		St. Pederburg
	The Castle	1926		Vest Vest Count's castle property, Court office, The junk room in K's workplace
	The Trial	1925		
	America	1927		USA, Germany, New York port, Hotel Occidental
André Malraux	The Human Condition	1958	1927	Shanghai
John Steinbeck	The Grapes of Wrath	1939	1930's	Oklahoma, Route 66, Arizona Desert, Rocky Mountains, California
William Faulkner	The Sound and the Fury	1929	1910, 1928	Harvard University, Cambridge Street, Mottson Town
	Light in August	1932	20th century	Alabama, Mississippi
Henry Miller	Tropic of Capricorn	1939	20th century	New York, Brooklyn
	Tropic of Cancer	1934	1930's	France
芥川龍之介[Akutagawa Ryūnosuke]	羅生門 [*Rashōmon*]	1917	The Heian period	Kyoto
	鼻 [*Hana*: The Nose]	1917	The Heian period	Ikeo, Uji, Kyoto
	河童 [*Kappa*]	1927	20th century	Hotakadake [Mt. Hotaka]
樋口一葉[Higuchi Ichiyō]	たけくらべ [*Takekurabe*]	1986	The Meiji era	Yoshiwara, Senzoku Shrine, Otori Shrine (Tokyo)
	にごりえ [*Nigorie*]	1895	The Meiji era	Fukuyamacho, Matuyama
	十三夜 [*Jūsan'ya*: The Thirteenth Night]	1895	The Meiji era	
志賀直哉 [Shiga Naoya]	城の崎にて [*Kinosaki Nite*: At KinosakiI]	1917	The Taishō era	Kinosaki Onsen (Hyōgo)
	小僧の神様 [*Kozō no Kamisama*]	1967	The Taishō era	Kanda (Tokyo)
	暗夜行路 [*An'ya Kōro*: A Dark Night's Passing]	1921	From the Meiji era to Taishō era	Onomichi, Kyoto
William Shakespeare	Hamlet	1603		Denmark
	Macbeth	1623	1040–1057	Scotland, Glamis
	Othello	1622		Venice
	King Lear	1606		
Albert Camus	The Stranger	1942	20th century	Algiers, Algeria
	The Plague	1947	20th century	Oran city, Algeria
	The Fall	1956	20th century	Amsterdam
	Exile and the Kingdom	1957	20th century	
山本有三[Yamamoto Yuzō]	路傍の石 [*Robō no Ishi*: A Stone by the Roadside]	1940	The mid Meiji era	Nezu, Hongō (Tokyo)
	真実一路 [Shinjitsu Ichiro: Towards the Truth]	1936	The Meiji era	
石坂洋次郎[Ishizaka Yojirō]	若い人 [*Wakai Hito*]	1937	The Shōwa era	A port town in a northern area (Japan)
石川達三[Ishikawa Tatsuzō]	私ひとりの私 [*Watakushi Hitori no Watakushi*]	1964	The Shōwa era	
Margaret Mitchell	Gone with the Wind	1936	1860's	Southern Georgia, Atlanta
Erich Remark	Arch of Triumph	1945	1938–1939	Paris

continued on following page

Table 1. Continued

Author	Title	Publication	Period(s) within the work	Location(s) within the work
Emily Brontë	Wuthering Heights	1847	19th century	Yorkshire (England), Haworth (Yorkshire, England)
Lev Tolstoy	Resurrection	1899	19th century	Russia, Siberia
	Anna Karenina	1877	1870's	Russia, Moscow, St. Peterburg
	War and Peace	1867	19th century	Russia, Austria
	Childhood	1852	19th century	Moscow, Russia
	Boyhood	1854	19th century	Moscow, Petrovskoy, Russia
	Youth	1856	19th century	Petrovskoy, Moscow, Russia
Fyodor Dostoevsky	Poor Folk	1846	19th century	St. Petersburg
	Crime and Punishment	1866	19th century	St. Petersburg, Siberia
	The Idiot	1874	19th century	Russia, Switzerland
	Demon	1873	1869	Russia
	The Brothers Karamazov	1880	19th century	Russia
	The Raw Youth	1875	19th century	Russia
	The House of the Dead	1862	19th century	Siberia, Russia
	Notes from Underground	1864	19th century	St. Petersburg
	Humiliated and Insulted	1861	19th century	St. Petersburg
	The Village of Stepanchikovo	1859	19th century	Stepanchikoovo Village
Ivan Turgenev	First Love	1860	1833	Moscow, Lake Neskunui
	Fathers and Sons	1862	May, 1859	Heidelberg
	Rudin	1857	19th century	Russia
	On the Eve	1860	19th century	Venice, Bulgaria, Balkans
	A Nest of the Gentry	1859	19th century	Baden-Baden, Germany
	Smoke	1867	19th century	Russia
	Virginity	1877	19th century	Russia
Alexander Pushkin	The Captain's Daughter	1836	1773–1775	Russia, Don River Basin, Bella Gorx fortress
	Eugene Onegin	1833	1819–1825	St. Petersburg
	The Queen of Spades	1834		Russia
Maxim Gorky	My Childhood	1914	19th century	Russia
	In the World	1916	19th century	Russia
	My Universities	1923	19th century	Russia
	Mother	1907	19th century	Russia
	The Lower Depths	1901-1902	19th century	Russia
Nikolai Ostrovskii	How the Steel Was Tempered	1932	20th century	Russia
武者小路実篤[Mushanokōji Saneatsu]	友情 [*Yujō*: Friendship]	1920	The Meiji era	Japan, Europe
	愛と死 [*Ai to Shi*: Love and Death]	1939	The Taishō era	Japan, Paris
伊藤左千夫 [Itō Sachio]	野菊の墓 [*Nogiku no Haka*: The Wild Daisy]	1906	The Meiji era	Matsudo (Chiba)
Hans Theodor Storm	Immensee	1851	19th century	Schleswig-Holstein
	Viola Tricolor	1874	19the century	Germany
Johann Wolfgang von Goethe	The Sorrows of Young Werther	1774	18th century	Germany
	Faust	1808, 1833	15th century–16th century	Germany
Jacques-Henri Bernardin de Saint-Pierre	Paul and Virginia	1787	18th century	Mauritius Island
Abbé Prévost	Manon Lescaut	1731		France, USA
Madame de La Fayette	La Princesse de Cleves	1678	16th century	Louvre Palace
George Sand	Little Fadette	1849	19th century	Franch countryside
Colette	Green Wheat	1923	20th century	Paris
Françoise Sagan	Hello Sadness	1954	20th century	Cote d'Azur
Alain-Fournier	The Great Meaulnes	1913		France, Sologne
James Hilton	Goodbye, Mr. Chips]	1934	From the end of 19th century to the beginning of 20th century)	England
Hans Christian Andersen	Billedbog uden Billeder	1839	19th century	Denmark
Saint-Exupéry	The Little Prince	1839		Sahara desert
	Night Flight	1931	20th century	Lisbon, Patagonia
Charles Lindbergh	The Spirit of St. Louis	1953	1923	Over Atrantic Ocean, Le Bourget airport (Paris)
Charles Dickens	Christmas Carol	1843	19th century	London
Annelies Frank	The Diary of a Young Girl	1947	Dering World War II	Amsterdam, Netherlands

continued on following page

Table 1. Continued

Author	Title	Publication	Period(s) within the work	Location(s) within the work
西川一三[Nishikawa Kazumi]	秘境西域八年の潜行 [*Hikyō Saiiki Hachinen no Senkō*]	1967	1945, 1950	Japan, Mongolia, Gobi Desert, Tibet
岸上大作[Kishigami Taisaku]	もうひとつの意志表示 [*Mouhitotsu no Ishihyoji*]	1973	Shōwa 30's	Tokyo
永山則夫[Nagayama Norio]	無知の涙 [*Muchi no Namida*]	1971	The Shōwa era	Prison
高野悦子 [Takano Etsuko]	二十歳の原点 [Hatachi no Genten]	1971	Jan/2–Jun/22, 1969	Kyoto
Hermann Hesse	Beneath the Wheel	1905	19th century	Germany
	Demian	1920	20th century	Latin school in a small town in Jarmany
	Narcissus and Goldmund	1957		Germany, Europe
	Steppenwolf	1927	20th century	Germany
Thomas Mann	Tonio Kroger	1913	Early 20th century	Lübeck, Germany, Gymnasium, Munich, Denmark)
	Death in Venice	1912	Early 20th century	Munich, Adriatic Sea, Venice
	Buddenbrooks	1901	19th century	Lübeck, Germany
	The Magic Mountain	1924	World War I	Swiss Alps, Davos
	Royal Highness	1909		Grand Grimburg
	Doctor Faustus	1947		Germany
	The Holy Sinner	1951		Flanders, Artois
	Lotte in Weimar: The Beloved Returns	1939		Germany
	Confessions of Felix Krull	1954		River Rhine, Romania
	Joseph and His Brothers	1933	Ancient times	Egypt, Babylon
	Little Herr Friedemann	1897	19th century	Germany
	Mario and the Magician	1930	20th century	Italy
Friedrich Nietzsche	Thus Spoke Zarathustra	1885		
Rainer Maria Rilke	The Notebooks of Malte Laurids Brigge	1910		Paris, Denmark
Hans Carossa	A Childhood	1922	After 1878	Bavarian Kingdom
	Das Jahr der schonen Tauschungen	1941	1897–1898	Munich
Jean-Paul Sartre	Nausea	1938	1932	France
	The Wall	1937		Spain
	The Flies	1943	Ancient Greece	Ancient Greece
André Gide	Strait is the Gate	1909	19th century	France
	The Pastoral Symphony	1919		France
	The Fruits of the Earth	1897	19th century	Africa
	Atarashiki Kate	1935	20th century	Africa
	If It Die, an Autobiography	1921	19th century–20th century	France
	The Vatican Cellars	1914	20the century	Rome, Bucharest, Paris
	The Counterfeiters	1925	20th century	Paris, France
Jean Genet	Our Lady of the Flowers	1944	20th century	Paris, Prison
	Miracle of the Rose	1946	20th century	Metley
	The Thief's Journal	1949	20th century	Spain, Italy, Austria, Czechoslovakia, Poland, Germany, Belgium)
	The Maids	1947	20th century	
Guy de Maupassant	Pierre and Jean	1887-1888	19th century	Le Havre
	A Woman's Life	1883	19th century	Les Pouples
	Ball of Fat	1880	19 th century	Rouen
Gustave Flaubert	Madame Bovary	1857	19th century	Rouen, Tost
	Sentimental Education	1869	1840–1867	Paris
	Bouvard et Pecuchet	1881	19th century	Paris, Proudhon Street
Stendhal	The Charterhouse of Parma	1839	Napoleonic's era	Parma, Lake Como, Farnese Tower, Italy
	The Red and the Black	1830		Besancon, France
	Vanina Vanini	1829	1820's	Italy
de Laclos	Dangerous Liaisons	1782	18th century	France
David Herbert Lawrence	Sons and Lovers	1913	19th century	Wales
	Women in Love	1920		Wales
	The Rainbow	1912		Wales
	Lady Chatterley's Lover	1928	During World War I	Venice
	The White Peacock	1911		
	St. Mawr	1925	After World War I	England, New Mexico

continued on following page

Table 1. Continued

Author	Title	Publication	Period(s) within the work	Location(s) within the work
	The Man Who Died	1929	Ancient times	Gorgoda Hill
Mikhail Sholokho	*Deshabarikko*	1925	1917	Russia
	And Quiet Flows the Don	1928-1940	During World War I	Tatar Ski Village near Byoshchenskaya in Don area
	Virgin Soil Upturned	1932, 1959	Jan., 1930	Gremyachi Lok Village in the River Don Basin
Romain Rolland	Jean-Christophe	1904	19th century	Germany, River Rhine, Berlin, Paris, USA, Switzerland, Rome
梶井基次郎 [Kajii Motojirō]	檸檬 [Lemon]	1931	Early 20th century	Kyōgoku, Maruzen, Teramachi Street, Kyoto
佐藤春夫 [Satō Haruo]	田園の憂鬱 [*Den'en no Yūutsu*: Melancholy in the Country]	1919	Early 20th century	Nakazato-mura, Tsuzuki-gun, Kanagawa Prefecture, Japan
	都会の憂鬱 [*Tokai no Yūutsu*: Melancholy in the City]	1922	Early 20th century	Musashino, Tokyo
中島敦 [Nakajima Atsushi]	山月記 [*Sangetsuki*]	1942	The Tō period (China)	Rousei, China
坂口安吾[Sakaguchi Ango]	白痴 [*Hakuchi*: The Idiot]	1947	During World War II	Kamata, Tokyo
古井由吉 [Furui Yoshikitichi]	行き隠れ [*Yukigakure*]	1972	The Shōwa era	Japan
川端康成[Kawabata Yasunari]	伊豆の踊り子 [*Izuno Odoriko*: The Dancing Girl of Izu]	1927	The Taishō era	Izu, Yugasima, Amagi, Shimoda
	花のワルツ [*Hana no Warutsu*: Waltzu of the Flowers]	1936	The Shōwa era	
	雪国 [*Yukiguni*: Snow Country]	1937	The Taishō era-The Shōwa era	Yuzawa Onsen
	古都 [*Koto*: The Old Capital]	1962	The Shōwa era	Kyoto, Nakagyō, Yasaka Shrine
	千羽鶴 [*Senbazuru*: Thousand Cranes]	1952	The Shōwa era	Engaku temple, Kamakura, Takeda, Kyushu
	掌の小説 [*Tenohira (Tanagokoro) no Shōsetsu*: Palm-of-the-Hand Stories]	1971		
	禽獣 [*Kinju*]	1935	The Shōwa era	Hibiya, Zen temple
	抒情歌 [*Jojōka*]	1934	The Shōwa era	Hanatake Peach Blossoms on the Coast
	愛する人達 [*Aisuru Hitotachi*]	1941	The Shōwa era	
	日も月も [*Hi mo Tsuki mo*]	1953	Shōwa 20's	Kamakura
	名人 [*Meijin*: The Master of Go]	1954	Jan/18–Dec/4, 1940	Atami
	川のある下町の話 [*Kawa no Aru Shitamachi no Hanashi*]	1954	Shōwa 20's	Tokyo
	美しさと哀しみと [*Utsukushisa to Kanashimi to*: Beauty and Sadness]	1965	The Shōwa era	Kyoto, Kita Kamakura, Saihoji temple
	眠れる美女 [*Nemureru Bijo*: The House of the Sleeping Beauties]	1961	The Shōwa era	Japanese-style hotel
	みづうみ [*Miduumi*: The Lake]	1955	The Shōwa era	Tokyo, Shinsyu (Nagano), Karuizawa
	片腕 [*Takaude*: One Arm]	1965	The Shōwa era	
	たんぽぽ [*Tanpopo*: Dandelions]	1972	The Shōwa era	Ikutachō, Jokoji temple, Miidera temple, Nishiizu
	東京のひと [*Tokyo no Hito*: Women in Tokyo]	1955	1950's	Tokyo
	女であること [*Onna de Aru Koto*]	1956-1957	The Shōwa era	Tokyo
三島由紀夫[Mishima Yukio]	美しい星 [*Utsukushii Hoshi*: A Beautiful Star]	1962	The Shōwa era	Han'no (Saitama), Mount. Rakan, Kanazawa Uchinada, Sendai
	盗賊 [*Tōzoku*: Thieves]	1948	1930's	San'nomiya (Kōbe), Murasakino (Kyoto)
	仮面の告白 [*Kamen no Kokuhaku*: Confessions of a Mask]	1949	Before/During World War II	Tokyo, Karuizawa
	純白の夜 [*Jumpaku no Yoru*]	1950	1948	Azabu, Tokyo, Kamakura
	愛の渇き [*Ai no Kawaki*: Thirst for Love]	1950	The Shōwa era	Umeda, Toyonaka, Osaka
	青の時代 [*Ao no Jidai*]	1950	1923	Chiba, Tokyo, Toriizaki
	禁色 [*Kinjiki*: Forbidden Colors]	1951-1953	1950–1951	Tokyo, Izu, Kyoto
	潮騒 [*Shiosai*: The Sound of Waves]	1954	The Shōwa era	Utashima
	沈める滝 [*Sizumeru Taki*]	1955	The Shōwa era	Tokyo, Nigata
	金閣寺 [*Kinkakuji*: The Temple of the Golden Pavilion]	1956	After World War II	Kinkakuji temple, Naryu misaki, Nanzenji temple, Maizuru, Ryuhoji temple
	美徳のよろめき [*Bitoku no Yoromeki*]	1957	The Shōwa era	Tokyo
	鏡子の家 [*Kyōko no Ie*: Kyoko's House]	1959	The Shōwa era	Yotsuya Higasi-Shinanocho, New York, Mexico

continued on following page

Table 1. Continued

Author	Title	Publication	Period(s) within the work	Location(s) within the work
	宴のあと [Utage no Ato: After the Banquet]	1960	The Shōwa era	Tokyo
	獣の戯れ [*Kemono no Tawamure*: The Frolic of the Beasts]	1961	The Shōwa era	Nishi-Izu, Ginza, Tokyo, Hamamatsu
	午後の曳航 [*Gogo no Eikō*: The Sailor Who Fell from Grace with the Sea]	1963	The Shōwa era	Yamate, Tomioka, Yokohama, Yokohama port
	音楽 [*Ongaku*: Music]	1965	The Shōwa era	Hibiya, Kōfu, Izu), Namida-bashi
	絹と明察 [*Kinu to Meisatsu*: Silk and Insight]	1964	The Shōwa era	Ōmi (Shiga), Konkai Kōmyōji temple
	遠乗会 [*Tōnorikai]*	1958	The Shōwa era	
	真夏の死 [*Manatsu no Shi*: Death in Midsummer]	1965	The Shōwa era	Izu
	卵 [*Tamago*: Eggs]	1966	The Shōwa era	Tokyo
	憂国 [*Yūkoku*: Patriotism]	1961	1936	Yotsuya, Tokyo
	三熊野詣 [*Mikumano Mōdo*: Acts of Worship]	1953	The Shōwa era	Kumano, Kumano Nachi Taisya Shrine, Kumano Hayatama Taisya Shrine, Kumano Hongu Shurine
	橋づくし [*Hashi Dukushi*]	1956	The Shōwa era	Tokyo, Shinbashi, Miyoshibashi, Tsukijibashi, Irifunebashi, Akatsukibashi, Sakaibashi, Bizenbashi
	癩王のテラス [*Raiō no Terasu*]	1969	Late 12th century	Cambodia
	鹿鳴館 [*Rokumeikan*]	1957	1886	Rokumeikan, Hibiya
	豊饒の海(春の雪) [*Hojo no Umi* (*Haru no Yuki*): The Sea of Fertility (Spring Snow)]	1969	The Meiji era–The Shōwa era	Tokyo, Yamato (Nara)
	豊饒の海(奔馬) [*Hojo no Umi* (*Honba*): The Sea of Fertility (Runaway Horses)]	1969	May, 1932–Dec, 1933	Osaka, Ōgami Jinja Shrine, Mt. Miwa, Isagawa Jinja Shrine, Yanagawa, Yamanashi, Manchuria, Ise Jingū Shrine, Izu
	豊饒の海(暁の寺) [*Hojo no Umi* (*Akatsuki no Tera*):The Sea of Fertility (The Temple of Dawn)]	1970	1941–1945, 1952–1967	Bangkok, India, Gotemba
	豊饒の海(天人五衰) [*Hojo no Umi* (*Ten'nin Gosui*):The Sea of Fertility:The Decay of the Angel]	1971	1970–1975	Miho no Matsubara (Shizuoka), Tokyo, Gesshūji (Yamato)
Raymond Radiguet	Devil in the Flesh	1923	1917	France
	Count d'Orgel's Ball	1924	1920	Robson, Basque Country
Oscar Wilde	The Picture of Dorian Gray	1891	19th century	London
	Salome	1893	Ancient times	
森鷗外 [Mori Ōgai]	舞姫 [*Maihime*: The Dancing Girl]	1907	Late 19th century	Germany, Saigon, Klosterstrasse, Berlin, Victoria Theater, Russia
	ヰタ・セクスアリス [Vita Sexualis]	1909	The Meiji eara	Tokyo
	青年 [*Seinen*: Young Men]	1913	The Meiji era	Tokyo
	雁 [*Gan*: The Wild Geese]	1913	1880	Muenzaka, Tokyo
	渋江抽斎 [*Shibue Chūsai*]	1916	The Edo period	Edo (Tokyo)
	高瀬舟 [*Takasebune*]	1916	The Edo period	River Takase, Kyoto
泉鏡花 [Izumi Kyōka]	歌行燈 [*Uta-andon*: A Song by Lantern Light]	1912	The Meiji era	
	高野聖 [*Kōya-hijiri*: The Holy Man of Mount Koōya]	1908	The Meiji era	Wakasa, Echizen, Shinshu, Amou Tōge
	婦系図 [*On'na Keizu*]	1908	The Meiji era	Yushima, Tokyo
	日本橋 [*Nihonbashi*]	1914	The Meiji era	Nihonbashi, Tokyo
尾崎紅葉 [Ozaki Kōyō]	金色夜叉 [*Konjijki Yasha*: The Golden Demon]	1898-1903	The Meiji era	Atami, Shiobara, Tokyo
太宰治 [Dazai Osamu]	斜陽 [*Shayō*: The Setting Sun]	1947	1945	Nishikata, Tokyo, Izu
	晩年 [*Ban'nen*: The Late Years]	1936	The Shōwa era	
	新釈諸国噺 [*Shinshaku Shokoku-banashi*: A New Cersion of Countries' Tales]	1945		Japan
	津軽 [*Tsugaru*]	1944	The Shōwa era	Kanagimachi, Kodomari, Tsugaru (Aomori)
	人間失格 [*Ningen Shikkaku*: No Longer Human]	1948	The Shōwa era	Tokyo
	ヴィヨンの妻 [*Viyon no Tsuma*: Villon's Wife]	1947	The Shōwa era	Asakusa park, Koganei, Nakano Station, Tokyo

continued on following page

Table 1. Continued

Author	Title	Publication	Period(s) within the work	Location(s) within the work
René Descartes	The World	1633		
	Principles of Philosophy	1644		
	Discourse on the Method	1637		
	The Passions of the Soul	1649		
道元 [Dōgen]	正法眼蔵随聞記 *[Shōhō Genzō Zuimonki*: The Treasury of the True Dharma Eye: Record of Things Heard]	1235-1238		
世阿弥 [Zeami]	風姿花伝 [Fūshikaden]	Early 15th century		
親鸞[Shinran]	歎異抄 [*Tan'nishō*]	The end of the Kamakura era		
	Old Testament			
	New Testament			
	古代歌謡集 [*Kodai Kayō Shū*]		Ancient times	Ancient Yamato area
	中世・近世歌謡集 [*Chūsei Kindai Kayō Shū*]		The Kamukura period–The Edo period	Many areas in Japan
	万葉集 [*Man'yōshū*]	The end of the Nara period	Ancient times	Many areas in Japan
	古今和歌集 [*Kokin Wakashū*]	c. 912		Kyoto, many areas in Japan
	新古今和歌集 [*Shin Kokin Wakashū*]	c. 1210		Kyoto, many areas in Japan
	金塊和歌集 [*Kinkaiwakashū*]	The Kamuraku period		
	和泉式部歌集 [*Izumi Shikibu Kashū*]	The Heian period		
	義経記 [*Gikeiki*]	From the Northern and Southern Dynasties to the early Muromachi period	The Heiji rebellion	Kyoto, Ishiyama-dera Temple, Echizen, Rokujogawara, Kyoto, Mino
	狭衣物語 [*Sagoromo Monogatari*]	The Heian period	The Heian period	Tsukushi, Kyoto
	夜の寝覚 [*Yoru no Nezame*]	The end of the Heian period	The Heian period	Kyoto
	平家物語 [*Heike Monogatari*: The Tale of the Heike]	The Kamakura Period	The Kamakura period	Kyoto, Kamakura, Setouchi, Dan'noura
紫式部 [Murasaki Shikibu]	源氏物語 [*Genji Monogatarri*: The Tale of Genji]	The Heian Period		Kyoto, Suma, Akashi, Tsukushi, Many places in Japan
	紫式部日記 [*Murasakishikibu Nikki*: The Diary of Lady Murasaki]	The early 13th century	1008–1010	Kyoto
与謝野晶子[Yosano Akiko]	源氏物語 [*Genji Monogatari*: The Tale of Genji]	1911-1940	The Heian period	Kyoto, Suma, Akashi, Tsukushi, Many places in Japan
上田秋成 [Ueda Akinari]	雨月物語 [*Ugetsu Monogatari*]	1776	The Edo period	Sanuki, Kōyasan, Dōjōji
	春雨物語 [*Harusame Monogatari*]	1808	The Edo period	Kyoto, Sagami, Ōmi, Settsu
式亭三馬 [Shikitei Sanba]	浮世風呂 [*Ukiyoburo*]	1809-1813	The Edo period	Edo (Tokyo)
	浮世床 [*Ukiyodoko*]	1809	The Edo period	Edo (Tokyo)
井原西鶴 [Ihara Saikaku]	好色一代男 [*Kōshoku Ichidai Otoko*: The Life of an Amorous Man]	1682	The Edo period	Kyoto, Osaka, Kyūshū, Ohara, Sado, Shimabara, Sano, Murotsu, Shinmachi, Nyogogashima
	好色五人女 [*Kōshoku Gonin On'na*: Five Women Who Loved Love]	1686	The Edo period	Himeji, Osaka, Komagome Kisshōji teimple, Satsuma
	日本永代蔵 [*Nippon Eitaigura*: The Eternal Storehouse of Japan]	1688	The Edo period	Edo, Kyoto, Izumi, Nanto, Ōtsu, Kinokuni, Sakata, Bungo, Fushimi, Suruga, Chikuzen, Echizen, Nagasaki, Yamazaki, Yamato, Hitachi, Sakushū, Bushū, Senshū, Yamashiro
	世間胸算用 [*Seken Munezan'yo*: Reckonings that Carry Men Through the World or This Scheming World]	1692	The Edo period	Nagamachi, Sakai, Edo
	西鶴織留 [*Saikaku Oridome*]	1694	The Edo period	
藤原道綱母[Fujiwara no Michitsuna no Haha]	蜻蛉日記 [*Kagerō Nikki*]	974	957–974	Kyoto
和泉式部 [Izumi Shikibu]	和泉式部日記 [*Izumi Shikibu Nikki*]	1008	1003–1004	Kyoto
菅原孝標女[Sugawara no Takasue no Musume]	更級日記 [*Sarashina Nikki*]	The Middle of the Heian period	1020–1059	Kyoto
藤原長子[Fujiwara no Nagako]	讃岐典侍日記 [*Sanuki no Suke Nikki*]	The End of the Heian period	The time of Emperor Horikawa and Emperor Toba)	Kyoto
清少納言 [Sei Shōnagon]	枕草子 [*Makura no Sohi*]	The Middle of the Heian period	The Middle of the Heian period	Kyoto

continued on following page

Table 1. Continued

Author	Title	Publication	Period(s) within the work	Location(s) within the work
吉田兼好[Yoshida Kenkō]	徒然草 [*Tsurezuregusa*]	The End of the Kamuraku period	The Kamura period	Ninanji temple, Kyoto
鴨長明 [Kamo no Chōmei]	方丈記 [*Hōjōki*]	1212	1170's–1180's	Ninanji temple, Kyoto
為永春水[Tamenaga Shunsui]	梅暦 [*Ume Goyomi*]	1833-1835	The End of the Edo period	Edo
源信 [Genshin]	往生要集 [*Ōjōyōshū*]	985		*Jigoku* [Hell]
夏目漱石[Natsume Sōseki]	吾輩は猫である [*Wagahai ha Neko de aru*: I Am a Cat]	1905-1907	The Meiji era	Tokyo
	坊ちゃん [*Botchan*]	1907	The Meiji era	Tokyo, Matsuyama
	虞美人草 [*Gubijinsō*: The Poppy]	1908	The Meiji era	Tokyo, kyoto, Ueno Park
	坑夫 [*Kōfu*: The Miner]	1908	The Meiji era	A cole mine
	三四郎 [*Sanshirō*]	1909	The Meiji era	Tarin, Tokyo
	門 [*Mon*: The Gate]	1911	The Meiji era	Tokyo, Kamakura
	それから [*Sorekara*: And Then]	1910	The Meiji era	Tokyo
	こころ [*Kokoro*]	1914	The Meiji period	Kamakura, Yuigahama, Tokyo, Zoōhigaya
	行人 [*Kōjin*:The Wayfarer]	1914	The Meiji era	Tokyo, Osaka
	道草 [Michikusa]	1915	The Meiji era	Hongō, Tokyo
	彼岸過迄 [*Higansugimade*: To the Spring Equinox and Beyond]	1912	The Meiji era	Ogawachō, Tokyo, Kansai
	明暗 [*Meian*: Light and Darkness]	1917	The Meiji era	Tokyo, Izu
永井荷風 [Nagai Kafū]	あめりか物語 [*Amerika Monogatari*: American Stories]	1908	Meiji 36–40	USA
	ふらんす物語 [*Furansu Monogatari*: Frence Stories]	1909	Meiji 40	France
	断腸亭日乗 [*Danchōtei Nichijō*]	1962-1965	1917–1959	Ushigome, Tokyo
	すみだ川 [*Sumidagawa*: River Sumida]	1911	The Meiji era	Asakusa, Tokyo
	腕くらべ [*Udekurabe*: Geisha in Rivalry]	1918	The Meiji era	Tokyo, Tōhoku
	日和下駄 [*Hiyori Geta*]	1915	The Meiji era	Tokyo
	おかめ笹 [*Okame-zasa*]	1920	The Meiji era	Tokyo
	つゆのあとさき [*Tsuyu no Atosaki*: During the Rains]	1931	The Meiji era	Ginza, Tokyo
	ひかげの花 [*Hikage no Hana*: Flowers in the Shade]	1946	The Meiji era	Tokyo
	墨東綺譚 [*Bokutō Kitan*: A Strange Tale from East of the River]	1937	The Meiji era	Tamanoi, Mukōjima, Tokyo
	踊り子 [*Odoriko*]	1948		Tokyo
	雨瀟々 [*Ame Shōshō*]	1922		Ginza, Tokyo
二葉亭四迷[Futabatei Shimei]	浮雲 [Ukigumo: The Drifting Cloud]	1887-1889	The Meiji era	Tokyo
	其面影 [*Sono Omokage*: An Adopted Husband]	1907	The Meiji era	Tokyo, Tenshin
	平凡 [*Heibon*]	1908	The Meiji era	Tokyo
徳田秋声 [Tokuda Shūsei]	新世帯 [*Shinsetai*]	1909	The Meiji era	Tokyo
	足迹 [*Ashiato*]	1910	The Meiji era	Tokyo, Asakusa, Yushima
	黴 [*Kabi*]	1912	The Meiji era	Kyushū, Osaka
	爛 [Tadare]	1913	The Meiji era	Tokyo
	あらくれ [*Arakure*: Rough Living]	1915	The Meiji era	Tokyo, Shitaya, Hongō
	仮装人物 [*Kasō Jinbutsu*]	1938	The Meiji era	Tokyo
	縮図 [*Shukuzu*: Epitome]	1944	The Meiji era	Tokyo, Hakusan, Fujimi
島崎藤村[Shimazaki Tōson]	破戒 [*Hakai*: The Broken]	1906	The End of the Meiji perido	Shinshū (Nagano)
	春 [*Haru*: Spring]	1908	The Meiji era	Tokyo
	家 [*Ie*: The Family]	1910	The Meiji era	Kiso (Nagano), Nagoya
	新生 [*Shinsei*]	1919	The Meiji era	Tokyo, Paris
	夜明け前 [*Yoake Mae*: Before the Dawn]	1932, 1935	1853–1886	Kisodani (Shinshū), Tokyo
谷崎潤一郎[Tanizaki Jun'ichirō]	恐怖時代 [*Kyōfu Jizai*]	1920	The Edo period	
	饒太郎 [*Jōtarō*]	1914		
	お艶殺し [*Otsuya Goroshi*]	1915	The Edo era	Surugaya, Fukagawa, Mukōjima
	お国と五平 [*Okuni to Gohei*]	1922	The Edo era	Nasunogahara
	痴人の愛 [*Chijin no Ai*: Naomi]	1925	1917-1922	Asakusa, Ōmori, Yokohama
	細雪 [*Sasameyuki*: The Makioka Sisters]	1944–1948	1936–1941	Osaka, Ashiya, Ōgaki, Tokyo

continued on following page

Table 1. Continued

Author	Title	Publication	Period(s) within the work	Location(s) within the work
舟橋聖一[Funahashi Seiichi]	雪夫人絵図 [*Yuki Fujin Ezu*: Portrait of Madame Yuki]	1948	The Shōwa era	
	ある女の遠景 [*Aru On'na no Enkei*]	1963	The Shōwa era	Tokyo
	好きな女の胸飾り [*Sukina On'na no Munakazari*]	1967	The Shōwa era	Tokyo
	木石 [*Bokuseki*]	1938		
	裾野 [*Susono*]	1952	The Shōwa era	
	とりかえばや秘文 [*Torikaebaya Hibun*]	1966	The Taishō era	Tokyo
瀬戸内晴美[Setouchi Harumi]	かの子繚乱 [*Kanoko Ryōran*]	1979	The Meiji era–The Shōwa era	Tokyo, France
岡本かの子[Okamoto Kanoko]	金魚繚乱 [*Kingyo Ryōran*: A Riot of Goldfish]	1937	The Taishō era	Tokyo
堀辰雄 [Hori Tatsuo]	燃ゆる頬 [*Moyuru Hoho*]	1939		Karuizawa
	聖家族 [*Sei Kazoku*: The Holy Family]	1932		Karuizawa
	風立ちぬ [*Kazetachinu*: The Wind Has Risen]	1938		Karuizawa
福永武彦[Fukunaga Takehiko]	風土 [*Fūdo*]	1952	1923–1939	Japan
	草の花 [*Kusa no Hana*: Flowers of Grass]	1954	Around 1945	Japan
	忘却の河 [*Bōkyaku no Kawa*]	1964	Shōwa 30's	Japan
	死の島 [*Shi no Shima*]	1971	1953–1954	Hiroshima, Tokyo
	廃市 [*Haishi*]	1960	The Shōwa era	A quiet, rural town with an abandoned loneliness
	海市 [*Kaishi*]	1968	The Shōwa era	Izu, Tokyo
伊藤整 [Itō Sei]	典子の生きかた [*Noriko no Ikikata*]	1953	The Shōwa era	Japan
	若い詩人の肖像 [*Wakai Hito no Syōzō*]	1956	1928	Otaru, Tokyo
	鳴海仙吉 [*Narumi Senkichi*]	1950	Around 1945	Hokkaido
	氾濫 [*Hanran*]	1958	1940's	Japan
安部公房 [Abe Kōbō]	砂の女 [*Suna no On'na*: The Woman in the Dunes]	1962		Dunes
	燃えつきた地図 [*Moetsukita Chizu*: The Ruined Map]	1967		City
	箱男 [*Hakootoko*: The Box Man]	1973		Ueno, Shinjuku
	方舟さくら丸 [*Hakobune Sakuramaru*: The Ark Sakura]	1984		Underground quarry cave
大江健三郎[Ōe Kenzaburō]	万延元年のフットボール [*Man'en Gan'nen no Football*: The Silent Cry]	1967	1960's	Shikoku
	みずから我が涙をぬぐいたまう日 [*Mizukara Waga Namida wo Nugui Tamau Hi*: The Day He Himself Shall Wipe My Tears Away]	1972		
	洪水はわが魂に及び [*Kōzui ha Waga Tamashii ni Oyobi*: The Flood Invades My Spirit]	1973	The Shōwa era	Tokyo
高橋和巳[Takahashi Kazumi]	悲の器 [*Hi no Utsuwa*: Vessel of Sorrow]	1962	Around World War II	Japan
	捨子物語 [*Sutego Monogatari*]	1968	The End of the Pacific War	Osaka
	我が心は石にあらず [*Waga Kokoro ha Ishi ni Arazu*]	1967	1950's–1960's	Japan
	憂鬱なる党派 [*Yūutsu naru Toha*: A Melancholy Faction]	1965	The Shōwa era	Japan
	邪宗門 [*Jashūmon*: Heretical Faith]	1965	Early Shōwa era	Japan
	日本の悪霊 [*Nihon no Akuryō*]	1980	The Shōwa era	Japan
	堕落 [*Daraku*]	1969	Around World War II	Japan, Manchuria
	散華 [*Sange*]	1963	Around World War II	Japan
	白く塗りたる墓 [*Shiroku Nuritaru Haka*]	1971	The Shōwa era	Japan
	黄昏の橋 [*Tasogare no Hashi*]	1968-未完	The Shōwa era	Japan
	わが解体 [*Waga Kaitai*]	1971		
	文学の責任 [*Bungaku no Sekinin*]	1963		
真継伸彦[Matsugi Nobuhiko]	光る声 [*Hikaru Koe*]	1966	The Shōwa era	Japan

continued on following page

Table 1. Continued

Author	Title	Publication	Period(s) within the work	Location(s) within the work
柴田翔 [Shibata Shō]	されどわれらが日々 [*Saredo Wareraga Hibi*]	1964	1955	Tokyo
	贈る言葉 [*Okuru Kotoba*]	1966	1960's	Tokyo
倉橋由美子[Kurahashi Yumiko]	わたしの中の彼へ [*Watashi no Naka no Kare he*]	1970		
	スミヤキストQの冒険 [*Sumiyakisuto Q no Bōken*: Adventures of Sumiyakisto Q]	1964		Sensitivity center on an isolated island
	アマノン国探訪記 [*Amanon-koku Tanbōki*: Journey to Amanon]	1986		Phantom Land Amanon
金井美恵子[Kanai Mieko]	愛の生活 [*Ai no Seikatsu*: Love Life]	1968		
	夢の時間 [*Yume no Jikan*: Time of Dreams]	1970		
	岸辺のない海 [*Kishibe no Nai Umi*: The Shoreless Sea]	1974		
	夜になっても遊び続けろ [*Yoru ni Nattemo Asobitsuzukero*]	1974		
小田実 [Oda Makoto]	アメリカ [*America*]	1962	The Shōwa era	USA
丸山健二[Maruyama Kenji]	夏の流れ [*Natsu no Nagare*: Summer's Passage]	1967	The Shōwa era	Japan
	正午なり [*Shōgonari*]	1968	The Shōwa era	Japan
石川淳 [Ishikawa Jun]	白頭吟 [*Hakutōgin*]	1957	The Shōwa era	Japan
	焼け跡のイエス [*Yakeato no Yes*]	1946	After World War II	Ueno
村上龍[Murakami Ryū]	限りなく透明に近いブルー [*Kagirinaku Tomei ni Chikai Būu*: Almost Transparent Blue]	1976	The Shōwa era	Fussa, Tokyo
外岡秀俊[Sotooka Hidetoshi]	北帰行 [*Hokkikō*]	1976	The Shōwa era	Train, Morioka
稲垣足穂 [Inagaki Taruho]	一千一秒物語 [*Issen Ichibyō Monogatari*: One Thousand and One Second Stories]	1923		
小栗蟲太郎[Oguri Chutarō]	黒死館殺人事件 [*Kokushikan Satsujinjiken*]	1935		Kōza, Kanagawa
野間宏 [Noma Hiroshi]	暗い絵 [*Kurai E*: Dark Pictures]	1947	The End of the Pacific War	Osaka, Kyoto
	崩壊感覚 [*Hōkai Kankaku*]	1948	The End of the Pacific War	
	真空地帯 [*Shinkū Chitai*: Zone of Emptiness]	1952	1944	Japanese army
	わが塔はそこに立つ [*Waga Tō ha Soko ni Tastu*]	1962	The Shōwa era	Kyoto
	さいころの空 [*Saikoro no Sora*]	1959	The Shōwa era	Kabutochō, Tokyo
	青年の環 [*Seinen no Wa*]	1949, 1950, 1966, 1970	Jul–Sep, 1939	Osaka
武田泰淳 [Takeda Taijun]	風媒花 [*Fūbaika*]	1952	After the Pacific War	Japan, China
	貴族の階段 [*Kizoku no Kaidan*]	1959	The age of the 2.26 incident	
	森と湖の祭り [*Mori to Mizuumi no Matsuri*]	1958	The Shōwa era	Hokkaiso, Lake Akan
	富士 [*Fuji*]	1971	The Shōwa era	A Mental hospital at Mt. Fuji
	快楽 [*Keraku*]	1972	Shōwa 10's	Tokyo
	司馬遷 [*Shibasen*]	1943	Ancient China	China
	ひかりごけ [*Hikarigoke*: Luminous Moss]	1953	The Shōwa era	Rausu, Hokkaido
島尾敏雄 [Shimao Toshio]	出発はついに訪れず [*Shuppatsu ha Tsuini Otozurezu*]	1964	Durging Pacific War	Kakeromajima
	死の棘 [*Shi no Toge*: The Sting of Death]	1977	The Shōwa era	
中上健次[Nakagami Kenji]	枯木灘 [*Karekinada*: The Sea of Withered Trees]	1977	The Shōwa era	Kishū (Wakayama)
	地の果て 至上の時 [*Chi no Hate Shijō no Toki*: Supreme Time at the End of the Earth]	1983	The Shōwa era	Singu, Kishū (Wakayama)
	千年の愉楽 [*Sen'nen no Yuraku*: A Thousand Years of Pleasure]	1982	The Shōwa era	Singu, Kishū (Wakayama)
	熊野集 [*Kumanoshū*]	1984	The Shōwa era	Kishū (Wakayama)
	岬 [*Misaki*: The Cape]	1976	The Shōwa era	Kishū (Wakayama)
James Joyce	Ulysses	1922	Jun/16, 1904	Dublin, Ireland
	Dubliners	1914		Dublin, Ireland

continued on following page

Table 1. Continued

Author	Title	Publication	Period(s) within the work	Location(s) within the work
	A Portrait of the Artist as a Young Man	1916		Dublin, Ireland
	Finnegans Wake	1939		
Virginia Woolf	To the Lighthouse	1927		Ramsey, Isle of Skye
	Mrs Dalloway	1925	After World War I	London
Henry James	The Bostonians	1886	19th century	Mississippi, Boston, Harvard University
	The American	1877	19th century	Europe, Paris
	The Wings of the Dove	1902	19th century	Europe
	Roderick Hudson	1875	19th century	Europe
	The Turn of the Screw	1898	19th century	England
Émile Zola	Therese Raquin	1867	19th century	Oreluan, Algeria
	L'Assommoir	1877	19th century	Paris
	Nana	1880	Second Empire	Paris
	The Joy of Living	1884	1863–1873	Bonneville Village, France
	Germinal	1885	1860's	Coal mine town monsue
	La Terre	1887	19th century	France
Honoré de Balzac	The Lily of the Valley	1836		France
	Eugenie Grandet	1833		Country town Saumur, Paris, India
	The Quest of the Absolute	1834-1835	French royal period	France
Ivan Goncharov	Oblomov	1859	19th century	Russia
Marquis de Sade	美徳の不幸 [*Bitoku no Fukō:* The Misfortunes of Virtue]	1791	18th century	France
	悪徳の栄え [*Akutoku no Sakae*: Juliette]	1797–1801	18th century	France
	The 120 Days of Sodom	1785	18th century	Schwarzwald old castle, Schilling castle
	Philosophy in the Bedroom	1795	18th century	France
	悲惨物語 [*Hisan Monogatari*]	1800	18th century	France
Georges Bataille	Madame Edwarda	1937	20th century	Paris
	Story of the Eye	1928	20th century	Spain
	The abbot C	1971	During World Wor II	France
	The Accursed Share: The Notion of Expenditure	1951		
	Erotism	1961		
Guillaume Apollinaire	The Eleven Thousand Rods	1907		Paris
沼正三 [Numa Shōzō]	Kachikujin Yapū	1970	A future age	Future Empire EHS
John Cleland	Fanny Hill	1748-1749	18th century	London
Jean-Jacques Rousseau	Emile	1762	18th century	France
	The New Heloise	1761	18th century	France
	Confessions	1,781	18th century	France
	Reveries of a Solitary Walker	1778	18th century	France
Michel de Montaigne	Essays	1580		
Niccolò Machiavelli	The Prince	1532		
Georg Wilhelm Friedrich Hegel	The Phenomenology of Spirit	1807		
Vladimir Nabokov	Lolita	1955	1947–1949	Europe, USA
	The Real Life of Sebastian Knight	1941	1899–1936	Paris
小島信夫 [Kojima Nobuo]	抱擁家族 [*Hōyō Kazoku*: Embracing Family]	1965	The Shōwa era	Setagaya, Tokyo
	島 [*Shima*: Island]	1956		An island
Homer	Iliad	8th century BC	Ancient Greece	Troy
François Rabelais	Gargantua and Pantagruel]	1532-1564		Poitiers, Oreluan, Paris, Utopia, Theleme Abbey
Victor Hugo	The Hunchback of Notre-Dame	1831	1482	Paris
Gottfried Keller	Green Henry	1879-1880	19th century	Munich
Aldous Huxley	Point Counter Point	1928		
Roger Martin du Gard	The Thibaults	1922	During World War I	Europe
Samuel Beckett	Malone Dies	1951		
Andrei Bely	The Silver Dove	1910		Russia
Louis-Ferdinand Céline	Journey to the End of the Night	1932	During World War I	Paris
Cesare Pavese	The Fine Summer	1949	20th century	Italy
Günter Grass	The Tin Drum	1959	1954(1899–1954)	Danzig Freedom City
Saul Bellow	Dangling Man	1944	20th century	Chicago
Jerome David Salinger	Nine Stories	1953	20th century	USA

continued on following page

Table 1. Continued

Author	Title	Publication	Period(s) within the work	Location(s) within the work
Jean-Marie Le Clézio	Terra Amata	1967	20th century	Nice
	The Interrogation	1963	20the century	France
Philippe Sollers	The Park	1961	20th century	France
Drieu La Rochelle	Gilles	1939	20th century	France
Jorge Vargas Llosa	Conversation in the Cathedral	1969	20th century	Bar La Cathedral
José Donoso	The Obscene Bird of Night	1970	20th century	
Julio Cortázar	Hopscotch	1963	20th century	Paris
Manuel Puig	Kiss of the Spider Woman	1976	20th century	Prison in Buenosaires
Gabriel García Márquez	The Autumn of the Patriarch	1975		fictional country facing the Caribbean Sea
石原慎太郎[Ishihara Shintarō]	化石の森 [*Kaseki no Mori*: Petrified Forest]	1970	The Shōwa era	Tokyo
高橋源一郎[Takahashi Genichirō]	ジョン・レノン対火星人 [*John Lennon Tai Kaseijin*: John Lennon vs the Martians]	1985	The Shōwa era	Tokyo Detention Center
埴谷雄高 [Haniya Yutaka]	死霊 [*Shirei*: Death Spirits]	1945-95	Shōwa 10s	Tokyo
	闇のなかの黒い馬 [*Yami no Naka no Kuroi Uma*: Black Horses in the Darkness and Other Stories]	1970		
Aleksandr Solzhenitsyn	The Gulag Archipelago	1973-1975	Former Soviet Union	Concentration Camp Glag, USSR
	Cancer Ward	1968-1969	1955	USSR
	One Day in the Life of Ivan Denisovich	1962	1950's	USSR
	Red Wheel	1993-1997		Eastern Prussia, Tannenberg

Chapter 7
Haiku Generation From Narratological Perspective:
A Circulation Between Haikus and Stories

Jumpei Ono
Aomori University, Japan

Takashi Ogata
https://orcid.org/0000-0001-9181-8876
Iwate Prefectural University, Japan

ABSTRACT

Haiku—a form of unrhymed poetry—is popular among the Japanese. A typical haiku is composed of 17 moras and three phrases. A haiku has the possibility of scratching the surface and uncovering a hidden message through an expression of events. According to Masaoka Shiki, a haiku is a kind of literature and has high affinity with our research on generating stories. In this chapter, the authors implemented the prototype system that has two functions: first, to produce multiple haikus from a single story, and second, to engender multiple stories from haikus. The system prototyped in this chapter is based on haiku theory, which is used by the authors in their research, and is rooted in the concept's co-occurrence information and frequency information used to generate a haiku. The method uses statistical information for selecting the words and creating the word network in the haiku. Through the aforementioned methods, the authors created a framework for a system of circulating haiku and stories and proposed a kind of narrative generation with narrative as an input.

INTRODUCTION

A *haiku* is a type of poem. This art form emerged in Japan and has been familiar to Japanese people for many years. A haiku poem is composed of 17 *moras* and 3 lines: the first line has 5 moras, the second 7, and the third 5. However, this standard definition is only a reference and does not define all types of haiku. Furthermore, many poets have attempted to describe haiku, with Matsuo Basho and Masaoka

DOI: 10.4018/978-1-7998-4864-6.ch007

Shiki being some of the famous haiku poets. In *Haikai Taiyo*, Masaoka Shiki (1966) mentions that haiku is a literary form, pointing out that what makes it different from other types of literature is the number of moras, although haiku does not necessarily consist of 17 moras. In *Haiku no Tukuriyo* ("*How to Make a Haiku*"), Takahama Kyoshi (2009), an apprentice of Masaoka Shiki, comments on the need for seasonal words and cuts based on 17 moras. However, there is a view that haiku does not have a strict formal style. Ishihara (2012) thus asserts that a haiku style is defined on the basis of a poetic mindset (emotions, scenery, etc.).

As maintained by Masaoka (1966), haiku is a literary form, and it shows a high affinity for our research on generating stories. In this chapter, we propose a system for the following two functions and implement a prototype. The first function is to produce multiple haikus from a single story. The second function is to generate multiple stories from haiku.

The authors have approached the haiku production using two methods. One of them is grounded in the structural aspects of haiku, while the other are based on scrutinizing the transitions of the parts of speech in haiku.

Ito and Ogata (2019) consider that a haiku consists of parts of speech, and they created haiku using the transition patterns of parts of speech in each line. Specifically, this approach utilizes the transition patterns of nouns and verbs only (Ito, Igarashi, & Ogata, 2018). The target of analysis to acquire the transition pattern is 63 phrases published in *Oku no Hosomichi* (Matsuo, 1702; Imoto, Hisatomi, Hori, Yamashita, & Maruyama,2008). First, each phrase is subjected to a morphological analysis, and only the nouns and verbs of the obtained parts of speech are examined. The haiku is a fragmentary set of words and is an experimental effort based on the expectation that nouns and verbs make up the majority of it. In this paper, we focus only on the number of occurrences of nouns and verbs. If other parts of speech (e.g., adjectives) appear first, we begin counting from the first noun or verb that appears in the line. The second strategy for constructing a haiku uses a time series based on deep learning and long short-term memory (LSTM). The third strategy uses *umeji*, which harnesses the method of devising a story. The authors concentrate on a strategy of dividing the haiku into word and character units, grounded in the approach of deep learning employed by Ono and Ogata (2018, 2019). In each type of learning, the 63 phrases in *Oku no Hosomichi* were used to study the data. Igarashi harnessed a "chainer" for the framework of deep learning and produced a haiku by learning to construct the LSTM. The learned pattern is the appearance pattern of the division unit. The target haiku was given a start symbol and end symbol. The mechanism was taught 10,000 times under the above conditions. The authors tried deep learning for the appearance pattern of words in a haiku; the appearance pattern of characters in the haiku for the learning method is described. With these learning data, a haiku was formed by concatenating words or characters from the start symbol to the end symbol. The basic procedures are as follows:

1. Begin using the start symbol.
2. Select the element based on the learning data for the element that appears next to the start symbol.
3. If the end symbol is selected, the haiku is complete. Otherwise, go to Step 4.
4. Choose elements based on learning data for elements that appear next to the selected element. Return to Step 3.

With the above technique, if the number of times needed to learn is small, both very short and very long haikus are generated. When the number of times needed to learn has reached 1,000, the results that satisfy the form of the haiku are generated. At 2,000 times, the same haiku was created as input data. Most haikus constructed after 5,000 times became the same as the input data. First, this subsection explains the experimental outcomes. Second, we discuss the results. Through this experiment, we found that the comprehensibility of the haiku and the degree to which it is interesting tend to be inversely proportional. Although there is no clear difference, this finding will serve as a foothold for experiments on a larger scale. In this research, by assessing the generated haiku based on two types of evaluation criteria, we discovered that comprehensibility and interest in the learning process tend to be inversely proportional.

This chapter aims to develop a system that establishes a story summary and new stories by producing haikus.

BACKGROUND

Haiku Generation Based on *Umeji*

Another approach is to model how poets describe the process of writing haiku. The method of generating one statistically has a limit in exceeding a certain level. Therefore, it may be significant to derive a computer science model from the poet's approach to writing haiku. We studied a computational science technique to constructing a haiku based on *umeji*, as described by Takahama (2009). He demonstrated in *Haiku no Tukuriyo* the first approach to haiku creation by "arranging 17 moras randomly." Thus, statistically random creation is not an incorrect way of forming haiku. However, deep learning for haikus is not a suitable strategy because it is based on existing haikus.

Here, we explain the architecture of the proposed system. There are two large modules. One produces a haiku using a story. The other generates a story from a haiku. For the module that produces a haiku from a story, we used the module that we mentioned previously. This module focuses on a sentence in the story, collects and edits related sentences to generate three lines, and combines them into one haiku. The module extracts elements such as people, places, and verbs from the observed sentences and gathers sentences with the same components. In this paper, we extend the method and comprehensively establish a haiku using the module. Instead of devising a single haiku, we create one from every angle. With this approach, the elements that constitute the story are thoroughly extracted. The discourse concept can be used as an exhaustive extraction technique. In particular, the notion of the viewpoint corresponds to the method. Haikus shaped by this method can be perceived as a network of story components.

When producing a story from a haiku, all or a part of the generated haiku can be input. The components of the story that have been broken down into weak connections are once again given a strong connection by writing the story. This process uses an integrated narrative generation system.

The technique suggested in this paper includes the above-mentioned work by Takahama. However, we propose not only a haiku but also a cyclical creation strategy, which includes generating a story from a haiku. This cyclical procedure is used in the process of forming a story described by Ogata and is based on the haiku theory outlined by Nitta (2018). Nitta stated that the recipient can transform the haiku context (narrative) and refer to the transformation as a strolling ontology. Recipients have their own context and accept the haiku context (narrative) in that context. The cyclical generation set forth in this paper represents the transformation. Generating a haiku by using a story as an input means creating a haiku, and generating a story using a haiku as an input embodies acceptance.

We approached the haiku generation from two aspects. The first one involves focusing on haiku's structural features. One entail analyzing the part-of-speech transitions in the haiku; the other requires deep learning to produce the haiku. These attempts yielded particular results.

Another technique is to model the haiku generation method described by the poet. We consider the first approach of forming a haiku statistically, to have a limit in exceeding a certain level. Therefore, the second method can be beneficial in creating a computer science model from the haiku creation approach commented on by the poet. We studied a haiku generation method based on computational science grounded in *umeji*, referred to by Takahama. In *Haiku no Tukuriyo* (2009), Takahama showed that randomly arranging 17 mora is the first haiku creation method. Hence, it is not wrong to produce a statistically random haiku. However, the deep learning of a haiku is not suitable because it uses an existing haiku.

As mentioned above, this paper uses the approach by Takahama. However, we propose not only haiku generation but also cyclical formation, including devising a story from a haiku. This cyclical generation is used in the story process described by Ogata, and it refers to the haiku theory outlined by Nitta (2018). Nitta stated that the recipient transforms the haiku context (narrative) and expresses the transformation as a strolling ontology. Recipients have their own context through which they accept the haiku context (narrative). The cyclical generation set forth in this paper represents the transformation. Shaping a haiku by using a story as an input is the process of creating a haiku, and generating a story by using a haiku as an input is a process of acceptance.

From the perspective of narratology, creating a haiku has a discourse aspect. Discourse is the concept presented by Genette (1972). Thus, haiku plays the role of a discourse technique. For example, it is possible to make a haiku by cutting a story from a particular viewpoint. Alternatively, a specific event is missing. A haiku can be established by employing various discourse-technique-like operations, such as reversing the time series of events.

The following paragraph exemplifies the architecture of the proposed system, which uses two large modules. One generates a haiku from the story, while the other produces stories from the haiku.

The module that creates a haiku from the story uses the module that we have previously published. It considers the sentences in the story, collects and edits sentences related to it, generates three lines, and combines these lines into one haiku. The module extracts elements such as people, places, or verbs from the observed sentences and collects sentences with the same elements. In this chapter, the technique is expanded to comprehensively form haikus by employing modules. Rather than creating a single haiku, it generates haiku from every perspective. With this type of haiku generation, the components that make up the story are thoroughly extracted. The concept of discourse can be used as a comprehensive extraction approach. In particular, the concept of the viewpoint corresponds to the method. The haiku generated using this strategy can be regarded as a network of story components.

When generating a story from a haiku, all or a part of the haiku consists of input. A story component that has been broken down into weak links can once again be given strong links as a story through story production. This story creation uses an integrated narrative generation system.

We suggest a system for generating stories and haikus using the above technique. When the haiku is defined on the basis of Masaoka's standards, the usual works using ordinary words are not good haikus. Based on this, using the deep-learning method only can lead to a type of haiku generation that looks like natural language. Moreover, this method has the possibility of approaching an expression of events via the surface, as well as a hidden message by expressing those events. In another study, we conducted the commercial message generation, and we believe in using the knowledge of this paper in the generation.

We carried out experiments to produce sentences (like those in a haiku) through deep learning, with haikus used as learning data. Even in this generation, as the number of times needed to learn increased, the comprehensibility of the sentences tended to comprise the output. However, when considering the fun of writing a haiku, it is hard to say that its comprehensibility was not necessarily interesting. The hypothesis was that haikus that seemed rather difficult to understand might be more interesting. Therefore, by evaluating the comprehensibility and fun at each constant learning number, we investigated the relationship between the two values.

Haiku Generation Based on the gap Technique

Ito, Ono and Ogata (2019) proposed haiku generation based on the gap technique. Regarding this method of production, we consider the following two techniques: (1) selecting two sentences that have a character as a designated agent, from a narrative, and (2) generating a haiku using the elements from these sentences. We expect to form a haiku to express an overview of the character's behavior in the narrative. What follows is a generation example using the following narrative:

Abstract sentences have a designated agent, and two sentences are chosen from this abstraction: When "wolf" is selected, we subtract "The wolf attacks sheep in a stock farm" and "The wolf escapes."

Figure 1. Input story

父が家屋で狼に困る.
A father is embarrassed a wolf in his house.
息子のアランが家屋で父を励ます.
His son, Alain encourages his father in his house.
父が「アランが外出する」ことを禁止する.
The father forbids Alain from going out.
アランが家屋で泥を洗う.
Alain washes off mud in his house.
アランが牧場で衣服を乾かす.
Alain dries clothes in a stock farm.
狼が牧場で羊を襲う.
The wolf attacks sheep in a stock farm.
アランが狼を待ち構える.
Alain waits for the wolf.
アランが狼を襲う.
Alain attacks the wolf.
アランが狼を殴る.
Alain hits the wolf.
狼が逃げる.
The wolf escapes.
アランが狼を逃がす.
Alain misses the wolf.
父がアランを称える.
The father praises Alain.

An abstract verb case and a purpose for the sentences are chosen in (1) Extract "attacks" and "sheep" from the first sentence. Abstract "escapes" from the second sentence. "The wolf" is S, and "attacks" is V1. "Sheep" is P1, and "escapes" is V2.

Then, we make a format and arrange the words in a sequence for each selected sentence: Here, we use "S V1+P1 V2," setting elements according to this format: "狼 (the wolf) 襲う(attacks) 羊を (sheep) 逃げる (escape)."

Adjustments are made to form a proper haiku: "the wolf" is added as a mora *Kirezi*, because "the wolf" consists of four moras. Similarly, "escapes" is three moras, so two moras are added to *Kirezi*. Finally, we end up with the following haiku: "狼 (the wolf) や (ya) 襲う(attacks) 羊 (sheep) を 逃げる (escape)かな (kana)."

GENERATION SYSTEM THROUGH CIRCULATION BETWEEN HAIKUS AND STORIES

Our main goal here is to create a haiku and a story cyclically. *Cyclic* in this case implies a mechanism whereby mutual inputs can be inserted into a production system, and one input can output the other. The simplest form of this cycle is when a haiku is formed from a story, and the original story is produced from the generated haiku. This is an example of haiku generation as a narrative creation technique.

In addition, we consider a method to shape a story, unlike the original one, based on the generated haiku. This is an attempt to form a new story through the mutual generation of the haiku and the story. A haiku is generated from a certain story, and we assess whether the original story can be produced on the basis of the generated haiku. We propose a haiku/story flow, as shown in Fig. 2. The haiku and the narrative are connected cyclically by two processes: (1) narrative generation from the haiku and (2) the creation of the haiku from the narrative.

Figure 2. Mutual generation by translating the haiku and the story

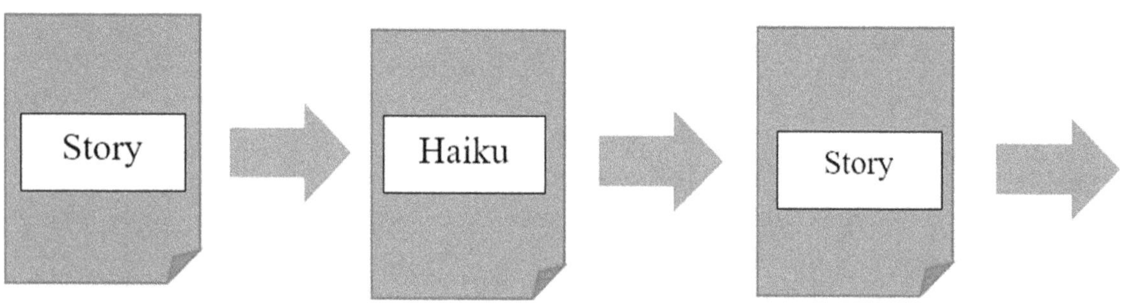

When starting a story, first, a story is given as input to the system. Such a story is transformed into a haiku that focuses on some point in the process of producing a haiku from the story. The haiku obtained through this process is developed into a story once again through story generation from the haiku. The story obtained at this time is elaborated around the elements that the input haiku is paying attention to, so we expect that the center of the original story and the topic will change.

To start a haiku, first, we input a haiku into the system. At this time, the input haiku considers both the existing haiku and the haiku produced using the haiku generation method proposed in this system. Next, we move on to creating stories from a haiku. One generation method is selected to form a story. Next, we move on to create a haiku from a story. The generated narrative is converted back into the haiku format using the chosen method. At this time, when a sentence is converted from a story to a haiku, the output changes depending on the technique used, and a resemblance to the first input haiku may be generated.

Interconversion uses not only superficial expressions but also structural aspects. Figure 3 depicts an example of one of the readers' interpretations of the haiku shown in Ito et al. (2017), which was written and analyzed. To create a story from a haiku, we asked the reader to read and interpret it. We performed an experiment in which the reader produced a story about the haiku. We examined the relationship between the haiku and the narrative regarding the development of the expression from the narrative obtained from the experiment. From the results, we devised a haiku and a semantic network of stories, as presented in Fig. 3. A story generation approach using a semantic network and an integrated story

Figure 3. Semantic network of the haiku and the story (Source: Ito, Arai, & Ogata, 2017)

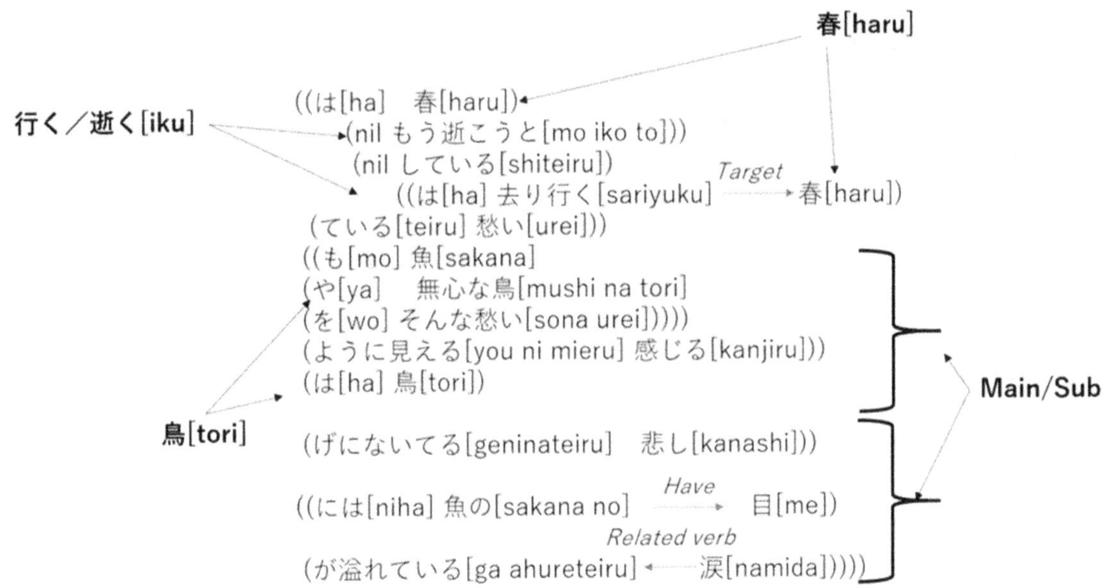

system was also developed. Specifically, we tried to create a new story by replacing words in the haiku in the interpretation of the haiku itself, with different words using a concept dictionary.

Here, the haiku's structure is generated at the level of the case structure, without analyzing the meaning up to the level shown in Fig. 3. Attempts were made to create a haiku using all four elements of the story (in this case, people, things, places, and events).

IMPLEMENTATION OF THE GENERATION SYSTEM

This section covers the methods of generation for haikus and stories. The first subsection portrays a haiku generation flow. The next subsection describes the story generation flow. Moreover, the last subsection contains a discussion. The system was made using Common Lisp.

METHOD FOR GENERATING A HAIKU

The section explains the following steps to produce a haiku:

1. Select one or more events in the input story.
2. Construct haiku structures.
3. Make an empty story structure.

Step 1.

This step includes selecting a sentence that has a character as a designated agent and choosing a sentence that appears just before the selected sentence or immediately afterward. A haiku is produced using the elements from these sentences. We expect that it will be created so as to describe one scene from the narrative. We demonstrate a generation example using a narrative. An abstract sentence has a designated agent, and a sentence is selected that appears just before the chosen sentence or immediately afterward: When “the wolf” is chosen and abstracted, the sentence “the wolf escapes” appears, followed by “Alain hits the wolf” and “Alain misses the wolf.” In this example, “Alain hits the wolf” was selected.

Figure 4. Example of input story

父のトムが家屋で狼に困る。
息子のアランが家屋で父を励ます。
アランが家屋で衣服を洗う。
アランが牧場で衣服を乾かす。
狼が牧場で羊を襲う。
アランが狼を襲う。
狼が逃げる。
トムがアランを称える。

Step 2.

Abstract sentences contain a word representing a designated location. Two sentences from this abstraction are selected: When "his house" is chosen, abstract sentences contain "his house": "A father is embarrassed by a wolf in his house," "His son, Alain, encourages his father in his house," and "Alain washes off his clothes in his house." In this example, we selected "Alain washes off his clothes in his house." A designated agent, a verb case, and a purpose from the sentences were selected. "Alain," "washes off," and "his clothes" come from "Alain washes off his clothes in his house." "Alain" is S, "washes off" is V, and "his clothes" is P, while "his house" is L. The format is made, and the words are arranged in the sequence for each of the selected sentences. Here, the format "L S+V2P" is used, and the elements are arranged accordingly: "家屋 [his house] アラン [Alain] 洗う[washes off] 衣服 [his clothes]."

The elements are adjusted to properly form a haiku: "アラン [Alain] 洗う [washes off]" is added to a mora kireji (Table 1), because it comprises six moras. Similarly, "家屋 [his house]" and "衣服 [his clothes]" are three moras, so two moras are added to kireji. Finally, we obtain the haiku: "家屋かな アラン洗うや 衣服なり." In particular, we make a rule using two moras kireji. "けり(keri)," "たり (tari)," and "なり(nari)" are auxiliary verbs in classical Japanese grammar. "けり" indicates the past and mourning, and it is connected to the conjunctive form of the conjugated word by utilizing the La variant. "なり" denotes assertion/existence, and it is used in the adjective verb *Nari* inflectional type to be attached to nouns, adnominal forms, adverbs, and particles. "たり" means completion and survival, and it can be used as a continuous form of an inflection word by utilizing the La variant. "たり" has a more conclusive meaning, depending on the context, and can be put into a physical form by employing the adjective verb *tari* inflection type. Here, the nouns are *nari* and *tari*, and the verbs are *keri* and *tari*.

Table 1. Kireji table

Mora	Kireji
One mora	や ら し ぞ か よ せ や つ れ ぬ ず に へ け じ
Two moras	かな けり たり
Three moras	もがな

Figure 5. Example of generated haiku structure and sentences

(1st-phrase "家屋" "かな") (2nd-phrase "アラン" "洗う" "や") (3rd-phrase "衣服" "なり"))
"家屋かな アラン洗うや 衣服なり"

Step 3

For the events, the input story structure is removed. The system obtains an empty story structure, and outputs generate haikus and an empty story structure. Figure 6 shows an example of the output data.

Figure 6. Haiku structure and a story structure

```
((((1st-phrase "家屋" "けり") (1st-phrase "アラン" "洗う" "や") (1st-phrase "泥" "もがな")))
(
$継起
 ($継起 (event "empty0")
       (event "empty1"))
 ($継起 (event "empty2")
      (event "empty3"))
 ($継起
   (event "empty4")
   (event "empty5")
   (event "empty6")
(event "empty7"))))
```

METHOD FOR STORY GENERATION USING A HAIKU

This section explains the following generation steps.

1. Select one verb (a noun) from the input haiku.
2. Obtain a verb with a similar frequency of information from the frequency information of the selected verb (noun).
3. Insert the generated events into an empty tree.

Step 1

A verb or noun is chosen using frequency information (Appendix 1). The c verb (noun) is used for the first generation of the event.

Step 2

The system chooses verbs for event production. The verb is selected on the basis of the number of empty events in the empty story structure. Figure 6 outlines examples of the generation functions. Figure 7 depicts the verb frequency data and verb corresponding data. (Appendix 2 contains more examples.)

Step 3

The events generated in Step 2 are inserted into an empty tree. Figure 9 shows an example of a generated story.

Figure 7. Examples of functions

```
(defun bH (av)
  (nth (- (position (intern v) *vlis* :test #'equal :key #'car)
             (- 3 (random 6))
             ) *vlis*)
  )

;***
(defun vRel (v &aux (temp nil)(temp2 nil) (temp3 nil))
  (setq temp (format nil "./v-data/~a.txt" v))
  (if (file-exist-p temp)
     (with-open-file (in temp)
            (setq temp2 (read in nil nil))
            )
    )
  (setq temp3 (nth (random (length (cdr temp2))) (cdr temp2)))
  (nth (random (length temp3)) temp3)
  )
```

Generation Experiments

This section covers two experiments, such as those in Figure 10. The first experiment demonstrates a cycle between a story and a haiku. The second experiment depicts continuous transformation between stories and haikus. Two experiments used a common story (Figure 4) as input data. Input stories, generated haikus, and generated stories are calculated as corresponding parameters. The parameter is used to compare the stories and haikus. If the two parameters are separated, the content in the story translates to haiku generation.

Regression Experiment

Regression experiments generate a haiku from a story and use the generated haiku to check whether a story that is close to the original can be generated. Figure 11 shows examples of the generated story. The co-occurrence information score of the haiku generated from the story and the co-occurrence information score of the story generated from the haiku were compared, and a story with a difference within 0.2 was judged to be a regressed story.

Figure 8. Example of corresponding information

(洗う4 ((女子 1.000769) (知人 1.009719) (親友 1.061499) (プロレタリアート 1.095198) (介 1.095923) (坊さん 1.115862) (使者 1.118878) (画家 1.120392) (隊長 1.134205) (人種 1.137321) (宮 1.137321) (家人 1.137321) (代表 1.152091) (乳母 1.162879) (水兵 1.210482) (人民 1.214922) (大衆 1.21641) (一行 1.221194) (夫人 1.236111) (駒 1.239475) (王子 1.297758) (独身 1.314237) (浪人 1.318423) (刑事 1.322635) (地主 1.352891) (事務 1.357327) (長男 1.361792) (母上 1.361792) (民衆 1.375365) (友 1.37903) (車夫 1.379949) (味方 1.383023) (同志 1.398597) (国 1.400806) (師匠 1.40215) (読者 1.405244) (信者 1.412917) (強盗 1.417756) (奴隷 1.432473) (法師 1.444133) (部長 1.447499) (姉さん 1.464569) (犯人 1.469769) (大臣 1.481162) (処女 1.491837) (先輩 1.49453) (民族 1.506769) (不良 1.508141) (弁護士 1.510893) (息 1.512901) (ママ 1.514578) (連れ 1.522003) (判事 1.538977) (師 1.544719) (乞食 1.556334) (尼 1.568124) (女王 1.574087) (名人 1.607708) (代わり 1.610836) (権威 1.610836) (兄妹 1.642797) (貴族 1.642797) (論者 1.649342) (一味 1.655939) (人間 1.656602) (坊ちゃん 1.669294) (坊や 1.696659) (記者 1.722067) (父 1.730861) (作家 1.740557) (孫 1.746771) (星 1.751699) (弟 1.753348) (兄貴 1.761638) (曲者 1.761638) (恋人 1.761638) (候補 1.776759)・・省略

Transformation Experiment

Figure 9. Example of a generated story

```
($継起
 ($継起
  (event 父のトムは地価を衣服に負けました)
  (event 青年のアランは本店で父のトムの心を洗いました))
 ($継起
  (event アランが衣服を洗う。)
  (event 菌類がケーブルで咲きました))
 ($継起
  (event 青年のアランは帆影で父のトムの心を洗いました)
  (event 父のトムは泥沼で衣服を結びました)
  (event 拡声器が乗り物で疲れました)
  (event 菌類が本店で咲きました)))
```

Figure 10. Example of a generated story

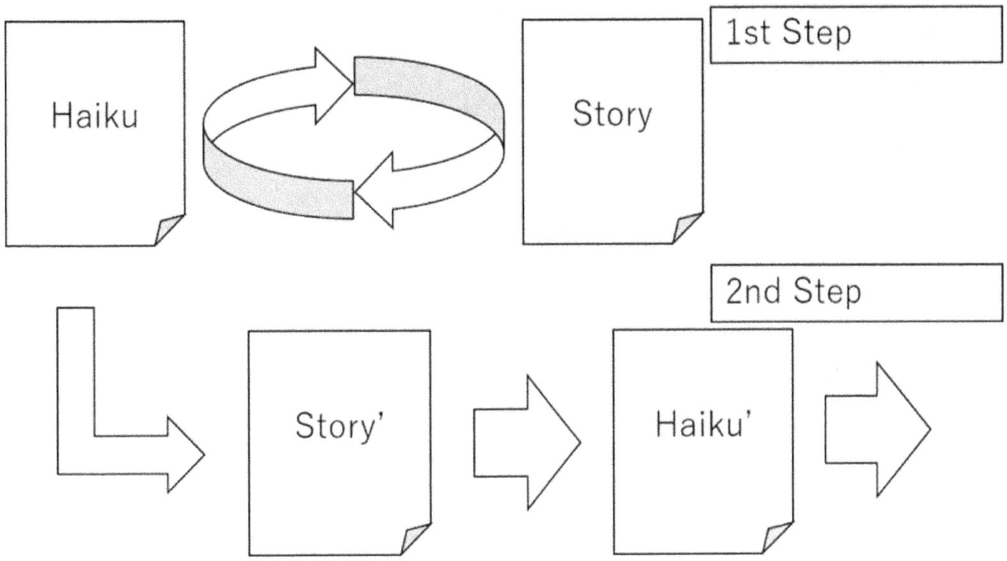

Regression experiments generate a haiku from a story and use the generated haiku to examine whether a story that is not close to the original can be generated. Figure 12 shows examples of the generated story. The co-occurrence information score of the haiku generated from the story and the co-occurrence information score of the story generated from the haiku were compared, and a story with a difference within 0.2 was judged to be a regressed story.

RELATED WORKS

Haiku generation research has been conducted since 2016, and researchers at Microsoft Development Ltd. and Hokkaido University have proposed a method using deep learning. In both cases, learning was attempted by employing the LSTM method with Matsuo Basho's or Kobayashi Issa's haikus as input. Furthermore, researchers at Hokkaido University have attempted to generate haiku from input images by matching images with haikus. Both methods are based on statistical data. Furthermore, researchers at the Kyoto Institute of Technology have proposed haiku generation using generative adversarial networks, which are widely used in the field of image generation. In this study, the standard used for the evaluation was whether the generated haiku looked like a human-created haiku.

Researchers at Ben Gurion University of the Negev approached creativity in haiku generation, with the word associative norm as the axis of generation. When broadening their horizons and looking at the study of poetry generation, researchers at Tsinghua University tried to create Chinese poetry. Their study controls sentiment. Therefore, a sentiment controllable poetry generation model was defined, and the researchers evaluated the sentiment of the generated haiku.

The core idea of AI Issa-kun (Chibi), which the research team of Hokkaido University first worked on, was learning, and learning with hiragana haiku data that could measure the number of moras to reliably form the five–seven–five structure. Thus, about 20,000 Kobayashi Issa's Hiragana Haiku character

Figure 11. Examples of a generated story in the regression experiment

```
($継起
 ($継起
  (event 青年のアランは便殿で内貨しに努力を払いました)
  (event 雹が家屋で便殿を叩きました))
 ($継起
  (event アランが衣服を洗う。)
  (event 真相が牧場で唯一 続きました))
 ($継起
  (event 父のトムは都下で衣服を結びました)
  (event 最高限が湾岸で唯一 続きました)
  (event 青年のアランは水源で父のトムの心を洗いました)
  (event 真相が牧場で最高限 続きました)))
```

```
($継起
  (event 父のトムは地価を衣服に負けました)
  (event 青年のアランは本店で父のトムの心を洗いました))
 ($継起
  (event アランが衣服を洗う。)
  (event 菌類がケーブルで咲きました))
 ($継起
  (event 青年のアランは帆影で父のトムの心を洗いました)
  (event 父のトムは泥沼で衣服を結びました)
  (event 拡声器が乗り物で疲れました)
  (event 菌類が本店で咲きました)))
```

strings were divided into morphemes by the morphological analysis engine MeCab. The haiku divided into morphemes was input into the Recurrent Neural Network (RNN), and Learn, which Yoneda, Yokoyama, Yamashita, and Kawamura (2018) proposed as the methods that learn relations between an image and a haiku and generate haikus showing relationships with a motif in an image. Wu, Klyen, Ito, and Chen (2017) proposed generating haiku-like sentences with elements that were uttered during the talk with Chatbot. Hirota, Oka, Araki, and Tanaka (2018) proposed methods that learn a large quantity of haiku data and character data strings containing five or seven syllables, and generate haikus using those data. In these previous studies, haiku was generated using elements that were strongly interrelated. As a result, the generated haiku is a common haiku that can be expected. The authors think that it is difficult to reach "surprising" results compared to previous studies. In addition, the authors think that the previous haiku generation methods are only a combination of elements. morphemes appear.

Figure 12. Example of a generated story in the transformation experiment

```
($継起
 ($継起
  (event アランが船長を加える。)
  (event アランが抜け道を残す。))
 ($継起
  (event アランが衣服を洗う。)
  (event アランが生涯を下る。))
 ($継起
  (event アランが同郷を限る。)
  (event アランが五月雨を結ぶ。)
  (event アランが足袋を叩く。)
  (event アランが土手を結ぶ。)))
```

```
($継起
 ($継起
  (event 青年のアランは便殿で内貨しに努力を払いました)
  (event 雹が家屋で便殿を叩きました))
 ($継起
  (event アランが衣服を洗う。)
  (event 真相が牧場で唯一 続きました))
 ($継起
  (event 父のトムは都下で衣服を結びました)
  (event 最高限が湾岸で唯一 続きました)
  (event 青年のアランは水源で父のトムの心を洗いました)
  (event 真相が牧場で最高限 続きました)))
```

CONCLUSION

In this paper, the authors propose a haiku generation system based on mutual transformation between a narrative and haiku-like sentences. In particular, this paper presents a method for generating haiku-like sentences from a narrative fragment. In addition, the authors have been developing an integrated narrative generation system that automatically generates narratives. Thus, this paper aims to combine this mechanism with the above-mentioned system in the future. Basic elements for generating a narrative and a haiku-like sentence are commonly the selection, arrangement, and modification of the elements. To generate haiku-like sentences from a narrative, the proposed method selects particular elements from a narrative and arranges them in the form of a haiku-like sentence. Further, it provides variations to the haiku-like sentence. This paper shows several forms of haiku-like sentence generation to experimentally generate haiku-like sentences from a narrative. The authors discovered that by evaluating the generated haiku based on two types of evaluation criteria, the understandability and interest in the learning process tend to be inversely proportional. In addition, although the number of learning times that make understandability and interestingness increase in a well-balanced manner was not found in this experiment scale, the authors would like to discover the number of learning that harmonizes interestingness and understandability in the future.

To realize the framework described, the following issues should be considered:

Morphological analysis of haiku: Since haiku is a set of fragmentary words, ordinary sentences, and parts of speech, it varies greatly. In addition, the notation sway is intense, especially when including former pseudonyms. The generation of events plays an important role in the concept described in this paper, and the acquisition of elements such as verbs is essential for the generation of the event. It is conceivable to extend the language notation dictionary of the integrated narrative generation system and manually correct the elements.

Correspondence with sentence pattern: In this study, an element is inserted into a sentence pattern by forcibly generating an event (sentence) and ignoring the constraint condition. However, if the event is generated by ignoring the constraint condition, unnatural events (sentences) will be created, which may affect the application of the subsequent gap technique. Elements that can be obtained from haiku are limited, and there is a high possibility that an element will not correspond with the current sentence pattern. Therefore, it is necessary to expand the technique to be able to flexibly deal with combinations of various elements.

Correspondence to haiku without a verb: For this study, the authors have assumed that haiku has one or more verbs as a premise. However, haiku without verbs also exists. When dealing with such haiku, it is necessary to first add a verb in some form.

REFERENCES

Genette, G. (1972). Discours du recit, Essai de methode, Figures III. Seuil.

Hirota, A., Oka, N., Araki, M., & Tanaka, K. (2018, March). Haiku generation by seqGAN that divided learning data set. *Proceedings of the Twenty-fourth Annual Meeting of the Association for Natural Language Processing*, 1292-1295.

Imoto, N., Hisatomi, T., Hori, N., Yamashita, K., & Maruyama, K. (2008). Nihon no koten wo yomu: Vol. 20. *Okuno hosomichi Basho / Buson / Issa syu*. Shogakukan.

Ishihara, Y. (2012). *Haiku no tukurikata* [How to make a haiku]. Meiji Shoin.

Ito, T., Arai, T., & Ogata, T. (2017). Narrative generation from the semantic networks of a haiku. *Proceedings of the 31st Annual Conference of the Japanese Society for Artificial Intelligence*.

Ito, T., Igarashi, K., & Ogata, T. (2018). A consideration on a multiple approach for haiku generation *Proceedings of the 32nd Annual Conference of the Japanese Society for Artificial Intelligence*.

Ito, T., & Ogata, T. (2019). A Framework for Haiku Generation from a Narrative. *Journal of Robotics. Networking and Artificial Life*, *6*(1), 23–26. doi:10.2991/jrnal.k.190531.005

Ito, T., Ono, J., & Ogata, T. (2018). Haiku generation using gap techniques. *Proceedings of 2018 International Conference on Artificial Intelligence and Virtual Reality, V-004*. 10.1145/3293663.3293666

Ito, T., Ono, J., & Ogata, T. (2018). Haiku generation using gap techniques. *Proceedings of 2018 International Conference on Artificial Intelligence and Virtual Reality.*

Masaoka, S. (1966). *Haikai taiyo* [Important points of Haikai]. Iwanami Shoten.

Mastuo Basyo. (1702). *Oku no hosomichi* [Narrow Road to the Interior]. (https://kotenseki.nijl.ac.jp/biblio/100000092/)

Nitta, Y. (2018). Haiku-like aesthetic sentence composition through the narrative process. In T. Ogata & S. Asakawa (Eds.), *Content generation through narrative communication and simulation* (pp. 286–309). IGI-Global. doi:10.4018/978-1-5225-4775-4.ch009

Ono, J., & Ogata, T. (2018). Considering a haiku generation process using deep learning. *Proceedings of the 28th Annual Conference of the Japanese Neural Network Society*, 166-167.

Ono, J., & Ogata, T. (2019). A method of haiku generation using deep learning for advertising generation. *Proceedings of the 2019 International Conference on Artificial Life and Robotics*, 578-580. 10.5954/ICAROB.2019.OS24-1

Takahama, K. (2009). *Haiku no tsukuriyou.* Kadokawa Shoten.

Wu, X., Klyen, M., Ito, K., & Chen, Z. (2017). Haiku Generation Using Deep Neural Networks. *Proceedings of the Twenty-third Annual Meeting of the Association for Natural Language Processing*, 1133-1136.

Yoneda, K., Yokoyama, S., Yamashita, T., & Kawamura, H. (2018). Generated haiku distinctions by deep learning. IPSJ SIG Technical Report Vol. 2018-ICS-192 No. 8.

KEY TERMS AND DEFINITIONS

Cyclic Story Generation: A concept for generation of story and haiku. Cyclic in this case implies a mechanism whereby mutual inputs can be inserted into a production system, and one input can output the other.

Haiku: A poetic form common in Japan. It consists of three lines. The first line contains 5 moras, the second 7, and the third 5, for a total of 17 moras).

Mora: A sound unit in haiku. In most cases, the authors do not use the term *syllable*. For example, the word *book* consists of a single syllable and two moras.

Narrative Discourse: Narrative discourse is a concept presented by Genette (1972). This chapter deals with narrative discourse as a technique for transforming the structure of a story. This chapter considers haiku as a discourse technique. For example, it is possible to make a haiku by cutting a story from a particular viewpoint. Alternatively, a specific event is missing, etc.

Umeji: A method of composing haiku. First, the first and third phrases are prepared. Then, the second phrase is inserted between them. The second phrase is based on the relationship with other two phrases. For example, the relationship is word familiar, dissimilar, and so on.

Chapter 8
On Logical Literary Work Generation and More

Akinori Abe
Chiba University, Japan

ABSTRACT

In several papers, the author has discussed the possibility of the computational literary work generation. That was based on Julia Kristeva's the concept of the intertextuality. This very simple application was achieved by random combination of the selected words/phrases/letters. It is a simple technique such as the cutup method. The author showed even such a simple method could generate literary works that can be regarded as not so bad literary works. However, for more intelligent generation, the author introduced abduction. By abduction, it is possible to guarantee the logical consistency and coherency. However, it does not care the very cognitive aspects such as beautiful, sentimental, and valuable, which will be very important factor in the literature. In this chapter, the author will discuss to add such cognitive aspects to the logical generation of literary works.

INTRODUCTION

In several papers, the author has discussed the possibility of the computational literary work generation. The generative process was based on Julia Kristeva's concept of the intertextuality (Kristeva, 1980). Kristeva defined intertextuality as follows:

The word's status is thus defined horizontally (the word in the text belongs to both writing subject and addressee) as well as vertically (the word in the text is oriented towards an anterior or synchronic literary corpus)... each word (text) is an intersection of words (texts) where at least one other word (text) can be read... any text is constructed as a mosaic of quotations; any text is the absorption and transformation of another.

The important point is the phrase "each word (text) is an intersection of words (texts) where at least one other word (text) can be read... any text is constructed as a mosaic of quotations.''

DOI: 10.4018/978-1-7998-4864-6.ch008

This means that it is possible to generate new literary works by the reconstruction/combination (mosaic of quotations) of the frictions of the previously written literary works.

In (Abe, 2005; Abe, 2910a; Abe, 2019), the author has discussed the possibility of the automatic waka (Japanese poem) generation. In the works, the main objective was to show the possibility of using the concept of intertextuality to generate literary works. Accordingly the author used a random combination of words from the widely known Japanese waka (Hyakunin-issyu (百人一首)). For instance, the following wakas could be generated (in Japanese[1]).

鳴く鹿の 尾上の桜 命にて ながながし夜を 山の奥にも

A bleating deer, cherry blossoms in Ogami, it may be our fatal, in a long long night, deep in the mountain.

初霜の 衛士のたく火の 奥山に つらぬきとめぬ 紅葉なりけり

The first frost, fire given by guardians, in the secluded mountain, cannot penetrate read leaves.

The very simple application was achieved by random combination of the selected words/phrases/letters. It is a simple technique such as the cutup method. Phrases are previously divided to 5, 7, 5, 7, 7. The phrases were extracted from the original wakas in Hyakunin-issyu. Accordingly, all phrases are the part of the original wakas. Then those phrases are selected randomly and combined according to the rule of waka[2]. The author showed even such a simple method could generate literary works which can be regarded as readable literary works. Actually computers cannot evaluate the quality of the generated waka. The quality of the generated waka can be determined by human. In addition, the random combination method cannot consider consistency coherency, nor meaning etc. during the generation as it is. Accordingly, for more intelligent generation, the author introduced abduction.

By abduction, it is possible to guarantee logical consistency and coherency. However, the generative process does not consider the cognitive aspects such as beauty, sentimentality and value such as preciousness or preference, which are very important factors in literature. In this chapter I will discuss adding such cognitive aspects to the logical generation of literary works.

The main purpose of this chapter is building a bridge between a computer and affective aspect of human. This is very significant trial and I will discuss from both computational aspects and authentic aspects. For that, the author explains abduction which can deal with a strict consistency and coherence maintenance. In addition, as a non-intelligent but exciting method, the author will introduce random generation such as cutup method. Then by using the result of previously conducted experiment, the author will show human's abductive story generation. This may show a certain emotional aspects of human as well. In addition, for the emotional aspects, the author will address the importance of analogical mapping as a pastiche like method.

ABDUCTION

In the previous section, the author mentioned literary work generation based on Julia Kristeva's concept of the intertextuality. Although wakas are constructed as a mosaic of quotations, it was almost a random generation. Though readable literary works could be generated, it is not a logical and intelligent generation. For more logical and intelligent generation, one of strategies is an introduction of abduction. Abduction is one of the human thinking styles and regarded as a creative activity. In this section, the author will briefly show the mechanism of hypothetical reasoning (Theorist) which can be regarded as a computational abduction.

Computational Abduction

Peirce illustrated abduction as follows (Peirce, 1955):

...abduction is an operation for adopting an explanatory hypothesis, which is subject to certain conditions, and that in pure abduction, there can never be justification for accepting the hypothesis other than through interrogation.

The important keyword of the definition is *explanatory hypothesis*. That is, abduction is performed as an explanation of something. Peirce would not have assumed the performance of computational abduction.

The first achievement of computational abduction was proposed by Pople (Pople, 1973). Pople explored the abductive reasoning process and developed a model for its mechanization, which consists of an embedding of deductive logic in an iterative hypothesis and test procedure. After several years, more logical abduction systems were proposed such as Theorist (Poole et al., 1987) and ALP (Abductive Logic Programming) (Kakas et al., 1992).

Theorist is considered as a computational hypothetical reasoning system. A hypothetical reasoning is an explanatory reasoning mechanism that generates (collects, selects) a consistent hypothesis set from hypothesis candidates to explain the given observation. The generated hypothesis set can be regarded as an answer (solution) which can explain the observation. The inference mechanism of Theorist is described as follows:

$F\ O$. (O cannot be explained only by F.) (1)

$F \cup h \vdash O$. (O can be explained by F and h.) (2)

$F \cup h\ \square$. (F and h are consistent.) (3)where F is called fact which is always true. On the other hand, h is called hypothesis which is not always true and included in the hypothesis base H ($h \subseteq H$). O is an observation to be explained. $\square$ is an empty set. When $F \cup h \vdash \square$, F and h are not consistent. In the knowledge base, inconsistent information is also included.

Though details were shown in (Abe, 2019), the author will show a brief application of abduction to the generation of poems. If an *observation* is "satisfy hungry,'' when the situation is ¬*knife*, *bitter summer orange*(*road side*), H will be a hypothesis set such as {*pick up*(*bitter summer orange*(*road side*)), *peel*(*nail, bitter summer orange*(*road side*)), *eat*(*bitter summer orange*(*road side*))}. Then the outline of the poem shown in the section 5.2 can be generated.

STORY GENERATION AS A RANDOM GENERATION

In the previous section, the author introduced his previous works, for instance, an automatic waka generation. In this section, the author will illustrate a literary work generation by a type of random generation method.

In the previous section, the author mentioned the possibility of the automatic (computational) waka (Japanese poem) generation. By using the concept of intertextuality, the generative system was able to generate wakas. These were acceptable in a certain sense. There was an almost random generation or combination of phrases. In the field of the modern literature, this type of generation was conducted as the cutup method. William S. Burroughs mentioned that ''[t]he method is simple. Here is one way to do it. Take a page. Like this page. Now cut down the middle and cross the middle. You have four sections: 1

2 3 4... one two three four. Now rearrange the sections placing section four with section one and section two with section three. And you have a new page. (Burroughs)'' In addition, Burroughs stated "Sometimes it says much the same thing. Sometimes something quite differentócutting up political speeches is an interesting exercise in any case you will find that it says something and something quite definite.''

For instance, the opening of the " *The Soft Machine*'' (1961) is as follows:

Dead on Arrival

I was working the hole with the Sailor and we did not do bad fifteen cents on an average night boosting the afternoons and short timing the dawn we made out from the land of the free but I was running out of veins ... I went over to the counter for another cup of coffee ... in Joe's Lunch Room drinking coffee with a napkin under the cup which is said to be the mark of someone who does a lot of sitting in cafeterias and lunchrooms ... waiting on the Man ... `What can we do?' Nick said to me once in his dead junky whisper...

It is a rather nonsensical sentence, but it can be acceptable as an English sentence. This "novel'' has been lauded as having literary value after the cutup.

David Bowie also said that "I use it [Burroughs' cutup method] for igniting anything that may be in my imagination... You write down a paragraph or two describing several different subjects, creating a kind of `story ingredients' list, I suppose, and then cut the sentences into four or five-word sections; mix 'em up and reconnect them.'' (SONGWRITING TIPS: Try David Bowie's `cut-up' method of writing lyrics, https://thehitformula.com/2015/09/21/songwriting-tips-try-david-bowies-cut-up-method-of-writing-lyrics-2/, September 21, 2015 (22, Nov., 2017 retrieved)) The important matter is that "for igniting anything that may be in my imagination.'' That is, by the cutup, a potential expression or emotion might appear.

In (Abe, 2005; Abe, 2010a; Abe, 2019), the author also cut the original waka into four-letters clusters (kanji and kana are included) and combined randomly. In this case, all the fragments are not always Japanese words and do not have a meaning. The generated wakas are:

るらむ誰 に雲がく る人にせ に朽ちな るも別れ
(Ruranu dare Ni kumogaku Ru hitonise Ni kuchina Rumo wakare)
かりほの 白妙の衣 がむれば 暮るるも がれつつ
(Karihono Shirotaeno koromo Gamureba Kururumo Garetsutsu)
あらねど 龍田の川 あぢきな 猪名の笹 あだ波は
(Aranedo Tatsutano kawa Ajikina Tyonano sasa Adanamiha)

A strange division such as "るらむ誰'' appears, but the last generated waka can be regarded as a waka composed by a (novice?) human. It happens to be composed of rather meaningful phrases. Actually the phrases in the waka are computationally divided. The original wakas were divided in four letters (including both kana and kanji) cluster, and combined with five clusters. Therefore, the generated waka is rather difficult to understand and seems like nonsense, compared with a waka composed of meaningful phrases. However, it may be obviously composed with very exciting "Japanese'' phrases. By the cutup method, a potential expression or emotion might appear. The acceptability of the generated works can be evaluated by human. Currently it is rather difficult to build a computational evaluation system. To build it, it will be necessary to model the human acceptance of artworks. In addition, the combination without consciousness and intention can generate exciting works. This might be related to the "Descriptions automatiques,'' which were performed in the age of Dada and Surrealism (Jenny & Trezise, 1989).

Although it will be difficult to extend the above method to the generation of general literary work, it is possible to generate literary work using the feature of intertextuality. In fact, it is a chance operation mechanism conducted by, for instance, John Cage and Karlheinz Stockhausen. It is addressed that "[t]he basic principle [of chance operation] is to remove one's own intention from the work and hand that over to the oracle. Intention is always to some extent circumscribed by one's own tastes and personality, whereas non-intention moves beyond like and dislike and becomes something more resembling an act of nature (https://www.biroco.com/yijing/cage.htm (retrieved 2 June, 2020)).'' Thus by chance operation, very exciting works can be occasionally generated.

Read pointed out that "[a]t the core of their work, OuLiPo considers language not as a veil of truth but as words, arranged within a logical and arbitrary structure of formal constraint. From this perspective, both traditional poetic forms and modern free verse can be seen as systems of composition based not in expressive honesty, but in the art of language. Writing poetry is a word game played according to certain rules (Read, 1988).'' OuLiPo`s member proposed several argorithms for poem generation. Some are interesting, but rather experimental.

Of course this type of generation can be easily conducted by a computer. However, this type of generation will not be intelligent. As an intelligent generation, I previously proposed an abductive generation (Abe, 2019). The author means "intelligent'' as consistency and coherent. In the following sections, the author will briefly introduce the procedure and the method of the abductive generation. First, the author will show the abductive story generation during an art appreciation. Results from this experiment will show the process of the abductive story generation.

STORY GENERATION DURING THE ART APPRECIATION

Previously Tadaki and Abe conducted several experiments ((Tadaki & Abe, 2017), (Tadaki & Abe, 2018), (Tadaki & Abe, 2019), (Tadaki & Abe, 2020) etc.) in order to know how we can appreciate art works. So that Tadaki and Abe controlled the quality and quantity of information, for instance, title, painter, history, and explanation of the art works, during the art appreciations.

The Effect Of The Title.

In (Tadaki & Abe, 2018), Tadaki and Abe wondered "would people use different strategies to cate the painting for each abstract painting and representational painting?.'' Our experiment was to control the description of captions. For instance, Tadaki and Abe hide the title of the abstract painting (Figure 2) (Forêt by Jean Fautrier (Figure 1)). The painting would be rather difficult to understand or create story without the title. We thought it would be difficult to understand it for novice. We would like to know the natural image to the paintings. However, when the title (Forêt (forest)) was given to viewers (they could open the hidden title), one of them created a story "I found a fairy in a forest.'' Thus a certain hint or guidance can support person to understand a difficult associations of meaning. However, the caption may also be regarded as a leading statement that over-anchors the meaning. Therefore, such a hint should be properly offered. But it should not offer too much information to viewers. In the following experiment, the quality and quantity of the information offering strategies can be shown. In addition, in the next paper (Fukushima et al., 2019; Abe et al., 2020), the authors will explore the change of preference and sense of values according to information.

Even simple information such as a title, we can generate a story. It can be regarded as abduction. The participant tries to explain the title to generate a story.

Figure 1. Forêt by Jean Fautrier

Captions With Several Levels Of Explanation.

In the previous experiment, a title is hidden. Especially for an abstract painting if the title is missing, it is very difficult to guess what is drawn. When a title is given, a new image will appear in participant's brain. For the modern painting some artists do not add the title to works. Instead they add a title such as "untitled.'' Of course a title sometimes will disturb our thinking. Without a title, images for the painting may sometimes be extended.

In order to determine the effect of information of artworks in art appreciations, Abe and Tadaki conducted an experiment (Abe & Tadaki, 2019). In this experiment, their question was how the description of caption would influence or help the viewer's thinking. In the above experiment, only one pattern of caption was offered to one participant. In this experiment, all patterns of caption were offered to all participants. That is, all participants took the same experiment. From the previous experiment, we observed even a title could help viewer's understanding of artworks. Accordingly, the authors' hypothesis was the more amount of information about an artwork is offered to viewer, the more deeply viewer can understand the artwork.

Abe and Tadaki used the painting shown in Figure 3. The experiment was conducted as a question on a paper. Since we think the artist was not famous in Japan, first no information was shown, then the information for the painting (title) was shown, at last, information of the artist was shown. On papers, a question was given as well, when participants finished their answers, the next paper with question was given. We prepared a question with three levels of information as follows[3];

1. Please imagine and write the story occurring in the painting.
2. The title of this painting is "New Day,'' please imagine and write the story occurring in the painting. If something are different from what you imagined in the previous question, please write them.
3. The painter who drew this painting is a female painter living in Beograd, Serbia. Her name is Ivana Živić. She was born in 1979 in Sarajevo. She drew this painting in 2018. Please imagine and write the story occurring in the painting. If something[s] are different from what you imagined in the previous question, please write them.

That is, all participants answered the similar questions three times. Information on the painting would increase as they proceeded the experiment. Participants were given time as long as they liked.

One of the answer[4] sets was:

1. Since she is ill or is confined in the house, she cannot go out of her house. She dreams to go out to the outside world and go out from the window in her mind.
2. The reason she did not go out of her house was not unavoidable but intentional. Since today, she strongly thinks of going out of her house.
3. She remembers her girlhood. She remember that since outside world was dangerous she wanted to go out of house but could not, and that she strongly dreamed that she could go out of house. Before shown the statement I think the painter drew current situation but after reading it I changed my mind to think that she drew in her reminiscence. The reason why she did not go out of house might be the Yugoslav Wars?

Their common understandings are:

1. Phrases such as "the girl has been confined'' and "she is floating'' are appeared. Many participants mentioned that she was in water. In fact, from the last three years she has been working on series of paintings called "Rooms of Water.'' Unlike the other painting, though this painting does not explicitly paint water, they felt a girl in water. Perhaps the painting shows a certain floating feeling.
2. Phrases such as "go to the new world'' and "go out'' are appeared.

Figure 2. Caption with hidden title

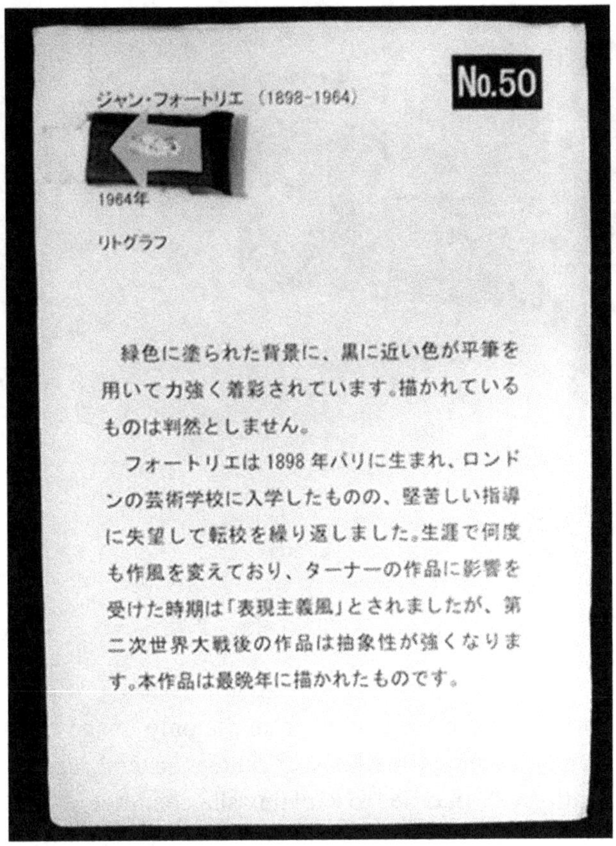

3. Since in the artist's introduction, such a description as "Ivana Živić was born in 1979 in Sarajevo (BIH)'' is included, they tend to think the story in the painting from the context of the Yugoslavia conflict. Phrases such as "escape'' is appeared.

As shown above, according as information increase (a title, a profile of the artist), the participants could add stories. The participants seem to be able to gradually understand the artwork. Even such a representational painting, the level of understanding is gradually changing. For abstract paintings this type of caption functions more like the Jean Fautrier's case.

From the phrase "Ivana Živić was born in 1979 in Sarajevo,'' it may be a misunderstanding to read a painting from the context of the Yugoslavia conflict. However, such thinking can be allowed as a free association.

In the following, the author will explain the process of the abductive story generation during art works appreciations.

Abductive Story Generation During An Art Appreciation

From the previous experiment, the author will model the abductive art appreciation. Since most of the participants created the same or similar stories, the author will use the above result.

Figure 3. Experiment

In the answer 1., the participant created a story from the only image in the painting. As an experiment for the phrase "she cannot go out of her house,'' quite a general reason such as "since she is ill'' was given. That is, for equation (2), in order to explain "she cannot go out of her house'' (*O*), "she is ill'' (*h*) was generated. The hypothesis will be generated, because when we cannot go out, usually the reason will be illness. The other reason "physically inconvenience'' was used.

It will be the same reason as illness. In the next session, as a title, when "New Day'' was given, the participant generated an explanation as "[s]ince today, she strongly thinks of going out of her house.'' That is, the imagined situation "she did not want to go out of her house'' (*O*) changes to "she could not go out of her house.'' As a solution, as *O*, "goes outside'' was obtained, then *h* as "thinks of going out of her house'' was generated. For the term "New,'' the phrase "[f]rom today'' was used, which can mean "new.'' Such a phrase as "early in the morning'' was also used. All participants created story by using a term which can explain "new.'' In addition, the profile of the painter was offered.

The painter who drew this painting is a female painter living in Beograd in Serbia. Her name is Ivana Živić. From the fact that "the painter was born in 1979 in Sarajevo,'' the participants remembered the Yugoslavia conflict which occurred during 1991 and 1999. In order to explain *O* as "the Yugoslavia conflict,'' *h* such as "since outside world was dangerous she wanted to go out of house but could not'' was generated.

All the participants created their stories in the context of the Yugoslavia conflict.

Thus when we create a story, we will abductively make an inference. In general, the art appreciation can be modelled by abduction. Of course, we will not perform the art appreciation by such a simple process. Kandel pointed out that "Artists often use reduction to serve a different purpose. By reducing figuration, artists enable us to perceive an essential component of a work in isolation, be it form, line, color, or light. The isolated component stimulates aspects of our imagination in ways that a complex image might not... (Kandel, 2018)'' Then we will separate (disassemble) the art work to basic components during the art appreciation. Accordingly it is necessary to discuss the process of the art appreciation based on the above situation.

STORY GENERATION AS ABDUCTION

In the previous works, the author modelled the story generation during an art appreciation as abduction.

In (Abe, 2019), I discussed an abductive story generation. The author will briefly show the study.

Natural Language Processing By Abduction

Many researchers proposed several natural language processing strategies by abduction. (e.g. (Stickel, 1988), (Stickel, 1989, (Hobbs et al., 1993), (McRoy & Hirst, 1993), (McRoy & Hirst, 1995)). They are systems that can conduct a natural language interpretation and understanding, and correct the misunderstanding language. For instance, an abductive interpretation by Hobbs (Hobbs et al., 1993) conducts the text interpretation based on the concept that the interpretation is the minimal explanation that the text is true. The interpretation of the text is abductively conducted by proving that the text is coherent. In addition, it is more interesting to abductively conduct a natural language processing to the dialogue (discourse) generation. Lascarides and Oberlander proposed the simple discourse generation method by abduction (Lascarides & Oberlander, 1992). In this method, the interpretation of the laconic discourse is abductively conducted by using defeasible rules. The discourse generation is performed by nonmonotonic deductive check and abduction.

Story Generation By Abduction

In (Abe, 1998), the author proposed a conversation generation method by abduction. In this framework, the structure of the previous conversation[5] is referred to to generate a new conversation. In addition, a humourous conversation is generated by AAR (Abe, 2000). This application was developed for the conversation with joke generation. In addition, the above application is based on a situational inference. A suitable situation change can lead our thinking in the better direction. Accordingly, if we use this type of situational inference, not only for making joke but also for giving a certain stimulation, an innovative activities can be encouraged.

A discourse is the process in which several persons perform story telling to each other. Accordingly, it will be possible to extend this scheme to narratology. The author proposed the following literary-work generation method (Abe, 2019). A source of literary-work can be divided into small units. Each small units will have own (partial) conclusions. Fragments of a literary-work can be generated as an abductive hypothesis set by explaining the partial conclusion. For instance, it is possible to formalize the literary-work generation by the description of the hypothetical reasoning (Theorist (Poole et al., 1987)) as follows:

$$\mathit{Rule\_set\_such\_as\_the\_connection\_rule\_of\_sentence} \cup h \vdash \mathit{partial\_conclusion}. \quad (4)$$

By this strategy, at least, an outline of a literary-work can be generated. That is, a set of generated hypothesis can be regarded as an outline of a literary-work. From the *partial_conclusions*, many partial stories (outline) will be generated. After generated partial stories (outline), a whole story will be constructed by the combination of them.

Actually, for the story generation, Ogata et al. have been proposed many methods. For instance, in (Ogata, 2007), Ogata showed the generated story such as the follows (for some sentences, English translations are given by the author.):

蛇が皇女を誘拐する(A snake abducts a princess.)。
老婆が嘆きの歌を歌う(An old woman sings the doleful song.)。
イワンが皇女の探索を決意する(Ivan decides to search for the princess.)。
イワンが蛇の国へ出立する(Ivan departs for the snake land.)。
蛇がイワンを闘う(The snake fights Ivan.)。
イワンが腕を負傷する。イワンが蛇を勝つ(Ivan defeats the snake.)。
イワンが皇女を誘拐する(Ivan rescues the princess.)。
イワンが皇女を故国へ帰路に着く。蛇が飛ぶ。蛇がイワンを追いかける。
イワンが岩を隠れる。皇女がイワンの傷を認知する。
小人達が宮殿を建てる。宮殿でイワンが住む。
イワンが皇女を結婚する(Ivan marries the princess.)。

He considered Vladimir Propp's narrative study from the viewpoint of "de-composition and re-composition.'' Based on the attempt of Propp's de-composition and re-composition, he discuss the positioning of Propp model as the narrative rhetoric in narratology. Then he introduced the Propp model into his narrative generation system. Actually he pointed out that "by a set of specific "function'' and the pattern of chain of it, a specific genre in story can be determined.'' According to Propp, "function'' means character's action from the viewpoint of the result. This type of generation (construction) will be regarded as a deduction based generation.

For the abduction based generation, the abduction will be performed to explain the *partial_conclusion*. For instance, one of *partial_conclusions* will be "marry the princess.'' And the other *partial_conclusion* will be ¬*princess* (the princess is not here.). If the *partial_conclusion* is ¬*princess*, in order to explain ¬*princess*, "search for the princess'' will be generated by abduction. Then "(Ivan) searches for the princess.'' will be created as a short story (outline). If the princess is in the snake land, in order to explain *see princess in the snake land*, "depart for the snake land'' will be generated by abduction. Then "(Ivan) departs for the snake land." will be created as a short story (outline). Of course in order to perform abduction, it is necessary to prepare the hypothesis base including necessary hypotheses and the fact base including several causalities. In this chapter, I will not show all knowledge (facts and hypotheses). As shown in (Abe, 1998), the combination of deduction and abduction can generate rather long story. Thus by abduction, a set of outlines can be generated. These outlines are reasonable, because they are logically generated.

However, a phrase with a literary flavor cannot be generated only by abduction. For the problem in (Abe, 2019), the author proposed a method that with referring to a certain literary work, abduction can be performed to generate a new literary work. This process can be regarded as pastiche. It could generate such a poem as follows:

月夜の晩に、夏みかんを一つ (On a moonlight night, a bitter summer orange)
道端に、見つけた。(I found it on the road side.)
それを採って、役立てようと (I picked it up to utilise it…)
僕は思ったわけでもないが (I did not think so…)
なぜだかそれを捨てるに忍びず (I did not know why… I could not help to throw it away.)
僕はその皮を、ナイフが見つからず、爪で剥いた。(For the bitter summer orange's rind, since I could not find a knife, I peel it with my nail.)

It seems good, because it was generated based on the poem of Chuya Nakahara (中原中也). Thus, it will be rather difficult to generate emotional phrases by only the logical method.

A BRIDGE BETWEEN LOGIC AND EMOTION (KANSEI)

In the previous sections, the author discussed the process of an abductive story generation. We usually create a story by an abductive way. Because a story is usually created based on an existing fact such as a history or news. If the story is exactly based on a history or news, the created story will be characterized as a non-fiction. If a certain modification is applied, it will be regarded as a fiction.

The author discussed the creation of story from the aspects of both computer and human. For a human generation, as shown in the section 4, the participants created stories according to the offered information. This is an abductive creation. Of course they have a plenty of background knowledge. Therefore they could rather easily create a story based on the history. For a computer generation, as shown in the section 5, very simple outline of the story can be created by an abductive way by the explanation of the observation (sub-goal). In fact, it is necessary to prepare a lot of knowledge to make abduction.

For the deductive way, we can generate a certain story by adding several phrases. By computer, as shown in the previous section, Ogata proposed the story generation system based on Propp's theory (Ogata, 2007). By the system, unlimitted story (outline) can be generated.

Thus by both abduction and deduction, a story can be automatically generated. However quality of the story is not sufficient. This is because current computers do not have a sufficient programming skill of sense or emotion.

For the word selection, a computer cannot have a character but we human have a definite character. In (Abe, 2010b), the author analyzed the usage of parts of speech by the Kamishibai KeyGraph (Ohsawa et al., 2008). Two authors' patterns of the usage of parts of speech are visualized in the same figure. If the both pattern are similar in their shape (structure), both writers show the similar usage of parts of speech. For instance, Figure 4 shows the comparison of the pattern of Hisaki Matsuura's "Ayame (あやめ)'' with that of Hiromi Kawakami's "Fuuka (風花)''. Matsuura's pattern is not similar to Kawakami's pattern. This result might be caused by the difference of plot. On the other hand, Figure 5 shows the similar pattern. Both novels are written in narration style. Accordingly the usage of parts of speech might be similar. Or because both writers are female (Matsuura is male.). Both writers are in the different generation (born in 1976 and 1947). In general, Yamazaki's novel is regarded as a popular literature and Kashimada's novel is regarded as a pure literature. It is quite difficult to conclude the usage of parts of speech shows writer's potential character. But the result can be applied to the automatic creating story system for the preferable word selection assistance.

Asakawa tried to apply recurrent neural network models to generating narratives (Asakawa, 2015). This work was inspired by recent advances of recurrent neural networks. A story like Hakusyu Kitahara's work was generated. Actually the usage of adjective or Kana is similar to the Hakusyu's usage. A part of generated story is as follows:

まあるいお月さまの中には 、いつも兎が杵《きね》をもつて餅を搗《つ》いてといふのがありませう。ほんとに金屏風でなくては、あの若い小母さまには似合はないでせうね。いかにも昔のお江戸の子供が謡つたやうでせう。...

The pattern of words used in the above story is learned by deep learning. Thus by the learning the style of a certain writer can be obtained.

Figure 4. Comparison of the pattern of Hisaki Matsuura's "Ayame (あやめ)'' with that of Hiromi Kawakami's "Fuuka (風花)''

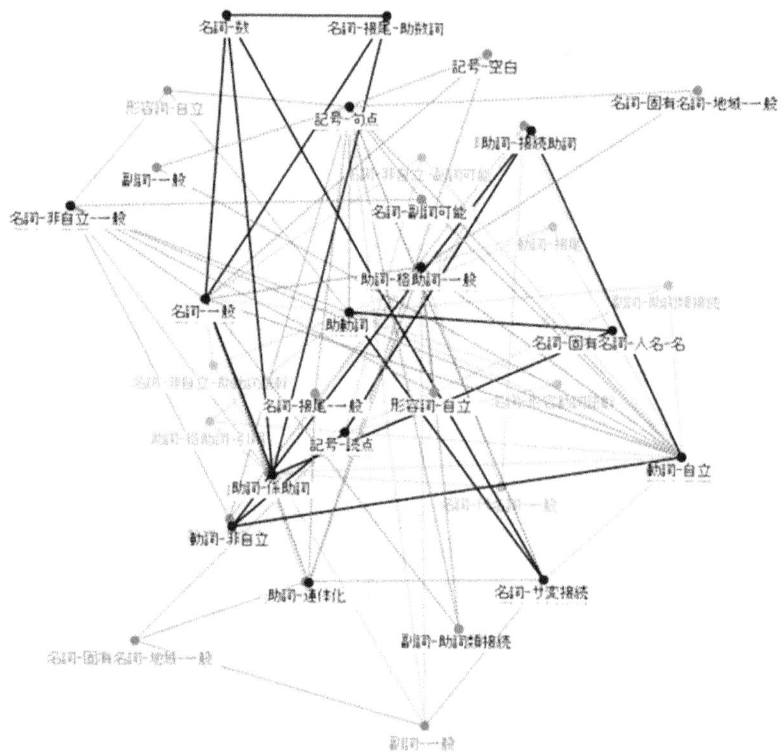

In general, an automatically generated story will not have an emotional adjective. Such a pattern of adjective can be shown as a hint of additional words. For instance, the "A snake abducts a princess.'' can be decorated as "A cunning snake abducts an innocent princess.'' If the writer prefers a formal way, the phrase will become "A guile snake abducts an ingenuous princess.''

The pastiche like strategy shown in (Abe, 2019) can be applied to create full of lyricism and more excellent work. The pastiche like strategy can generate a certain work as if it were written by a certain exist writer.

In addition, an analogical mapping strategy can be applied to create more excellent work. If the story is generated by referring to the previous work, the generated story will become as good as the original work. In the previous works, the author used an analogical strategy for the supplement of lacked hypothesis. For instance, in (Abe, 2015), the author discussed as follows:

"In addition, when the generated hypothesis set cannot be used as it is, an analogical mapping is introduced to generate new and usable hypothesis set in the situation.

AAR can generate more complex hypothesis set than the minimal hypothesis set. Accordingly, the obtained hypothesis set can be more reasonable hypothesis set by referring to the other existing techniques. Of course, during an analogical mapping, the consistency of hypothesis set should be considered.'' If the concept is applied to the literary work generation, at least more complex story than outline can be

Figure 5. Comparison of the pattern of Yoko Yamazaki's "Salome in the sea (海のサロメ)" with that of Maki Kashimada's "Kingdom: e of Zero"

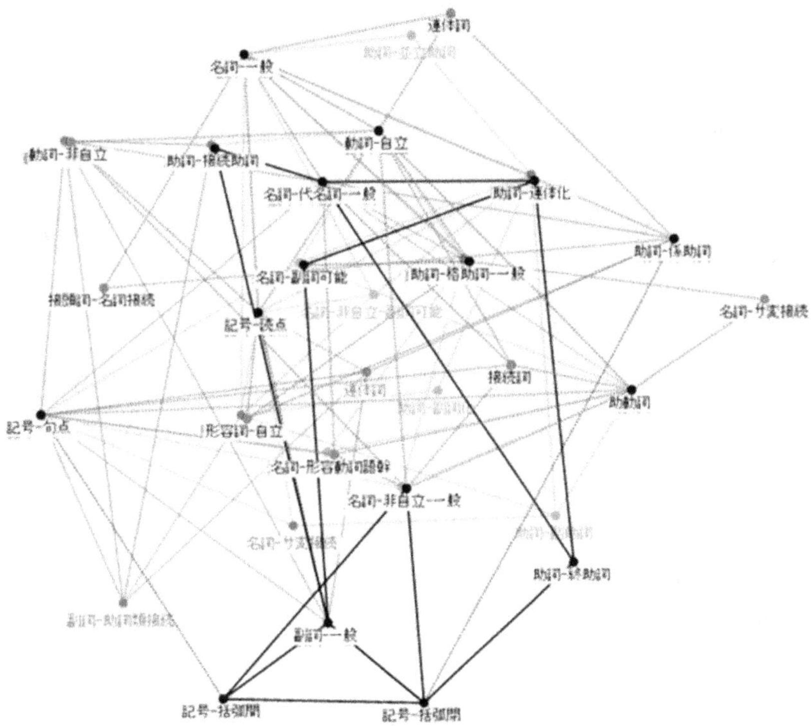

generated. In addition, if an analogical mapping functions well, a full of lyricism and more excellent work could be generated.

CONCLUSION

In this chapter, the author discussed the literary work generation.

First the author discussed a random combination of the selected words/phrases/letters, which is a simple generation technique such as the cutup method. The generated works can be sufficient as literary works. However the method is not so logical and intelligent. The generated works are products of chance.

For the more intelligent generation, the author introduced abduction. By abduction, it is possible to guarantee the logical consistency and coherency. Furthermore, the author observed human viewers use of abduction when they create a story during an art appreciation experience. The human viewer could create almost a full story by abduction. This is because human has a lot of back ground knowledge and emotion. By computational abduction, it is not possible to consider the cognitive aspects such as beauty, sentimentality and aesthetic value. These are very important factors in the literature.

Then the author introduced an analogical mapping strategy to create the better story. If the analogical strategy is applied to the literary work generation, at least more complex story than outline can be generated. In addition, if an analogical mapping functions well, a full of lyricism and more exellent works could be generated.

REFERENCES

Abe, A. (1998). Applications of abduction. *Proc. of ECAI98 Workshop on Abduction and Induction in AI*, 12-19.

Abe, A. (2000). Abductive analogical reasoning. *Systems and Computers in Japan*, *31*(1), 11–19. doi:10.1002/(SICI)1520-684X(200001)31:1<11::AID-SCJ2>3.0.CO;2-E

Abe A. (2005). *Computer and KANSEI (III).* Technical Report of JSAI, SIG-LSE-A502-5-2, 33—36.

Abe A. (2010a). *About the generation of novels.* Technical Report of JSAI, SIG-LSE-A903-16-8, 129—134.

Abe A. (2010b). *The style of novelists' writing.* Technical Report of JSAI, SIG-LSE-B002-4, 23—32.

Abe, A. (2015). The role of abduction in IMDJ. *Proc. of IJCAI2015 International Workshop on Chance Discovery, Data Synthesis and Data Market*, 59—64.

Abe, A. (2019). *The possibility of the generation of literary works by abduction.* In T. Ogata & T. Akimoto (Eds.), *Post-Narratology through computational and cognitive approaches* (pp. 144–163). Information Science Reference.

Abe, A., Fukushima, K., & Kawada, R. (2020). *How will sense of values and preference change during art appreciation? Information 2020, 11(6), 328.*

Abe, A., & Tadaki, K. (2019). Captions with several levels of explanation. *Proc. of KES2019, Procedia Computer Science*, 2335—2344.

Asakawa S. (2015). *A neural narrative model.* Technical Report of JSAI, SIG-LSE-B501-2, 17—24.

Burroughs, W. S. (n.d.). *The cut-up method of Brion Gysin.* http://tbook.constantvzw.org/wp-content/cutup_gysin_burrough.pdf

Fukushima, K., Abe, A., & Kawada, R. (2019). How will sense of values change during art appreciation? *Proc. of 2019 IEEE International Conference on Data Mining Workshops (CDEC2019)*, 5—12. 10.1109/ICDMW.2019.00009

Hobbs, J. R., & Sticke, M. (1993). Interpretation as abduction. *Artificial Intelligence*, *63*(1-2), 69–142. doi:10.1016/0004-3702(93)90015-4

Jenny, L., & Trezise, T. (1989). From Breton to Dali: The adventures of automatism. October, 51, 105-114.

Kakas, A. C., Kowalski, R. A., & Toni, F. (1992). Abductive logic programming. *Journal of Logic and Computation*, *2*(6), 719–770. doi:10.1093/logcom/2.6.719

Kandel, E. R. (2018). *Reductionism in art and brain science.* Columbia University Press.

Kristeva, J. (1980). *Desire in language: a semiotic approach to literature and art.* Columbia University Press.

Lascarides, A., & Oberlander, J. (1992). Abducing temporal discourse. *Proc. of 6th. int'l Workshop on Natural Language Generation (LNAI-587), 167—182.*

McRoy, S., & Hirst, G. (1993). Abductive explanation of dialogue misunderstandings. *Proc. of EACL93*, 277—286.

McRoy, S., & Hirst, G. (1995). The repair of speech act misunderstandings by abductive inference. *Computational Linguistics*, *21*(4), 435–478.

Ogata, T. (2007). To the rhetoric of story from Propp: Centering around the rhetoric of de-composition and re-composition. *Cognitive Studies*, *14*(4), 532–558.

Ohsawa, Y., Ito, T., & Kamata, M. I. (2008). Kamishibai KeyGraph as scenario map visualizer for detecting transient causes from sequential data. *Proc. of PAKDD 2008 Working Notes of Workshops on Data Mining for Decision Making and Risk Management*, 272—283.

Peirce, C. S. (1955). Philosophical writings of Peirce. Academic Press.

Poole, D., Goebel, R., & Aleliunas, R. (1987). Theorist: A logical reasoning system for defaults and diagnosis. In The Knowledge Frontier: Essays in the Representation of Knowledge (pp. 331-352). Springer Verlag.

Pople, H. R. Jr. (1973). On the mechanization of abductive logic. *Proc. of IJCAI73*, 147—152.

Read, A. G. (1988). OuLiPo, architecture, and the practice of creative constraint. *Proc. of the 86th ACSA Annual Meeting and Technology Conference*, 190—194.

Stickel, M. E. (1988). A Prolog-like system for computing minimum-cost abductive explanations in natural-language interpretation. *Proc. of Int'l Computer Science Conf.*, 343—350.

Stickel, M. E. (1989). Rationale and methods for abductive reasoning in natural language interpretation. *Proc. of Int'l Scientific Symp. "Natural Language and Logic" (LNAI-459), 233—252.*

Tadaki, K., & Abe, A. (2017). Museum visitors' behavioral change caused by captions. *Proc. of the 2nd. Int'l Workshop on Language Sense on Computer in IJCAI2017*, 53—58.

Tadaki, K., & Abe, A. (2018). The influence of story writing worksheets on art appreciation. *Proc. of ICAROB2018*, 687—690. 10.5954/ICAROB.2018.OS17-7

Tadaki, K., & Abe, A. (2019). *The potentiality of first person views: Analysing narratives during art appreciation.* In T. Ogata & T. Akimoto (Eds.), *Post-Narratology Through Computational and Cognitive Approaches* (pp. 375–396). Information Science Reference. doi:10.4018/978-1-5225-7979-3.ch010

Tadaki, K., & Abe, A. (2020). The influence of story creating activities while appreciating abstract artworks. Advances in Artificial Intelligence, 1128, 126-135.

ENDNOTES

1 I offer English translations, but they are not complete. Because they are not complete poems. Accordingly, they will be difficult to understand.

2 The order of phrases is 5, 7, 5, 7, 7.

3 Before the questions, questions about the user's profile (How often do you go to (art) museums? etc.) were conducted. But in this paper, the answers to the questions were not in consideration.

4 Answer sets are translated from Japanese to English by the authors.

5 For simplification in the example, clauses are simply saved in the knowledge base. Actually, they do not have a structural feature.

Chapter 9
An Analysis of Cognitive Elements and Effects of Video Commercials:
Toward Effective Indexing and Video Production

Yoji Kawamura
Kindai University, Japan

ABSTRACT

This chapter starts with an analysis of differences between the indexes attached to still images in the image database and cognitive elements of still images extracted from the cognitive experiment. It then analyzes the relationship between cognitive elements of each video shot generated based on still images and cognition of the video commercial produced by combining them. Lastly, it discusses how to index the video shot and production methods of video commercials based on the analyses. The image database tends to attach the records of the timing and place of shooting and features of the persons as indexes. In order to produce an attractive video commercial, it is necessary that video shots convey rich cognitive elements that are not too simple. This ensures that when combining several video shots, their cognitive elements have some consistency and consistent cognitive elements are woven into the video commercial to constitute a rich semantic network.

INTRODUCTION

The Need to Produce Customized Video Commercials for Individuals

Conventionally, in producing video commercials using mass media, the advertising planner sets/creates the advertising concept and storyboard drawing from the overall story and atmosphere of the video. Then, based on these elements, the advertising director shoots/edits video shots (concrete screen components)

DOI: 10.4018/978-1-7998-4864-6.ch009

to produce video commercials (the expression). What is being strived for is advertising communication in which the advertising planner and director create a video commercial with a high degree of aesthetic impact and creativity that in turn delivers an impact on many people via mass media, who will then recognize it. In this instance, the overall story and atmosphere of the video commercial and concrete expressive elements tend to be implicitly processed and agreed upon between the advertising planner and director through their personal communication; however, this has rarely been examined explicitly. In addition, production methods for video commercials with a high degree of aesthetic impact and creativity are largely seen as an area of implicit knowledge that is difficult to capture explicitly.

However, as media that target individuals have started to spread in recent years, the production of video commercials customized for individuals has become a challenge. It is impossible for the advertising planner and director to produce a video commercial that speaks to all viewers individually. Systematizing production methods of video commercials to automatically produce a video commercial that is customized for each individual has become an urgent issue. While it may be difficult to automate the production of video commercials with a high degree of aesthetic impact and creativity immediately, for the medium and long term, the systematization of knowledge the advertising planner and director have about producing such videos by means of artificial intelligence and other means is necessary.

The Video Shot/Video Commercial and Index

The author has been engaged with the development of an interactive video commercial production system that generates video commercials in accordance with information input by the consumer (or user) to realize advertising communication that is effectively customized to each consumer. Kawamura (2019) has attempted to develop a system that generates video commercials adjusted to the consumer based on the information input by him/her by creating a video database of video shots from existing video commercials (TV commercial videos) (Figure 1). The system edits/generates video commercials by attaching indexes (constitutive elements such as people, things, space, and meaning/atmosphere of the video shot) to video shots as text information and by searching/combining text information.

In this case, the generated video commercial changes depending on what kind of index is attached to the video image shot. The key to an effective customization is accurate indexing.

The components (the overall story and atmosphere of the video commercial and concrete expressive elements) that were implicitly processed and agreed upon between the advertising planner and director through their personal communication need to be examined as explicit indexes. However, as the video shot is a complex body of information, its interpretation differs from viewer to viewer. The challenge here is it is difficult to have uniform specifications for what kind of framework should be used to capture the content of the video image shot and the kind of index that is to be attached.

Furthermore, most video commercials are produced by combining multiple video shots. The effect of a series of different video shots also differs from viewer to viewer. When indexing the video shot, the video commercial made up of multiple video shots can be seen as a combination of indexes. The combination then constitutes the overall story and atmosphere of the video commercial. A rule for accurate combination of indexes is the key to achieving an effective customization.

Figure 1. Interactive Video Commercial Production System Plan (Kawamura, 2003)

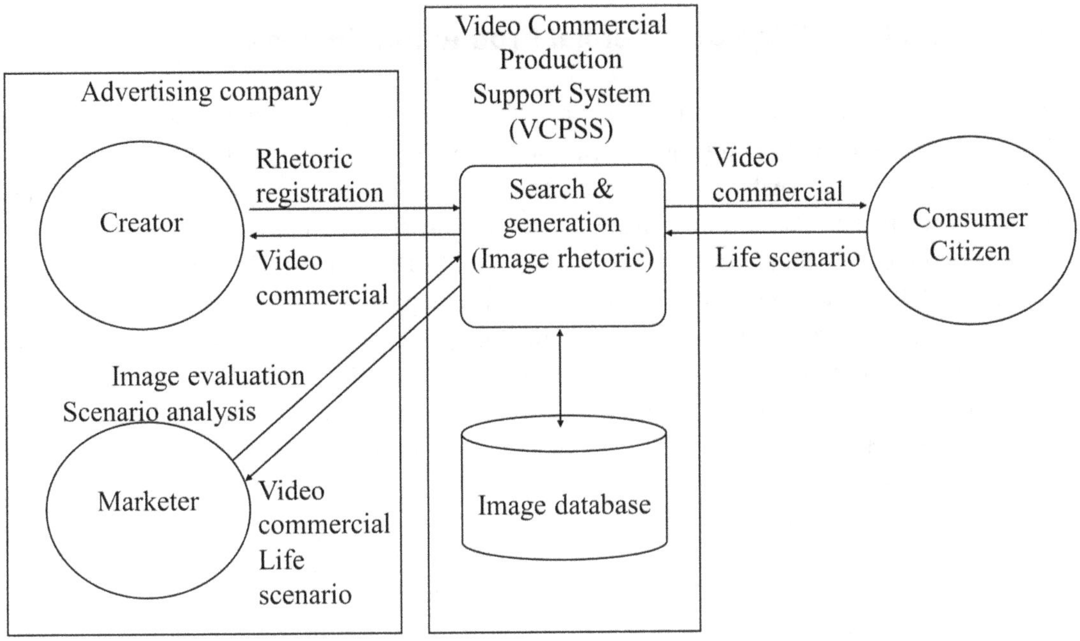

The Database of Still Images/Videos and Indexes

Recently, as content production has diversified, services to create and provide databases of still images/videos of various scenes and situations (hereafter "image database") have emerged (Aflo, 2020; Amana Images, 2020). Still images/videos these services provide are diverse and useful in producing diverse video commercials.

In these services, still images and videos are searched/extracted based on the index the service provider has attached. However, those indexes are merely information to enable a search there is no examination of whether they are accurately set for the purpose of creating content using still images and videos. This chapter examines if the index used in the image database can be utilized in producing a video commercial.

Chapter Objective

This chapter starts with an analysis of differences between the indexes attached to still images in the image database and cognitive elements of still images extracted from the cognitive experiment. It then analyzes the relationship between cognitive elements of each video shot generated based on still images and cognition/effect of the video commercial produced by combining them. Lastly, it discusses how to index the video shot and production methods of video commercials based on the analyses.

BACKGROUND

Film Theory and the Video Commercial Production Process

There has been extensive research into film in image theory. Drawing on his experience as a filmmaker, Whitaker (1970) reflects on the language of film and explains various techniques related to the elements that constitute video, such as visual content, audio content, editing and montage. Metz (1968) attempts to systematize montage, while Arijon (1976) and Katz (1991) provide detailed explanations of filmmaking techniques. Film researchers, Monaco (1977), Deleuze (1983; 1985), Bordwell (1985) and Asanuma (1990), have shown that cinematic expressions possess both denotative and connotative meanings. They considered specific examples of film and prior studies as to how denotative and connotative meaning is constructed through the combination of various techniques. There is a wealth of research on the frameworks and explanations of these visual techniques. However, in this chapter, based on the aforementioned studies, we consider how denotative meaning in film is constructed based on expressive techniques (mise-en-scène, sound and editing), and the construction of connotative meaning (story, atmosphere) by combining these techniques.

In relation to video commercials, Shimp (1976) endeavored to categorize the structures they had identified as individual-oriented, story-oriented, product-oriented, and technique-oriented. Subsequent research into video commercials has shifted from the structure of film techniques to how they are perceived by the viewer. In this, the discussions explored the relationship between creative elements and effects (message comprehension, related recollection, persuasion) (Stewart & Furse, 1986), as was the classification and effects of video commercials (Laskey, Day, & Crask,1989; Crask & Laskey, 1990; Laskey, Fox, & Crask, 1994, 1995). Moreover, these classifications were discussed in the creation of video commercials (Uejo, 1993; Abe, 2003) and in the classification and commentary of television commercial (O'Barr, 2010). In addition, video commercial creators have advanced the theory of video commercial production (Uejo, 1993) and compiled the video commercial production manual (Abe, 2003). In reference to the foregoing studies, Kawamura systematically structured the video commercial production process according to filmmaking techniques and attempted to systematize film techniques and experiment with its effects. Given this background, this chapter considers the video commercial production process to comprise the following stages:

1. **Advertising History Recall:** The advertising creator recalls an event population based on the advertising concept to produce the video commercial.
2. **Setting the Advertising Content (Story):** The advertising creator selects events to be turned into the video commercial from the recalled events and the flow.
3. **Setting the Advertising Atmosphere (Mood):** The advertising creator sets the overall atmosphere of the video commercial when producing it.
4. **Mise-en-scène:** Setting visual elements (screen components) and composition that constitute the video shot and shooting them.
5. **Editing:** Setting the duration/sequence of the video shots and ordering them.
6. **Adding Sound Effect:** Setting and adding audio elements to be added to the video shot and the video commercial as a whole.

This chapter deals with the meaning of the video shot in setting the advertising story (the flow of events), the atmosphere of the video shot in setting the advertising atmosphere, and the screen components (e.g., people, things, and space) that serve as mise-en-scène techniques for the indexes of video shots.

In our discussion of a system for creating video commercial based on fragmentary still images and videos, we define the following terms:

- Still images: Images stored in image databases
- Videos: Videos stored in image databases
- Video shots: 3-second videos containing partially enlarged still images extracted in the production of video commercials
- Video Commercial: A video commercial that combines 7 video shots

Indexes in the Image Database

Figure 2. Index example of image database (partial excerpt)

Image	Index
 Image provided by Amana Images	2014, August, 21st, Summer, Afternoon, Foreign, United States, Hawaii, Oahu, Waikiki, Outdoor, Coast, Landscape, Sheraton Waikiki Hotel, Pool, Poolside, Hotel, People, Human, Multiple, Leisure, Landscape main photo, Refreshing, Fun, Lively

Pattern Recognition Technology for Indexing of Still Images and Videos

In recent years, advances in pattern recognition technology have been made in the use of artificial intelligence technology to automatically analyze the components of still images and videos (Fauzi & Belkhatir, 2014). As well as the addition of indexes using pattern recognition of still images or video itself, attempts have been made to extract high-accuracy indexes by combining text information in images, adjacent text information and capture information (Leong, Mihalcea, & Hassan, 2010; Bhute & Meshram, 2014). However, while accuracy in the analysis of specific objects (e.g., vegetation, animal, building, places etc.) has improved, it has yet to reach the level of analyzing real world images, and it has also been indicated that image recognition technology without combining text information has not advanced (Leong, Mihalcea, & Hassan, 2010). Against this background, Vinyals Toshev, Bengio, & Erhan (2015) reports that accuracy was improved by using neural networks to extract text from images based on large volumes of supervised data. If such efforts become fully advanced in the medium- to long-term, it may be possible to accurately extract representative elements as an index featured at the level of image databases. However, at this juncture, we have not reached the level of simultaneously analyzing the various elements that constitute an advertisement video shot.

The authors carried out an analysis by manually attaching indexes about the constitutive elements and atmosphere to the video shot (Kawamura, 2003, 2004, 2006, 2007, 2009). The indexes analyzed are diverse including characters, things that are used, backgrounds and atmosphere, and its precision can be increased compared to the image/video processing technology. However, the challenge here is that the analysis requires a large amount of labor and time.

Denotative and Connotative Meanings (Concepts) of Video Shots

We discuss three issues in this chapter. The first concerns our view of the elements and concepts that comprise mise-en-scène. The information elements that constitute a video commercial have diverse and ambiguous connotations. As an example of the above, an index of image database is extensive. However, the effectiveness of such an index in the production of video commercial has not been studied at present.

In the field of cognitive science, research into story cognition has largely been conducted on texts.

Proposals include the formulation of story grammar (Rumelhart, 1975) and narrative inference using scripts (Schank & Abelson, 1977). Moreover, Lakoff (1987) proposes a conceptual framework for constructing diverse narratives, and their concepts are considered to be applicable to artificial intelligence (Kuznetsov, 2013). However, video shots (events that construct the narrative) are not only words and sentences, but also include various screen components (denotative meaning) expressed in the video shot, as well as connotative meaning (event content, atmosphere) incorporated into the video shot as a whole.

We also examine single words (summarizing the video shots used to construct the story), atmosphere (the elements used to build the mood), and screen components (the representational elements of the video shot) and discuss how these are connected and structured. We refer to the index of an image database as an example of the various information elements associated with video shots, and extract and compare cognitive elements (text, atmosphere, screen components) through cognitive experiments on still images. In this chapter, we consider the connotative meaning (concept) of video shots and the indexes that contribute to the production of video commercials.

Film Grammar and Cognition

The second issue concerns our understanding of the rules of composition when combining video shots to compose a video commercial. When creating a video commercial, the advertising creator designs not only the representational elements (denotive meaning), but also the connotative meaning (story, atmosphere) imbued by combining them.

In addition to story grammar (Rumelhart, 1975) and scripts (Schank & Abelson, 1977), video cognition requires a discussion that takes into consideration the various screen components expressed in the video shot, the ambiguity of connotative meaning embodied in the video shot, and the processing of information by the viewer in response to the temporal constraints of the sequence of video shots. Talmy (2000) suggests that narrative cognition changes not only according to the cognition of the narrative work itself, but also according to the author of the story, the recipient of the story, the space and time, as well as the cultural context in which it is received. Moreover, studies that focus on information processing by the viewer, present findings into the relationship between the degree of confusion in the narrative order and recollection (Stein & Nezworski,1978; Yussen, Mathews, Huang, & Evans, 1988), the multiple resource model in television video cognition (Basil, 1994), the validity of schema in image comprehension (Collins & Wellman, 1982; Pingree, 1986), and video event summary theory (Schwan & Garsoffky 2004).

In considering the future automation of video commercial production, the video commercial used in the viewing experiment in this chapter was produced using an experimental system (Kawamura, 2019) featuring a video database of video shots converted from still images in an image database, and an index. It was created to target the travel services genre (concept, space, and context) and to set keywords relating to the three types of audience orientation (experiences at the travel destination, comfortable mood at the destination, and mood of the accommodation). The three types of video commercials make use of a typical advertising story (consumption situation -> product function -> product acceptance -> consumption effect -> effect on company) as the script, and a slice of life (service) format showing the scene of the travel service (product). While there is no difference in advertising story, the atmosphere differs between video commercials. We then consider the elements and composition rules that appeal to viewers by using a cognitive experiment involving advertisement images with different atmospheres.

Evaluating a Video Commercial

The third issue concerns the method of evaluating the video commercial created by us. In cognitive science, a framework for the cognitive process is assumed to comprise attention, understanding, and memory. In advertising, the AIDMA (Strong, 1925) consumer model has been proposed and used for advertising goal management (DAGMAR, Colley, 1961). However, with the development of the Internet in recent years, the AISAS (Akiyama & Sugiyama, 2004) model and iDEACCycle (Kawamura, 2013; Kawamura, Kato, Tachibana, Okai, & Kinoshita, 2014; Kawamura, 2018) model, which engages consumers in advertising, has been posited. These models point out that it is important to model the stages of active consumer engagement (search, share, acquisition, confidence, expression). Additionally, with the growth of social media, engagement effect indicators, namely those in which consumers are actively engaged (Click, Likeable, Comment, Share), have been emphasized (Gavilances, Flatten, & Brettel, 2018). In this chapter, we set "attractive" as the evaluation indicators for video commercials.

In the study of advertisement images (Scott, 1994; McQuarrie & Mick, 1996, 1999, 2003; Phillips, 1997; Mothersbaugh, Huhmann, & Franke, 2002; McQuarrie & Phillips, 2005; Stathakopoulos, Theodorakis, & Mastoridou, 2008; Huhmann & Albinsson, 2012; Theodorakis, Koritos, & Stathakopoulos, 2015), it has been pointed out that the viewers recognize/recall not only screen components of the advertisement image/poster but also the overall meaning, opinion and feeling of the advertisement due to effects of expressive techniques and rhetoric based on research using various advertisement images/posters. These studies measure the effect of techniques and rhetoric from a variety of perspectives. McQuarrie & Mick (1999) has evaluated the effect of rhetoric with the following six items:

1. "provokes"/"does not provoke imagery"
2. "vivid"/"dull"
3. "interesting"/"boring"
4. "I had many thoughts in response"/"I had few thoughts in response"
5. "The ad has multiple meanings"/"The ad has one meaning"
6. "The ad has rich, complex meaning (s)"/"The ad has simple meaning (s)"

Here, an attractive advertisement is one that leads to cognition/recall of a variety of meanings and feelings (atmosphere). Following this, this chapter uses the following as indicators of still images and video commercials:

- Diversity of cognitive elements in meaning (The size of variety)
- Diversity of cognitive elements in atmosphere (The size of variety)
- Complexity of semantic networks (the number of links)

In this chapter, attractive and its reasons are set as the evaluation indicators for video commercials, and we analyze the diversity of cognitive elements in meaning and atmosphere and complexity of semantic networks in attractive reasons.

METHOD

Setting the Video Shot and Video Commercial

In this chapter, a video commercial for travel to Hawaii was produced. This is because travel services are enjoyed by a relatively large number of people and disliked by very few. Further, Hawaii is often ranked toward the top of the list as a preferred destination, regardless of age. There are many aims for a travel: some would look for experiences at the destination, some would seek a relaxed atmosphere at the destination, and some would value the atmosphere of accommodation. This chapter produced three video commercials comprising seven video image shots (Figure 3, one shot of a consumption situation, two shots on product function, two shots on product acceptance, one shot on consumption effect and one shot on effect on company), and their examined indexes/cognition. The video shot is a video of three seconds in which a still image is manipulated with slight zooming. The video commercial is a slide show combining seven video image shots.

Video commercial A (hereafter "A") is a video commercial focusing on the experience at the destination and is produced by extracting still images from the image database using keywords such as beach, hotel, meals, nature, and adventure as indexes.

Video commercial B (hereafter "B") is a video commercial focusing on the relaxed atmosphere of the destination and is produced by extracting still images from the image database using keywords such as sea, beach, hotel, meals, peace, and plurality as indexes.

Video commercial C (hereafter "C") is a video commercial focusing on the atmosphere of accommodation and is produced by extracting still images from the image database using keywords such as sea, beach, hotel, meals, luxury, calm, elegance, relaxed and grace as indexes. Figure 3 shows the outline of the video shots and some of the images used.

Tallying Indexes of the Image Database

The still images extracted are still images from an image database. Each still image is added text information as an index for search purposes. We tallied the text information of the still images used to produce a video commercial by the video shot and extracted commonalities and differences among A, B and C.

Figure 3. Outline of the video shots and some of the images

Stage	A	B	C
Consumption Situation	A bird's-eye view of the mountains	Beach with sailboat	Overlooking the island floating in the sea
Product Function 1 ＊: Image provided by Amana Images	Balcony with white umbrellas  ＊	Overlooking the rocky area where the waves splash from the terrace seat  ＊	Sea view from room with sofa  ＊
Product Function 2	Palm-roofed hut among palm trees	Overlooking the coast from a balcony with a sofa bed	A glowing cottage on the sea during twilight
Product Acceptance 1	Woman holding a surfboard  ＊	Pool that seems to be integrated with the sea  ＊	Ocean view restaurants  ＊
Product Acceptance 2	A man holding a torch	Seats for two on the sea	Flowers and wine
Consumption Effect	Coast seen from the car window  ＊	Relax in a hammock  ＊	Park at sunset  ＊
Effect on Company	Overlooking Gulf facilities from airplane	Blue sea with horizon looking down from airplane	Blue sea with horizon looking down from airplane

The Cognitive Experiment Using the Still Images and Tallying of the Cognitive Elements

A total of 21 still images were used to produce the video commercials. As the seventh video image shot for B and C was the same, we showed twenty kinds of still images to the experiment participants and asked them to answer the following questions in the form of questionnaire.

1. Please explain this image in one word (hereafter, "one word")
2. Please describe the atmosphere and image you feel from the image as a whole (preferably using three or more keywords) (hereafter, "atmosphere")
3. Please write down information (elements) about what you can see in this image (preferably using three or more keywords) (hereafter, "screen components")

Collected responses were broken down/extracted into morphemes (cognitive elements) using Text Mining Studio (TMS), a morpheme analysis software (Mathematical Systems, 2011). Coding processes (identification of synonyms and deletion of abstract words) were applied in the process of breaking down/extracting/tallying.

The extracted cognitive elements were tallied by the video commercial and by the still image and commonalities and differences among A, B and C were extracted.

The Cognitive Experiment on the Video Commercial and Tallying of the Reasons for Selection

We showed the three video commercials (A, B, C) to experiment participants and collected their answers to the following questions in the form of a questionnaire about effects.

1. Which among A, B and C did you find most attractive? (hereafter "evaluation")
2. Why was it the most attractive? Please explain in more than 40 characters (hereafter "reasons for selection")

The video commercials the experiment participants found most attractive were tallied and presented in a figure. The collected reasons for selection were broken down/extracted into morphemes (cognitive elements) by TMS. Coding processes were applied in the process of breaking down, extracting, and tallying relationships.

The extracted relationships were tallied by the video commercial and commonalities and differences among A, B and C were extracted.

Comparison Between the Index of the Image Database and Cognitive Elements

We compared the tallied results of the index and the cognitive elements. Further, correlation analysis between the main words in the index of image database (100 words) based on the twenty still images and the main cognitive elements (100 elements) from the cognitive experiment was carried out to investigate the relationship between the two.

Discussion of the Relationship between the Cognitive Elements of the Video Shots and Cognition of the Video Commercial

Following the tallied results of the cognitive elements of video shots and video commercial, characteristics of cognitive elements of the video shots and the relationship with evaluation/cognition of the video commercial were examined.

RESULT

The Index of the Image Database

We tallied the index for image search that was attached to the image database. Table 1 shows words that appear three or more times. What are common among A, B and C are "coast," "foreign countries," "sea," "photograph focusing on the landscape," "outdoors," "landscape," "no people," and "calming." Characteristic keywords for A are "persons," "tropics," "adventure," and so on; for B, they are "sky," "tourist spots," "refreshing," and so on; and for C, they are "buildings," "restaurants," and "luxury."

Table 1. Main index keywords of A, B and C

	A	B	C
Common to A, B, and C	coast, foreign countries, sea, photograph focusing on the landscape, outdoors, landscape, no people, calming		
Common to any two of A, B, and C	USA, nature, vacation, Hawaii, coastline, escape, travel, shore, water, sunny, freedom, waterside		-
	-	resort, hotels, quiet	
	islands	-	islands
Difference between A, B, and C	persons, tropics, adventure, north America, 3rd generation, brown hair, Honolulu, black hair, young people, authenticity, adults, adventure lover, candid	sky, tourist spots, blue sky, refreshing, relaxing, leisure, rest, peace, gaze, flee, scenic, copy space, Waikiki	buildings, restaurants, luxury

Cognitive Elements of the Still Images

We showed twenty still images to a total of 39 experiment participants (26 men and 13 women) aged between 19 and 23 and collected their answers to one word, atmosphere, and screen components. The responses were analyzed by TMS and coding processes were applied and tallied. Most frequently appeared cognitive elements down to 100th in the ranking were extracted and a total of 3,357 elements were identified for tallying. Table 2 shows the breakdown of cognitive elements to be tallied.

Table 2. Breakdown of cognitive elements to be tallied

	Number of Types	A	B	C	Total
One word	20	188	181	254	623
Atmosphere	35	186	337	282	805
Screen components	45	650	696	583	1,929
Total	100	1,024	1,214	1,119	3,357

Table 3 shows cognitive elements that appear ten or more times by the video commercial. What are common among A, B, and C are "resort" in one word; "travel" and "refreshing" in atmosphere; and "sea," "trees," "sky," "airplanes," and "palm trees" in screen components. In terms of characteristic cognitive elements, for A, they are "towns," "islands," "surfing," and "drive" in one word and "cars," "mountains," "clouds," and "fire" in screen components; for B, they are "the seaside" and "visible" in one word, "vacation," "cool," "comfortable," and "relaxing" in atmosphere, and "hammocks" and "pools" in screen components; for C, they are "hotels" and "dinner" in one word, "tropical," "fashionable," "calming," and "atmosphere" in atmosphere, and "houses" and "the setting sun" in screen components.

Table 4 shows the tally of cognitive elements that appear three or more times by the video commercial and still image. The table shows the meaning, atmosphere, the sequence of screen components of the still image (video shot). The video commercial is made up of seven video shots and when there is a connection among many video shots, the viewer will feel consistency. In the table, the elements that

Table 3. Cognitive elements of the still images of A, B and C

<table>
<tr><th></th><th>Items</th><th>A</th><th>B</th><th>C</th></tr>
<tr><td rowspan="3">Common to A, B, and C</td><td>One word</td><td colspan="3">resort</td></tr>
<tr><td>Atmosphere</td><td colspan="3">travel, refreshing</td></tr>
<tr><td>Screen components</td><td colspan="3">sea, trees, sky, airplanes, palm trees</td></tr>
<tr><td rowspan="5">Common to any two of A, B, and C</td><td>One word</td><td>-</td><td colspan="2">restaurant</td></tr>
<tr><td rowspan="2">Atmosphere</td><td colspan="2">nature, open feeling, hot</td><td>-</td></tr>
<tr><td>-</td><td colspan="2">beautiful, fun, good</td></tr>
<tr><td rowspan="2">Screen components</td><td colspan="2">persons</td><td>-</td></tr>
<tr><td>-</td><td colspan="2">tables, chairs</td></tr>
<tr><td rowspan="3">Difference between A, B, and C</td><td>One word</td><td>towns, islands, surfing, drive, surfer, ethnic</td><td>the seaside, visible</td><td>hotels, dinner</td></tr>
<tr><td>Atmosphere</td><td>-</td><td>vacation, cool, comfortable, relaxing</td><td>tropical, fashionable, calming, atmosphere</td></tr>
<tr><td>Screen components</td><td>cars, mountains, clouds, fire, woman, buildings, parasols, surfboards</td><td>hammocks, pools, beach, beds, yacht</td><td>houses, the setting sun, buildings, sofa</td></tr>
</table>

are recognized in three or more still images (hereafter "consistent cognitive elements") in the video commercial (a group of seven still images) are marked with +. And the elements that are recognized in one or two still images in the video commercial are marked with *. The consistent cognitive elements common to A, B, and C are "coast" in one word and "sea" and "sky" in screen components. For A and B, consistent cognitive elements are "travel" and "refreshing" in atmosphere and "persons" in screen components; for B and C, they are "resort" and "tropical" in one word, "beautiful," "fun," and "vacation" in atmosphere, and "trees" in screen components. In addition, for B, "landscape" in one word and "nature" and "cool" in atmosphere are consistent cognitive elements, and for C, they are "fashionable," "good," and "spacious" in atmosphere and "houses" in screen components.

Figure 4 shows the graph of the amount of consistent cognitive elements. As for A, while it is consistent in reference to screen components, it has fewer consistent cognitive elements in reference to one word and atmosphere compared to B and C.

Evaluation of the Video Commercial and Reasons for Selection

We showed three video commercials to a total of 386 experiment participants (259 men and 127 women) aged between 19 and 23 and collected their evaluation on which video commercial they found attractive and the reasons for selection.

Figure 5 shows the tally of evaluation of video commercials. C received the highest evaluation (169 persons) followed by B (149 persons), and A (68 persons) in this order. An χ-square test was conducted against the number of persons evaluating the video commercial, which yielded $\chi 2=44.46$, df=2, $p<.001$, indicating that A was rated significantly lower than B and C. As for gender, women tended to evaluate A lower.

We carried out morpheme analysis of reasons for selection with TMS, applied the coding process, and tallied the result in reference to relationship. We extracted most frequently appearing relationships

Table 4. Tally of cognitive elements by still images of video commercial

Items	Word	A							B							C						
		1	2	3	4	5	6	7	1	2	3	4	5	6	7	1	2	3	4	5	6	7
One word	coast			+	+		+		+	+	+	+	+	+		+			+	+	+	
	resort		*	*					+		+	+	+			+	+	+	+		+	
	hotels		*								*		*				*	*				
	restaurants									*			*						*	*		
	tropical			*					+	+			+			+				+	+	
	landscape						*				+	+			+	*						*
	sunset				*								*								*	
Atmo-sphere	beautiful		*		*				+		+				+	+	+	+			+	+
	travel		+				+	+	+		+			+	+		*					*
	nature	*									+			+	+	*						*
	fun				*				+		+	+					+		+	+		
	fashionable		*		*								*					+	+	+	+	
	vacation		*						+		+		+			+	+				+	
	refreshing		+	+	+						+		+		+							*
	good										*				*		+		+			+
	calm down												*					*			*	
	hot			*					*	*						*						
	cool								+	+		+	+									
	spacious	*												*	*		+		+			+
	blue								*						*							*
	elegance		*										*	*								
	freedom													*	*							*
Scre-en comp-onents	sea	+	+	+	+	+	+	+	+	+	+	+	+	+	+	+	+	+	+	+	+	+
	trees			*	*				+		+		+	+	+	+			+		+	+
	sky	+		+		+		+	+	+	+			+	+	+		+				+
	airplanes							*							*							*
	persons	+	+		+	+	+		+			+		+							*	
	palm trees			*					*					*		*					*	
	towns							*			*				*	*						*
	tables									*			*						*	*		
	chairs									*			*				*		*			
	houses			*												+	+	+				
	beach				*				*				*									
	buildings		*	*							*							*				
	clouds	*				*									*							*
	feather							*							*							*
	veranda		*								*						*					
	wings							*							*							*
	green	*													*							*

Figure 4. Size of variety of consistency cognitive elements

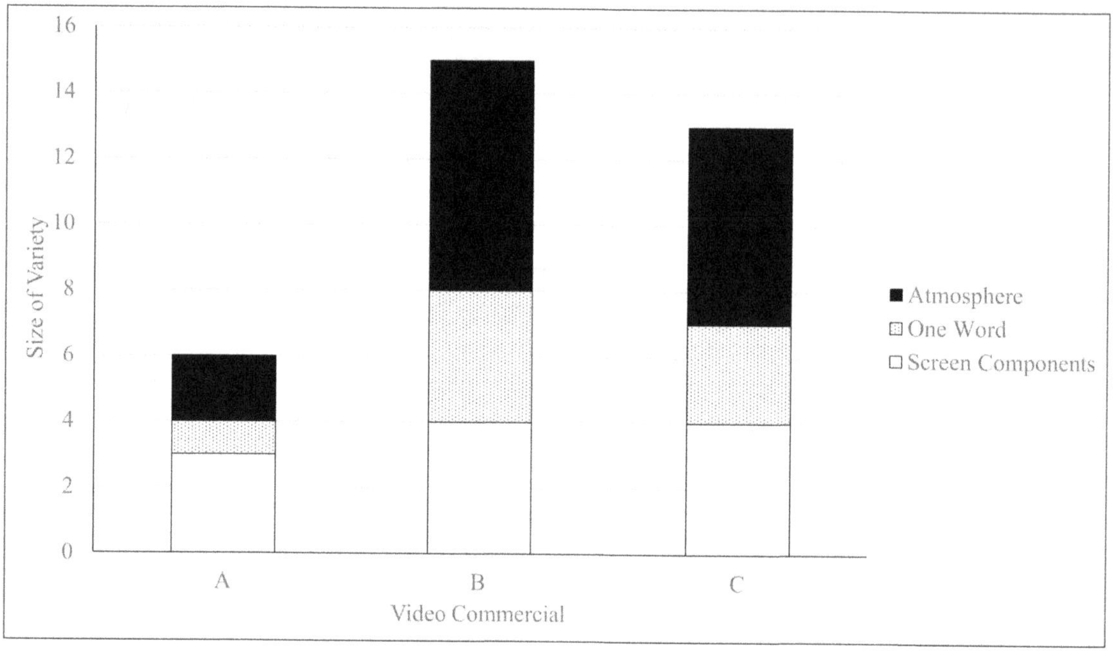

Figure 5. Evaluation of video commercials

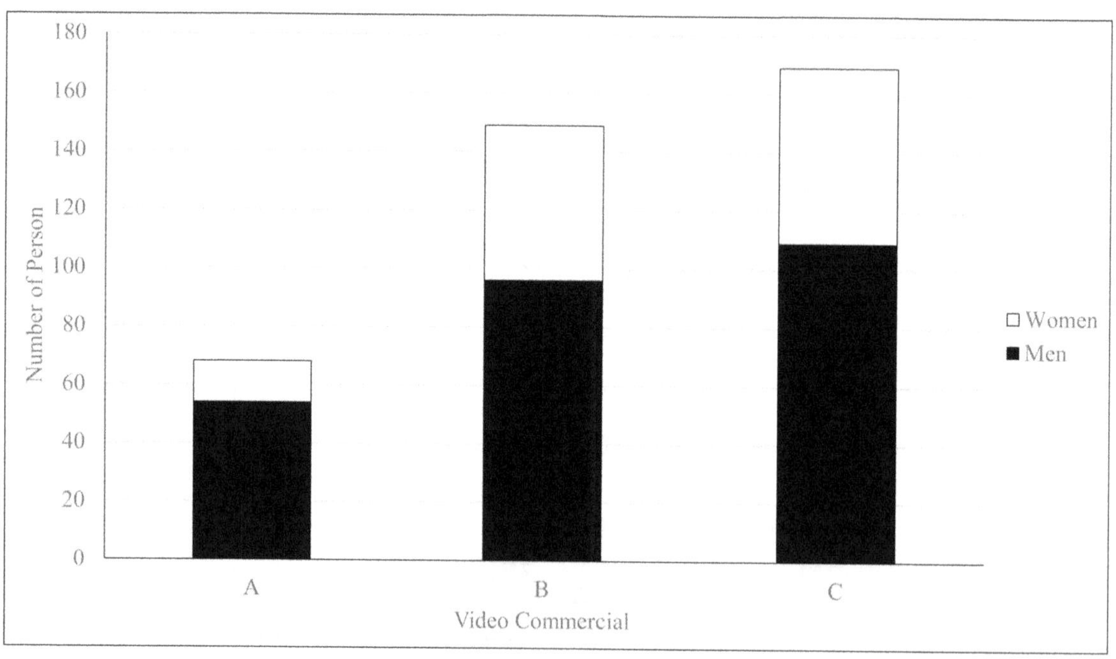

down to the 100th in the ranking and a total of 485 were analyzed. Table 5 shows the relationships that

Table 5. Tally by video commercial of the reason for selecting the video commercial

<table>
<tr><th></th><th>A</th><th>B</th><th>C</th></tr>
<tr><td>Common to A, B, and C</td><td colspan="3">attractive-feel, sea-beautiful</td></tr>
<tr><td rowspan="2">Common to any two of A, B, and C</td><td>impression-lasting, landscape-beautiful</td><td>-</td><td>impression-lasting, landscape-beautiful</td></tr>
<tr><td>-</td><td colspan="2">Hawaii-want to visit, photo-many, sea-travel, myself-like</td></tr>
<tr><td>Difference between A, B, and C</td><td>impression-strong</td><td>sea-like, sea-take a photo, sea-reflected, sea-image, sea-visible, like-feel, travel-want to visit, myself-want to visit, picture-many</td><td>Hawaii-visit, travel-visit, landscape-see, landscape-good, atmosphere-good</td></tr>
</table>

appeared four or more times in each video commercial by the video commercial. Reasons for selection that are common among A, B and C are the relationships of "attractive-feel" and "sea-beautiful." In C, we found a relationship of "landscape-beautiful; see" and there were relationships to do with attitudes: "landscape-good" and "Hawaii-visit; want to visit." In B, we found relationships of "sea-take a photo; reflected; image; visible" and relationships to do with attitudes of "sea-like" and "Hawaii-want to visit." On the other hand, in A, although we found a relationship of "impression-lasting; strong," we did not find a relationship to do with attitudes.

DISCUSSION

Comparison of the Index of the Still Images Database and Cognitive Elements

The number of keywords extracted in Table 1 is 55 and the number of cognitive elements extracted in Table 3 is 53. While the number of items extracted is similar, the number of common keywords is 14 (sea, nature, vacation, travel, coast, resort, hotels, islands, persons, sky, refreshing, relaxing, buildings, restaurants). Differences are found in that while in the index of the image database there are many keywords to do with concrete place names (USA, state of Hawaii, Honolulu and so on) and features of person (brown hair, black hair, young people and adults), among the cognitive elements from the cognitive experiment, there are many keywords to do with senses (hot, cool, fun and calming), judgement (good, pleasant), and physical facilities (cars, tables, chairs, parasols, surfboards, hammocks, pools, beds and sofa).

The still images database tends to attach the records of the timing and place of shooting and features of persons as indexes, but it appears necessary to attach keywords that are related to the viewers' emotional involvement as indexes in order to make the most of the database in the advertisement.

Table 6 shows correlation analysis of the main indexing words of the image database (100 words) and the main cognitive elements in the cognitive experiment (100 elements). The table shows the result of grouping of correlating words and cognitive elements with above 0.4 correlation coefficient (excerpts).

Table 6. Result of grouping of correlating words and cognitive elements (excerpts)

Group	Cognitive elements			Image database index words
	One word	Atmosphere	Screen components	
Tropical countries	tropical, Hawaii	warm, couple, calm down	horizon, setting sun, sunset, sun	trees, palm trees
Vacation related	ocean view, summer vacation, weather	vacation, relax	room, veranda, hotel, resort hotel, sofa, beds, summer houses	idyllic, Mexican, Cabo san Lucas, tropical climate, spa, furniture, seated, courtyard, wonderful, flashy, escape, scenic, rest, quiet, peaceful, relax, sightseeing, freedom, waterside
Travel related	travel, scenery, landscape	free, blue, beautiful, refreshing	window, wing, feather, airplane, sky, green	blue sky, airplane, wing, vehicle, early summer
Ethnic related	people	strong, scary	man, fire	beautiful, sunset, dark, one person, foreign
Dinner related	dinner	fashionable, delicious, luxury	flowers, rice, dishes	restaurant, Caribbean, indoor, meal
Restaurant related	restaurant		table, roof	restaurant, Caribbean, indoor, meal, chair
Cafe related	cafe	waves	chair	tropical climate, spa, furniture, seated, courtyard, wonderful, flashy

If we can build comprehensive knowledge about the grouping of the combination of cognitive elements and indexes, we can utilize it well in developing advertising concepts with expressive elements. For example, if we want to express the concept of "tropical countries," we think of using "horizon," "sunset," and "palm trees" as expressive elements. Then we search the image database using these keywords to extract still images of tropical countries. If we want to express the concept of "relaxing vacation," we can think of using "verandas," "sofa," "beds," "summer houses," "spa," and "courtyard" as expressive elements. By searching the image database using these as the keywords, we can extract still images of a relaxing vacation.

The Relationship between Cognitive Elements of the Still Images and Evaluation/Relationship of Cognition of the Video Commercial

Drawing from the results of cognitive elements of still images by the video commercial and evaluation of the video commercial, we examine the relationship among them. When we compare Figures 2 and 3, we find the number of consistent cognitive elements decreases from B to C to A and the evaluation of attractive video commercial decreases from C to B to A. While the positions of B and C are swapped, it appears the small number of consistent cognitive elements of A and its low evaluation is related.

Figure 6 to 8 visually represents the consistent cognitive elements shown in Table 4 and the semantic network of the relationship shown in Table 5 by the video commercial. The number of links to the semantic network is the highest in B at 15, followed by C with 13 and A with 5. A has a small number of consistent cognitive elements and is not linked to the semantic network of the reasons for selecting the video commercial. B has a large number of consistent cognitive elements, and among which "sea," "beautiful," and "travel" are the nodes in the semantic network, and "sea" and "travel" nodes are linked to various nodes. C also has a large number of consistent cognitive elements, of which "sea," "beauti-

Figure 6. The consistent cognitive elements and the semantic network of A

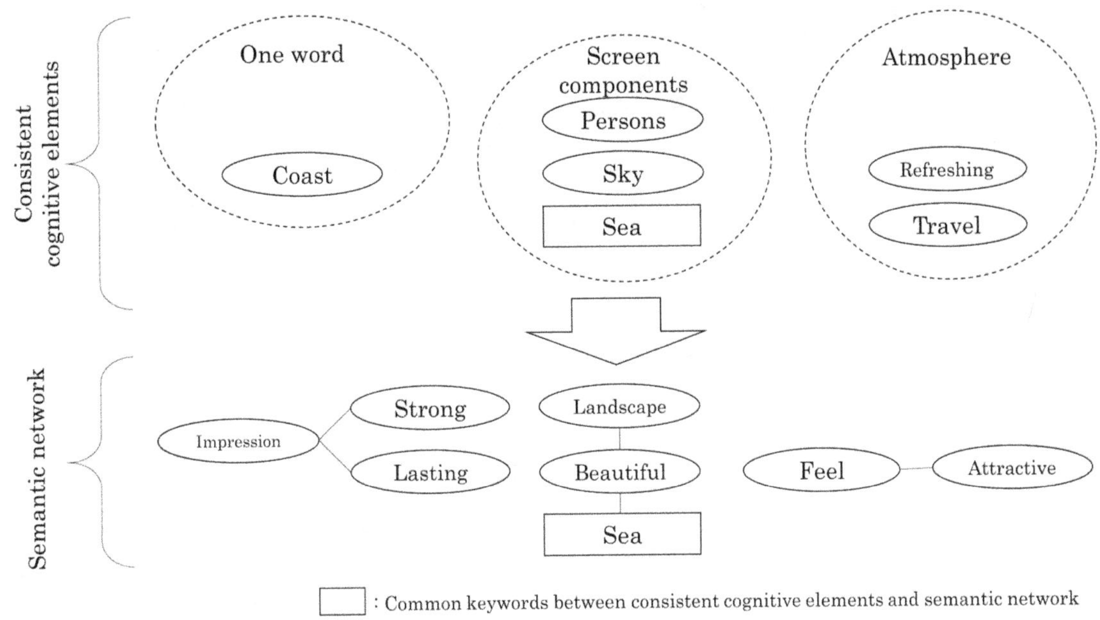

Figure 7. The consistent cognitive elements and the semantic network of B

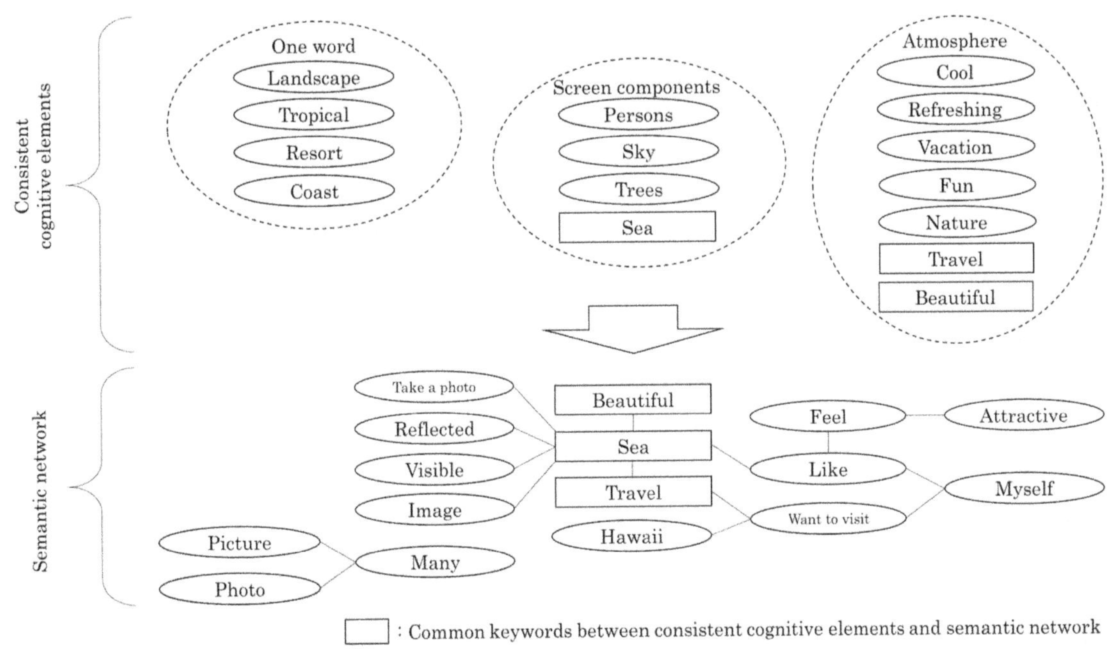

Figure 8. The consistent cognitive elements and the semantic network of C

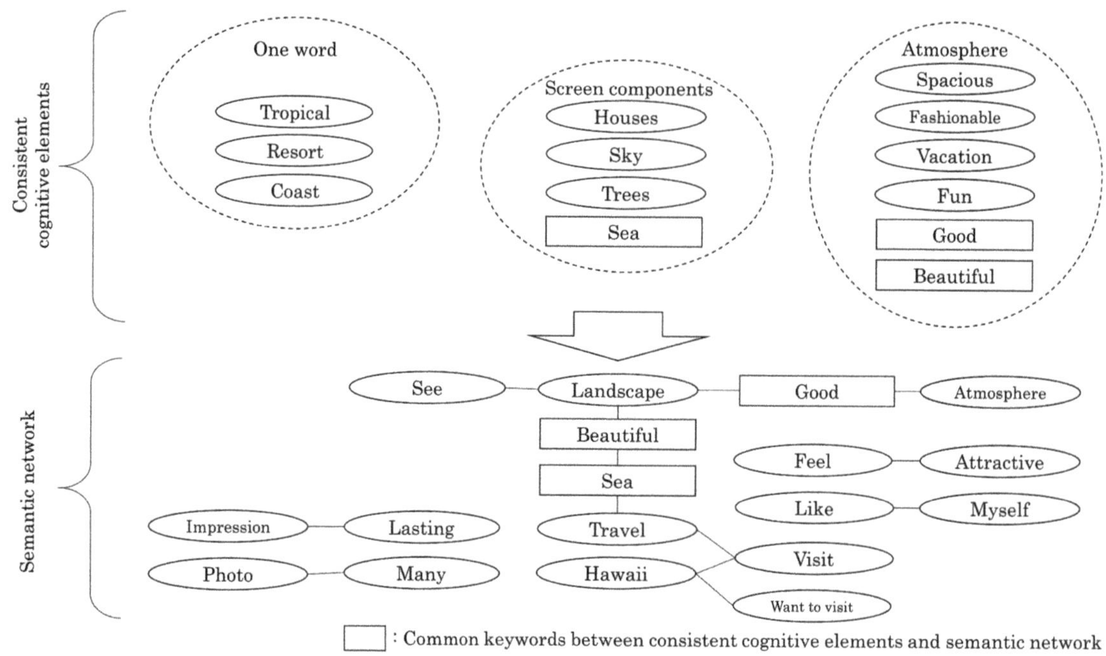

ful," and "good" are the nodes in the semantic network, and "sea," "beautiful," and "good" nodes are linked to various nodes.

To the right of the semantic network in Figure 6 to 8 we positioned keywords related to positive feelings ("feel [attractive]," "like," "good") and keywords related to desire ("visit," "want to visit"). In B (Figure 7), these form a network with "sea," "travel," and "Hawaii," while in C (Figure 8) these form a network with "landscape," "travel," and "Hawaii." Thus, a sense of liking and desire for "Hawaii" and "travel" is evoked, because they are associated with the connection of video shots.

In order to produce an attractive video commercial, it is necessary that video shots convey rich cognitive elements that are not too simple. This ensures that when combining several video shots, their cognitive elements have some consistency and consistent cognitive elements are woven into the video commercial to constitute a rich semantic network.

SOLUTIONS AND RECOMMENDATIONS

Use of Still Image Database in Advertising

This chapter has first analyzed the differences between the index attached to the still images in the still images database and cognitive elements of the still images extracted by the cognitive experiment. Next, it has analyzed the relationship between cognitive elements of each video shot created based on the still image and cognition/effect of the video commercial produced by combining them.

The still images database tends to attach the records of the timing and place of shooting and features of the persons as indexes. It appears necessary to attach keywords that are related to the viewers' emotional involvement as indexes in order to make the most of the database in advertisement. If we can build comprehensive knowledge about the grouping of the combination of cognitive elements and indexes, we can utilize it well in developing advertising concepts with expressive elements.

Evaluating Video Commercial

The results showed that video commercial A had been assessed as being lesser than B and C. Schwan & Garsoffky (2004) shows that a lack of a coherent sequence of events have cognitive repercussions. Therefore, A, in which the cognitive elements are not organized, was assessed as being lesser than B and C, which are organized. In this chapter, we discussed video commercials in the slice of life (service) format depicting travel service scenes, and in video commercials with similar stories, we found that the consistency and diversity of the series of events (text, atmosphere) were associated with attractive video commercials.

Story, Atmosphere, and Screen Components of the Experimental Video Commercial

Table 4 summarizes the structure of the video shots used in conducting the experiment based on three video commercials in this chapter. The outline of the story of these video commercials is "Enjoy the hotel, restaurants and scenery at a tropical beachside resort," and the atmosphere was consistent as a whole: A, "travel" and "refreshing"; B, "beautiful," "travel," "nature," "fun," "vacation," "cool," and "refreshing"; C, "beautiful," "fun," "fashionable," "vacation," "good," and "spacious." Considering that the evaluation for video commercial A was low, the screen components present in B and C and not in A were extracted, namely "tables" and "chairs." These were extracted as a result of analyzing the cognitive elements, and based on these findings, we consider the direction of the method of production in the video commercial generation system below.

How to Index

Where we assume that the diversity of events and atmosphere and their chronological consistency, it necessitates the addition of keywords to the index of still images and video shots for both screen components and events and atmosphere. Although keywords relating to atmosphere are also indexed in conventional image databases, more keywords need to be added. However, an additional index has cost and effort implications, and images are ambiguous in that their interpretation changes from viewer to viewer. By referencing the conceptual base as shown in Table 6, we infer an index of image layout components from events (one word) and atmosphere, while taking into account the characteristics of the individual viewing the video. There is also requirement for a method of inferring an index of events (one word) and atmosphere from screen components. If we attempt an explanation by taking Figure 6 as an example, we find "vacation" to be the characteristic atmosphere of B and C, from which we infer "room," "veranda," "resort hotel," "sofa," "beds," and "summer houses" as the screen components. In addition, "tables" and "chairs" are also characteristic screen components in B and C, from which we infer "restaurant" and "cafe" as events (one word) and "waves" as atmosphere. If it is possible to construct a rich conceptual base using images and their related captures published on the Internet, it is possible to construct a rich index of images from a variety of perspectives.

The Video Commercial Production Method and the Settings and Results of the Viewing Experiment in this Chapter

Based on the information obtained, we explain the method for producing video commercials. In addition, we present the settings and results of the video commercial production for the cognitive experiment conducted in this chapter. Although the settings and results of the video commercial production for the cognitive experiment are limited in scope and lack the versatility for application in the production of diverse video commercials, this represents a step toward the automation of video commercial production. By taking full advantage of the cognitive experiments conducted in this chapter and constructing a comprehensive knowledge base that exploits recent advances in artificial intelligence technology, we believe it is possible to achieve more accurate video commercial production. It should be noted that there are other production methods to the one discussed:

1. **Advertising History Recall:** The construction of a large image database (still images, video shots) with various indexes corresponding to the target segment. An index is added to videos relating to the meaning of that video (event), the atmosphere created by it, the representational elements that comprise the scene, and the temporal and sound elements of the video. We also construct a knowledge base related to the index. The knowledge base has a conceptual base (dictionary) that defines (quantifies) the relationship (hierarchy, causation) between items such as the viewer, events, atmosphere, screen components, temporal elements and sound elements, and features a function (rule base) to set search rules (target search items, the weighting thereof and so on) according to the viewer. The knowledge base also features functions to supplement any missing items in the index and to provide feedback on the video commercial.
2. **Setting the Advertising Content (Story):** A request is made to a rule-based system for content relating to the relationship between the viewer and a chain of events, and thereby the genre (advertising concept, life keywords and so on) and chain of events (number of events, script) are set. In this chapter, the following were set as genres of events: experience (with sea, beach, hotel, meals, nature, adventure, and activities as specific keywords), vacation (with sea, beach, hotel, meals, peace, and plurality as specific keywords), and accommodation (with sea, beach, hotel, meals, luxury, calm, elegance, relaxed and grace as specific keywords). Consumption situation, product function, product acceptance, consumption effect and effect on company were set as indexes related to the script in the following chronological order: consumption situation -> product function -> product acceptance -> consumption effect -> effect on company.
3. **Setting the Advertising Atmosphere (Mood):** A request is made to a rule-based system for content relating to the relationship between the viewer and atmosphere, and thereby the genre of the atmosphere and the development of expressive elements (mise-en-scène, editing, sound) are set. In this chapter, the following were set as genres of atmosphere: vivid (with nature, adventure, activities as specific keywords), poem (with peace and plurality as specific keywords), and gorgeous (with luxury, calm, elegance, relaxed, and grace as specific keywords). The development of expressive elements was limited to mise-en-scène.
4. **Mise-en-scène:** The video database was searched based on the genre of events and atmosphere set, and the video shot extracted. For still images, these were converted to video shots. In this chapter, seven video shots were extracted by searching the database using keywords for the aforementioned genres of events (1 consumption situation shot, 2 product function shots, 2 product acceptance shots,

1 consumption effect shot, and 1 effect on company shot). In conducting the cognitive experiment in this chapter, we found that diversity and consistency of the event sequence (one word, atmosphere) were important to the production of attractive video commercials. It is possible to conceive the creation of a knowledge base of keywords related to various atmospheres, and to systematize rules for creating a consistent atmosphere when combining video shots.

5. **Editing:** The running time for the extracted video shots was set, the temporal sequence set, and the video shots connected. In this chapter, the duration of a single video shot was set at 3 seconds. The temporal sequence was taken to be chronological.
6. **Adding Sound Effect:** Dialogue and sound effects are added as sound elements related to the screen components of the video shot, and narration, sound effects and music are added as sound elements to be overlapped with connected video shots. In this chapter, the cognitive experiment was conducted without adding sound.

FUTURE RESEARCH DIRECTIONS

As the participants of the experiment in this study were young, it is necessary to examine if a similar result can be produced with middle-aged or elderly participants, and if there is any influence of the experience in terms of vacationing in Hawaii. While this study has carried out an experiment on consistency based on a basic advertising story to make the case for travel services, it is also necessary to examine more complex advertising stories and mise-en-scène techniques using rhetoric.

This chapter has conducted a cognitive experiment with still images. However, in the internet environment, the number of people who publish still images and videos with a capture through social media is exploding. A system that can take in such information and attach indexes to still images and videos would enable more efficient and effective production of video commercials. We would like to explore this in the future.

CONCLUSION

This chapter has examined suggestions for using the image database in the production of video commercials (how to index them and how to make sense of the knowledge base) and basic knowledge about production methods of video commercials (how to direct video shots, methods of drawing up an advertising story, and methods to create an atmosphere for an advertisement). The research outcomes are still at a basic level, but it is important to keep working on these endeavors.

ACKNOWLEDGMENT

This work was supported by JSPS KAKENHI Grant Number 22500102.

The cognitive experiment was conducted with the cooperation of Synergy Marketing Inc., Amana Inc., Dentsu Inc. and AOI TYO Holdings Inc. Thanks to everyone involved for their cooperation.

REFERENCES

Abe, S. (2003). *Saishin CM seisaku no kisochishiki: puranningu kara dejitaru seisaku made* [Latest basic knowledge of the commercial production: From planning to digital production]. Sendenkaigi.

Aflo. (2020). *Service site operated by Afro*. https://www.aflo.com/ja

Akiyama, R., & Sugiyama, K. (2004). *Horisuthikku komyunikeishon* [Holistic Communication]. Sendenkaigi.

Amana Images. (2020). *Service site operated by Amana Images*. https://amanaimages.com/home.aspx

Arijon, D. (1976). *Grammar of the film language*. Silman-James Press.

Asanuma, K. (1990). Eizo no tekusuto no giho [Techniques for image text]. In Eiga no tameni II [For movie II] (pp. 130-247). Tokyo, Japan: Suiseisha.

Basil, M. D. (1994). Multiple resource theory I: Application to television viewing. *Communication Research, 21*(2), 177–207. doi:10.1177/009365094021002003

Bhute, A. N., & Meshram, B. B. (2014). Text based approach for indexing and retrieval of image and video: A review. *Advances in Vision Computing, 1*(1), 27–38. doi:10.2139srn.3430312

Bordwell, D. (1985). *Narration in the fiction film*. University of Wisconsin Press.

Colley, R. H. (1961). *Defining advertising goals for measured advertising results*. Association of National Advertisers.

Collins, W. A., & Wellman, H. M. (1982). Social scripts and developmental patterns in comprehension of televised narratives. *Communication Research, 9*(3), 380–389. doi:10.1177/009365082009003003

Crask, M. R., & Laskey, H. A. (1990). A positioning-based decision model for selecting advertising messages. *Journal of Advertising Research, 30*(4), 32–38.

Deleuze, G. (1983). *Cinema' Image-mouvement.* Paris, France: Les Editions de Minuit.

Deleuze, G. (1985). *Cinema' Image-temps*. Paris, France: Les Editions de Minuit.

Fauzi, F., & Belkhatir, M. (2014). Image understanding and the web: A state-of-the-art review. *Journal of Intelligent Information Systems, 43*(2), 271–306. doi:10.100710844-014-0323-6

Gavilances, J. M., Flatten, T. C., & Brettel, M. (2018). Content strategies for digital consumer engagement in social networks: Why advertising is an antecedent of engagement. *Journal of Advertising, 47*(1), 4–23. doi:10.1080/00913367.2017.1405751

Huhmann, B. A., & Albinsson, P. A. (2012). Does rhetoric impact advertising effectiveness with liking controlled? *European Journal of Marketing, 46*(11/12), 1476–1500. doi:10.1108/03090561211259943

Katz, S. D. (1991). *Film directing shot by shot visualizing from concept to screen*. Michael Wise Productions.

Kawamura, Y. (2003). An analysis of the rhetoric of commercial film: Toward the building of a commercial film production support system based on image rhetoric. In *Proceedings of the 2003 IEEE International Conference on Systems, Man and Cybernetics* (pp.993-1000). Washington, DC: IEEE. 10.1109/ICSMC.2003.1243944

Kawamura, Y. (2004). An analysis of the rhetoric of commercial film: Toward the building of a commercial film production support information system. *Journal of Advertising Science, 45*, 122–139.

Kawamura, Y. (2006). An analysis on rhetoric of the commercial film of beer and personal computer: A symbolic approach on rhetoric of commercial film. *Journal of Advertising Science, 47*, 33–48.

Kawamura, Y. (2007). A study on the technique-rhetorics and effects of commercial film. *Cognitive Studies, 14*(3), 409–423.

Kawamura, Y. (2009). An analysis on the story and editing techniques of commercial film: Toward the building of a commercial film production support information system. *Journal of Advertising Science, 50*, 16–32.

Kawamura, Y. (2013). Junkangata shohisha kodo moderu iDEACCycle no teian: Shohisha nikki chosa/intavu chosa wo motoni [A proposal of cyclical consumer behavior model iDEACCycle: Based on consumer dairy survey and interview survey]. *Japan Marketing Journal, 33*(2), 95–109.

Kawamura, Y. (2018). Practice and modeling of advertising communication strategy: Sender-driven receiver-driven. In T. Ogata & S. Asakawa (Eds.), *Content generation through narrative communication and simulation* (pp. 358–379). IGI Global. doi:10.4018/978-1-5225-4775-4.ch013

Kawamura, Y. (2019). An attempt of the commercial film production support system based on the image rhetoric of commercial film. In T. Ogata & T. Akimoto (Eds.), *Post-narratology through computational and cognitive approaches* (pp. 292–317). IGI Global. doi:10.4018/978-1-5225-7979-3.ch007

Kawamura, Y., Kato, T., Tachibana, K., Okai, G., & Kinoshita, S. (2014). Kokoku kyanpen no jirei ruikei no kokoromi: Junkangata shohisha kodo moderu iDEACCycle no kanten kara [An attempt of case categorization of advertising campaigns: From the view point of cyclical consumer behavior model iDEACCycle]. *Bulletin of Nikkei Advertising Research Institute,* (273), 10-16.

Kuznetsov, O. P. (2013). Cognitive semantics and artificial intelligence. *Scientific and Technical Information Processing, 40*(5), 269–276. doi:10.3103/S0147688213050067

Lakoff, G. (1987). *Women, fire and dangerous things: What categories reveal about the mind.* The University of Chicago Press. doi:10.7208/chicago/9780226471013.001.0001

Laskey, H. A., Day, E., & Crask, M. R. (1989). Typology of main message strategies for television commercial. *Journal of Advertising, 18*(1), 36–41. doi:10.1080/00913367.1989.10673141

Laskey, H. A., Fox, R. J., & Crask, M. R. (1994). Investigation the impact of executional style on television commercial effectiveness. *Journal of Advertising Research, 34*(6), 9–16.

Laskey, H. A., Fox, R. J., & Crask, M. R. (1995). The relationship between advertising message strategy and television commercial effectiveness. *Journal of Advertising Research, 35*(2), 31–39.

Leong, C. W., Mihalcea, R., & Hassan, S. (2010). Text mining for automatic image tagging. In *Proceedings of the 23th International Conference on Computational Linguistics* (pp.647-655). Beijing, China: Academic Press.

Mathematical Systems. (2011). *Text mining studio operating manual version 4.0*. Mathematical Systems.

McQuarrie, E. F., & Mick, D. G. (1996). Figures of rhetoric in advertising language. *The Journal of Consumer Research, 22*(4), 424–438. doi:10.1086/209459

McQuarrie, E. F., & Mick, D. G. (1999). Visual rhetoric in advertising: Text-interpretive, experimental, and reader-response analysis. *The Journal of Consumer Research, 26*(1), 37–54. doi:10.1086/209549

McQuarrie, E. F., & Mick, D. G. (2003). Visual and verbal rhetorical figures under directed processing versus incidental exposure to advertising. *The Journal of Consumer Research, 29*(4), 579–587. doi:10.1086/346252

McQuarrie, E. F., & Phillips, B. J. (2005). Indirect persuasion in advertising: How consumers process metaphors presented in picture and words. *Journal of Advertising, 34*(2), 7–19. doi:10.1080/00913367.2005.10639188

Metz, C. (1968). *Essais sur la signification au cinéma* (Vol. 1). Klincksieck.

Monaco, J. (1977). *How to read a film: the art technology, language, history and theory of film and media*. Oxford University Press.

Mothersbaugh, D. L., Huhmann, B. A., & Franke, G. R. (2002). Combinatory and separative effects of rhetorical figures on consumers' effort and focus in ad processing. *The Journal of Consumer Research, 28*(4), 589–602. doi:10.1086/338211

O'Barr, W. M. (2010). The rise and fall of the TV commercial. *Advertising & Society Review, 11*(2).

Phillips, B. J. (1997). Thinking into it: Consumer interpretation of complex advertising images. *Journal of Advertising, 26*(2), 77–87. doi:10.1080/00913367.1997.10673524

Pingree, S. (1986). Children's activity and television comprehensibility. *Communication Research, 13*(2), 239–256. doi:10.1177/009365086013002005

Rumelhart, D. E. (1975). Notes on a schema for stories. In D. G. Bobrow & A. Collins (Eds.), *Representation and understanding: studies in cognitive science* (pp. 211–236). Academic Press. doi:10.1016/B978-0-12-108550-6.50013-6

Schank, R., & Abelson, R. P. (1977). *Scripts, plans, goals, and understanding: An inquiry into human knowledge structure*. Lawrence Erlbaum Associates, Inc.

Schwan, S., & Garsoffky, B. (2004). The cognitive representation of filmic event summaries. *Applied Cognitive Psychology, 18*(1), 37–55. doi:10.1002/acp.940

Scott, L. M. (1994). Images in advertising: The need for a theory of visual rhetoric. *The Journal of Consumer Research, 21*(2), 252–272. doi:10.1086/209396

Shimp, T. A. (1976). Methods of commercial presentation employed by national television advertisers. *Journal of Advertising*, *5*(4), 30–36. doi:10.1080/00913367.1976.10672661

Stathakopoulos, V., Theodorakis, I. G., & Mastoridou, E. (2008). Visual and verbal rhetoric in advertising: The case of 'resonance'. *International Journal of Advertising*, *27*(4), 629–657. doi:10.2501/S0265048708080177

Stein, N. L., & Nezworski, T. (1978). The effects of organization and instructional set on story memory. *Discourse Processes*, *1*(2), 177–193. doi:10.1080/01638537809544433

Stewart, D. W., & Furse, D. H. (1986). *Effective television advertising: A study of 1000 commercials*. Lexington Books.

Strong, E. K. (1925). Theories of selling. *The Journal of Applied Psychology*, *9*(9), 75–86. doi:10.1037/h0070123

Talmy, L. (2000). Toward a Cognitive Semantics: Vol. 2. *Typology and process in concept structuring*. The MIT Press. doi:10.7551/mitpress/6848.001.0001

Theodorakis, I. G., Koritos, C., & Stathakopoulos, V. (2015). Rhetorical maneuvers in a controversial tide: Assessing the boundaries of advertising rhetoric. *Journal of Advertising*, *44*(1), 14–24. doi:10.1080/00913367.2014.930679

Uejo, N. (1993). *Kokoku kopi gairon* [Advertising copy overview]. Sendenkaigi.

Vinyals, O., Toshev, A., Bengio, S., & Erhan, D. (2015). Show and tell: A neural image caption generator. In *Proceedings of 28th IEEE Conference on Computer Vision and Pattern Recognition* (pp. 3156-3164). Boston, MA: IEEE. 10.1109/CVPR.2015.7298935

Whitaker, R. (1970). *The language of film*. Prentice-Hall, Inc.

Yussen, S., Mathews, S., Huang, S. T., & Evans, R. (1988). The robustness and temporal course of the story schema's influence on recall. *Journal of Experimental Psychology. Learning, Memory, and Cognition*, *14*(1), 173–179. doi:10.1037/0278-7393.14.1.173

KEY TERMS AND DEFINITIONS

Advertising Atmosphere: Advertising creators set the atmosphere (mood) of the entire video commercial when making the video commercial. The advertising atmosphere technique is the rule of which atmosphere elements are used to express the entire atmosphere, based on the advertising concept. At this time, the advertising atmosphere is set based on the advertising concept and the positioning of the product, including visual elements, temporal elements, sound elements, and the combination of these elements.

Advertising History: Advertising creators recall an event population for advertising visualization based on the advertising concept. At this time, they recall events (state, action, atmosphere, etc.) in which words, sentences, images, videos, sounds, etc. are mixed. These events depend on the personal characteristics of the advertising creators, and it is difficult to define the structure of the population centrally. However, the advertising creator recalls the time flow (history) of those events and the causal relationship between the events.

Advertising Story: Advertising creators set the events to be converted into video commercial and the flow from the recalled events. The advertising story technique is the rule of which event to focus on, based on the advertising concept. At this time, the advertising story is set based on the advertising concept and the positioning of the product by extracting events among the advertising history.

AIDMA Model: A model that consumers' psychological/behavioral stages consist of Attention, Interest, Desire, Memory, Action.

AISAS Model: A model that consumers' psychological/behavioral stages consist of Attention, Interest, Search, Action, Share.

DAGMAR Model: A model that consumers' psychological/behavioral stages consist of Unawareness, Awareness, Comprehension, Conviction, Action. DAGMAR is an abbreviation for Defining Adverting Goals for Measured Advertising Results.

Editing Technique: Advertising directors set the duration and connection of the video shots and rearranges the order. The editing technique is the rule of determining which editing elements make up the time sequence. For example, as editing elements, the duration of a video shot (up tempo: short time, slow tempo: long time, constant tempo: constant time, unstable tempo: random time), connection of scenes (time series type, manipulation type), and so on. The editing of video commercial in this chapter was time series type (consumption situation->product function->product acceptance->consumption effect->effect to company).

iDEACCycle Model: A model that consumers' psychological/behavioral stages consist of Awareness, Acquisition, Acceptance, Consideration, Confidence, Ideation, Expression, and Action. The unique characteristic of this model is the fact that it is a model of the process that begins with a consumer receiving information stimuli (awareness, acquisition) from the media and then processing that information (acceptance, consideration, confidence), to the process of conceiving of an idea based on the processing of that information and then expressing (communicating) that information in the media. The cyclical consumer behavior model does not trace the ideas and behaviors of consumers in a linear fashion but rather provides a structure to the process by which their ideas and behaviors proceed in cyclical stages (in some cases, amplified or radiating).

Image Database Service: The image database service is a service that captures and accumulates still images and videos in advance and extracts and sells them based on search keywords. There are companies that sell up to 120 million photos, illustrations, and videos.

Image Rhetoric: The combination of image techniques such as mise-en-scène (direction) of shots, editing shots, and so on.

Image Techniques: The techniques used to express images: advertising story, mise-en-scène technique, editing technique, and sound technique, among others.

Index: The index of the image database includes the situation (date, time, place) at which the still images and videos were taken, the expression element (facility, background, person), the overall configuration (purpose, photographing, atmosphere), and so on. The user searches and extracts still images and videos based on these indexes.

Interactive Video Commercial Production System: Interactive video commercial production system is a system that converts a video shot of a video commercial into an image database and generates a video commercial adapted to the consumer, based on the input information of the consumer. The researches on customization systems that provide video commercial corresponding to consumer groups (segments) based on various video shots created by advertisers are being pursued. In addition, the research on a co-creation system that generates video commercial for consumers individually based on an image database of daily and fragmentary life video shots created by consumers is also being attempted.

Mise-en-Scène Technique: Advertising directors set and shoot the visual elements that make up the video shot. The mise-en-scène technique is the rule of which visual elements are used to compose the screen. For example, visual elements include people (product providers, consumers), objects (products), information (products), backgrounds (space, time), and shooting method (camera position, movement).

Sound Technique: Advertising directors set and add sound elements to be added to the video shot or the entire video commercial. The sound technique is the rule of which sound element are used to add. For example, as sound elements, dialogue/narration, sound effect, music, and so on. No sound was added in the cognitive experiments in this chapter.

Video Commercial: The video commercial in this chapter was composed of seven video shots (consumption situation 1shot, product function 2shots, product acceptance 2shots, consumption effect 1shot, effect on company 1shot).

Video Shot: For the video shot in this chapter, the image of the still image database was displayed for 3 seconds, and a slight zoom process was performed.

Section 3

Social Perspectives of Narratological and Narrative Generation Methods

Chapter 10

Narrative Generation Using Psychological Value Variables: Probabilistic Model of Language Expressions and Values

Yasuo Tanida
Kotonoha Research Laboratory, Japan

ABSTRACT

In terms of story generation, the author proposes a method wherein stories can be rewritten using psychological value variables to ensure positive evaluation from readers. This method includes two phases wherein (1) a story likely to be chosen by readers from a selection of stories in the story generation system is chosen and (2) the selected story is rewritten using words that readers can relate to. It is imperative to understand the psychological values of the recipient and the relation between linguistic expressions and psychological values to complete the two phases of this study. In this chapter, the author examines the definition of psychological values and how to obtain them as well as the verification of the relation between linguistic expressions and values. This study aimed to rewrite the output of a story generation system using psychological values, and this paper is a research report on the preparation for rewriting stories. A proposal is offered for a study to construct and verify the system.

INTRODUCTION

The goal of this research is to rewrite the output of a story generation system (Ogata & Kanai, 2010) using psychological values. Ogata's narrative research attempted to translate the output of a narrative into universal words using automatic translation and evaluate it. However, different regions have different values in evaluating automatically generated stories, and the individual values of a region are dependent on cultural and environmental differences. As a result, different regions have different stories and movies.

First, the concept of value should be clarified. Everything that humans feel is controlled by universal strategies. A strategy is a measure of the magnitude of value, and positive and negative values indicate

DOI: 10.4018/978-1-7998-4864-6.ch010

individual differences. This is our general definition of a universal value. However, a universal value cannot include all universal values in the world. However, conceptual values can be evaluated using references to various viewpoints and past studies, and causes for individual differences can be considered. In the **BACKGROUND** section, we consider a universal value and discuss Tanida's concepts thereof.

We consider the following approaches, using differences in individuality with respect to a universal value:

- Rewriting selected stories with linguistic expressions that match the universal value of the reader
- Selecting a story that the reader is likely to select from multiple stories in the story generation system

For the two phases of research, it is necessary to define psychological values (strategies for universal values) and examine the relationship between psychological values and linguistic expressions. The value perceived by humans is quantified as a psychological variable, and individual differences are expressed by the magnitude of the variable. To rewrite a story with psychological value variables, it is necessary to develop a quantitative model of the psychological value variables and a stochastic model that expresses the relationship between the quantitatively defined value variables and linguistic expressions. In the **MAIN FOCUS OF THE CHAPTER** section, for each of them(Quantitative definition and Stochastic model), we consider the approach, data selection, collection methods, and verification methods. In addition, in the process of rewriting the expression of the story and selecting the story itself, we discuss possible challenges, such as the expected value and preference for a story beyond the superficial expression.

This chapter focuses on rewriting the output of a narrative generation system using personality differences for universal values. In the **SOLUTIONS AND RECOMMENDATIONS** section, we describe solutions for quantifying universal values, clarify the contents of questionnaires on specific psychological values, and describe how to collect utterance data. We extract linguistic expressions related to universal value variables and examine the extracted examples to confirm the validity of the relationships. In the **FUTURE RESEARCH DIRECTIONS** section, we discuss the concept and verification method of the system using the relationship between the value variables obtained from this experiment and linguistic expressions. We also present a paradigm for future research on achieving and verifying ideas for the story selection. In addition, we discuss a research approach that involves selecting a story that the reader is likely to select from multiple stories of the story generation system.

BACKGROUND

Why humans feel value is not discussed here. Although personality differences exist, this discussion proceeds on the premise that humans feel similar values and share them using a common language. For example, words that have lost value disappear from a community in which humans live. The premise is that words themselves have the universality of sharing value, and there are individual differences in how they are used. If individuality presupposes universal value (Brown, 1991), it may be possible to identify the structure of value by providing a typical example of universal values and examining differences in individuality. This section considers common examples of universal values.

From a biological point of view, a *universal value* reflects the goal of transmitting one's own genes. In the society, a competitive advantage can be considered a universal value. In the context of the preservation of species throughout the human society, altruistic and selfish values are important. Economics measures a universal value with the value of *utility*. If humans act selfishly to maximize the economic value, then selfish economic value is a universal value; however, in this case, altruistic benefits can also become a universal value. From an emotional point of view, it is important to determine whether a person feels positively or negatively. The emotional value of a certain stimulus may be a universal value or there may be individual differences.

A universal value is experienced by most people who live in a cultural community of a certain size, regardless of whether the values are positive or negative for an event. In addition, a universal value is a value that can be felt intuitively before discussing personality differences. The following are examples of values that are considered universal, and we discuss differences in individuality for these values.

We first discuss several values that arise from primitive human needs. Sleeping, eating, drinking, and preserving health are physiological desires for sustaining one's life. The desire for sleep, for example, is a universal value shared by all humans. However, there are individual differences regarding the amount and timing of sleep, which is caused by various circumstances and ways of thinking. In addition to sleep, eating and drinking are also essential to human survival. The desire to eat and drink is a universal value; however, the variety of human nutrients and cultures has created individual differences whose aim is no longer to only support life. Preferences for eating and drinking vary depending on one's environment, experience, and culture, and there are individual differences in taste. Nutrition sources, such as alcohol, that are unnecessary for survival are also subject to taste and desire.

Health is another source and universal value aimed at avoiding the fear of death and promoting survival. However, the concept of health that is shared by humans is not only a life-sustaining source. Athletic and medical activities are values that are related to goal orientation, such as maintaining health and mental health, maintaining an active life, and creating a sense of accomplishment. However, individual differences increase depending on one's culture, environment, physical individual differences, gender differences, and views on life. Although not directly related to one's own survival, humans also feel strong emotional and sexual pleasure toward sexual objects due to the reproductive purpose of gene preservation. Sexual desire has gender and individual differences, and gender targets also vary based on the influence of the environment and one's experience. Sexual objects that all people share a desire to have universal values. Humans share the concept of sexual objects among various communities as a value that has evolved beyond gene conservation.

Competition is essential for gene propagation and requires power, knowledge, ability, self-expression, and cooperation with others. However, during competitions, various desires and feelings are generated, such as a desire for approval, anxiety, and dependence, the elimination of anxiety and dependence, and a sense of self-satisfaction. These are conceptualized as words and recognized as universal values.

The above-mentioned forces are not only powerful but also competitive in small communities. Power is the ability to retain resources and social attention, and it is necessary for promoting competitions within a community (Oohira, 2010). Strategies, such as showing off power, dominating others, emphasizing equal relations with others, and making concessions to others, depend on the community, setting, and one's individuality. However, everyone uses a certain power strategy. Therefore, power is a type of strategy and a universal value.

Aside from power, self-advantage within the community and a competitive advantage over other communities are also important. In addition, there is a universal value of information, knowledge, and competence. In the hunter-gatherer era, this value may have been strategic information, knowledge, and the physical ability to survive. However, these values vary greatly depending on the environment and culture. From this universal value, emotions, such as ambition, the desire for knowledge, and curiosity, have emerged to acquire knowledge and abilities and use information. In addition, these emotions generate other emotions, such as longing and praise, self-satisfaction, and a sense of achievement. Universal values, such as knowledge, competence, and information, have evolved from strong links to resource retention and social attention retention.

Similarly, the economic value of income and assets is a universal value derived from competition. Economic value can be calculated in terms of money, a fictitious unit created by humans that exists as a result of humans' self-organizing ability. A universal value is the desire for something that can be exchanged for money; however, the concept of money has a universal value in itself. An income level creates a sense of security in the human mind. The same holds for assets, such as savings, real estate, and securities. Economic value creates a competitive advantage and not only involves competition with others but also satisfaction with one's status. The feeling of self-satisfaction stems from obtaining a sense of superiority. Perspectives on economic value and the required size vary across individuals depending on their environment and situation. If there are no conflicts, it may suffice to only have enough reserves for survival.

Here, we discuss the concept of time and space, which are universal values that have already been described, and values that are related to universal values, which will be described later in this section. Time separates various concepts and actions, such as seasons, morning, day, night, New Year, Christmas, dawn, sunset, meeting from time to time, date from time to time, supermarket opening hours, profit for the current term, next year's target, and precipitation for this month. Words and time-based communication are infinite and indispensable to human life. The value of time varies depending on one's environment and personal circumstances. In addition, whether there is enough time affects mental states, such as stress. Furthermore, spaces, such as places, scenes, and environments, create infinite possibilities for language and communication. Language and communication are some of the most important attributes that determine the perspective from which universal values are experienced. Because humans can communicate using concepts of time and place, they can organize and share knowledge and information on survival and competition, dangers, laws, territories, and borders, among others, and can pass on their genes.

Next, we consider art, aesthetics, and religion. In this section, we consider art to be the product of personal self-expression, in other words, something that can help one sympathize with others. Art is also a competition because it involves showing off to others, expressing oneself, and being complacent. Naturally, the individuality of artistic sensibility, sensitivity, and direction greatly varies depending on one's environment, experience, and personal abilities. There are a variety of values for beauty, which are related to self-containment, showing off, sexual objects, art, and a view of life. In addition, there are considerable differences in individuality due to cultural and environmental differences, and these cannot be uniformly defined but are derived from competition, as in art. In addition, religion exists for various reasons, such as to alleviate anxiety and fear, to avoid something, or to find the meaning of living. For some, religion may relate to knowledge, information, aesthetics, and arts, or it may relate to family, reciprocity, community, and communication. Religion is undoubtedly a universal value given the existence of Christianity, Islam, Buddhism, and religions worldwide.

Finally, we touch upon human altruistic behavior. Families have genetic and selfish values, but they also have altruistic values. Moral values are dependent on culture and environment. Due to differences in personal circumstances, family values also vary considerably. There are various subjects, such as parents, children, siblings, entire families, ancestry, and blood relations. All altruistic behaviors have a strong relationship with emotions and is a universal human value. Humans often help others who are not their family members or spouses or who they have a sexual relationship with. Such a system of reciprocity is, however, vulnerable. This is because an individual who does not feel value in reciprocal altruism leads to decreased chances for other individuals to survive. Therefore, humans should regard altruism as a universal value and consider it to be equipped with a mechanism for eliminating foreign individuals or creating fear and aggressive feelings about foreign individuals.

Reciprocal altruism appears to be an evolutionary value or social sentiment for species conservation, not individual genetic conservation. This value presents itself in many forms. Likes and dislikes and friendships with others contribute to maintaining a system of reciprocal altruism. Those who experience friendship behave altruistically and receive emotional rewards instead of altruistic benefits; however, this involves excluding traitors and foreign elements. Attacks on foreign elements trigger emotions, such as moral concepts and a sense of justice. In addition, appreciation serves as a strong emotional reward, and sympathy also arises for the motivation to take altruistic actions. Although these values are universal, there are large individual differences due to one's environment, experience, personality, and other factors.

Humans generally do not live alone but in a community. Communities are also associated with reciprocal altruism and communication, and humans sometimes prioritize community interests over those of individuals. The smallest communities are sexual subjects and families. However, regardless of genetic preservation, such as sexuality and kin, humans belong to multiple communities, such as companies, hobby circles, SNS connections, and local activists. Strategies, such as competition and cooperation within a community and outside a community, are sources of universal values.

Communication within the community shares the concept of universal values. Humans communicate with others in many ways, including non-verbal communication, which involves facial expressions and gestures, in one-on-one conversations, in conversations with multiple people, and in e-mail, chat, and SNS communication. At the heart of communication is language. Although various values arise from communication itself, such as rewards for empathetic emotions, arguments, and desires for approval, the words and concepts used in communication are universal values. However, it is difficult to grasp the individual differences in the value of words and concepts because they are shared and empathized.

Here, we describe a way of thinking about the value of activities, interests, and opinions (AIO) (Ikeo, 2010), which is relevant in the field of marketing. *Activities* refer to the idea that time-consuming actions are valuable in terms of what individuals spend their time on. *Interests* represent the types of interests, indicating a preference before the action. *Opinions* refer to individuals' feelings about various events, such as politics and social issues. This is related more to social interests that depend on culture, environment, and experience than to strategies in the community. Personality differences also play an important role. AIO is an approach that seeks to measure one's general lifestyle or a particular area of life or product category by asking questions about three aspects.

The well-known Big Five personality traits (Tanno, 2003) include openness to experience, conscientiousness, extraversion, agreeableness, and neuroticism. This system is used to express personality using five factors, and it is used in many studies and businesses. In addition, its correlation with combinations, actions, and languages has been reported.

Maslow's five-stage theory of desire (Maslow, 1987) outlines humans' needs for living toward self-realization from the most primitive to most advanced: physiological, safety, social, approval, and self-fulfillment.

Schwartz's value theory (Schwartz, 2012) is categorized into self-direction (related to the desire to think and decide for oneself and to support one's own life), stimulation (relevant to exciting experiences), desire, hedonism, achievement (related to competition and approval), power (related to ability to retain resources, ability to retain social attention), security, conformity (social compatibility), tradition (culture and religion), benevolence (related to coordination, reciprocal altruism), and universalism (development of reciprocal altruism that embraces philanthropy in a broader sense than benevolence and love of nature and peace for mankind). In addition, the 10 types are displayed and arranged on a map expressed by two axes: personal–cooperative and adventurous–conservative. In addition, Schwartz's value theory is also used in the World Value Survey, in which approximately 100 countries participate, and data from approximately 400,000 individuals are distributed free of charge (Kano, 2015).

Cloninger's personality theory (Kijima, 2014) divides human behavior into temperament and personality. Temperament is a habit system (associative conditioning) using unconscious procedural memory, whereas personality is a conscious system and a cognitive set using propositional memory (conceptual insight). Aside from the expressions of temperament and personality, this concept is the same as the dual structure theory of cognitive function. However, whether temperament is hereditary and personality is experiential is unclear. Dispositions include damage avoidance as passive avoidance, pursuit of novelty as an incentive activity, remuneration dependence as a social contact point, and persistence as partial strengthening. The personality is self-oriented in an individual and defines coordination and self-transcendence as the universe.

The value of the psychological values used to select and rewrite a story appears to be an overstated examination; however, we examine various ideas of value. A story is a secondary content contact, not a primary original experience. However, the space of human perception in that contact is wider than the real world, and there is a measure from the feeling of closeness to the original experience to the point at which a fictional world is felt. The examination of universal values presented by the author, past values, and studies of values can all be combined. When attempting to summarize the values felt by humans into one system, we believe that there are roughly two ways to summarize the system: one is to interpret humans as vehicles of genes (Dawkins, 1976) and define values objectively and the other is to interpret how humans feel in terms of information stimuli. Beginning from the Buddhist idea of interpreting information stimuli as *kuu* (none exists), we focus on the path where all is *mu* (zero) is *ujou* (infinite). This is a method of defining the concept of words as a value that replaces individual ideas. However, *hannya* (wisdom) is not so easily completed. We have selected the former method of defining the value of a human being, the vehicle for a gene, based on the past definitions of value. The latter idea is revisited at the end of the chapter.

In considering humans as genetic vehicles, two paths can be observed: a strategy for survival and gene preservation. All of the values mentioned thus far can fall under this control. Neither selfish desire or value nor altruistic desire or value is the only strategy for survival or gene transfer. The first of the most primitive strategies is the choice between adventure and risk taking (adventure strategy) or risk aversion (maintenance strategy). The second approach is to take an egoistic strategy (individual strategy) or altruistic strategy (cooperative strategy). The third approach is whether to stick (stick strategy) or not (flexible strategy), and the fourth is whether to endure (endurance strategy) or succumb to temptation (pleasure strategy). Humans combine and use these strategies for survival and gene conservation. However, when individuals encounter a certain situation, their tendency to select a certain strategy is innately dependent on their personalities. Personality traits (curiosity, self-growth, coordination, criticism, self-love, delicateness, leisure) and acquired values (family, work, marriage, money, time, friends, and stress) can be explained by the four strategies. These strategies are based on the double-process theory of cognitive function. In addition, in the definition by Cloninger, the four basic strategies are intuitive, whereas the others(characters) are logical.

The author has described a *universal value* in order to use a language-value relationship model for story rewriting systems. The author's past studies have already defined the value model (Tanida, 2013), which is the relationship model between linguistic expressions and values (Tanida, 2013). The value model is quantitatively defined from a hypothesis based on the idea of a *universal value* as described above. The relationship model between linguistic expressions and values has also been verified. In the following, the author will explain the two models based on existing studies.

MAIN FOCUS OF THE CHAPTER

Issues, Controversies, Problems

There are two problems upon rewriting a story with psychological values. The first is how to define psychological value variables quantitatively, and the second is how to build a model that expresses the relationship between quantitatively defined value variables and linguistic expressions. Each of these challenges involves an overall approach, data selection and collection, and validation.

There is also the problem of selecting a better story for the readers, as suggested in this chapter. The problem is that the optimal story cannot be selected only by the relationship model between linguistic expressions and values. This is because the action of rewriting the expression of the story and selecting the story itself includes an expectation value and a preference for the story beyond the surface expression.

1. **Problems with story rewriting**

    a. Quantitative definition of psychological value variables

By considering universal values, we obtained the hypothesis of psychological value variables. However, considering the link with the story generation system, it is necessary to first quantify these variables. The focus of this chapter is to identify the relationship between psychological value variables and words to rewrite a story. For this reason, the problems and methods of quantifying psychological value variables are outlined here and are not discussed later.

i. Overall approach

Although there are hypotheses about universal values, data are needed to quantify them. The first approach is to collect data through surveys, and the second approach is to create teacher data in which various behavioral data are tagged with a universal value. In the former approach, the scrutiny of the contents of the questionnaire is important. We must consider what information can be gained from the survey respondents to determine the universal value of our hypothesis. The latter approach requires searching for behavioral data that encompass universal values. However, there are no such behavioral data that can be conveniently obtained. Attaching a universal value tag to each action is also a costly task. Thus, we attempted to quantify psychological values using the former approach through surveys.

ii. Data selection and collection method

We first considered the challenges of creating a questionnaire and determining what to ask the respondents. First, to grasp the actual behavior, we conducted a survey to identify what people care or are worried about in daily life. Although the number of participants was small (7), we attempted to extract their behaviors and values by conducting an in-depth interview of approximately 2 hours after obtaining the detailed daily behavior and consumption in a week. The findings suggest four primitive strategies of behavior, personality traits, and acquired values.

To explore these hypothetical values, we mass-produced alternatives using combinations of thoughts and behavior patterns in various scenes (approximately 1,300). The question sentences were constructed by grouping the options of a behavior pattern. The questionnaire instructed the participants to select the ones that applied, all in a multiple selection system. The survey was repeated several times using multiple survey companies to narrow down the options. Finally, we created a quantitative survey proposal that could identify the psychological attributes of the hypothesis by answering 60 options.

iii. Verification method

It is impossible to prove that the respondents answered the questionnaire correctly, and numerous reports have indicated that survey responses are often inaccurate. However, to model behavioral data by assigning value tags to people, it is necessary is to trust the questionnaire responses. The only solution is to use the responses as a baseline to examine their relevance to the behavioral data and make corrections. The true values of the survey respondents can only be considered halfway between the survey responses and the respondents' actions.

Even so, it is necessary to eliminate fraudulent answers, which can be eliminated mechanically. To verify this, in addition to the 100,000 questionnaire responses collected, we created a large quantity of response data that were randomly answered mechanically. Then, from the distribution of human and random answer trends, a mechanism was created to eliminate the answers of respondents who appeared to answer at random. As this is not related to the main subject, further details are omitted. Using the developed mechanism, the data of approximately 87,000 respondents, excluding those with suspected random answers, were used to create a correct answer for the value tag. The principal component obtained from these answers is a psychological value tag. Using the data, a probabilistic model based on a stochastic decision tree and a Bayesian network was created, and a closed test and a cross-validation test were performed. The results confirm that almost 100% correct answers can be obtained for any value tag if there are 60 answer options.

b. Stochastic model that expresses the relationship between quantitatively defined value variables and linguistic expressions

i. Overall approach

The purpose of creating a probabilistic model that expresses the relationship between quantitatively defined value variables and linguistic expressions is to rewrite a story using that information. However, the question is to determine whether the features of the linguistic expression related to the value of the reader relate to the linguistic features related to the value of the writer. In other words, the goal is to determine whether the characteristic linguistic expression with the value that the writer wishes to convey to the reader possesses a similar value. Matching would be difficult if the writer and reader had multiple choices of linguistic expressions to express a value. Thus, for the same linguistic expression that is related to a value, the value perceived by the writer and the value perceived by the reader are not identical. A question to be considered is whether readers who choose a value also choose that value as a writer.

In the fundamental definition presented in **BACKGROUND**, a value is a strategy, and language is a typology of that strategy. There are individual differences in the tendency to select certain values. The difference in the position of the writer and reader is one of the personality differences. However, because the reader agrees on the value that the writer wishes to convey, if the value selected by the writer matches the value perceived by the reader, then the reader is more likely to sympathize. In other words, if the relationship between the value selected by the writer and the characteristics of the linguistic expression is used, it is simple to obtain sympathy with the reader who easily selects the same value as the writer.

ii. Data selection and collection method

Here, we evaluate the approach that should be taken to create a probabilistic model that represents the relationship between the writer's easily selected values and distinctive linguistic expressions. A large quantity of expressions, such as novels, essays, stories, dialogs, blogs, and SNS, are transmitted throughout the world. There are two approaches: adding value to stories and novels or adding value to writers. The former is a task that requires a large cost of adding values to a large quantity of data. Although the latter has a quantified value framework, the problem is that the novelist cannot answer the questionnaire. It is necessary to collect data that are easily collected as text and that reflect the values of the text writer (e.g., by answering a questionnaire).

iii. Verification method

Suppose that a certain number of a writer's linguistic expression texts and values are collected. It is important to determine what type of verification should be performed with the language model created.

First, the verification must be able to extract at least characteristic linguistic expressions that are correlated with certain values. It is important for this result to be intuitive. In addition, it is necessary to confirm the estimation accuracy of a stochastic model that estimates a writer's values from a plurality of linguistic expressions. However, this has little to do with the problem of rewriting a story and is thus revisited later.

2. Problems in selecting a story

Here, we discuss the challenge of choosing an improved story for the reader. An optimal story cannot likely be selected using only the relationship model between a linguistic expression and values. We believe that at least using the relationship between the value of the writer and the characteristics of the linguistic expression can help empathize with readers who select the same value as the writer; however, the writer should convey this concept throughout the story. The concept or story seeks to convey an atmosphere that cannot be conveyed by features of partial linguistic expression, such as the style or language. The question of what creates a feeling is a difficult problem. It is necessary to clarify various factors that separate the concept and the story, such as semantic features rather than surface expression features, the overall story structure, and the expected value at the beginning of the story. Rather than using the linguistic expression–value relationship model used for rewriting, an entirely different approach of deconstructing stories and concepts is required.

Story authors provide content to target audiences who react to specific values. For example, let's say you provided a story that makes you feel "impressed by the efforts of the hero or heroine". It feels the same when a poor-performing player in a baseball team makes an effort, or when a salesman who has a poor sales record makes an effort. However, readers of the type "impressed by the efforts of the hero or heroine" do not respond to both stories. That is why this problem is difficult.

SOLUTIONS AND RECOMMENDATIONS

To study psychological value variables quantitatively, data can be collected through questionnaires, specifically, a questionnaire with a total of six questions and 60 choices.

In the first question, the respondents are instructed to answer questions regarding their personality traits. This question is primarily useful for identifying inborn strategies and personalities; however, it also involves family and friendship and asks about altruistic trends.

Q1. Tell us about yourself. Select all that apply.

- It excites me to try new things.
- I like to explore the unknowns.
- I have a strong sense of curiosity.
- I care about how people see me.
- I am occasionally distressed.
- I lose sleep over trivial things at times.
- I prefer not to do things that are too much of a bother.
- I prefer indoor activities rather than outdoor activities.
- I like to kick back and relax.
- I consider myself pretty good at remembering anniversaries.
- I am cooperative.
- I want to make my family and friends happy.

In the second question, respondents are instructed to report scenes that feel positive and those that feel negative. This question is closely related only to innate strategy tendencies and personalities; however, it also helps identify acquired values by imagining specific situations, such as work.

Q2. Tell us about yourself. Select all that apply.

- It makes me happy to wear clothes that I like.
- It makes me happy when I make a tasty dish.
- I enjoy being complimented on my appearance, including how I look and how I am dressed.
- It makes me happy to receive recognition at work.
- I enjoy being appreciated at work by a customer.
- I enjoy the sense of personal growth.

- It is unbearable when I get criticized on my personality or inner self.
- It is unbearable to have my ego hurt.
- I do not like to be criticized about my appearance, including how I look and how I am dressed.
- It frustrates me when someone displays bad manners in public.
- It frustrates me when someone does not apologize for their mistake or wrongdoing.
- It frustrates me to be with people who are uncooperative and who can't read between the lines.
- It disappoints me when a restaurant that I go to serves food that doesn't taste good.
- It disappoints me to find that another store has something I bought for a cheaper price.
- It frustrates me to have to wait for a long time.

The third question is related to family. Families have universal values that are unique to human beings, and their individual differences are large. This question is primarily useful for identifying acquired values related to family and for identifying fundamental strategies and personality trends.

Q3. Tell us about your view on "you" and "family." Select all that apply.

- I want to get married someday.
- I want to have a child someday.
- I feel as though I may never get married.
- It is annoying when family members who I do not live with come and visit me.
- I do not get along very well with my parents.
- My parents and I respect one another's space.
- I am currently focused on being a "mother" and a "wife".
- I am dependent on my family.
- I contact my parents frequently.

The fourth question is related to work and money. It is mainly useful for identifying acquired values, such as the ability to retain resources and social attention, and it is also related to fundamental strategies and personalities.

Q4. Please tell us about your work and finance. Select all that apply.

- I feel that my work right now is challenging.
- I am proud of my work.
- There are times when I can sense that I am growing professionally.
- I feel pressured at work.
- I feel a lot of stress.
- The first work day of the week is depressing.
- I do not have enough money.
- I am not in a financial bind.
- I save a fixed amount every month.

The fifth question is related to friendship. This question examines universal acquired values, and it is also related to fundamental strategies and personalities.

Q5. Tell us about your social relationships. Select all that apply.

- At times, I get tired of human relationships in general.
- At times, I feel that it is bothersome to have to be considerate toward others.
- It is a hassle sometimes to meet up with people.
- I don't have many, but I do have a few close friends.
- I believe that my current relationship with my close friends will continue on forever.
- I have someone to call a "best friend".

The sixth question is related to time, which is an important universal value. This question is also related to acquired values, fundamental strategies, personalities, and stress.

Q6. Tell us about your time. Select all that apply.

- I have a lot of free time.
- I have plenty of time to have fun.
- I live a comfortable lifestyle.
- I don't have time to be alone.
- I feel that I don't have much freedom.
- I have less free time now compared to before.
- Higher priority is placed on my family and child(ren)'s plans over my work.
- I generously give my time and effort to my family.
- My life is centered around household chores, managing the family budget, and taking care of my child(ren).

In the contents of the questionnaire, various answer results are related to the value trend tag. When the answer result of the questionnaire is inputted to the value tag assignment system, the value tag of the respondent is outputted (Tanida, 2013).

As mentioned earlier, we needed to collect data in which the linguistic expressions were easily obtained as text and which indicated the value of the writer who expressed it (writer answers the questionnaire). To estimate a writer's values, a value tag should be associated with a characteristic linguistic expression using a quantitative survey. The questionnaire system described above was used for estimating values, and Twitter was used as the target of data to be analyzed as the data capture associated linguistic expressions as text. With Twitter, it is easy to obtain a statement text, and the realness and nature of human behavior in actions are likely to appear due to the immediacy and anonymity of the statement. A model that expresses the relationship between quantitatively defined value variables and linguistic expressions was achieved by associating Twitter utterance data with value variables. The details are described below.

The participants were gathered on the condition that they had published their Twitter ID, had written more than 1,000 tweets in the past, gave permission for their Twitter data to be analyzed, and completed a questionnaire survey on the above values. In addition to acquiring a participant's Twitter ID, we conducted a value questionnaire survey and collected valid data for approximately 1,200 individuals. Of the vocabulary obtained from the utterance text, linguistic expressions in which the number of uttered speakers was 10% or more and less than 90% were analyzed for their relationship with the value component. The reason for this is that if the speaker's usage rate is too high, the vocabulary may not represent the characteristics of the participant. By contrast, a vocabulary whose usage rate is too low may disproportionately reflect the characteristics of the speaker. This is because when a conditional probability model is used, there is a high possibility that the learning data become sparse (Tanida, 2013).

In the following, a characteristic linguistic expression that possesses a large correlation with the value tag is considered. We examine the distinctive linguistic expressions that are highly correlated with the first of the most primitive strategies: adventure and taking risks (adventure strategy) or risk aversion (conservative strategy). First, those who are likely to select the adventure strategy use highly abstract expressions and topical keywords related to news on politics and social issues. By contrast, those who are more likely to choose the conservative strategy use entertainment news, buzzwords, and common and peaceful words. *Fukushima* is a place in Japan where a nuclear accident occurred due to an earthquake. *Chu* is an onomatopoeic word for a kiss and is primarily used with children.

Adventure Strategy

ask, sympathy, history, United States, research, continuation, stimulation, suggestions, warnings, markets, sustain, City, territory, full, es, nuclear, economy, surveillance, cost, analysis, reconstruction assistance, operation, Fukushima, nuclear, 3.11, University, college students, head office, Mac, nuclear power plant, competition, China, accuracy, after the earthquake, nuclear accident, TEPCO, ldnews, Shinjuku, Apple, J-CAST News

Maintenance Strategy

Cute, benefits, combinations, entertainers, spoilers, pick-up, browsing, girl power, Chu

The following are function words that have a high correlation between the adventure strategy and maintenance strategy. Because the Japanese language is Romanized, subtle nuances may not be transmitted to non-Japanese speakers. For example, the function word *teirukara* described in the adventure strategy is used to present past evidence and state inference results. All other function words are used to promote a logical conversation. As described above, the expression used by a person who easily selects the adventure strategy is an expression for imparting logic to a statement. By contrast, the "*" in *ohohoh** described in the maintenance strategy is an exclamation expression indicating that "*" has a repeated meaning and that *oh* may be continued. The expression used by a person who easily selects the conservative selection is an exclamation expression that acts like a surprise, a bad ride, empathy, joy, or rejection and is exaggerated, cute, and acts badly.

Adventure Strategy

teirukara, toshiteno, teimasune, dattaga

Maintenance Strategy

*oh *, ah *, uwa-, suru-, ka-, gya-, wa-i, wa, ba-, Yoh-, yada, www **

Next, we examine characteristic linguistic expressions that are highly correlated with the second choice of value strategy, selfish strategy (individual strategy), or altruistic strategy (cooperative strategy). *Suki-nandakedo* in the individual strategy does not only express "I love you." The nuance of *nandakedo* has a cool meaning. *Yatsu* and *aitsu* represent *he* and *she*; however, the nuances in Japanese contain connotations of disgust. *Omaera* is a plurality of *you*, which is also directed at a person with disgust. *one piece (OP)* is an abbreviation for One Piece's anime title. Other content words listed in the individual strategy also include the *meaning of outlaws*, *retreating thoughts*, *giving up*, *nerds*, and *escaping from reality*. The content words presented in the cooperation strategy have unknown nuances in English; however, in Japanese, there are expressions that express gentleness, warmth, concern, and modest altruism. *Obaachan* does not simply mean *grandmother* but a grandmother that is kind and friendly.

Individual Strategy

sukinanndakedo, break, neat, file, loop, painful news, scam, Item, mackerel, OP, match, unexpected, illustration, brain, yatsu, aitsu, omaera, yellow

Cooperative Strategy

Sincerely, medical examination, obaachan, encouragement, warming, staying, otsukaresama, feast, One day, last night, half a day, lunch, hot, cherry blossoms, dentist, toast, 6 people, GET, beginning of the year, aftershocks, National highway, chestnut, hot, forecast, classmate

By examining function words that are highly correlated with the individual strategy and collaboration strategy, the features appear more clearly. Examining the expressions listed in the individual strategy shows that it is difficult for non-Japanese speakers to understand the nuances; however, *nandaze* is a slightly negative expression when boasting, whereas *tekurenai* means to look down at the opponent. The expression of time, *teitai*, expresses selfishness while looking down, *kah-, kana-* is a lazy hammer, *shiro* is an order, and *koe* is very scary. It is an emotional expression used to playfully despise others. As regards the cooperation strategy, *batabata* is an adverb expressing busyness; however, it is a humble expression used when asking for sympathy from an opponent. *Chauyone, dekimasu, masuyouni, masen-nyouni*, and *masho* presuppose cooperation with the opponent, seek sympathy, communicate that it is possible, or make a wish. They are expressions of putting in, praying for something, and soliciting. *Iyah, yoh*, and *nnah* are interjection-like expressions for easing communication.

Individual Strategy

nanndaze, tekurenai, teitai, kah-, kana-, koe, shiro

Cooperation Strategy

batabata, chauyone, dekimasu, masenyouni, masenyouni, masho, iyah, yoh, nnah

Next, we examine the expressions of content words that are highly correlated with the third primitive strategies, the persistence strategy and flexible strategy. The persistence strategy appears to be closer to the individual strategy, but they also have differences. *Ore* is *I* used by men, but *I* sticks to himself. *Yatu* and *omae*, which are correlated with the individual strategy, represent individuals who consider others rather than themselves. *Yaba (i)* has the same meaning as *koe*, which is correlated with the individual strategy. However, *koe* refers to something that happens to others, whereas *yaba(i)* is seen as something that happens to the speaker. Expressions that are correlated with the persistence strategy are all negative expressions; however, they all concern themselves and are not as far away from the society as expressions that are correlated with the individual strategy.

By contrast, the flexible strategy is close to the adventure strategy. Similar to the adventure strategy, expressions with a high level of abstraction and topical words related to news on politics and social issues are correlated, and some expressions are identical. The only difference is *Kamakura, eldest son, 0 years old, children, weather, okonomiyaki, syujin, top, Chinese cabbage, cold,* and *Sakura*. There are many expressions with a sense of life, and there are many soft expressions. *Okonomiyaki* is a Japanese-style pizza-like food. *Disaster prevention* is a topical story that is based on daily life. *Syujin* literally means *master* but can be used to refer to a husband. Therefore, the topic of current affairs and life in the flexible strategy plays the role of small talk to facilitate communication.

Persistence Strategy

ore, yaba (i), sorry, jealousy, byte, depression, boobs, talk, play, Love, last night, learn, part-time, handsome, girl power

Flexible Strategy

United States, disaster prevention, nuclear accident, electricity, power generation, nuclear power, NHK, manufacturers, experts, Nuclear power plant, pollution, East Japan, U.S. military, electric power company, electricity rate, denuclearization, 3.0%, release, Eldest son, 0-year-old, children, full face, weather, okonomiyaki, disaster prevention, master, leader, Chinese cabbage, release, cold, cherry blossoms

Examining function words that are highly correlated with the persistence strategy shows that the persistence strategy appears to be similar to the individual strategy. *Ka-* and *kana-* are also correlated with the individual strategy. However, there is no expression that makes the other person feel disgust. The last term, *kuso*, represents regret and disgust for oneself but not others. A difference between the flexible strategy and the adventure strategy clearly appears when observing function words. Expressions that correlate with the flexible strategy are expressions that display concern for others and facilitate communication.

Persistence Strategy

ka-, kana-, wah-, ne-, maji, dayo-, anma, su-, na-, da-, fa-, dayone-, kuso

Flexible Strategy

deimasu, yosasou, nanto, dattanda

For the fourth primitive strategies, the durability strategy and maintenance strategy, the content word expression and functional word expression are displayed together. The endurance strategy is a composite strategy of three strategies: adventure strategy, cooperative strategy, and flexible strategy. To endure may represent being adventurous at times, cooperative at times, and flexible at times. The pleasure strategy is a composite strategy of three strategies: conservative strategy, individual strategy, and persistence strategy. Losing oneself to temptation may be the result of protecting oneself and relying on oneself.

Endurance Strategy

sincerely, hurry, energy, pool, important, bright, encourage, wisdom, organization,

tell, sleep over, lunch, learn, today, all day, last night, learn, go live, US, bureaucratic, nuclear, articles, nuclear, domestic, abolition, nuclear accident, crisis, TEPCO, electricity, safety, media, concerns, NHK, China, Fukushima, 0% (more than content words)

more and more, tadesune, kome, good!, teiruka (the above are functional words)

Pleasure Strategy

Love, thank you, good night, yaba (i), suffer, twink, niisan, senpai, hair, voice actor, mahjong, animate, omaera, main story, figure, kiss, aitsu, win, boxing (more than content words)

*dayo-, ka-, anma, ne-, tayo-, waka, maji, www *, ku-, yo-, kuso-, su-, kayo, kakke-, guraisika (the above are functional words)*

Humans have been modeled on the hypothesis that the four strategies are combined and used for survival and gene conservation. However, the hypothesis has been raised that the fourth strategy is a superordinate concept of the first three strategies. This hypothesis must be reviewed. This book does not discuss personal characteristics or acquired values. It is poorly organized, but at least it convinces expressions correlated with each value tag. This completes the verification that characteristic linguistic expressions correlated with values can be extracted.

Although not very relevant to the problem of story rewriting, there are experiments that have confirmed the accuracy of a probabilistic model that estimates the writer's values from multiple linguistic expressions. The verification experiment does not verify the correct answer rate of individual value tags but estimates 12 types of values (Tanida, 2013). In other words, it is a hypothesis test of whether a value type can be estimated from a linguistic expression. If this is effective, the prospect of rewriting linguistic expressions in accordance with the values will also hold. The following summarizes the results of the verification experiment:

- The correct answer rate is a coincidence rate between the type number of the inference result when the linguistic evidence is provided and the correct answer type number of the teacher data. The random correct answer rate was 8.3%.
- The accuracy rate of the estimated model was 31.4% for more than 1,000 tweets. Because the estimation accuracy of the value type was 32.6%, it almost reached the upper limit. This model

is thus a powerful estimation tool that considers the value tag if there is a remark text. Therefore, inferring the type of values through linguistic components is effective.

The conclusions are that it is at least possible to infer a value strategy from the textual information of microblogging speakers, such as Twitter users. In addition, we can rewrite the text based on the value strategy. The following sections describe how to apply this knowledge to story rewriting. In addition, they describe ideas and research methods on the topic of story selection.

FUTURE RESEARCH DIRECTIONS

In the previous section, we examine linguistic expressions that are correlated with psychological value variables. Of the four primitive value selection strategies, three are consistent with intuition, indicating that each strategy works independently. The other complex values, the fourth primitive strategy, personality traits, and acquired values, are not used for rewriting a story because they increase complexity. Therefore, in this research, we focus on the three value strategies and consider how to apply them to story rewriting.

First, by using the Twitter data from the previous section, we can create the data presented in Figure 1. This is because there is tweet data from people whose value strategies are known. Hereinafter, for simplicity, the first primitive strategy in the figure and text (adventure strategy–maintenance strategy) is referred to as g1, the second strategy (individual strategy–cooperation strategy) is referred to as g2, and the third (persistence strategy–flexible strategy) is referred to as g3. The state values of the variables for these strategies are represented by three values: 1, 0, and −1. If g1 is 1, it is easy to select the individual strategy, whereas if it is −1, it is easy to select the cooperative strategy. If it is 0, neither of the strategies is selected. The values of each ID are obtained from the questionnaire responses.

Next, a characteristic word list, as illustrated in Figure 2, is created from the data using a plurality of statistics indicating the characteristics of the expression, such as the correlation. The original default

Figure 1. How to build train data using Twitter data

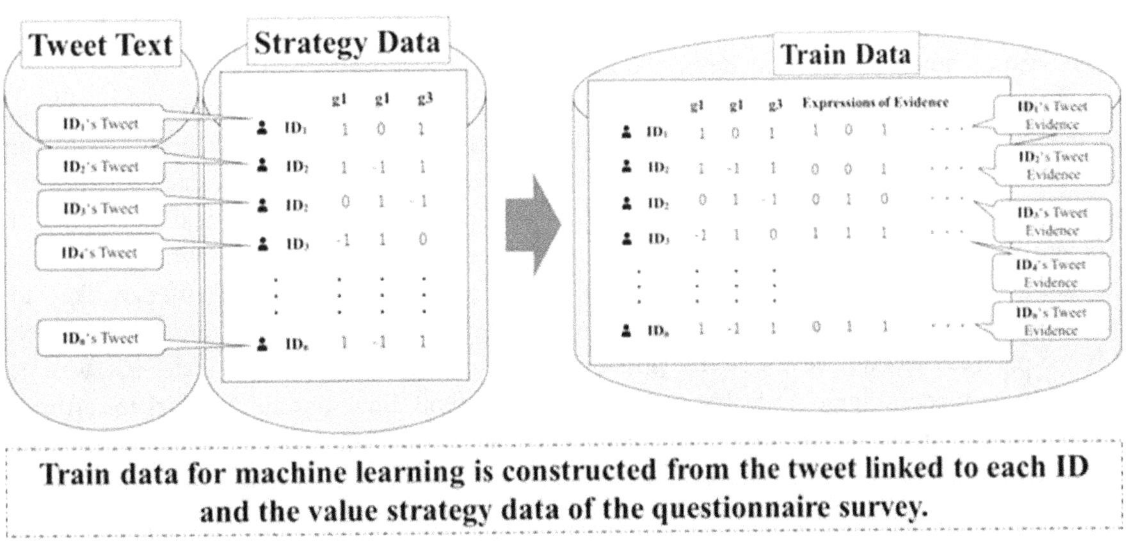

value of this automatically generated list is likely to be original if there is an inflected form. Therefore, this list is modified manually, for example, *aitsu* (*he* or *she* for humans, expressed with disgust), a value strategy like *kare* (*he* in English), or *kanojo* (*she* in English) is rewritten to the default expression only for information transmission.

Figure 2. List of expressions

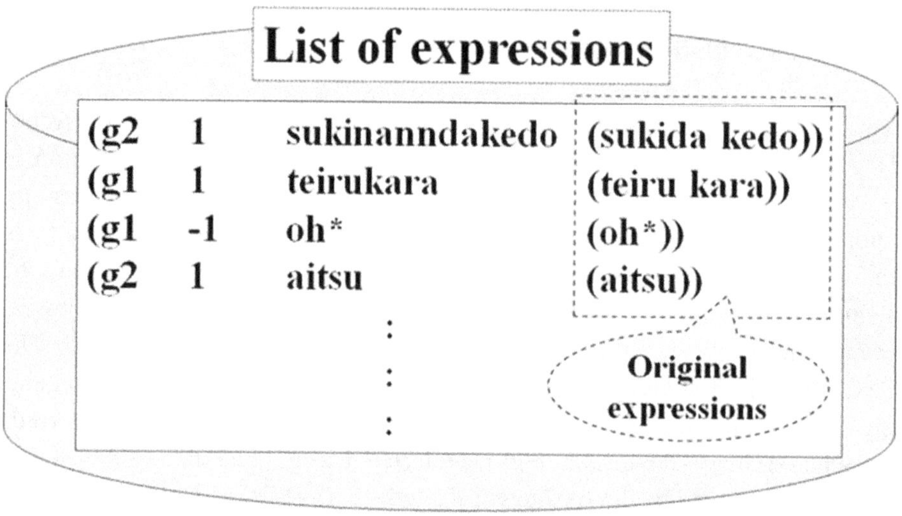

Finally, the list is rewritten using the default linguistic expression as a key, as illustrated in Figure 3. However, a linguistic expression that is characteristic for one strategy may not be found in the data for another strategy, and the space must be filled manually. In the previous example, ((kare kanojo) => ((g2. 1) aitsu)) and the output list as ((kare kanojo) => ((g2. 1) aitsu) ((g3. 1) yatsu)). If there is no correspondence with the strategy in the list, it is not rewritten. In terms of the system, minor problems remain, such as changing the result of the morphological analysis to the default expression and syntactically rewriting; however, this is not described in this chapter.

A bigger challenge is described in this chapter. Although a simple rewriting mechanism can be constructed by the method described above, there are problems with the quantity of data and the task area. Only 1,200 Twitter users were linked to Twitter data and values. We examined the characteristics of the linguistic expressions of the users and discussed the relationship with psychological values. This is a sufficient number to verify that there is a relationship; however, our focus is on rewriting. We believe that the quantity of data is insufficient. In addition, the task areas of the story and Twitter are likely to be too different, and the expressions used may not greatly overlap. Possible solutions are described below.

In the previous section, we discuss the possibility of inferring values from the text. First, we extend the estimation model. Figure 4 illustrates the extension method. First, an initial model for estimating value, which is limited to g1, g2, and g3, is created. The Tweet used here is linked to the values obtained from the questionnaire. Up to this point, to avoid interference with other psychological value variables and complications, a new estimation model of only g1, g2, and g3 is created. Then, using this model,

Figure 3. Key list of expressions

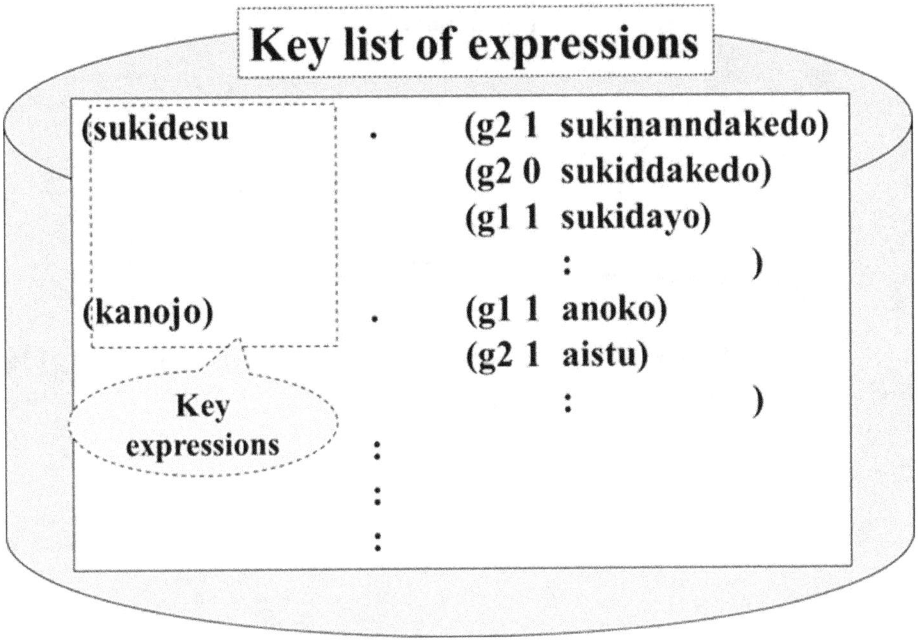

we assign value tags to the same amount of text (for each Twitter ID) as used in the experiment. Then, while manually verifying the linguistic expression linked to the psychological values, the learning data of the estimation model are extended, and the estimation model is extended. This cycle continues until near saturation.

With the above, a value estimation model extended by a large quantity of Twitter data can be created. Regarding the difference between the story and the domain, we use the value estimation model extended by Twitter data as the initial model and use the story data and fill the task area of the story, as shown in Figure 5. We then perform the model extension. The story data are processed separately for each story, author, country, and age with the same amount of text as used in the experiment. Then, in the same way as the extension with the Twitter data, we assign value tags and manually verify linguistic expressions linked to psychological values and expand the learning data of the estimation model and extend the estimation model. This cycle continues until near saturation.

The learning data of the model correspond to Figure 2 of the linguistic expression list related to psychological values. Figure 3 is thus created based on Figure 2. This demonstrates the rewrite rule that is the basis for rewriting stories using psychological values, which is the purpose of this research.

The structure of the story rewriting system is illustrated in Figure 6. The story is rewritten on the assumption that the reader's values of the story are understood. It is necessary to verify whether the rewriting results are successful. The story evaluation experiment for this verification is performed in a simple manner. We prepare moderate stories about the psychological values (g1, g2, and g3) to be built. The method is to randomly prepare approximately half of the target story that has been rewritten using the system and half that has not been rewritten at all. Then, a large number of participants evaluate the story. With this, the effect of rewriting can be verified from the story side and the human side.

Figure 4. How to extend the rewrite model using Twitter data

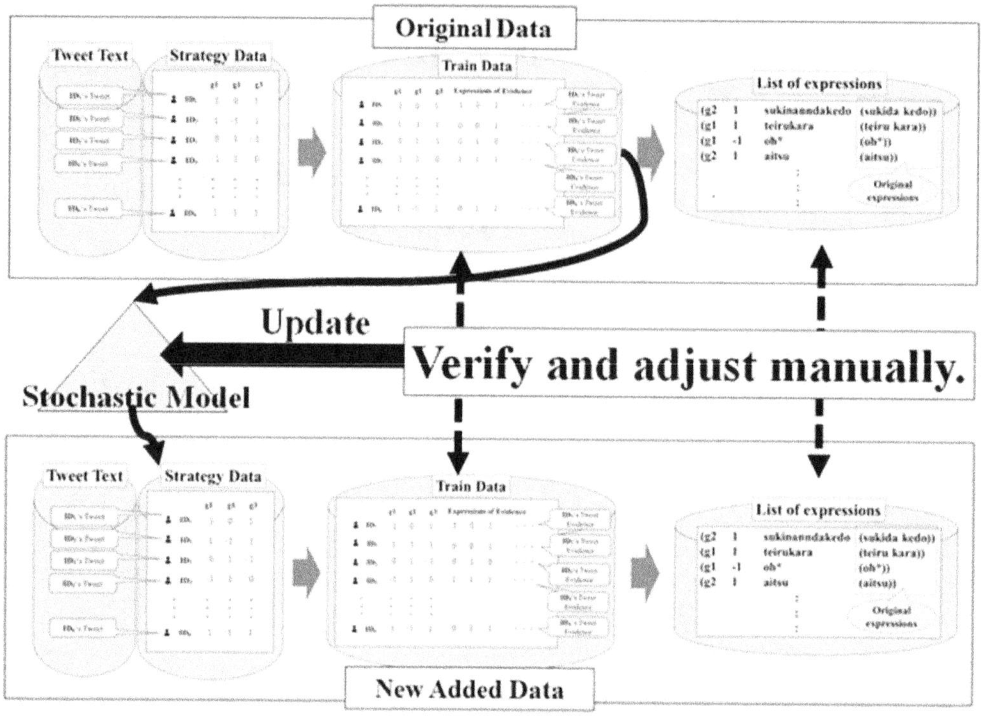

Next, we discuss how the idea of story selection can be achieved. The simplest method is to assign the psychological value tag used for rewriting to a plurality of input stories and recommend a story close to the reader's psychological tag. However, this may be an overly simple method because the rating of a story depends on the scenario, mood, and expectations. Simply because an individual likes fantasy does not mean that all fantasy is interesting. This problem cannot be solved by rewriting linguistic expressions, and there is a sense of conformity that only the reader can feel. However, it is impossible for the author to disassemble the concept of the story and all parameters starting from the story. The following points are thus considered.

The preference for making a story interesting is also consistent with a view of life and choices on how to live. Based on this idea, a connection with the strategy for living, that is, the three strategies described above, exists. However, instead of viewing the connection as a relationship with linguistic expressions, it is a way to look at the story as a whole, just as with life. First, we gathered survey data that included what types of stories the participants liked and how they rated it and obtained 19 evaluation keywords. We instructed 3,000 participants to select 19 metrics for their favorite story trends. Figures 7 and 8 present the rounded results of the survey. Figure 7 is a graph displaying the correlation coefficients for the two axes of the first and second components (temporarily named *warm* and *cold*) of the four principal components obtained from the 19 trend evaluation index responses. In addition, Figure 8 shows two axes of the third and fourth components (temporarily named *daily life* and *fantasy*).

The narrative may differ from one's view of life; however, this study claims that there are two large axes that cover the whole. The first axis is *warm* and *cold* (abbreviated as g4), and the second axis is *fantasy* and *daily life* (abbreviated as g5). We regard this as a story selection strategy and intend to apply

Figure 5. How to extend the rewrite model using stories

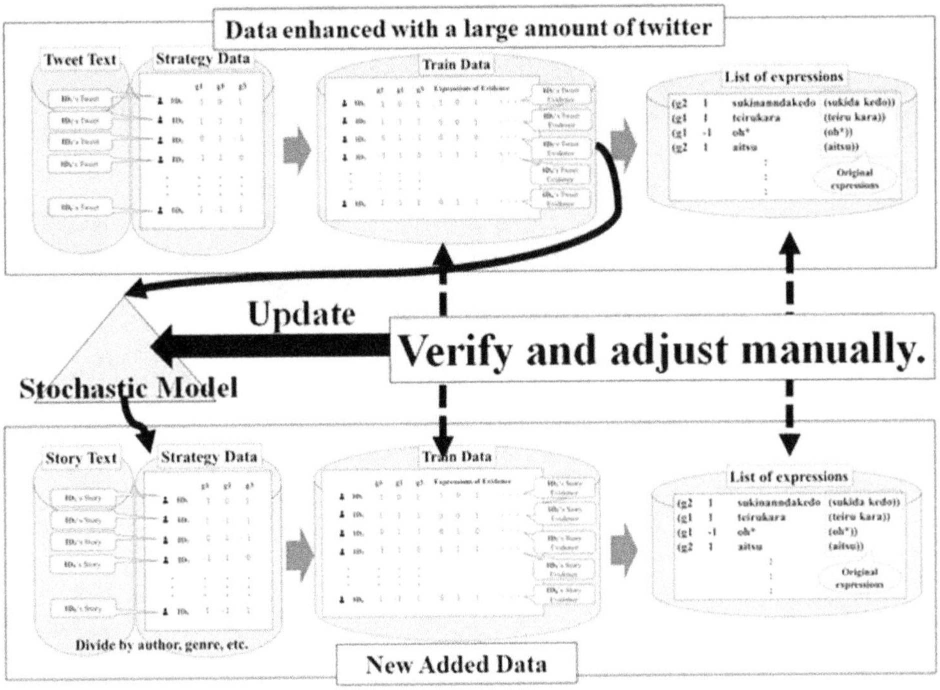

it to a story selection system. As illustrated in Figure 9, this is an estimation model that outputs a value of g4 and g5 with a story as the input. The outcome of the evidence of the estimation model is left for future works. However, if there exists such an estimation model, in addition to the values of g1, g2, and g3 of the reader, the values of g4 and g5 can be known in advance. Then, in addition to this evaluation result, it is possible to output the story closest to the strategy of the reader using the values of g1, g2, and g3 outputted by the estimation model described above. In terms of the method of verifying the selection of a story, as in the verification of rewriting the story, a method of randomly selecting the story using a large number of experimental participants can be used.

It is assumed that the target reader's value strategy, story preference, and outlook on life are known. Let's rewrite the story of a poorly performing salesman making an effort while keeping the type of story "impressed by the efforts of the hero or heroine". A poor mother cooks a lunch every day for her high school son, or a father who can't play the piano practices the piano for his daughter's wedding. Rewrite or select a similar story from existing stories. When selecting, it is advisable to combine the document summarization technique and the document similarity determination technique. When rewriting it, it will be necessary to solve the problem of filling in the semantic frames and story elements. The author has no specific idea of how to realize it systematically. However, the author thinks that a combination of existing technologies can provide such a rewriting image mechanism.

Figure 6. Story rewriting system

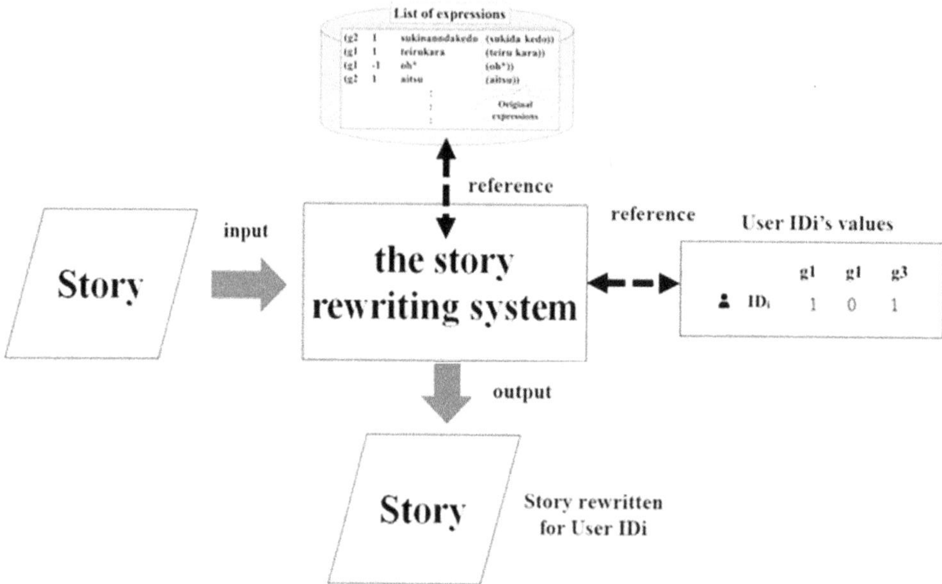

CONCLUSION

This chapter is a research report on the preparation for rewriting a story outputted from a story generation system, and it offers a proposal for a study to construct and verify the system. The following three are the main points of this chapter.

- The relationship model between linguistic expressions and values, validated in the author's previous work (Tanida, 2013), is used to rewrite the story.
- Among the many values defined in the author's past work, three value strategies that have been verified to fit intuitively are used.
- In order to deal with the value to the story, the author plans a study that assumes output that also considers the taste to the story.

This chapter begins with a discussion of the concept of universal value for humans, and as a criterion for quantifying universal value, we use choices based on strategies for survival and gene preservation. These are adventure strategy–human strategy–cooperative strategy, persistence strategy–flexible strategy, and endurance strategy–pleasure strategy. We quantify this hypothesis using questionnaire-based survey data and build a system that attaches a value variable tag from the questionnaire results.

Next, in exploring the rewriteability of the story, we discuss the question of whether the reader feels comfortable when the values of the writer and reader are similar. Based on the idea that value is a strategy and language is a typology of that strategy, rewriting a story using the relationship between values and the characteristics of linguistic expressions makes it easier to gain sympathy with the reader. This is the fundamental concept of a story rewriting system. For this purpose, the validity of the probability model of the relationship between the value variables and linguistic expressions was evaluated. Characteristic linguistic expressions that were highly correlated with the four primitive value selection

Figure 7. Story evaluation map (warm and cold)
Source: Tanida, 2019

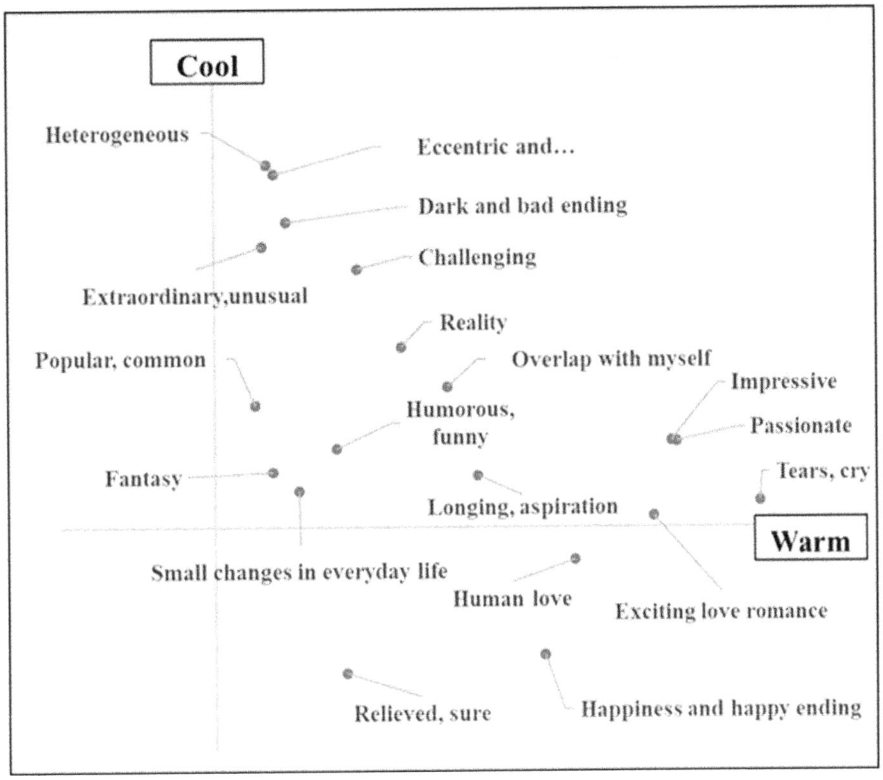

strategies were discussed. We also performed a verification of the estimation validity of the quantitative stochastic model. The results reveal that three of the four assumed primitive value selection strategies were related to characteristic linguistic expressions and had an independent role as a basic strategy for communication. Through such considerations, we propose a feasible approach that includes a future paradigm for story rewriting systems.

In addition, we discuss not only the rewriting of a story but also the problem of selecting the story. There may be a value criterion not only in linguistic expressions and basic value choices for living but also in the story choices themselves. As a result, rewriting a story uses the first three value selection strategies; however, story selection also requires a value selection strategy of warm–cold and fantasy–daily. We thus propose a system for selecting a story that suits the reader by using it.

We thus conclude the information presented in this chapter. However, an approach that is omitted in this chapter should be noted. To consider how humans perceive information stimuli, we can start with the Buddhist idea of *kuu*, which interprets information stimuli. Next, we focus on the path where all *mu* is *ujou* and define the concept of words that can be substituted for individual ideas as values. Although the challenges are great, this approach is important in determining the true relationship between language and human value. To determine the value that humans feel, the universal hypothesis that the way the human mind works is not bound by race or gender should be evaluated.

Figure 8. Story evaluation map (daily life and fantasy)
Source: Tanida, 2019

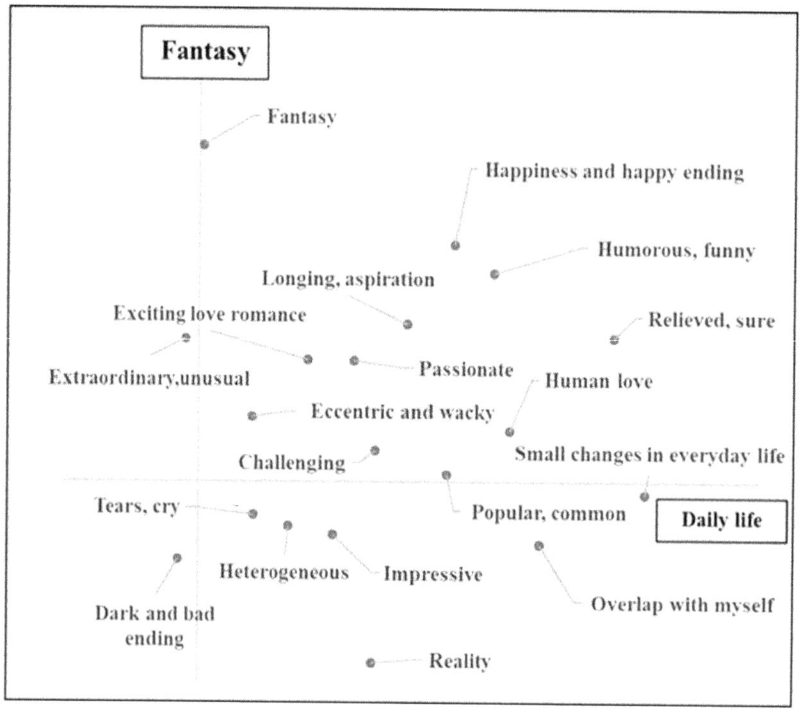

Figure 9. Story selection system

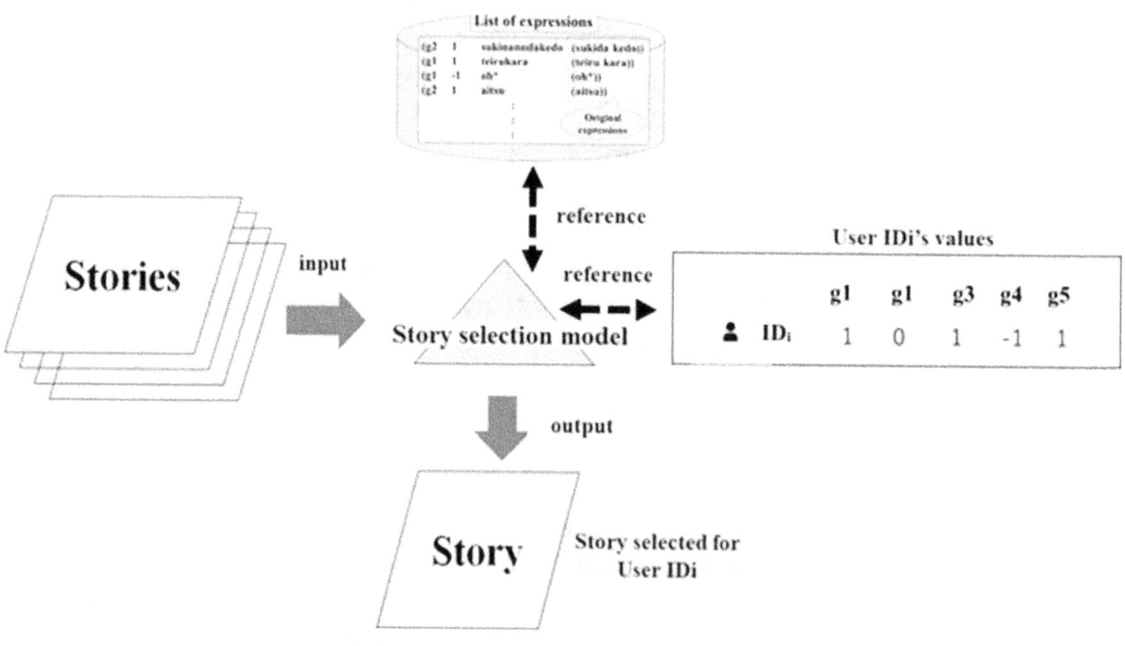

The first major hypothesis is that the evolution of genes is the reason why humans have the mental capacity that they do. In addition, the human mind is formed by the action of neurons, that is, nerve cells. What is obtained by the evolution of nerve cells is the current way in which the human mind works.

The second major hypothesis is that there are two types of ways in which the nerve cells that control the work of the mind work. One is to completely ignore the information and discard it, and the other is to build a mental network while editing the information to degenerate and replace it. Both functions are important mental functions for humans who receive a large amount of stimuli by sensory neurons. These actions already begin when the sensory neuron inputs the stimulus information. In particular, the ability to classify *similar* things, such as *categories*, *types*, and *associations*, is required to enable an analogy, which is the function of the latter mind. Classified *similar* can be reused. In other words, it is a valuable stimulus information, a value that humans perceive. This ability gives rise to human creativity (Levy-Strauss, 1976). In the following, this *similar thing* is referred to as *isomorphism* (Hayek, 2008) using terms from Gestalt psychology. *Homomorphism* refers to the relationship between structured sets. When there is an isomorphic relationship between the values of different categories, the two values are based on some criteria based on the human mind and can be regarded as similar. We believe that exploring this criterion can define an analogous criterion for the value of the human world. Both types of mental activities are cost reduction operations that discard information; however, the latter requires an associative structure and a place to store mental networks that allow analogy operations. These need to be organized to store mental networks. This mental activity requires the sensory neurons of the body and the nerve cells of the brain. Sensory neurons do not perform advanced sorting and storage even when classifying information. The latter half of the mental work is performed by the evolved brain. There are also two types of processes for organizing and storing: the emotional or semantic working style and the logical or syntactic working style.

The third hypothesis pertains to the relationship between humans' perceived value and language. The value perceived by humans is verbalized if there is a common value in either direction. The same applies for words for things and for emotions. Words whose emotional and semantic values are close to zero from a social point of view will disappear in the long term. However, humans can create new values and words through their self-organizing ability. Words shared among humans are conceptual structures of value and have universal value. Emotional and semantic values precede and gradually shift to logical and syntactic values using society as a filter. Humans have an instinctive tendency to speak (Darwin, 2016), and language is an instinct (Pinker 1995) that is universal and innate (Chomsky, 2011). The relationship between language and emotional and semantic values is closely related to the mental work that requires the manipulation of analogical information. To study human values, living words(actually used) should be observed. In other words, to systematize the proposed idea, associative networks of languages that are linked in a large analogy should be collected, which may be impossible, however. We believe that artificial intelligence does not need to enable any human thinking. Mechanical associative networks have the potential to produce stories that are entirely different from human thinking.

The first two of these hypotheses are not author-verified. However, the third hypothesis, which started on the assumption that these two hypotheses are correct, can be verified if there is a large amount of communication data of online society. An associative network centered on analogy is automatically acquired and organized by humans. Automatic acquisition proceeds again according to the organized way of thinking. Very simple, but not organized. The author believes that this is the only way to test the third hypothesis.

ACKNOWLEDGMENT

This chapter's research was supported by Prof. Takashi Ogata of Iwate Prefectural University and Synergy Marketing Inc.

REFERENCES

Brown, D. E. (1991). *Human universals*. McGraw-Hill.

Chomsky, N. (2011). The generative enterprise revisited (T.Fukui & N.Tsuji, Trans.). Tokyo, Japan: Iwanamibunko.

Darwin, C. (2016). Ningen no yurai (M. Hasegawa, Trans.). Tokyo, Japan: Kodansha.

Dawkins, R. (1976). Rikotekina idensi (T. Hidaka, Trans.). Tokyo, Japan: Kinokuniyashoten.

Hayek, F. A. (2008). Kankakuchitujo (S. Akiyama, Trans.). Tokyo, Japan: Syunjuusha.

Ikeda, K. (2016). *Nihonjin no kangaekata sekai no hito no kangaekata—Sekaikachikanchousa kara mierumono* [Japanese way of thinking World way of thinking: What we can see from the World Values Survey?]. Keisoushobou.

Ikeo, K., Aoiki, K., Manami, C., & Inoue, T. (2010). *Matketing*. Yuibikaku.

Kano, F. (2015). Comprehending consumers: Computing complexity of cultures. In *Proceedings of the 29th Annual Conference of the Japanese Society for Artificial Intelligence*, Hokkaido. Japan: The Japanese Society for Artificial Intelligence.

Kijima, N. (2014). *Cloninger no personality riron nyuumon* [An introduction to Croninger's personality theory]. Kitaoojishobou.

Levy-Strauss, C. (1976). Yasei no sikou (Y. Oohashi, Trans.). Tokyo, Japan: Misuzushobou.

Maslow, A. H. (1987). Ningensei no sinrigaku (T. Oguchi, Trans.). Tokyo, Japan: Kinokuniyashoten.

Ogata, T., & Kanai, A. (2010). *Monogatariron no johogaku josetu—Monogatari seisei no gijutu wo megutte* [An introduction informatics of narratology: Around the thoughts and technologies of narrative generation]. Gakubunsha.

Oohira, H. (2010). *Kanjoo shinrigaku nyuumon* [Introduction to emotional psychology]. Yuibikaku.

Pinker, S. (1995). *Gengo wo umidasu honno* (N. Mukuda, Trans.). Tokyo, Japan: NHKbooks.

Schwartz, S. H. (2012). An overview of the Schwartz theory of basic values. *Online Readings in Psychology and Culture, 2*(1). Advance online publication. doi:10.9707/2307-0919.1116

Tanida, T., & HU, Y. (2013). Twitter, nan tweet areba kachikan ga wakarunoka? [Twitter, how many tweets do you understand your values?]. In *The 34th kotobakougakukenkyuukai* [*Japanese Conference for Language Engineering Research*]. Hokkaido, Japan: The Japanese Society for Artificial Intelligence.

Tanno, Y. (2003). *Seikaku no sinri—Big five to rinshou kara mita personality* [Personality psychology: Big Five and Clinical Personality]. Tokyo, Japan: Saiensusha.

KEY TERMS AND DEFINITIONS

Associative Network: An associative network of linguistic value expression based on analogy.

Relationship Model Between Linguistic Expressions and Values: A probabilistic model that quantifies the relationship between linguistic expressions and values, which was constructed using the data of values and tweets of about 1200 people.

Rewriting Selected Stories: Rewriting selected automatically generated stories.

Selecting a Story: Selecting an automatically generated story.

Universal Value: An abstraction based on genetic strategies with positive and negative values that most humans feel.

Value Model: A value model of past study quantified in consideration of universal value.

Value Strategy: Three value strategies (g1, g2, g3) selected from the value model based on universal value and two value strategies (g4, g5) based on the expected value of the story.

Chapter 11
Effect of External Activate Factors Serving as Clue:
Creating New Products or Services Ideas With Storytelling

Jun Nakamura
Chuo University, Japan

ABSTRACT

Why is there a demand for consultants and outside directors? That is because what they are asked for objective advice and the clue—this chapter describes them as external activate factors—is precious. This chapter focuses on how the external activate factor affects human creativity, especially the process of creating new products or services, and the author conducted an experiment to compare the effect of external activate factors at a lecture of graduate school. As a result, when the external activate factor is provided, the emergence of abduction and new rules was observed, and the effect of storytelling was confirmed. Further research is to find what kind of external activate factor is preferable.

INTRODUCTION

The describing of decisions made by a human being is often influenced by others (Asch, 1951). To drive innovation, IBM embedded a mechanism called FOAK, or First of A Kind (Frederich & Andrews, 2009), which is an innovative process to swiftly bridge technology to the market. In other words, it can be said to be a mechanism for developing the first service to be put in the market while exclusively collaborating with a specific external client for a specified period. The mechanism allows IBM to attain first-hand data from the client, and the client can also enjoy the upfront gains of the new service in the market. Furthermore, both IBM and clients can cooperate in terms of marketing strategies for a certain period of time. The author notes that IBM not only behaves individually as a leader in research but also employs the FOAK mechanism, which incorporates external input from the client in the innovation

DOI: 10.4018/978-1-7998-4864-6.ch011

process. From this viewpoint, the author has focused interest in the effect of external inputs on creative activities and innovation.

A creativity support system is crucial for innovation. Finke et al. (1992) suggests that such a support system is important so as to be able to rearrange components and rejoin existing shapes and concepts in order to explore new concepts, thus demonstrating that the Geneplore model could lead to the restructuring of discoveries (Finke et al., 1992). Also, the further the distance between two concepts is, the more effective would be the creativity in a concept that has been formed by the combining of concepts (Klahr & Dunbar, 1988; Wilkenfeld & Ward, 2001; Taura et al., 2005). Some studies found that creativity can be triggered by ambiguous ideas (Gaver et al., 2003) and accordingly examined external restricting conditions that motivate creativity (Bonnardel, 2000). An occupation of subjects in these articles was, however, nothing by design so that the topic of the effects of external input in the process of creating new ideas of products or services with storytelling would be ignored.

Given the above, the author sets the research question of this article as follows:

How external activate factors serving as a clue affect an individual's way of thinking when dealing with ill-defined creative ideas, including concept design with the storytelling?

The next section covers the concept creation model and the logic behind the analysis tool, followed by the analytical method. The outcome of the analysis is discussed, and the conclusion is presented next.

Concept Creation Model on the Basis of an External Activate Factor

The supplying of the idea to address the research question is referred to as the effect of combining different elements (Wilkenfeld & Ward, 2001; Costello & Keane, 2000). Considering previous works on creating novel ideas, the author's interest herein focuses on the power of analogy (Holyoak & Thagard, 1995). Teachers' use of analogy can provide students valuable words for better understanding of programming to students (Burstein, 1986). The function of programming know-how equates to the "base" in cognitive science since know-how determines what functions need to be realized and activated. The output of programming would be equal to the "target." This study gives some keywords as "base"; however, the "target" is unknown because it refers to the very stories that subjects create during experiments. Gick and Holyoak (1980) point out that the effect of analogical thought weakens except for obvious external activate factor. This paper expects that analogical thinking is brought about as a clue of storytelling, rather than a straightforward causal relation in which clue itself triggers analogical thinking. In forming such a novel idea, this author has focused on outlier items in statistics. Outlier items are the very factors that can trigger a value creation (Nakamura & Ohsawa, 2009).

See Figure 1 for the model, which is a series of simple flows. In the first stage, some items are already clustered based on the existing concept as default (left side in Figure 1). Later, the cluster is reconstructed, leaving no single item alone (right side in Figure 1). In most cases, the more items that are easily clustered together, the more "other groups" are left unclustered (in the middle of Figure 1). At this time, testees or subjects notice the simple rule of "no one left behind" and has no choice but reconsider the clusters that have been built halfway. They try to reconstruct items all over again. It is demanding for the testees of an experiment, yet, some new, more comprehensive stories are created after the completion of the whole process as a result of the expending of considerable effort and energy. Several attempts lead to better outcomes, after all.

Forcing concept reconstitution may not altogether remove the initial bias from one's mind (Cosejo et al., 2009). However, the hypothesis behind the concept creation model is that the experimental subjects

Figure 1. A concept creation model revised from (Nakmaura & Ohsawa, 2009)

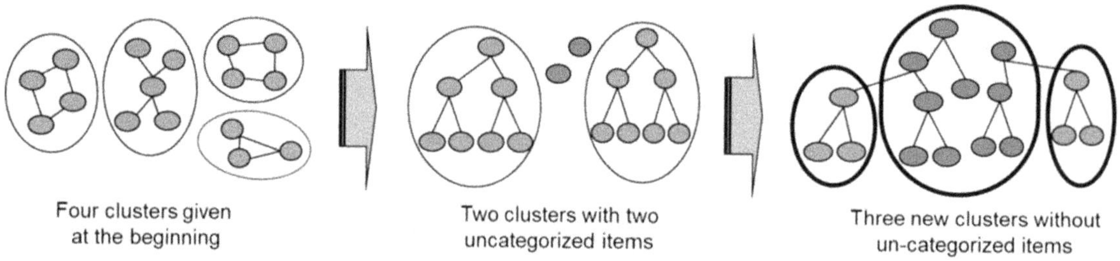

overcome the embedded biases, notice the shift of mindset, and find new combinations of words. This research aims to promote the switching of the problem space, notably by using the polysemy of words (Costello & Keane, 2000) and by using analogy as a link between polysemic words.

As similarly seen in the concept creation model described in this section, it is challenging for a subject to effortlessly break the cards bundled by the existing concepts. People easily get trapped in the prejudice and established ideas that have been embedded in their minds for some time, which is the initial situation we face. However, novel ideas may come to light as this established concept is broken and the combinations of different items originally categorized in other clusters are reconsidered. In this process, in the first heuristic, it is expected that there would be the chemical reaction. In the next step, a subject is asked to reconstruct the same given concept twice. The first time, subjects need to reconsider them with no clues, but the second time, they are given a couple of clues. This clue of "something" is thej second heuristic.

The possibilities of various innovative ideas have been examined by using this model. I've examined in particular what kind of story design could be developed if subjects were given items related to hardware products (Nakamura, 2014a). The article discussed the city and automotive design, and both topics can be related to each other with regard to surroundings. The city becomes a vital space for rediscovering the surroundings of the several buildings located in a particular block as a set of space (Sakamoto et., al. 2012). Automotive design requires consideration of three networks, i.e., traffic networks, telecommunication networks, and energy (electric) networks (Yamashita, 2013).

The logic behind the development of the model is presented below, with theoretical premises being based on the creative activity model (Hori, 2007) expressed in the following formula.

$$\dot{q}_\lambda = -k_\lambda q_\lambda + \sum_\mu g_{\mu_\lambda} \alpha_\mu + F_\lambda \quad (1)$$

$$\dot{\alpha}_\mu = -\gamma \alpha_\mu + \sum_\lambda h_{\lambda_\mu} q_\lambda + \Gamma_\mu \quad (2)$$

where q_λ is the differential calculus of creative concept's activation, $\alpha\mu$ is the component of concept factor, both $-k\lambda q\lambda$ and $-\gamma\alpha\mu$ are natural diminution and both $F\lambda$ and $\Gamma\mu$ are external activation.

This chapter talks particularly about $F\lambda$ and $\Gamma\mu$, exploring those features of the dependent variable. This researcher interprets them as inputs based on own experiences and knowledge, new information from interaction with various people, and new perspectives and viewpoints. Also, $g\mu\lambda$ and $h\lambda\mu$ are functions for concept creation, which can be said to be an innovative skill1 among human.

Based on the logic and concept creation model referred to above, this research is designed to enjoy the categorization of words ($\alpha\mu$, $\mu = 1,2,\cdots,20$) **in** consideration of not only similarity but also analogy in making a batch of words to produce new concepts ($q\lambda$, $0 < \lambda < 5$). As one word has plural interpretations, each testee was asked to consider the most suitable meaning to the relevant word.

The experimentation was conducted twice per testee. The first time, a subject was provided with no external input, which effects on $F\lambda$ and $\Gamma\mu$, and the second time, s/he was provided with the external input. Both $g\mu\lambda$ and $h\lambda\mu$ represent various motion capture in this research.

Changes in human behavior between with and without external activate factors have already been quantitatively observed (Nakamura & Ohsawa, 2008b). It was a preliminary experiment to explore the effect of external input. This chapter presents a qualitative analysis attempt to find out if external input makes a difference in storytelling, including concept design on products or services, utilizing the words given in classes at a graduate school.

As a result of the preliminary experiment, two types of subjects were categorized into two groups: Contemplation type, which concentrated on a particular word card over time, or Haste type, which spent only a short time on each card (Nakamura & Ohsawa, 2008a). External input was given to both types. Quantitative differences in the outcome between the case with the external input and the one without input were seen (See Figure 2 below for the result, paying closer attention to the scale of both axes). The left side shows the Haste type, and the right side illustrates the Contemplation type. When the external input was provided, the distance of one drag (x-axis) did not dramatically change. However, the distance of a drag of a card (y-axis) greatly slowed downy (Nakamura & Ohsawa, 2008b). In other words, it can be seen that there is an extreme difference in the mode of action between Haste type and Contemplation type. Although comparison is possible regardless of clues, the Contemplation type suggests that giving input as a clue in particular has an effect on the depth of thinking. The depth here is expressed by movement time and speed.

The outcome is consistent with that of IBM FOAK, showing the influence of the external input as having been quantitatively confirmed to a certain degree in the preliminary experiment. Nevertheless, the storytelling semantic experiment and qualitative analysis had been inadequate so these are to be addressed in this chapter.

The concept of analytical method is discussed below. Upon completion of the experiment, but before analysis, the created story was transformed into a first-order predicate logic expression. The number of individual constants and the length of formal logic regarding the output, were considered the depth of attributes. Then, those indicators were examined relative to how deeply participants considered the created products. To calculate the depth of attributes of individual constants, an expression of created products was formalized as follows, where P is either an individual constant or a variable, and t is a term to be quoted.

$$P(t1, t2, t3, \ldots, tn) \quad (3)$$

Here, the term (n) is interpreted as the depth of attributes. Whereas, the length of the formal logic regarding why the product created is interpreted as background or a differentiator. It is then counted with

Figure 2. Moving velocity in the preliminary experiment

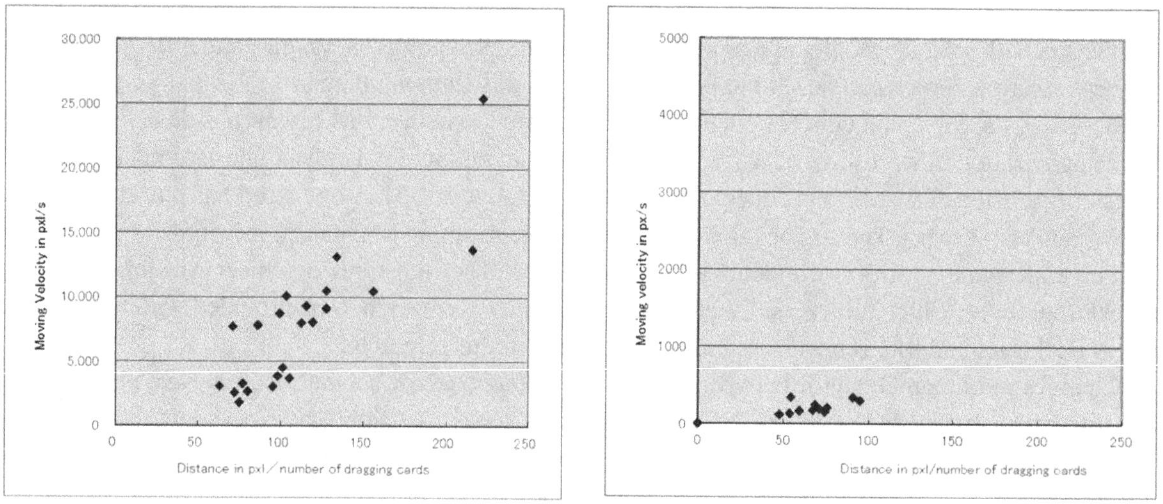

both the number of logical connections and the predicate symbols that are assumed as an appropriate level of satisfaction, assuming no contradiction has occurred. For example, following the former expressions and/or representations (Tsujii, 1987) can be better satisfied than the latter one by considering the number of terms, predicate symbols, and logical connections.

$$\forall x \exists y (Human(x) \rightarrow Eq(father(x), y) \wedge Human(y))$$

$$\forall x (father(x) \rightarrow Human(x))$$

The above concept of analysis on causal relation itself is not directly related to the origin of creation but the prerequisite way for the purpose of capturing the features of created stories. The features are discussed by comparing the stories with the ones with external input and the ones without external input. From the following section, let us introduce two experimental web-based systems as an enabler for users to create storytelling, as well as a more concrete method of analysis, results, and discussion.

EXTERNAL INPUT WTH A SIMPLE WORD THROUGH ANALOGY GAME

Regarding the logic noted above, the two web-based systems based on this concept creation model was designed, with the ways of storytelling with each system observed. The first system, herein called Analogy Game (hereinafter referred to as AG) and the second, Kotofactory (hereinafter referred to as KF). This section introduces AG, followed by details of experiments, analysis of results, and discussion. When it comes to AG, there are 20 words on the screen, with clusters of fruits as including oranges, strawberries, and apples, and clusters of nature as including mountains, rivers, deserts, and lakes, for instance. These clusters and items (word cards) are placed on the screen by default. Due to the constraint of screen space in PCs, the number of word cards was limited to 20. During the experiment, subjects dragged word cards and reconstructed cards on the screen, making sure any new categorization would convey the story. It sounds simple, yet some people find this type of storytelling difficult. Needless to say, some subjects could not break the established concepts while some enjoyed trying to conceptualize storylines with cards.

This section describes how to proceed with the analysis by using AG. Note that both input and output from/to AG used only unstructured data, i.e, a noun written on cards displayed on the screen. Why nouns? This is because the perception of a singular noun differs from people to people, which thus potentially allows broad options. In the case of an adjective or verb, the context would be identified accordingly, which may limit the creativity of ideas. Here is an example. Using the noun "Mt. Fuji," possibilities could be "Mt. Fuji dyed red that shines in a winter morning," "Snow-white Mt. Fuji captured from the satellite," or "Mt. Fuji witnessed from the climbing route," depending on one's imagination. On the other hand, what if a phrase given at the beginning is "Mt. Fuji reflecting red light of the sun in the cold winter morning"? The description is so specified and the imagination would be restricted too narrowly. There is little space left for one's creativity. Therefore, to make subjects recall a variety of scenes—it implies that the experiment requires enough knowledge and experience to be creative—the given words were thereafter limited to nouns. This researcher then designed AG in a way that asks subjects to think of combinations among words.

The system used provided two types of experimental output: one was a label for each cluster of cards signifying new ideas of products or services, and the other was an overall story covering all the clusters. For this purpose, participants moved and collected words by dragging a set of 20 given words. These words were combined and categorized into new groups according to participants' interpretation of new ideas. A new idea could be a product or service, with storytelling as previously mentioned.

The experiment consisted of two phases—named Experiment A and Experiment B. With regard to the first Experiment A, subjects were instructed to express the story by using presentation slides. In particular, subjects could set an arrow between clusters and put additional explanations, adding some illustrations in order to convey the storyline in detail. There were various designs of slides and information on them, depending on how the users considered the structure of the clusters. In other words, participants prepared a presentation slide as a result of breaking the given clusters and rearranging clustered cards, considering the structure of entire storytelling with newly built clusters. On the other hand, Experiment B asked to describe storytelling itself in a written form instead of a presentation slide. The intention on having two types of experiments was to observe whether the "how to give" the input triggered any change in the output. The subjects were asked to present with slides and/or to explain your idea through the text by mail or word documents depending on the situation in real business world. So the experiments were designed separately to presentation slides as opposed to a description of storytelling in text.

As mentioned at the end of previous section, analytical method is performed by transforming the created story into a first-order predicate logic expression. The formal logic is then analyzed in terms of rich*ness of knowledge and/or memory. The* level of richness of knowledge and/or memory is defined as shown in Eq.(4), where k is the sum of the number of nouns, verbs, adjectives, which is determined by morphological analysis, and the number of arrows and boxes for better expression in the presentation material regarding created product. m is the number of nouns and verbs ($k > m > 0$) that are the same as those shown on the cards remaining in the workspace, which are considered as the final composition of the combined cards. By using adjective and adverb within the context, it makes the contents enrich of being substantial, and bring it to better understanding, although it is not a sort of a rhetorical manner. This means that as the percentage increases, more external knowledge and/or memory is introduced in the context of the richness of expression. It is considered based on the assumption that frequent use of adjectives and/or adverbs, which is not a rhetorical manner, enrich the outcome in the context of storytelling, and the content of being substantial, and bring it to better understanding.

$$Knowlede\mu = {}_{\Sigma}(k - m) / \Sigma k \qquad (4)$$

Let us move on to the Experiment B. The given 20 cards and clusters were precisely the same as in Experiment A, yet in Experiment B, the first round had no external input, and the second round used Samsung as an external input. Besides, "how to give" of the external input was changed, asking subjects to conduct storytelling in a text format instead of a presentation. Those who participated in the experiment were all graduate school students from Shibaura Institute of Technology, but the group who took part in the Experiment B and Experiment A were not overlapped. A leader, a PC operator, and someone to take discussion minutes, were assigned to each group. The experiment required that all participants take part in discussions.

The overall category of experiments is depicted in the following Table 1, where the analysis method is described, i.e., Eq.(4) is indicated as the level of richness and/or knowledge, and logical connection means the analysis using the formal logic with predicate symbols, and description of story in text is the data to be analyzed.

Table 1. Outline of Experiment

	Experiment A **outcome: presentation slide**			**Experiment B** **outcome: written text**	
	Group 1	**Group 2**	**Group 3**	**Group 4**	**Group 5**
1st round **without** external input	Eq.(4) & logical connection	Eq.(4)		Description of story in text	
2nd round **with** external input					

Based on the methodology described above, the author conducted the two types of Experiments A and B in classes with fashioned four brands, such as Facebook, Starbucks, TripAdvisor, and Ikea, to attract and motivate participants as shown in Figure 3. Students as testees were asked to present five key words relevant to respective brand in consideration of their familiarity of word. To be fair on the number of word cards between clusters, each cluster should have the same number of word cards, resulting to 5. Keywords are listed below and shown on the screen at the beginning of the experiment. Then subjects were requested to think of new products or services in line with storytelling by dragging the word cards so as to be recombined on the screen. Three groups out of five participated in Experiment A, and the rest took part in experiment B.

Capital letter describes concepts representing Cluster.

· TRIP ADVISOR: discount ticket, foreign tour, backpacker, guide, and word of mouth
· FACEBOOK: search, friends, like!, share, and network
· IKEA LIFE: living room, do it yourself, easy to store, Europe, and wardrobe
· STABA LIFE: JPY400, third place, extra job, coffee, and steamer.

The first round had no external input, and in the second round, "Samsung" was given as an external input. Samsung is a world-famous South Korean company, specializing in smartphones and consumer

Figure 3. The initial ideas of combined words which are given for the experiment

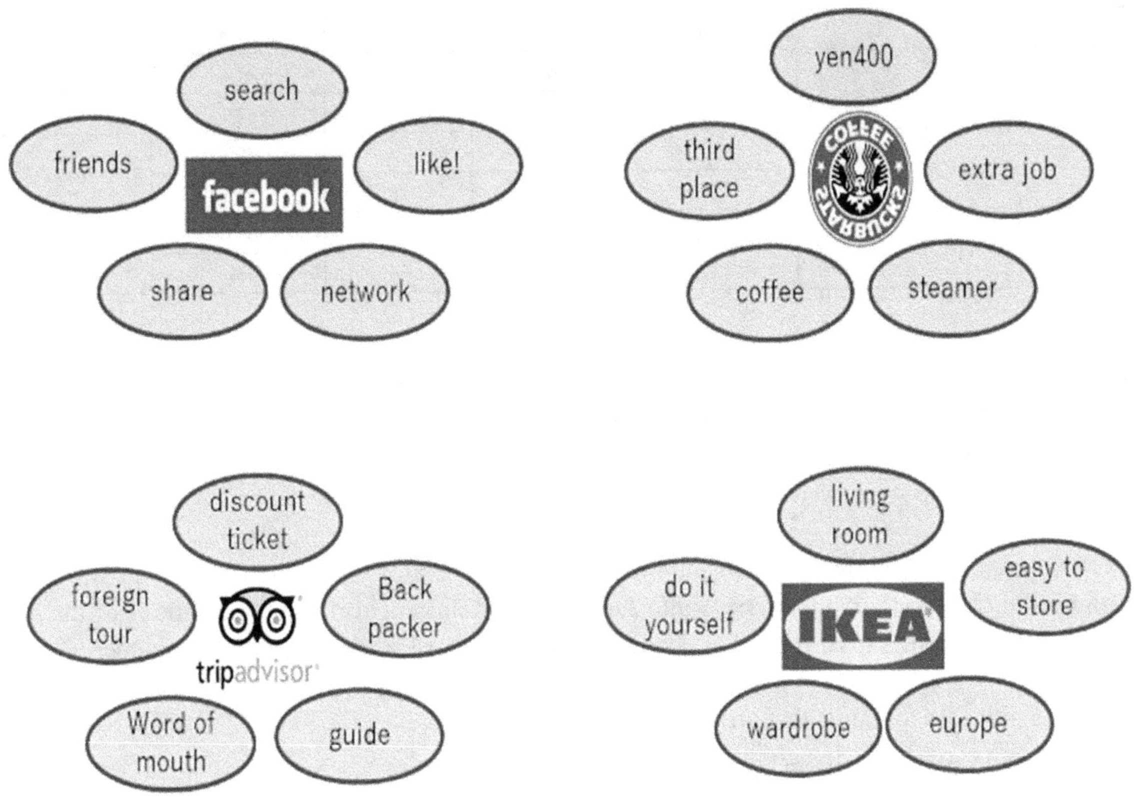

electronics. The focus of this chapter was on how external input affected the outcome, so any brands or manufacturers work for this experiment. However, experimental subjects needed to be familiar with the brand and can associate with the image of it, which is why the well-known company was appropriate. It is obvious that Samsung should not be a mandatory brand but APPLE can be possibly be used instead. This chapter, however, does not try to compare the outcome yielded by the type and/or content of external input itself which might entail an unlimited work of comparison, while the author explores here the effect in comparing the stories created with or without the external input. Also, Experiment B clarified whether the outcome was a product (s) or service (s) as well as its characteristics, by analyzing the storytelling text.

Let us start with the score as a result of quantitative analysis for Experiment A (presentation slide by Group 1 to 3) based in line with Eq.(4), continuing the semantic analysis on the generated story afterward. Table 2 shows the results of the calculation by the method Eq.(4). Every group indicated a higher score of Eq.(4) in the second round than the first one.

Let us discuss the Group 1 scenario in which the impact of the external input on the richness of knowledge and/or memory was the highest (115%). Firstly, the author compared the hypothesis reasoning of each output, with and without external activate factor. Clusters made without external input in the first round were as follows.

· SIM Card: discount ticket, network, do it yourself, and JPY400

Table 2. Result of calculation according to Eq.(4)

Group	Round	k	k-m	Eq.(4)	2nd/1st
Group 1	1st	47	22	47%	115%
	2nd	54	29	54%	
Group 2	1st	50	25	50%	106%
	2nd	53	28	53%	
Group 3	1st	55	30	55%	106%
	2nd	59	34	58%	

· SHARE HOUSE: backpacker, share, living room, and third place
· CAMPING: foreign tour, friends, Europe, and coffee
· SNS: word of mouth, guide, like!, and extra job
· IoT: search, easy to store, wardrobe, and steamer

Out of the clusters above, supplemental explanations came along with the first cluster was saying "Selling Sim Card to tourist that arrive into Japan." Other clusters also had explanations of the same volume. When it comes to SIM Card, it could be formally described in the following manner.

P_1: Selling sim card to tourist that arrive into Japan
Set = {r1: sim card→to be bought in Japan}
Fact = {sim card→discount ticket, network, do it yourself, JPY400, Using sim card in Japan → being able to communicate}
***Hypothesis*:** Tourist in Japan needs for SIM Card to communicate

Consequently, F U {Σ*r*, *h*} expects revenue of sim card, which means that we take hypothesis *h*, where selling sim card to tourist is satisfied with rule r1, the given fact, and hypothesis *h*.

As noted, hypothetical reasoning which explains relations of SIM Card as one of the clusters, and a word "tourist," were used. Nevertheless, four keywords composing a SIM Card cluster, discount ticket, network, do it yourself, and JPY400, were all just factors explaining sim card itself. In other words, additional 'logical rule' was not detected and the depth of attribute was still at low level.

Other clusters, SHARE HOUSE, CAMPING, SMS, and IoT, also indicated a similar tendency in logic as SIM CARD, which means that cards composing a cluster were less likely to be interpreted as additional "logical rule." Also, storytelling connecting clusters is simple, as shown below.

$$(\text{enjoy(camping)} \wedge \text{stay (shared house)}) \rightarrow (\text{needs for (SNS, sim card, IoT)}) \quad (5)$$

Like this, all the keywords were replaced with variable symbols. Moreover, a logical rule was evoked by connecting each cluster.

Next, the same subjects took part in another round. Subjects were provided with the keyword Samsung as an external activate factor in the second round, results of which are demonstrated in Figure 4. Each storytelling related to each cluster had much more new expressions than the first round. It is possible to interpret a new story as "Samsung washing machine makes customers" lifestyle active. On days off,

Figure 4. Presentation slide prepared by Group 1 for the Experiment A

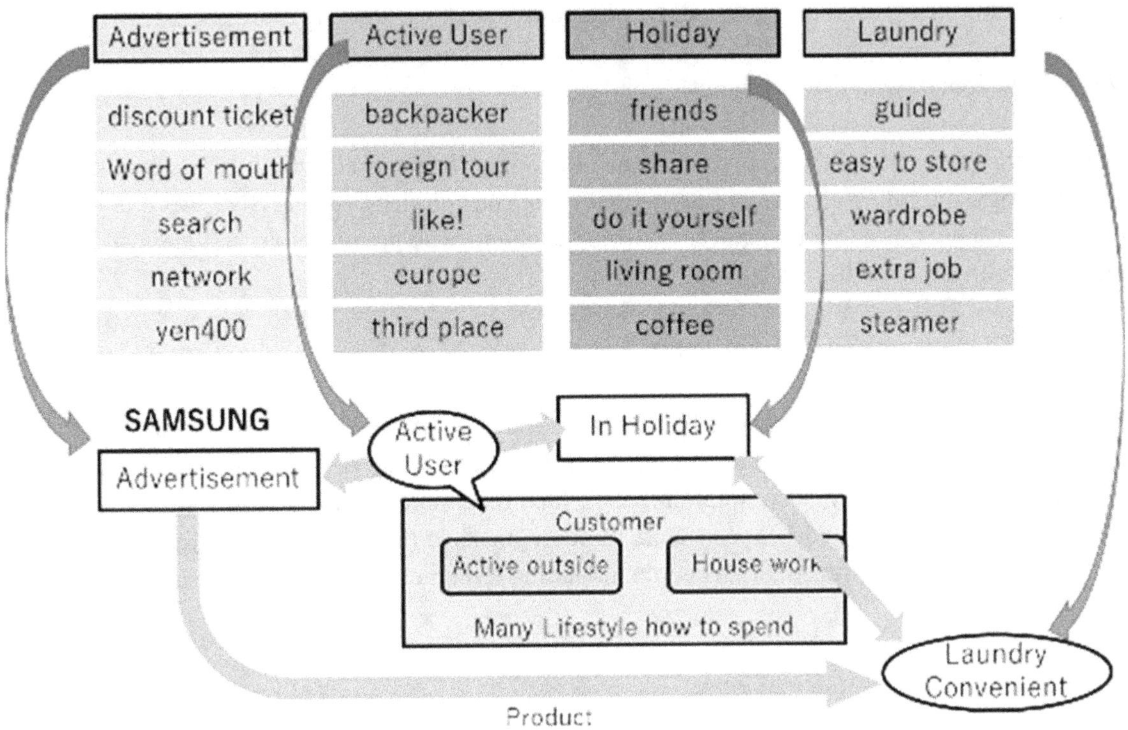

lifestyle can be both outdoors and house chores, while using Samsung washing machine, that makes users feel active. This interpretation can be formalized in the formal logic with predicate symbols.

$\forall x$ (human(x) $\wedge$ buy(laundry)) $\rightarrow$ (feel(active, holiday) $\wedge$ human(x) $\wedge$ (spend(place(outdoor $\vee$ house), lifestyle))) (6)

While (5) is consisting of created clusters where the given cards are combined with, (6) is consisting of created words like variables, e.g., active, lifestyle, house, and so on, that are not shown on the screen supporting created clusters where the given cards are combined with. Therefore, a comparison of (5) and (6) makes it clear that the external activate factor enhances the depths of the attribute. In other words, context becomes more complicated.

Now disclose the outcome of the experiment B. The only difference between the experiment A and B was the way of delivering stories. Group was asked to prepare slides for presentations in the experiment A, while people in the experiment B was supposed to write brief sentences. The stories of Group 4 and 5 followed:

Group 4: 1st round (without external input)

Our service provides a co-working space for freelancers. We provide a nice facility to enhance freelancer's productivity. Using our service, any freelancer can find a new friend and form a network to share a new job or help each other. Also, by using that network, they can organize their business trip much easier. In addition, a special discount will be offered users who share our service on SNS.

Group 4: 2nd round (with external input)

Our service is a smartphone system for general people. Clients can customize their own rooms with our technology and IoT. We focus on Europe to promote our service by using SNS, word of mouth, etc. In addition, we provide a lottery game as a gift to the clients who patronize our service. All of them get a chance to win a foreign tour as a prize.

Group 5: 1st round (without external input)

The speed dating service is going to be held in a coffee shop at Roppongi. The participants will have the opportunity to talk with possible partners for a while and enjoy their time. They also can borrow costumes in the shop so they can cosplay and make their time more fun and also get JPY400 of discount. After the official event, the couples can share more moments in the common space to get closer. These events are going to be promoted for all people, including foreigners and travelers, as a good opportunity to make friends!

Group 5: 2nd round (with external input)

Samsung is suitable for your life. The new screen technology suits your daily life routine and more. You can enjoy watching a movie in your house with a delicious coffee or go to the cinema and get JPY400 discount for Samsung users. If you are traveling, you can use our smartphone to see a show or find your way through the city. Samsung is always innovating, and now we are going to have a new store in France, Europe, please come to see our new products!

As a matter of course, the nature of the outcome changes with the way of giving external activate factor, from presentation to written text. However, results show that stories in both groups in the experiment B changed completely from without the external input to with the input. More interestingly, since stories were short and complete, it seemed that the external input itself had no effect on the nature of the outcome. One of the reasons for this might be based on the result of richness of expression by means of Eq.(4), i.e., the first round score (28%) is higher score than in the second round (18%) in Group 4, while that in the 2snd round (21%) is higher than in the 1st round (17%) in Group 5. Therefore, it is obvious that the outcome was not affected by the external input.

On the other hand, the characteristics of the contents of the external input affected the nature of the outcome. The second round used Samsung, a company that manufactures "goods" for home use, such as appliances and smartphones. Both groups relate *people* with service. That means, users mediate services and products, and products used is not limited to devices like CPU. On contrary, since the first round expanded the story without any input, it expressed "Service" narrating the story about *people* rather than *goods*.

With regard to service(s), Lush and Vargo (2014) advocate the Service-Dominant Logic that defines service(s) as follows:

Service(s)

Service-dominant logic uses the singular term "service" to reflect the process of doing something beneficial for and in conjunction with another actor, rather than units of output—immaterial goods—as the plural "services" implies. This distinction is important because it leads to the conclusion that there are no "services" in Service dominant logic. In-Service dominant logic, goods, and service are not alternative forms of products. Service represents the general and universal case, the common denominator, of the exchange process; service is what is always exchanged. Goods and money, when employed, aid the service process.

Tis difference between "service" and "services" to the results and see how storytelling was deployed is applied herein. When no external input was given, both groups developed a story preceding service. Usually, it was about people: the productivity of people and the sharing time about people, for example. In Group 4, the interesting point was that those who "can find a new friend," "make network to share new job," and "organize their business trip" were not business owners of service but participants of the beneficiary. Subjects in Group 5 was participants of the beneficiary, not the business owners, so that "participants have an opportunity to talk with possible partners," "can borrow costumes," "can cosplay and make their time more fun," "get JPY400 discount," and "can share more moments."

On the other hand, for Samsung, Group 4 mentioned services that connected the devices in the room with Samsung's technology device through IoT. Further, a possible marketing campaign was developed, in which people could win overseas air tickets by lottery. Group 5 presented a proposal that includes services such as watching a movie on a Samsung TV screen or distributing a JPY400 discount coupon to Samsung users. They also suggested that Samsung phones could be used to watch shows and help find directions when traveling, both of which bore a relation to services associated with Samsung products. Moreover, this storytelling presentation ended with the phrase, "Please come visit our new store opened in France." Both groups produced the impression that the participants associated storytelling for services as a function related to products.

A summary of Experiment B is presented in Figure 5.

As noted above, it was found that the nature of the output differed dramatically, depending on how external input was given. In particular, if no instructions were given to convey storytelling, the story produced was shorter but nevertheless complete, such that there was an influence from external input on ones' action. In addition, when the external input was content related to Samsung products, the presence or absence of external input was impacted. The outcome shifted to either process related to people or services relevant to goods.

After Experiment B, I asked participants which word cards were challenging to handle. Group 5 responded that, in the first round, the expression "easy to store" was difficult to integrate into clusters because the business focuses on *people* and not on *goods*. The word "wardrobe" also did not fit, for the same reason, but they said, "we had to fit it into the speed dating concept." Another expression that did not fit, "extra job," was interpreted to present it as relating to employee advantage. As the story was developed, the leftover word cards were translated to complete the ends and maintain story consistency. In the second round, Group 4 also struggled to associate the word cards "easy to store" and "wardrobe." They commented that "easy to store/wardrobe—our concept for the Samsung international market was based on product use, so we had to fit this word into the opening of a new project in Europe that would require storage and a wardrobe."

Figure 5. Summary of Experiment B (first round shown on left and second round shown on right) for Groups 4 and 5

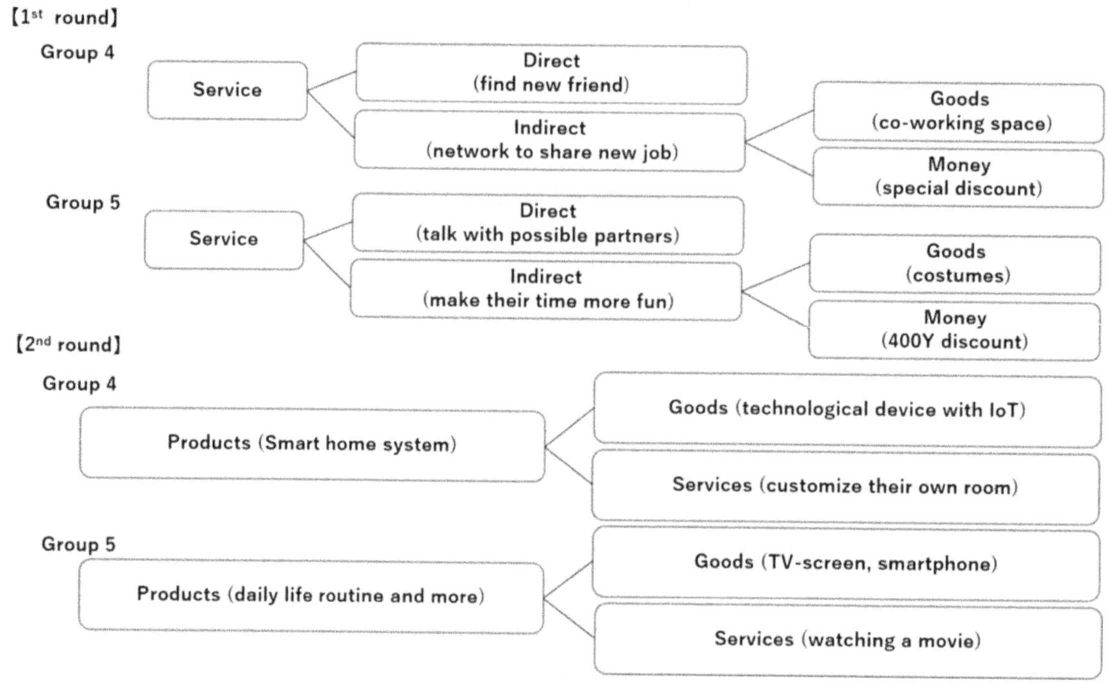

The students gave much feedback to the author on the experiment called the analogy game in the class, including the following: "It was my first time to take part in an experiment like this, and it went beyond my imagination. It is so helpful to generate a business idea without using a framework. During the experiment, I noticed that my brain was more active than normal. It was so interesting."

EXTERNAL INPUT WITH AN ARCHITECTURAL INFORMATION THROUGH KOTOFACTORY

The AG used in the activities described in the previous section is a tool for combining word cards. The Kotofactory (KF) introduced in this section is another tool for combining the same word cards. However, the word cards in AG were given clustered into groups, whereas in KF, two cards each for goods and services were divided into frames, based on architectural constitution. Then, word cards are given to each of these structures. The reason for focusing on frames is that although AG promotes flexible storytelling without a given frame, frames could enhance thinking.

Zackman's framework describes a taxonomy for a fundamental structure in enterprise architecture, which provides a formal and structured way of viewing and defining an enterprise (Zackman, 1987). With reference to Zackman's framework, this study presents an empirical environment, as shown in Figure 6. The expected outcome in this case is not strictly defined, but new ideas of storytelling with hard or soft products are the target output. Hard products are understood to be physical objects that can be touched, such as a building, printer, or bicycle. Soft products refer to actions, events, or services, regardless of the use of the hardware (or software) that are needed to play games or participate in any other experience.

This is the background to the idea of KF. When the author worked for an automaker, engines had multiple sensors, and sensor technology was understood to have come a long way. However, the author has noticed that certain technologies remained in the positions in the same industry, and automobile engineers did not consider applying technology to completely novel sectors, such as gaming or other parts of the entertainment industry. In other words, interdisciplinary usage of technologies were not envisioned. For instance, Nintendo's Wii also uses sensor technology, so it was certainly possible for games that incorporated sensor technologies similar to the best-selling Wii-Fit might have appeared. For this reason, the data for the product on the left in Figure 6 (for example, wearable devices) are equivalent to data relating to products belonging to one's own company. The products on the right exhibit data for irrelevant products. All of the data for the components of the products on the right are interpreted as external inputs in this section.

In Eqs. (1) and (2), the categorization of words with KF ($\alpha\mu$, $\mu = 1, 2, \ldots, 20$) can be placed by ($\alpha\mu$, $\mu \in (L \vee R)$), where L represents the set of words on the left and R represents the set of words on the right. New concepts ($q\lambda$, $0 \leq \lambda < 5$) can be placed by ($q\lambda$, $0 \leq \lambda$), which shows more flexibility in making clusters than AG.

In KF, two products or services are shown on the screen and broken down into a group of components, such as usage, application, technology, and other relevant issues. Participants attempt to create ideas of

Figure 6. Screen image of Kotofactory with captioned guide

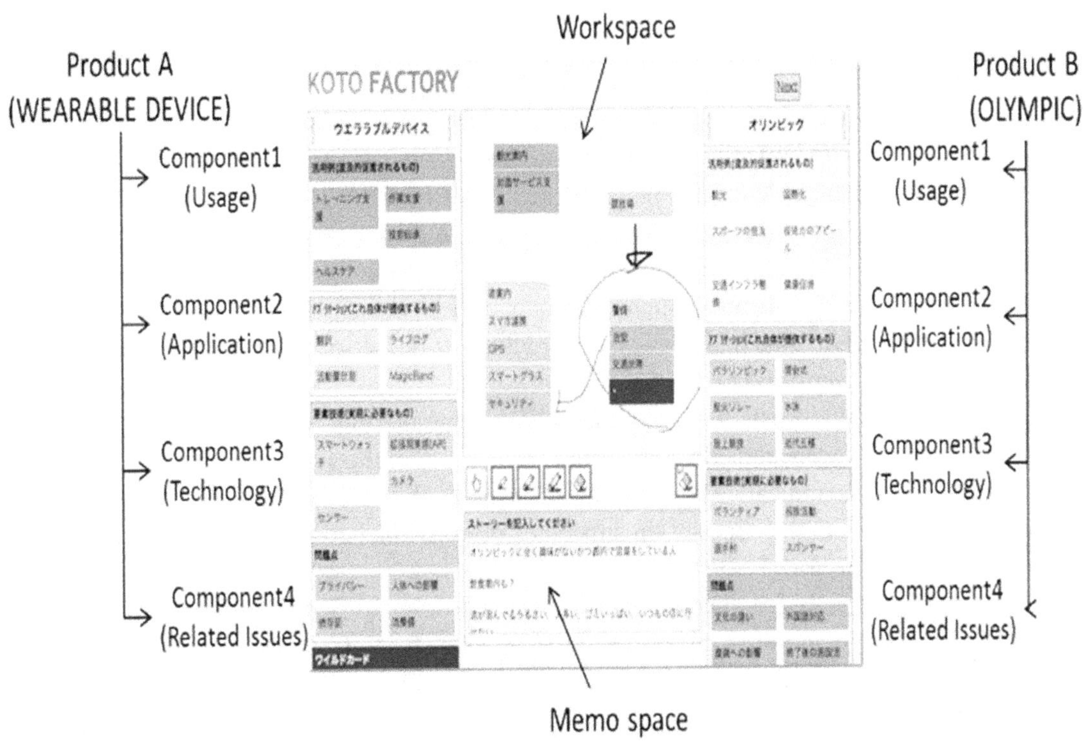

new things by combining or linking elements within components written on cards, using the rule of the appropriate hypothesis (Nakamura et al., 2012).

The previous findings are as follows. First, hard products are more often generated by participants who pay closer attention to usage at the beginning compared than by those who pay less attention. The latter group tend to generate soft product ideas. Second, by developing various types of synthesis of components populated by frames, the data itself do not directly support the generation of new ideas. Instead, it leverages cognition of ideas by linking elements with different components (Nakamura & Teramoto, 2013). Third, by formalizing generated ideas, an abductive inference model was developed to create an essential hypothesis and rules for product planning, and these act as clues for the design of a data market (Nakamura, 2014b). Abduction, which is a heuristic measure that represents chance discovery in formal logical research (Abe, 2003), was given evaluation criteria by Peirce for its outcomes, and these define plausibility, verifiability, simplicity, and economy (Yonemori, 2007). Note that these criteria are almost all resultant indicators for performance evaluation. For example, plausibility is considered to be a resultant indicator that is determined with questionnaires after an event, e.g., how plausible do you consider the idea, from 0 (not plausible) to 5 (entirely plausible)?

The author explored capabilities that required criteria for assessment of an ability to perform action, which entailed the precedent indicator as a reference. Such actions may include selection, motion, or clustering of cards. Another means of exploring performed capabilities is how a text was structured when expressing what you are aware of. The analytical method is demonstrated as follows.

Note that both inputs and outputs from/to KF were only unstructured data. That is, cards with nouns were displayed on the screen, as in AG. At the end of the experiment, the system provided two types of output. The first was a cluster of cards that signified new storytelling ideas. For this purpose, participants dragged cards from products A and B. These cards were combined according to the participants' own interpretation of new ideas of hard or soft products. Between the clusters, lines with arrows could be drawn to represent connections in the workplace. The use of a line signified a causal relation, purpose, or other relationship, indicating how participants considered the structure of the clusters. Another output was that the participants were asked to prepare a presentation by responding to questions about generated ideas that considered WHO, WHY, and WHAT to output an initial structured framework (Zackman, 1987). Upon completion of the experiment, the descriptions of the WHO, WHY, and WHAT content were transformed into a first-order predicate logic expression prior to analysis.

Analysis incorporated the following three steps. In step 1, the two types of output were evaluated by a third party, using resultant indicators, such as uniqueness and utility. High-scoring outputs were then compared with other outputs that had a lower rating, in terms of the semantic and formal perspectives. These perspectives are discussed later in this section.

Step 2 was a semantic analysis, conducted as follows. The number of individual constants and the length of the formal logic were considered in terms of the depth of the attributes. WHO was regarded as a target segment. Then, the indicators were examined in terms of how deeply participants considered the target segment. To calculate the depth of the attributes of individual constants, an expression of WHO can be formalized as in Eq. (4).

Step 3 was a formal analysis. The output was analyzed in terms of the level of concentration and the richness of knowledge and/or memory. The level of concentration was calculated using to Eq. (7), where p represents the number of cards that were clustered on the screen, and q is the number of cards consisting of clusters that were frequently moved during the Experiment ($q > 0$). γ is an invariable (0 or 1) dependent on α, which is a threshold relative to the frequency according to which cards are moved or placed. Here, γ is 1 for $\alpha \geq 5$ and 0 for $\alpha < 5$. Cards were, in most cases, moved about 10 times per experiment; thus, α was generally 5. As the percentage of frequently moved cards clusters increased in relation to the number of clustered cards, more of the same cards were moved in the workspace in the context of concentration.

$$Conc_{\mu} = \Sigma\Sigma\gamma q\mu_{l} p\mu \quad (7)$$

The level of richness of knowledge and/or memory was defined as shown in Eq. (4) for the AG experiments. The cards that were shown were set on the screen in advance, before discussion began among the participating students. As a result of this discussion, it was found that product A was a Wearable device, and product B was a device needed for the Olympic Games. The products were composed of four components: Usage for deployment, Application as an offering, Technology as an enabler, and related issues. Each component had six cards with relevant nouns. An experiment was conducted with 31 student participants from Tokyo Metropolitan University, who were arranged into eight groups (an average of four per group). The experiment lasted 90 minutes. During this time, the participants considered a synthesis of components from products A and B and combined cards by dragging them and drawing lines or diagrams to rearrange the clusters in the workspace. These tasks were performed on a PC display, and the following task instructions were given on a PowerPoint slide:

- describe WHO as a target segment or target customer
- describe WHY as only a background or differentiator
- describe several WHATs relative to hard or soft products in relation to the fixed WHO and WHY

After the experiment, each group gave a presentation, which included the final screen image and one PowerPoint slide, explaining WHO, WHY, and several WHATs.

An evaluation followed, conducted by a vote taken after all the presentations, including consideration of the resultant indicators (uniqueness and utility). The scores used to rank the eight groups were ranged from 1 (not significant) to 8 (significant) by 31 student participants. They have scored all groups respectively expect his/her own group.

Although WHY questions are the most important, WHO questions include criteria for considering stories in relation to personas (Cooper, 2004). Here, the persona must provide a sufficient level of detail for discussing WHY, as is relevant to a target segment or a customer segment. Repeatedly asking WHO and WHY together allowed the participants to assemble several WHAT stories. The author believes that determining WHO and WHY helped the participants generate WHATs without incorporating meta knowledge.

The results of the experiments are presented below in steps. Step 1 involved an evaluation the presentations. The eight presentations received scores ranging between 86 and 176 points. The two groups that received the highest evaluation were Group 2, with 176 points, and Group 5, with 147. The lowest scores were given to Group 8, with 114 points, and Group 7, with 86. These four were examined to determine these groups' differences in capability for focusing on WHO and WHY, as these questions are critical for generating WHATs.

Step 2 was a semantic analysis, referencing the output of WHO. We transformed natural language into a formal logical representation, calculating the number of terms (n), which represented the depth of attributes. The groups with the highest scores showed the following.

Group 2 (n = 4): human (foreigner) ∧ visits (Japan) ∧ work-for (Olympic Games), no-good-at (Japanese)
Group 5 (n = 7): human (students, engineering) ∧ stay (Tokyo) ∧ want (friends) ∧ attend (family home, university, one hour)

The two groups that received a low rating included only verbs and objects, which made it difficult to understand who the target segment was. Their results are given as follows:

Group 7 (n = 1): human (torch-runner)
Group 8 (n = 2): depend (smartphone) ∧ deploy (sports)

As is clear, a difference appears in the number of terms and the depth of attributes between the sets of groups. Table 3 presents the list of evaluation score and the average of *n* for all the groups. It is natural that the evaluation results increase, considering the definition of a target segment deepens.

Table 3. Average n per evaluation ranking cluster

	Evaluation score	*n*	Ranking cluster	Average *n* per cluster
G2	176	4	TOP2	5.5
G5	147	7		
G4	143	5	MID4	3.5
G1	134	3		
G3	129	4		
G6	124	2		
G8	114	1	LOW2	1.5
G7	86	2		

Note: *n:* number of terms, G: group.

Here, we examine the logical composition for WHY, in terms of appropriate levels of satisfaction. Groups that were more highly rated groups more proficiently incorporated predicate symbols and logical connections, and they also used more terms.

The high-scoring groups produced the following:

Group 2: (human (Japanese) no-good-at (English)) → (need-for (environment) ∧ communicate-with (foreigner))
Group 5: (human (students) ∧ want (friends)) → (difficult-to-visit (students, Olympic Games)) → (introduce (students, targets) ∧ search (friends))

The low-scoring groups produced the following:

Group 8: experience (sports) ∧ same eye-line-level of (player)
Group 7: promote (health-care) ∧ increase (population) ∧ activate (environment)

Figure 7. Results of formal analysis

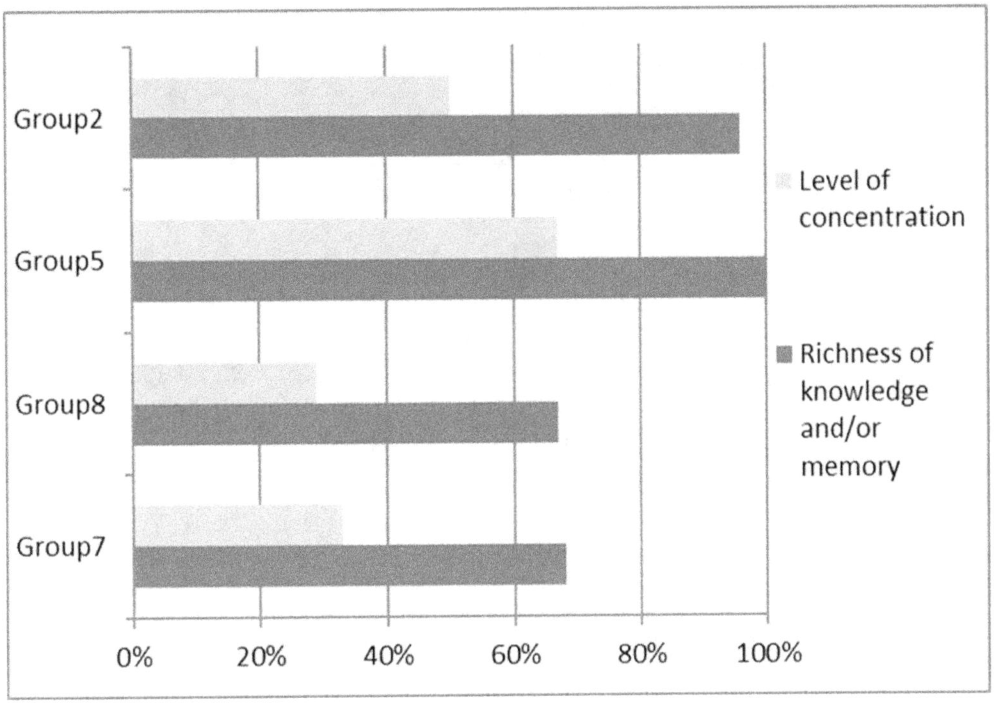

Thus, in relation to the level of satisfaction, the high-scoring groups created logical representations that indicated connectivity more frequently than the low-scoring groups. Nevertheless, this can be formalized without logical connections with the use of conjunctions.

Step 3 was a formal analysis. Figure 7 shows the results for level of concentration and richness of knowledge and/or memory.

The difference between the high-performing groups and the low-performing ones is obvious for both levels of concentration (t = 3.14929, df = 3, p < .10) and richness of knowledge and/or memory (t = 14.79467, df = 3, p < .05), i.e., the former groups had results of greater than 50% for level of concentration, and the latter obtained less than 25%. This suggests that the former groups tended to stick to specific cards, but the latter groups considered multiple, wide-ranging cards.

The former groups made more use of knowledge and/or memory within the context of ideas, beyond in addition to the given card data, than the groups that received lower evaluations.

The average percentage value for these two factors approximately correlates with the results of the evaluation (coefficient of correlation value: 0.607) for each of the eight groups. If the two factors are seen to be into precedent indicators and can guide participants to more easily understand how to perform actions as reference, the hypothesis given in section 3.1 would be supported. Guidance in relation to the two factors will be given to participants in the course of further experiments.

Now, let us consider the implications from the results. The results indicated three capability factors in merging data with awareness. First, capabilities in concept design merge data through a wide range of aspects relative to horizontal and vertical awareness. The more deeply you express or represent what your ideas are, the greater capability you have. This implication might be important, and it is supported with experimental evidence.

Figure 8. Possible learning process as relates to information richness and learner concentration, drawn from groups with higher and lower evaluations

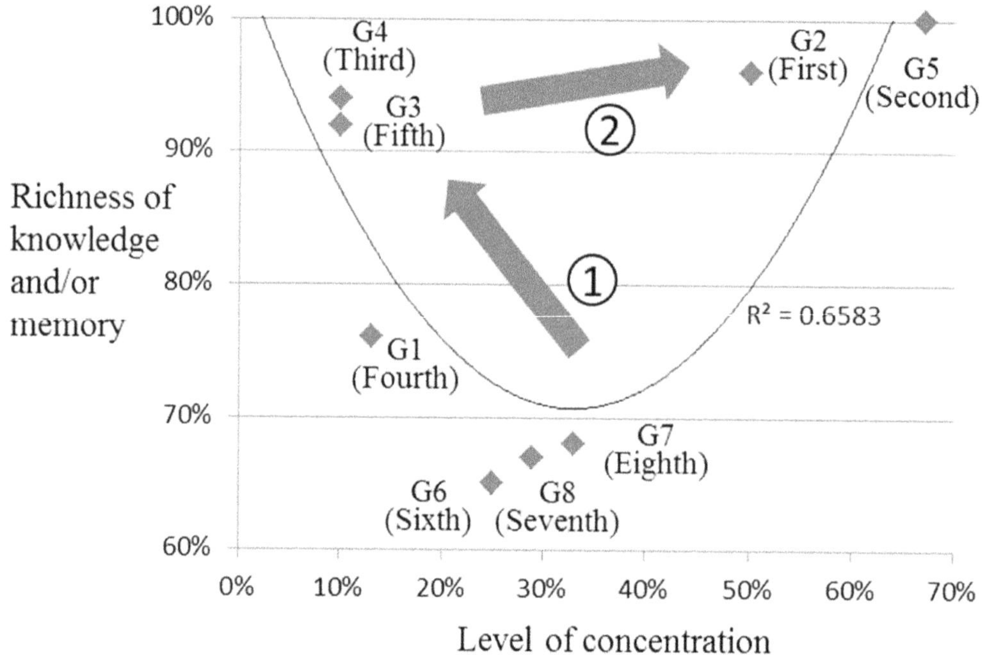

Second, in any explanation of WHY as a background or a differentiator, it is important to consider logical connections. This is because such connections imply scenarios that help understand what specific reasons are given.

Third, capabilities in concept design embody the ability to concentrate on specific data in addition to knowledge and/or memory. In Figure 8, these variables are represented by the x-axis and y-axis, respectively.

The groups with lower evaluations would be expected to enhance their capabilities by repeating the KF play in various cases. They could perform better, scoring better on the x–y axis, as shown in Figure 3. One possible process here would be to focus on the data first, and then the external factors in knowledge and/or memory can be inferred with the data in place (see ① in Figure 8). Certain specific cards are examined by rearranging the cards as a whole to create new clusters that were not anticipated at first. This eventually results in improving the level of concentration (see ② in Figure 8). The result appears to be a double helix representing a chance discovery cycle (Ohsawa & Nara, 2002), as the shape of the curves would be similar. However, the double helical chance discovery process is a model of one's thinking process as trial and error, while the finding from KF here could be a sort of learning process from lower score group to higher score one.

The externalization of knowledge and/or memory takes place through a truth maintenance system (TMS), which is a knowledge representation method for beliefs and their dependencies (Doyle, 1979), in the form of an inference engine that can explore alternatives, make choices, and examine the consequences of a choice in a given search space (Stanojevic et al., 1994). This study adopts a rather static approach; however, a dynamic approach, such as a TMS may be included in future research.

Capabilities in both semantic and formal perspectives create an essential hypothesis and rules of abduction (Nakamura, 2014b), as well as implying performance indicators that consist of the depth of attributes, level of satisfaction, concentration, and richness of knowledge and/or memory. These indicators may be associated with either precedent or resultant indicators, depending on whether such indicators help participants consider how to generate a concept design from scratch with only the given clues. For example, a concept design could be considered as deeply as possible in terms of the individual constant or variable attributes. Again, it would be possible to be guided to consider contextual stories that are plausible and logical by focusing on the external activation factor, along with your knowledge and/or memory, as much as possible. Again, this should be tested by further experiments.

DISCUSSION

Following the results of experiments conducted with AG, the effects of external input were discussed in terms of how it impacts the creating story. The effect of external input was different between Experiment A (presentation slide) and Experiment B (description of text) in terms of the use of AG. Following Eq. (4), which gives a calculation of the level of richness of knowledge and/or memory, it is obvious that the external input sufficient for creating storytelling in terms of a presentation slide, as shown in Table 2, while the outcome incorporates a unique feature relating to people through a service by writing brief sentences, regardless of whether there is any external input. The absence of external input leads to a mental picture oriented more toward service, while external input results in the imagination of people with a service that requires goods. It assumes that the outcome (service with goods) is influenced by Samsung as an external input, which is mainly goods manufacture. This could be changed to process-oriented services in storytelling, depending on the type of external input, such as the consulting company, which may be a non-manufacturer, in place of the manufacturing company like Samsung. The appropriate type of expression (slide or text) depends on the individual environmental status, but because an effect of external input was a clue in terms of the presentation slide. This implies that the expression through slides implies a higher degree of freedom in the creation of story in one's mind than is involved in writing a story.

The comparison of AG to KF shows that the level of constraint is clearly different, in terms of frame. KF is required to exhibit more constraint than AG, as KF introduces a frame that is used in both inputs (architectural components) and outputs (WHO, WHY, and WHAT). In this sense, AG can be more useful for generating new ideas from scratch relative to KF, which can be used to design more concrete storytelling, following after the use of AG. It should be noted, however, that both AG in terms of presentation slide and KF have an impact on the effect of external input, in relation to the level of richness and/or memory, because of the higher degree of freedom in the process of creating a story in one's mind, as previously discussed.

Throughout all of the experiments, the author found that a conceptual metaphor (Lakoff and Johnson, 1980) has been yielded in the process of creating story. In other words, the process of extraction of experimental results could be interpreted as a process of the construction of conceptual metaphors. For example, a subject constructed a short component of the context with a new cluster, as follows, "The speed dating service is going to be held in a coffee shop at Roppongi," with Group 5 at the first round in Experiment B, on AG. This reflected the construction of the conceptual metaphor. The execution of the conceptual blending (Fauconnier and Turner, 1998) could be used here as well to reflect the cognitive process of the blending the parts of conceptual metaphorical structures in mental space, as relates to the given clusters on the screen for both AG and KF.

For better understanding, let us examine a similar business case from the real world. The author conducted an interview with the head of IBM Research Japan and asked, "As a Head of Research, it is assumed that you do not conduct research yourself. Where do you see your value contribution at work?" He responded, "It is important to give some direction to researchers, and that is my responsibility. They tend to have biases on their own topics and are likely to get wrapped up in them. However, this will never bring about anything innovative. As head of the center, therefore, I provide the researchers with external stimulation. In that sense, my job is heuristic." I asked, "What type of information is your external input?" and learned that it should be as close to the market and as hot as possible. For example, no more than 20 years ago, when text mining did not yet exist, the head instructed the R&D team to develop a beer research target. The R&D team was looking for something loose and quick, and beer was gaining popularity in Japan at the time. The team focused on the catchy slogan of Asahi Breweries, "dry." They then used data from customers' telephone inquiries regarding competing companies and strove to find timely and valuable keywords like "dry." This series of attempts resulted in building the world's first text mining application. It is worth noting here that the beer itself was a product, but as a result of shedding light on the voice of customers and regarding it as an external input, the research's own output as the software and related services.

Here is another example. Suppose you are the business manager of a liqueur manufacturer. Your company's sales have been declining. What would you do first? What idea comes to your mind? You are more likely to think of straightforward sales approaches, such as pushing discount campaigns at stores, for instance. What if, in this case, you paid a visit to a factory that is not under your supervision. There, you see people on the assembly line fitting caps on liquor bottles. What is on your mind? The caps themselves are supplementary, external parts of the manufacture of the liqueur, not the liqueur itself. Could you view this in a storyline, where it is a scene in which you open the bottle at midnight by yourself? It can be after work, relaxing by yourself. Then you come up with a new strategy offering quality time to customers, not just a liquor. This strategy focuses more on the best time and space to make people happy and relaxed, rather than on traditional discount coupons. This campaign becomes the company's largest success. This might be your happy ending scenario.

This storyline was triggered by two external inputs. One was the opportunity to visit a manufacturing location that the subject of the story is not responsible for. Second, the manager pays attention to a simple cap, not the original liqueur. Thus, if a person remains closed in within the already present world, there is no going beyond understanding already present thoughts in a narrow way. Expanding the imagination and gain another perspective becomes possible. Innovative ideas often come from external stimulus.

Due to the embedded biases in the real world, it is never easy to overcome existing concepts. However, tools and external stimulus can help people go beyond their previously existing imagination and to think of unexpected narrative stories that can be applied to creating books, films, and television. Analogical thinking skill and the sense of metaphor should be made the most of.

At the same time, external stimulus has some shortcomings. External input might limit a person's ability to imagine freely and develop creatively because the content of the input has effects. Besides, it is nearly impossible to completely control the amount of information and bias built up with one's experiences. People are placed in different surroundings, so that the information they are exposed to differs from one to another. To exercise a jump of imagination (going beyond ordinary imagination), unpredictability, novelty, and uniqueness are needed, and unrestricted environment and some constraints should be ensured in the creation of a narrative. Providing external inputs is one way to give constraints to experimental subjects.

CONCLUSION

This paper executed a specific pattern in the effects of external input in the process of creating new products or services. AG helps create story by re-combining words from non-framed components, but KF creates story by re-combining words from framed components to the framed outcomes in text (WHO, WHAT, and WHY). The effects of external input were found in both experiments, but the particular features of the effects vary depending on the experimental design. This implies that the experimental design may be essential for practical counseling and consulting to generate new ideas of product or service in real business. This depends on the level of readiness, project phase, and problem setting. The number of subjects was, however, limited in this section, so that it is necessary to conduct repeated observations in further experiments. Further, what type of external input should be given was sufficiently explored.

As noted, if defining activities giving an external input as a heuristic, it would be challenging to identify factors. Social issues, including environmental and social welfare, have a wide psychological impact on people (Mead, 1934), meaning that questioning social issues could lead to a choice of external input. In other words, social issues might require deep insight and be dependent on interaction with society (Csikszentmihalyi & Sawyer, 1995), meaning that social problems like environmental issues could be an interpretation of external activation factors.

At any rate, to enhance the reproducibility of the factor effects, the author is willing to extract the characteristics of the external input itself other, beyond those of the manufacturers. To illustrate this, it was observed in the experiment that services were created with the concept people as mediation. However, the concept of people, in the context of hardware, such as Samsung, were only equated to retail sellers to end users. The use of external input that could shed more light on people in the sense of this concept may eventually enable an innovative story that will emerge and develop endlessly, overcoming established and embossed biases. This is a preliminary study of such a project.

ACKNOWLEDGMENT

The author thanks all the members of participants in the course of international marketing at Shibaura Institute of Technology in Tokyo, Japan, and Tokyo Metropolitan University, Japan, for their participation in the Experiments. This research was supported by the two grants from JSPS (JSPS KAKENHI Grant Number 17K03872 and JSPS KAKENHI Grant Number 19K03062). The authors would like to thank Enago (www.enago.jp) for the English language review.

REFERENCES

Abe, A. (2003). The role of abduction in chance discovery. *New Generation Computing*, *21*(1), 61–71. doi:10.1007/BF03042326

Asch, S. E. (1951). Effects of group pressure on the modification and distortion of judgments. In H. Guetzknow (Ed.), *Groups, Leadership and Men* (pp. 177–199). Carnegie Press.

Bonnardel, N. (2000). Towards understanding and supporting creativity in design: Analogies in constrained cognitive environment. *Knowledge-Based Systems*, *13*(7-8), 505–513. doi:10.1016/S0950-7051(00)00067-8

Burstein, M. H. (1986). Concept formation by incremental analogical reasoning and debugging. *Machine Learning*, *2*, 351–369.

Cooper, A. (2004). *The inmates are running the asulum*. Sams Publishing.

Cosejo, D. G., Oesterreich, J., & Ohlsson, S. (2009). Re-categorization: Restructuring in categorization. In *Proceedings of the Annual Meeting of the Cognitive Science Society (CogSci 2009)*. Red Hook, NY: Curran Associates.

Costello, F. J., & Keane, M. T. (2000). Efficient creativity: Constraint-guided conceptual combination. *Cognitive Science*, *24*(2), 299–349. doi:10.120715516709cog2402_4

Csikszentmihalyi, H., & Keith Sawyer, K. (1995). Creative insight: The social dimension of a solitary moment. In R. J. Sternberg & J. E. Davidson (Eds.), *The nature of insight* (pp. 329–364). MIT Press.

Doyle, J. (1979). A truth maintenance system. *Artificial Intelligence*, *12*(3), 231–272. doi:10.1016/0004-3702(79)90008-0

Fauconnier, G., & Turner, M. B. (1998). Conceptual integration networks. *Cognitive Science*, *22*(2), 133–187. doi:10.120715516709cog2202_1

Finke, R. A., Ward, T. B., & Smith, S. M. (1990). *Creative cognition: Theory, research, and applications*. MIT Press.

Frederich, M. J., & Andrews, P. (2009). *Innovation passport: The IBM first-of-a-kind (FOAK) journey from research to realty*. IBM Press.

Gaver, W. W., Beaver, J., & Benford, S. (2003). Ambiguity as a resource for design. In *Proceedings of the SIGCHI Conference on Human Factors in Computing Systems (pp.* 233-240*)*, New York, NY: ACM.

Gick, M. L., & Holyoak, K. J. (1980). Analogical problem solving. *Cognitive Psychology*, *12*(3), 306–355. doi:10.1016/0010-0285(80)90013-4

Holyoak, K. J., & Thagard, P. (1995). *Mental leaps: Analogy in creative thought*. MIT Press.

Hori, K. (2007). *Creativity support, theory and application of AI approach toward creativity*. Ohmsha.

Klahr, D., & Dunbar, K. (1988). Dual space search during scientific reasoning. *Cognitive Science*, *12*(1), 1–48. doi:10.120715516709cog1201_1

Lakoff, G., & Johnson, M. (1980). *Metaphor we live by*. The University of Chicago Press.

Lusch, R. F., & Vargo, S. L. (2014). *Service dominant logic, premises, perspectives, possibilities*. Cambridge University Press.

Mead, G. H. (1934). *Mind, self, and society*. University of Chicago Press.

Nakamura, J. (2014a). Discovering design concept for immovable objects. *Procedia Computer Science, 35*, 954–960. doi:10.1016/j.procs.2014.08.175

Nakamura, J. (2014b). Creating essential hypothesis and rules in product planning: Introducing abductive inference model. In *Proceedings of International Conference on Data Mining Workshop (ICDM2014)* (pp. 471-476), New York: NY: IEEE. 10.1109/ICDMW.2014.63

Nakamura, J., & Ohsawa Y. (2008a). Effect of vacilliation on analogical thought process for concept creation. *Journal of Transdicsiplinary Federation of Science and technology*, 2(1), 40-48.

Nakamura, J., & Ohsawa, Y. (2008b). *Kouji ninchi kinoo wo hikidasu kashikagiho* (in Japanese). Paper presented at the 25th Annual conference of Japanese Cognitive Science Society, Kyotanabe, Kyoto.

Nakamura, J., & Ohsawa, Y. (2009). A shift of mind – introducing a concept creation model. *Information Sciences, 179*(11), 1639–1646. doi:10.1016/j.ins.2008.11.036

Nakamura, J., Takagi, S., & Ohsawa, Y. (2012). *Reframing method for deployment of business domain*. Paper presented at SIG-KST, Hiyoshi, Kanagawa.

Nakamura, J., & Teramoto, M. (2013). Frame as a clue to intention of data: Toward new product ideas with framed components. In *Proceedings of International Conference on Data Mining Workshop (ICDM2013)* (pp. 720-725). New York, NY: IEEE. 10.1109/ICDMW.2013.28

Ohsawa, Y., & Nara, Y. (2002). Modeling the process of chance discovery by chance discovery on double helix. In *Proceedings of AAAI Technical Report FS-01-01*. Palo Alto, CA: AAAI Press.

Sakamoto, K., Iwaoka, T., Ogawa, J., & Nakai, K. (2012). *Kentikukouseigaku, kentiku design no houhou*. Jikkyo Shuppan. (in Japanese)

Stanojevic, M., Vranes, S., & Velasevic, D. (1994). Using truth maintenance systems: A tutorial. *IEEE Expert: Intelligent Systems and Their Applications, 9*(6), 46–56. doi:10.1109/64.363270

Taura, T., Nagai, Y., & Tanaka, S. (2005). *Design space blending: A key for creative design*. Paper presented at the 15th International Conference on Engineering Design (ICED05), Melbourne, Australia.

Tsujii, J. (1987). *Chikishi no hyogen to riyoo*. Shokodo. (in Japanese)

Wilkenfeld, M. J., & Ward, T. B. (2001). Similarity and emergence in conceptual combination. *Journal of Memory and Language, 45*(1), 21–38. doi:10.1006/jmla.2000.2772

Yamashita, M. (2013). Innovation of next automobiles. *Journal of Society of Automotive Engineers of Japan, 67*(6), 12–17.

Yonemori, Y. (2007). *Abduction*. Keiso Shobo.

Zackman, J. A. (1987). A framework for information systems architecture. *IBM Systems Journal*, *26*(3), 276–292. doi:10.1147j.263.0276

KEY TERMS AND DEFINITIONS

Analogy: A way of thinking that brings new idea is characterized as a cognitive process from the given base words to the different concepts.

Creativity: A phenomenon that is formed as a brand new and unique idea is one of human's capability.

External Input: Some keyword or instruction that is given to the Experiments would be expected to be a clue to new ideas of storytelling.

Products and Services: Something visible object or invisible offering that bring benefit is expected to be useful for users.

ENDNOTE

[1] How to improve the multiplier included in this function is not covered in this chapter, and its approach is the next challenge for further study.

Chapter 12
Design of Narrative Creation in Innovation:
"Signature Story" and Two Types of Pivots

Akimitsu Hirota
Kindai University, Japan

ABSTRACT

Among brand stories that affect brand equity, the importance of signature stories has been pointed out in particular. Casio Computer Co.'s G-SHOCK enjoys strong brand equity. The first model has created a new market. The development of G-SHOCK went through two years of trial and error. The trial and error that can exist only in the development of a market-creating product exert a great impact on brand equity. Stories about the trial and error of the developer and marketer during the development of a market-creating product are called "innovation narratives" in this chapter. This study examines the process of creating an innovation narrative from the viewpoint of a dialog and identifies the existence of a chain of two types of pivots: open pivots and closed pivots.

INTRODUCTION

Documentaries about product development amaze and inspire many people. People are amazed and inspired by the activities of the developer and marketer that cannot be known from the product itself. Many of the activities that elicit these feelings are stories of trial and error. People wonder how and why a certain product has been developed, thinking that they could not have developed it themselves. Those involved in the development of a product have gone through an experience that others cannot imagine by merely looking at, or using, the product. Many people are intrigued by the stories of trial and error of the developer and marketer behind a successful product. For this reason, companies have recently been providing information regarding the experiences of developers and marketers using their own media platforms or mass media. Such activities influence the product's brand equity (Aaker, 1991; Keller, 2013) and brand story (Aaker, 1995, 2018). In particular, many people have learned about the development process of Sony's Walkman and Apple's iPhone, which brought a fresh experience to

DOI: 10.4018/978-1-7998-4864-6.ch012

their daily lives, through online news, TV dramas, and books. The development process of such products becomes a topic of conversation in society and influences people's purchase decisions. Thus, the experiences of revolutionary products' developers greatly contribute to the products' brand equity and brand story. In the field of brand management, a story that influences brand equity and connects it with a brand vision, customer relations, organizational values, and business strategies is called a "signature story," which is distinguished from a strategy story aimed at short-term communications (Aaker, 2018). Signature stories are not a depiction of facts. Rather, they create an interpretation of facts in relation to the receiver (Aaker, 2018). Signature stories raise customers' interest, attract customers, have authenticity, and provide the product with recognition and vitality for a long period of time (Aaker, 2018). As for the creation of signature stories, the predominant view is that they should be based on facts (Aaker, 2018). Thus, this study examines the process of creating compelling facts for signature stories from the viewpoint of products and services, major elements of signature stories (Aaker, 2018). For this reason, the study focuses on the development process of G-SHOCK, which Casio developed, and for which the company created a market. Stories about the trial and error method of the developer and marketer during the development process of a market-creating product are called "innovation narratives" in this article. This study describes the creation process of an innovation narrative and discusses the creation of a "pivot" in the development process within the framework of a dialog with "projections". "Projection" is the act of projecting the sensation obtained in one scene to another (Suzki, 2016).

BACKGROUND

Product Development in an Uncertain Environment

This study examines the development process of a product that creates a new market. It identifies the problems that may occur in the early stages of the product-development process and examines how such problems can be solved. Generally speaking, there is a high level of uncertainty in the initial stages of the development process of a market-creating product. Thus, there is a situation in which the developers cannot clearly identify what exactly they should develop in the early stages. In these stages, the establishment of a clear objective is impossible even if the opinions of customers are carefully considered (Lester & Piore, 2004). Thus, the developer seeks to identify the problem through trial and error (Ishii, 1993; Lester & Piore, 2004).

This study examines "experimental behavior" in the process of trial and error (Dyer, Gergersen & Christensen 2011). It then presents the results of research focusing on experimental behavior through prototyping (the development and experiments using a prototype). The research target is the development process of the first model of G-SHOCK by Casio, which has created the market for shock-resistant watches (or "toughness watches"). Trial and error involve failure, as well as actions to correct the failure. A series of activities involving failure and efforts to correct the failure are called a "pivot" (Ries, 2011; Stainert & Leifer, 2012, Furr & Dyer, 2014). This study analyzes the development process of G-SHOCK using the concept of pivot. The study, based on the results of the analysis, discusses the contribution of a pivot to the pairing of "need-solution pairs" (Von Hippel & von Krogh, 2016).

Innovation and Behavior of Developers

Trial and error are essential in the creation of a new market or business. It has been pointed out that trial and error are necessary to identify problems that have not been previously discovered and to find a solution to these problems (Ishii, 1993; Lester & Piore, 2004). Similar views have been expressed in the area of research regarding star entrepreneurs (Sarasvathy, 2008; Ries, 2011), design thinking (Kelly & Kelly, 2013), frequent behaviors and corrections (Steinert & Leifer, 2012), and architecture (wide-area spatial design) (Alexander, 2013).

Trial and error are not a one-time act. The process involves multiple actions and corrections to create new customers, and values, as well as products and services to achieve these values. Thus, a behavior is revised based on the phenomenon produced by a behavior. Trial and error involve repeated actions and corrections. A "pivot" is an act of correcting a behavior based on the phenomenon produced by an experimental behavior (Ries, 2011; Stainert & Leifer, 2012; Furr & Dyer 2014).

Perspective of Pivots and the Development of Market-creating Products

"Linear" and "waterfall development" models forecast the possibility of moving to a next step in the development process under the assumption that the process will proceed to a next step without going backward. However, it has been pointed out that trial and error exist in product development in an uncertain environment, particularly in the early stages of development (Ishii, 1993; Lester & Piore, 2004). In such a situation, the developers cannot identify exactly what they should develop, or establish a clear development goal even if they heed the opinions of customers (Lester & Piore, 2004). However, even in such a situation, the developers must find a clue in certain information in order to establish a development goal. One of the solutions suggested for this problem is "experimental behavior" (Dyer et al., 2011). This behavior involves the development and use of a minimal prototype to repeatedly revise the current behavior to create a new behavior based on the information obtained. A pivot is a series of actions accompanying the experimental behavior.

Need-solution Pairs

Need-solution pairs (Von Hippel & Von Krogh, 2016) provide a framework for identifying the problems (needs) of customers and simultaneously providing solutions to these problems. The framework differs in concept from conventional product-development research, which mainly focuses on solving problems. The framework has three key features: 1) It points to the existence of an innovation mechanism that identifies the problems before they are defined and finds solutions to these problems at the same time; 2) The combination (pairing) of needs and solutions is chosen from the accumulated knowledge (landscape) regarding solutions in a mutually explorative fashion; 3) For pairing (combination), what is chosen is the combination of a need in the needs landscape that may achieve the greatest profit if a solution is found, and a solution in the solution landscape that entails the lowest cost. In addition, in need-solution pairs, 1) Activities for identifying problems and those for solving problems are not necessarily separated; and 2) There are cases in which problem identification does not precede problem-solving.

Management of the Combination (Pairing) of Needs and Solutions

Need-solution pairs are an excellent model that illustrates the combination of needs and solutions. However, the premise seems to be that potential needs and potential solutions can be listed beforehand. The pairing (combination) of potential needs and potential solutions is explained as follows. A combination is made between a potential need that could generate a large profit if a solution is found, and a potential low-cost solution. This indicates a dynamism in which a combination is made from multiple needs and multiple solutions with a careful examination of both. However, this does not mean that the size of the profit and the cost can always be estimated beforehand. In addition, the combination process or the process of change in the landscape is not mentioned. For this reason, additional explanations may be required with respect to the relationship between the creation of the need/solution landscape and the pairing.

The Role of a Pivot in Creating a Pairing

Under the premise that problems cannot be solved unless they can be expressed in a form that can be solved (Anzai, 1985), problem-identification activities must be repeated until the problems reach that stage. Furthermore, under the assumption that problem-solving activities do not occur until after problems are found, problem-solving activities do not occur unless there are problems that can be solved. However, in many product-development venues, there are cases in which a prototype is developed, and experimental activities and corrections are repeated even though problems have not been identified in such a way that they could be solved, and there are cases in which solutions are attempted with an erroneous understanding of the problems (Hirota, 1999).

Need-solution pairs (Von Hippel & von Krogh, 2016) incorporate the idea that problems may be found while solving problems. This type of development behavior is not necessarily inefficient. It has been confirmed that the development of a prototype to create a tangible object and the use of a prototype to actualize values may raise the chances of identifying the problems, even if the problems are wrongly perceived or understood. That would be more efficient than spending time discussing whether problems have been correctly recognized or understood (Hirota, 2018). This is where a pivot comes in. The importance of pivots has been pointed out in lean startups that require the development of a new market (Ries, 2011; Stainert & Leifer, 2012; Furr & Dyer, 2014). Pivots can be used to address uncertainties (Furr & Dyer, 2014). They learn and revise the original plan based on information obtained from the reality created by the experiment. The revisions, in addition to correcting the path toward the goal, may also involve changes in the initially established goal with the use of experiments and pivots (Stainert & Leifer, 2012). Changes involving a pivot do not reset everything each time. Instead, the process involves learning and revising based on the information obtained from the previous experimental behavior (Furr & Dyer, 2014). Pivots are useful in that they allow for the discovery of insights that cannot be found merely by changing various factors all at once (Furr & Dyer, 2014). In an uncertain environment, mistakes and failures always occur (Furr & Dyer, 2014). A pivot generates "mistakes" or "failures" so that a lesson can be learned for creating a new direction. A pivot is an important step toward repeatedly learning from experimental behavior and the resultant phenomena, and toward finding and solving the correct problems at the same time.

FORMATION OF SIGNATURE STORY AND TRIAL AND ERROR BEHAVIOR OF DEVELOPERS

Casio G-SHOCK

Casio put G-SHOCK on the market in 1983. The cumulative shipments of the product exceeded 50 million units in 2009. As of the end of August 2017, 34 years after the release of the first model, the cumulative shipments reached 100 million units. The recognition level of the product exceeds 95% in Japan. The product is now sold in 138 countries around the world.

G-SHOCK was not developed systematically with a clear goal in mind. Nor was it developed to meet a need discovered through market research. Instead, it was based on an intuition of the developer who wanted to create a watch that "does not break when dropped." It was born of the developer's experience and novel idea. The developer spent two years developing a prototype and conducting experiments through trial and error almost all by himself addressing a problem that was not at all clear: a watch that does not break when dropped. About 300 prototypes were developed. In the process, problems and solutions were discovered simultaneously.

Data Collection

Data regarding the development of G-SHOCK exist in the form of a book authored by the developer, as well as newspaper, magazine, and online articles based on interviews with the developer. There are also studies that recorded and examined the development process of G-SHOCK. The present study searched such data using a database of newspaper articles and magazine articles, and by conducting an online search. In addition, speeches delivered by the developer were obtained with the help of research partners. Furthermore, information on patents and utility-model registrations have been confirmed with the use of a patent application platform.

Based on this documented data, the activities of the developer and the resulting facts were listed in chronological order. An interview was then conducted with the developer, and the audio recording was converted to text, from which the facts that would lead to a pivot were listed. Finally, the information was classified using grounded theory.

The Signature Story of G-SHOCK

Watch products are worth the time. Therefore, the functional value of accurately indicating the time is important. Furthermore, lightness and compactness are also functional values. Watch product has other value. It is the status value. Therefore, watches are classified as precious metal products.

Since the 1970s, watch product have shifted from precision mechanical products to digital products. In particular, the parts called "movements" that support accurate time indication have been digitized and standardized. The product gap due to the accuracy of the time has disappeared. Until 1983, when the G-SHOCK was developed, the value of wristwatches was competing for the functional value of thin and light weight, and the emotional value of precious metal products. Until the first half of 1980, it was common knowledge that watches were vulnerable to shock. Based on that common sense, companies were competing for thin and light weight. G-SHOCK has created a new value that is a shock-resistant watch that is completely different from the value. However, as a result, the product has become larger in size and weight than conventional watches. In addition to the shock-resistant functional value of G-SHOCK, attention was paid to the behavior of its developers.

G-SHOCK has acquired emotional value such as "challenge", "innovator", and "toughness". Such values have been found by fashion designers, artists, fashion models and creators. The actor likes to wear G-SHOCK and appears in movies (Nobeoka, Kimura & Osanai, 2005). And the fans of the actors who saw the movie become fans of G-SHOCK. Many such cycles are born. In addition, many collaboration products with fashion companies and environmental protection groups have been created.

To celebrate the 25th anniversary of G-SHOCK, SHOCK THE WORLD, an event for fans of G-SHOCK that started in 2008, is held all over the world, and many fans gather. At each venue, G-SHOCK's first model developer will give a speech. At each venue, the 1st model developer of G-SHOCK is worshiped as "Father of G-SHOCK", and his enthusiastic fans gather. Also, from 2018, the 35th anniversary of birth, a mixed-genre event G-SHOCK FAN FESTA has been held such as sports, music and dance genres. At G-SHOCK FAN FESTA, along with many G-SHOCK fans, fans of each genre are packed and various exchanges are realized. In 2020, a limited model for collaboration between first model developers and fashion designers, which will be adopted in this research, has been developed and released worldwide.

The origin of such a market is the development of the G-SHOCK first model, and the development behavior of the developer.

In this research, we investigate the behavior of the first model developers who support the G-SHOCK signature story, and clarify the development behavior that leads to the signature story as a result.

A Watch That Does not Break When Dropped

In 1981, five years after joining the company, Mr. Ibe,Kikuo broke his treasured watch by accidentally dropping it on the floor upon contact with another person. He still remembers vividly how the watch fell and broke. At the time, members of his department were all required to submit a project proposal. Thereafter, he submitted his proposal based on this incident two or three months after he broke his watch. The project he proposed was clearly different in nature from those he had submitted before. He had previously designed light and thin watches, which dominated the market at the time. For this reason, his earlier proposals were about the shape of the back lid or a new way to mold watch cases. These proposals had never been adopted. This time, relying on his intuition, he submitted a one-line proposal: a solid watch that does not break when dropped. This proposal was approved. He himself was surprised that this intuitive, one-line proposal had been adopted.

Problem, Possible Solution, Creation of an Experimental Method

Initially, the developer thought that the problem would be solved by attaching shock absorbers to the four corners of existing watches. No shock absorbers had been used between the module and existing watches. The problem, he thought, was a lack of function in existing watches to absorb shocks. He thought that the problem could be solved by attaching shock absorbers in the four corners of the module. He also came up with an experimental method. Because of his experience of dropping and breaking a watch, he instinctively chose a method to simulate a natural fall. The developer, who is also an engineer, could have also chosen a method involving experimental devices. Nevertheless, he did not choose such a method. This is because shocks produced by a fall would vary. The conditions would differ depending not only on the height of the fall but also on which part of the watch would hit the ground. Even if any particular part is specified, there is still a vast number of different combinations with respect to angles. He could not possibly imagine all the combinations. Thus, he could not design any experimental devices. The developer could not have come up with the objective, "a watch that does not break," if he had conducted experiments using test devices.

Prototype of a Watch that Does not Break When Dropped

He repeated an experiment in which he developed prototypes (the act of manufacturing) and dropped them to the ground (the act of using). He would drop a prototype from a window of a third-floor restroom. He would then go down stairs to the location where the prototype fell to check its condition. If broken, he would climb upstairs to the third-floor lab wondering why. In the lab, he would disassemble the module to confirm which parts were broken. He repeated this routine every day. However, the module always broke even if he added more shock absorbers around it. Finally, he developed a prototype whose module would not break. This prototype, however, had too many protective layers, except for the liquid crystal display portion of the module. Moreover, it was the size of a softball.

Discovery of a New Problem and the Simultaneous Creation of a Solution

It would be difficult to commercialize a watch that had a module with a shock-absorbing function if it was the size of a softball. Even so, it was nonetheless a watch that would not break when dropped. For the developer, the problem that had to be addressed became clear. His challenge now was to make the shock-absorbing function, which was the size of a softball, small enough for a watch. At the same time, he created a new solution in the form of a five-step shock-absorbing structure. The experience of wrapping the shock-absorbing material around the prototype led to this new solution.

Discovery of Target Customers Brought About by a Different Environment

The five-step shock-absorbing structure made it possible to produce a smaller prototype. As he continued to develop prototypes and conduct repeat experiments, he encountered a group of people engaging in road-repair work, who were digging the road with rock drills, outside the company building. The developer noticed that none of the workers wore watches. He realized that they could not wear a watch because it would break from the vibration of the drill. This was the first time that the target users of a shock-resistant watch became clear to him.

Clue to the Solution Generated by an Environment Different from the Unsolved Problem

No matter how many experiments were carried out, a certain component in the module always broke. The results were the same even if the shock-absorption method was tweaked and the location and size of the shock absorber were changed. The developer decided to abandon the project because a solution could not be found.

On the day he decided to abandon the project, the developer nevertheless went to work to carry out an experiment even though it was a Sunday. He went outside the company building to buy lunch. He then saw kindergarten children playing with a ball in a park. While watching the ball, a certain picture came into his mind. This was a picture of a module floating inside a ball. Then, he also had another mental image: a wineglass floating inside a ball.

These mental images gave him the idea of a structure in which the module was supported by points, instead of being surrounded by a shock absorber. In other words, the module would float in the air. In contrast, the five-step shock-absorbing system would immobilize the module in such a way that an unabsorbed shock would reach the module. Thus, the mental picture of a module floating inside a ball made him realize the problem of the five-step shock-absorbing system, which would convey the unabsorbed shock directly to the module. This mental picture also presented a solution to this very problem. The situation where a module would float in the air could be created by supporting the module with points. Thus, if an unabsorbed shock hit the module, it would oscillate and evenly distribute the shock. A structure in which a module is supported by points and floats in the air, along with a five-step shock-absorption system, became a major characteristic of G-SHOCK, for which the company earned a patent.

DISCUSSION

Innovation and Problem-identification Behavior

Product development consists of two activities: identifying new problems and solving the right problems (Norman, 2013). Product-development research in the past primarily focused on solving problems (Clark & Fujimoto, 1991). In recent years, because of the uncertainties in the environment, it has been pointed out that there is a need for a new marketing approach (Kotler & Keller 2009, Kotler, & Kartajaya et al.2017), and a need to identify problems (Kotler & Takaoka, 2016), in addition to merely solving problems. Prior studies focused on solving problems were predicted on the assumption that the problems had already been identified. For this reason, even if problem-solving was in view, problem-solving activities were preceded by problem identification. The assumption was that problem-identification activities were separate from problem-solving activities, and that problem-identification activities preceded problem-solving activities.

Meanwhile, the problems that customers seek to have solved are not always the right problems. Thus, understanding the real problems is necessary in order to achieve innovation (Norman, 2013). For these problems to be solved simultaneously, they must be expressed in such a way that they could be solved (Anzai, 1985; Suzuki, 2016a). For this purpose, it is necessary to express in some form a situation where it is unclear how to go about addressing the problem. Understanding problems in this way is the most difficult step in the process of solving them (Anzai, 1985). Problems cannot be expressed appropriately unless the situation is understood. If one tries to express problems (to the extent possible) and solve them based on his or her current understanding, the areas where the expressions are inadequate or incorrect become clear and a better understanding of the situation can be obtained (Anzai, 1985). This insight is important for two reasons. First, understanding the situation that leads to problem-solving requires that one pay attention to the existence of a situation in which problems have been partially identified and a situation in which problems are gradually being identified, instead of making a distinction between a situation in which problems have been identified and a situation in which problems have not been identified. Second, in a situation in which problems have not been identified, attention must be paid to activities designed to solve the problems based on the understanding at the moment.

Pivots and Dialogs

The development of G-SHOCK began at a time when light and thin models dominated the wrist-watch market and the company was unaware that there was a demand for shock-resistant watches. A watch that "does not break when dropped" did not exist at the time. Therefore, no one was able to provide an adequate description of such a watch. In a situation where the goal is the development of a product to solve a need of which even customers were unaware, as in the case of G-SHOCK, the developer must address a problem that has not been identified. In the initial stages of the development, it is difficult to identify a problem that must be solved even if the voices of customers are considered. This is because the fundamental nature of uncertainty makes it difficult for the developer to evaluate the results or analyze and classify the problems (Lester & Piore, 2004). However, the developer must still identify the problem that must be solved even in such a situation. How the problem is defined greatly affects the outcome (Lester & Piore, 2004). Thus, the developer identifies a problem through trial and error in a haphazard way. The problem may be found accidentally, from a clue that occurs only in a particular situation (Ishii, 1993; Lester & Piore, 2004). The process of "dialog" is important when defining a problem and finding a solution (Ishii, 1993; Lester & Piore, 2004; Ishii, 2009). When there is a separation between customers and the developer, customers attempt to communicate certain ideas in their day-to-day activities and the developer seeks to interpret their meaning (Lester & Piore, 2004). The customers convey certain ideas related to the need that they experience in their daily lives through both verbal and non-verbal methods. The developer, a third party, interprets this dialog. However, there is no guarantee that the interpretation of the developer is always correct as long as the interpreter remains a third party. In this situation, in which the problem has not been identified, it may be effective to develop and use many prototypes frequently and repeatedly to achieve innovation. This was demonstrated with the use of a dialog framework (Hirota, 2016). Experimental behavior involves repetitions of prototype development (the act of manufacturing) and the use of prototypes in the field (the act of using) (Hirota, 2018). In such a behavior, there is a dialog between the act of manufacturing on the one hand, and the act of using or experiencing by the customers on the other (Hirota, 2018). From the perspective of cognitive science, the act of manufacturing and the act of using are a projection by the developer (Suzuki, 2016). The developer, through this projection, repeats a dialog between external representations and internal representations (Suzuki, 2016). This dialog involves repeated conversations regarding how the users will use the product (the procedure for the act of using) and how the developer should manufacture it (the procedure for the act of developing) (Hirota, 2018).

Information Creation Through Experimental Behavior and Pivots

A pivot, which targets areas where forecasting is impossible, is a framework created during a process of analyzing successful startup models (Ries, 2011; Blank & Dorf, 2012; Blank, 2013). A pivot is predicated on the occurrence of failure, under the assumption that repeated experiments would result in successful products and services. People cannot make accurate forecasts when faced with uncertainties. A pivot, based on information obtained from activities called experiments, creates a behavior that diverges from conventional behavior. It is based on the premise that action (experiment) inevitably begets failure in an uncertain environment (Furr & Dyer, 2014). Information obtained from failure is used to create a new target and take new action.

What is noteworthy about these studies is that trial and error are not just a one-time occurrence. Moreover, trial and error accompany experimental behavior, the results of which include failure. Startups that create a new market make many mistakes before they succeed (Blank & Dorf, 2012). Experimental behavior brings about a phenomenon called success, as well as a phenomenon called failure. Failure provides new information to the developer, who takes new action using information obtained from this phenomenon.

FUTURE RESEARCH DIRECTIONS

Discuss future and emerging trends. Provide insight about the future of the book's theme from the perspective of the chapter focus. Viability of a paradigm, model, implementation issues of proposed programs, etc., may be included in this section. If appropriate, suggest future research opportunities within the domain of the topic.

CONCLUSION

Two Types of Pivots: Closed and Open

The development process of G-SHOCK and the resultant pivot will be examined in this section within the framework of need-solution pairs. The focus is on the combination of the needs (problems) and the solutions with respect to the behavior of the developer immediately before and after the occurrence of a pivot. When the experimental behavior for the development of G-SHOCK is examined within the framework of a dialog, at least two types of pivots are confirmed: "closed type" and "open type." Innovation occurs based on a repetition and the combination of these two types of pivots.

A closed pivot is a development activity that does not doubt the problem to be solved. Therefore, development actions are limited to solving certain problems. It is described as "closed" because the problem is fixed and continuous. On the other hand, the open pivot includes an action of discovering and paying attention to a problem different from the conventional one due to some reason. It is described as "open type" because it includes actions that recapture a wide range of problems.

Two Types of Pivots and Dialog

An open pivot occurs when a convergence behavior toward a certain problem in the double-diamond model (Norman, 2013) shifts to a diffusion behavior for finding other problems. A closed pivot, on the other hand, maintains a convergence behavior and strengthens the combination of a problem and a solution. The difference between a closed pivot and an open pivot lies in the fact that, in a closed pivot, only the solutions change while the problems remain unidentified, whereas in an open pivot, the problems are identified, and the solutions change accordingly. The difference between the two can be explained by the concept of a dialog. A dialog is a framework through which the developer's thinking takes the form of a conversation with the object of development. The framework addresses what the problem is in pursuing development, and how this problem should be solved. The particular problem should be a problem that can actually be solved (Anzai, 1985). The structure of the dialog is a combination of a certain problem that can be solved and a solution to a certain problem. A closed pivot would repeat solution activities for a certain solvable problem. This is the type of pivot that emerges in relation to a solution if a certain problem persists. On the other hand, a different type of pivot can be observed in the development of G-SHOCK. This is an open pivot. It is the type of pivot that emerges when a new solution combination is created along with a problem.

Innovation Narratives and a Chain of Two Types of Pivots

A closed pivot changes the method of solving a certain problem and changes the activities for solving the problem based on experimental results. Changes in activities manifest themselves as a small pivot chain. The emergence of a small pivot chain confirms an external representation of phenomena, i.e., a projection in the form of the development of a prototype and a projection in the form of an experiment of the prototype (Suzuki, 2018; Hirota, 2018). This confirmation gives rise to an internal representation, through which a refractive conversation between cognition and expression is carried out (Schon, 1983). However, there are situations in which a solution cannot be found even when various measures are created to address a problem and when experiments are conducted. In such a situation, the emergence—both intentional and unintentional—of environmental changes that may alter the object of the developer's dialog would lead to the emergence of an open pivot. An open pivot is a phenomenon whereby a situation in which problems remain immobilized in a closed pivot leads to the discovery of a new problem because of changes in the environment. These two types of pivots mutually interact with each other and manifest themselves as a pivot chain.

This study has pointed out the existence of a development process that creates a new market as a signature story that is said to be effective in brand equity. The study called such a development process an "innovation narrative." Innovation narratives are stories about trial and error on the part of the developer. They are a combination of mutual interactions that emerge from various activities. Thus, innovation narratives require a chain of two types of pivots.

Innovation Narratives That Enable Signature Stories

In this study, we confirmed the behavior of product development that leads to a signature story.

G-SHOCK is the world's first product that created the shock-resistant watch market. However, the signature story cannot be realized only with the world's first product. Focusing on the behavior of G-SHOCK developers, the following points become clear.

First, create value by changing the common sense (schema) of existing products

Second, the existence of actions that change common sense (schema) in development

The signature story requires a transformation of these two common senses (schemas).

Signature stories are valid for brand equity (Aaker, 2018). For product brands that have a signature story, the trial and error behavior of developers and marketers in the market creation type product development process is called "innovation narrative" in this Study. Innovation narrative is a trial-and-error behavior of developers, which is a combination of interactions created by the behavior. The innovation narrative requires action to realize the transformation of these two common senses (schemas).

In this Study, the behavior of the developer was grasped from the viewpoint of the design double diamond (Norman, 2013). Then, it was clarified that the trial-and-error behavior that leads to the innovation narrative is linked by the combination of the two types of pivots.

In this Study, the development behavior in which two pivots occur in a chain is called "pivot chain". The first pivot is an action (closed-type pivot) that performs trial and error in solving a problem without changing the setting of the problem. In addition, a second pivot is needed to realize the innovation narrative. The second pivot is the action of creating a problem (open-type pivot). Each of these two actions is not a one-time action. Multiple pivots occur multiple times in succession. This is the pivot chain. Realization of an open pivot requires multiple closed pivots.

ACKNOWLEDGMENT

This study was supported by JSPS KAKENHI Grant Number 19K01974、19K01969.

REFERENCES

Aaker, D. A. (1991). *Managing brand equity*. Free Press.

Aaker, D. A. (1995). *Building strong brands*. Free Press.

Aaker, D. A. (2018). *Creating signature stories: Strategic messaging that energizes, persuades and inspires*. Morgan James Publishing.

Alexander, C. (2013). *The nature of order: An essay on the art of building and the nature of the universe*. Oxford University Press.

Anzai, Y. (1985). *Mondai kaiketuno shinrigaku — Ningenno jidaieno hassou*. Chukoushinsyo.

Blank, S., & Dorf, B. (2012). *The startup owner's manual: The step-by-step guide for building a great company*. Palo Alto, CA: K & S Ranch.

Blank, S. (2013). The four steps to the epiphany: successful strategies for products that win, Louisville, KY:Cafepress.com.

Clark, K. B., & Fujimoto, T. (1991). *Product development performance: strategy, organization, and management in the world auto industry*, Brighton, MA: Harvard Business Review Press.

Dyer, J., Gregersen, H., & Christensen, C. M. (2011). *The Innovator's DNA: Mastering the Five Skills of disruptive innovators*. Harvard Business School Press.

Furr, N., & Dyer, J. (2014). *The innovator's method: bringing the lean start-up into your organization*. Harvard Business Review Press.

Hirota, A. (1999). Only one market -Development of "Cyber Zero" by ASICS, Tokyo. *Japan Marketing Journal*, *19*(2), 62–74.

Hirota, A. (2016). The realization of prototypes in the product-design process - factors behind the realization of rapid prototyping. *Japan Society of Marketing & Distribution FY2016 Collection of Papers Presented at the National Research Conference, Japan Society of Marketing & Distribution*, (65), 6-98.

Hirota, A. (2017a). The prototype used in the implementation of the "Need-Solution Pairs" - The prototype use of single-user and multi-user to relate. In *Proceedings of 14th Open and User Innovation Society Meeting* (pp. 1-3). Innsbruck, Austria: Open and User Innovation Society Meeting.

Hirota, A. (2017b). The role of prototyping in creative moments (business insight). *Japan Society of Marketing & Distribution FY2017 Collection of Papers Presented at the National Research Conference, Japan Society of Marketing & Distribution, 66*, 81-87.

Hirota, A. (2018). Effect of "prototyping stage" for "Need-Solution Pairs" in design thinking, Stockholm. *ISPIM Innovation Conference*, 1-16.

Hirota, A. (2019). Pivot chain to create the pairing in Needs solutions pairs, Ferenze. *ISPIM Innovation Conference,* 1-17.

Hirota, A., Takemura, M., & Mizuno, M. (2017). Design prototyping in the "fuzzy front end" of product development – rapid prototyping at the stage of high uncertainty. *ISPIM Innovation Forum,* 1-7.

Ishii, J. (1993). *Marketing myths*. Iwanami Shoten.

Ishii, J. (2009). *Business insight – what is creation knowledge?* Iwanami Shoten.

Keller, K. (2013). *Strategic brand management* (4th ed.). Pearson.

Kelly, T., & Kelly, D. (2013). Creative confidence. New York, NY: Fletcher and Company.

Kotler, P., Kartajaya, H., & Setiawan, I. (2017). *Marketing 4.0: Moving from traditional to digital.* Gildan Media.

Kotler, P., & Keller, K. (2009). *Marketing management*. Pearson.

Kotler, P., & Takaoka, K. (2016). *Marketing no susume*. Chuoukouronsya.

Lester, R. K., & Piore, M. (2004). *Innovation the missing dimension.* Harvard University Press.

Nobeoka, K., Kimura, M., & Osanai, A. (2015). Creation design value. *Hitotubashi Business Review, 62*(4), 6–21.

Norman, D. (2013). *The Design of Everyday Things: Revised and expanded edition.* Basic Books.

Ries, E. (2011). *The lean startup: How today's entrepreneurs use continuous innovation to create radically successful businesses.* Currency.

Sarasvathy, S. (2009). *Eectuation: Elements of entrepreneurial expertise.* Edward Elgar Pub.

Schoen, D. A. (1983). *The reflective practitioner: How professionals think in action.* Routledge.

Steinert, M., & Larry, J. L. (2012). Finding one's way': Re-discovering a hunter-gatherer model based on wayfaring. *International Journal of Engineering Education, 28*(2), 251–252.

Suzuki, H. (2016a). *Cognitive science as an education.* Tokyo University Publishing.

Suzuki, H. (2016b). Perspectives on projection science. In *Proceedings of the Annual Meeting of the Japanese Cognitive Science Society* (pp. 20-25). Japanese Cognitive Science Society.

Von Hippel, E., & Von Krogh, G. (2016). CROSSROADS—Identifying Viable "Need–Solution Pairs"-Problem Solving Without Problem Formulation. *Organization Science*, *27*(1), 207–221.

KEY TERMS AND DEFINITIONS

Dialogue: Interaction of the creation of projection and the representation. Other than the talks with a person and the person, the talks in a process making prototype, the talks in the process using the prototype, the talks with the experience that wants to come true exist.

Need-Solution Pairs: Von Hippel and Von Krogh (2016) pointed out that in innovations, there are times when the needs and solutions are paired before the problem is defined. Needs and aggregate of the knowledge of the solution exist in the space. Innovation is realized by a combination of each information.

Pairing: A combination of a certain needs information that there is in a needs landscape and a certain solution information existing in a solution landscape.

Pivot: Actions to correct based on phenomena caused by experimental actions.

Problem Finding: Discover a problem and solution in an action of the trial and error simultaneously.

Procedural Knowledge: To investigate a combination of good combination through a trial by a combination of a certain needs, information, and a certain solution information.

Compilation of References

Aaker, D. A. (1991). *Managing brand equity*. Free Press.

Aaker, D. A. (1995). *Building strong brands*. Free Press.

Aaker, D. A. (2018). *Creating signature stories: Strategic messaging that energizes, persuades and inspires*. Morgan James Publishing.

Abe A. (2005). *Computer and KANSEI (III)*. Technical Report of JSAI, SIG-LSE-A502-5-2, 33—36.

Abe A. (2010a). *About the generation of novels*. Technical Report of JSAI, SIG-LSE-A903-16-8, 129—134.

Abe A. (2010b). *The style of novelists' writing*. Technical Report of JSAI, SIG-LSE-B002-4, 23—32.

Abe, A. (1998). Applications of abduction. *Proc. of ECAI98 Workshop on Abduction and Induction in AI*, 12-19.

Abe, A. (2000). Abductive analogical reasoning. *Systems and Computers in Japan*, *31*(1), 11–19. doi:10.1002/(SICI)1520-684X(200001)31:1<11::AID-SCJ2>3.0.CO;2-E

Abe, A. (2003). The role of abduction in chance discovery. *New Generation Computing*, *21*(1), 61–71. doi:10.1007/BF03042326

Abe, A. (2015). The role of abduction in IMDJ. *Proc. of IJCAI2015 International Workshop on Chance Discovery, Data Synthesis and Data Market*, 59—64.

Abe, A. (2019). *The possibility of the generation of literary works by abduction*. In T. Ogata & T. Akimoto (Eds.), *Post-Narratology through computational and cognitive approaches* (pp. 144–163). Information Science Reference.

Abe, A., Fukushima, K., & Kawada, R. (2020). *How will sense of values and preference change during art appreciation? Information 2020, 11(6), 328.*

Abe, A., & Tadaki, K. (2019). Captions with several levels of explanation. *Proc. of KES2019, Procedia Computer Science*, 2335—2344.

Abe, S. (2003). *Saishin CM seisaku no kisochishiki: puranningu kara dejitaru seisaku made* [Latest basic knowledge of the commercial production: From planning to digital production]. Sendenkaigi.

Adam, J. M. (1999). *Le récit* [The narrative]. Presses Universitaires de France.

Aflo. (2020). *Service site operated by Afro*. https://www.aflo.com/ja

Akimoto, T., & Ogata, T. (2013). Towards a discourse mechanism in narrative generation system: Proposal of a system introducing narrative discourse theory and reception theory. *Cognitive Studies*, *20*(4), 396–420.

Akiyama, R., & Sugiyama, K. (2004). *Horisuthikku komyunikeishon* [Holistic Communication]. Sendenkaigi.

Alexander, C. (2013). *The nature of order: An essay on the art of building and the nature of the universe*. Oxford University Press.

Amana Images. (2020). *Service site operated by Amana Images*. https://amanaimages.com/home.aspx

Amino, T., Kawamura, Y., & Ogata, T. (2002). Hierarchical generation of geinō-idol stories—Toward geinō information system and narrative marketing. In *Proceedings of 17th Congress of the International Association of Empirical Aesthetics* (pp. 549–552). Rome, Italy: University of Rome Tre.

Anzai, Y. (1985). *Mondai kaiketuno shinrigaku — Ningenno jidaieno hassou*. Chukoushinsyo.

Aoki, S., Ogata, T., & Ono, J. (2018). Narrative generation related to cognitive patterns seen in ASD -From the perspective of "Surprise". In *Proceedings of the 35th Annual Meeting of the Japanese Cognitive Science Society* (pp. 652-660). Academic Press.

Arai, T., Ono, J., & Ogata, T. (2016). Semi-automatic generation of events sequence knowledge for narrative generation: The use in an integrated narrative generation system. In *Proceedings of the 30th Annual Conference of the Japanese Society for Artificial Intelligence* (3P1-1in2). Tokyo, Japan: Japanese Society for Artificial Intelligence.

Arijon, D. (1976). *Grammar of the film language*. Silman-James Press.

Arishima, T. (1964). Umareizuru nayami. In Arishima Takeo shū (pp. 350–393). Tokyo, Japan: Chikuma Shobō. (Original work published 1918)

Arishima, T. (1995). Aru on'na. Tokyo, Japan: Shinchōsha. (Original work published 1919)

Aristotle. (1997). Poetics (M. Heath, Trans.). London, UK: Penguin Classics. (Original work published 4th century BC)

Aristotle. (1997). *Poetics* (N. Matsumoto & M. Oka, Trans.). Iwanami Shoten. (Original work published 335 B.C.)

Asada, I., & Namiki, S. (1995). Dōjōji genzai uroko. In Toyotakeza jōruri syū, 3 (pp. 7-100). Tokyo, Japan: Kokusho Kankōkai. (First performed in 1742)

Asakawa S. (2015). *A neural narrative model*. Technical Report of JSAI, SIG-LSE-B501-2, 17—24.

Asanuma, K. (1990). Eizo no tekusuto no giho [Techniques for image text]. In Eiga no tameni II [For movie II] (pp. 130-247). Tokyo, Japan: Suiseisha.

Asch, S. E. (1951). Effects of group pressure on the modification and distortion of judgments. In H. Guetzknow (Ed.), *Groups, Leadership and Men* (pp. 177–199). Carnegie Press.

Avrahami, J., & Kareev, Y. (1994). The emergence of events. *Cognition*, *53*(3), 239–261. doi:10.1016/0010-0277(94)90050-7 PMID:7842635

Bakhtin, M. (1984). *Problems of Dostoevsky's poetics* (C. Emerson, Trans.). University of Minnesota Press. (Original work published 1963) doi:10.5749/j.ctt22727z1

Bandō Tamasaburō, V. (2003). *Bandō Tamasaburō buyōsyū* [Motion Picture DVD]. Syōchiku Home Video.

Basil, M. D. (1994). Multiple resource theory I: Application to television viewing. *Communication Research*, *21*(2), 177–207. doi:10.1177/009365094021002003

Berliner, T., & Cohen, D. J. (2011). The illusion of continuity: Active perception and the classical editing system. *Journal of Film and Video*, *631*(1), 44–63. doi:10.5406/jfilmvideo.63.1.0044

Bhute, A. N., & Meshram, B. B. (2014). Text based approach for indexing and retrieval of image and video: A review. *Advances in Vision Computing*, *1*(1), 27–38. doi:10.2139srn.3430312

Bickmore, T., Bukhari, L., Vardoulakis, L., Paasche-Orlow, M. & Shanahan, C. (2012). *Hospital buddy: a persistent emotional support companion agent for hospital patients*. Academic Press.

Blank, S. (2013). The four steps to the epiphany: successful strategies for products that win, Louisville, KY:Cafepress.com.

Blank, S., & Dorf, B. (2012). *The startup owner's manual: The step-by-step guide for building a great company*. Palo Alto, CA: K & S Ranch.

Bonnardel, N. (2000). Towards understanding and supporting creativity in design: Analogies in constrained cognitive environment. *Knowledge-Based Systems*, *13*(7-8), 505–513. doi:10.1016/S0950-7051(00)00067-8

Bordwell, D. (2012). *Constructive editing in Robert Bresson's Pickpocket* [Video]. Vimeo. https://vimeo.com/52312154

Bordwell, D. (1985). *Narration in the fiction film*. University of Wisconsin Press.

Bordwell, D. (2007). *Poetics of cinema*. Routledge.

Borkman, T. (1976). Experiential knowledge: A new concept for the analysis of self-help groups. *The Social Service Review*, *50*(3), 445–456. doi:10.1086/643401

Braunberger, P. (Producer), & Renoir, J. (Director). (1936). *A Day in the Country* [Motion picture]. France: Panthéon Productions.

Brecht, B., & Willett, J. (1964). *Brecht on theatre: The development of an aesthetic*. Hill and Wang.

Bresson, R. (1959). *Pickpocket* [Film]. A. Delahaie (Producer). France: Compagnie Cinématographique de France.

Bresson, R. (1983). *Money* [Film]. J. M. Henchoz & D.T. Plantier (Producer).

Broekens, J., & DeGroot, D. (2004). Scalable and flexible appraisal models for virtual agents. In *Proceedings of the Fifth Game-on International Conference* (pp. 208-215). Academic Press.

Bronte, E. (2002). Wuthering heights. London, UK: Penguin Books. (Original work publishued 1847)

Brown, D. E. (1991). *Human universals*. McGraw-Hill.

Bruner, J. (1986). *Actual minds, possible worlds*. Harvard University Press.

Buchanan-Barker, P., & Barker, P. J. (2008). The Tidal Commitments: Extending the value base of mental health recovery. *Journal of Psychiatric and Mental Health Nursing*, *15*(2), 93–100. doi:10.1111/j.1365-2850.2007.01209.x PMID:18211556

Burroughs, W. S. (n.d.). *The cut-up method of Brion Gysin*. http://tbook.constantvzw.org/wp-content/cutup_gysin_burrough.pdf

Burstein, M. H. (1986). Concept formation by incremental analogical reasoning and debugging. *Machine Learning*, *2*, 351–369.

Caballero, R. (2007). Manner-of-motion verbs in wine description. *Journal of Pragmatics*, *39*(12), 2095–2114. doi:10.1016/j.pragma.2007.07.005

Caballero, R. (2017). From the glass through the nose and the mouth: Motion in the description of sensory data about wine in English and Spanish. *Terminology*, *23*(1), 66–88. doi:10.1075/term.23.1.03cab

Çağlayan, E. (2016). The aesthetics of boredom: Slow cinema and the virtues of the long take in once upon a time in Anatolia. *Projections*, *10*(1), 63–85. doi:10.3167/proj.2016.100108

Calleja, G. (2009). Experiential narrative in game environments in breaking new ground: innovation in games, play, practice and theory. *Proceedings of DiGRA 2009.*

Chatman, S. (1980). What novels can do that films can't (and vice versa). *Critical Inquiry*, *7*(1), 121–140. doi:10.1086/448091

Chatman, S. (1990). *Coming to terms: The rhetoric of narrative in fiction and film.* Cornell University Press.

Chomsky, N. (2011). The generative enterprise revisited (T.Fukui & N.Tsuji, Trans.). Tokyo, Japan: Iwanamibunko.

Clandinin, D. J. (Ed.). (2007). *Handbook of narrative inquiry: Mapping a methodology*. Sage Publications Inc. doi:10.4135/9781452226552

Clark, K. B., & Fujimoto, T. (1991). *Product development performance: strategy, organization, and management in the world auto industry*, Brighton, MA: Harvard Business Review Press.

Cohn, N. (2013). Visual narrative structure. *Cognitive Science*, *34*(3), 413–452. doi:10.1111/cogs.12016 PMID:23163777

Cohn, N. (2014). You're a good structure, Charlie Brown: The distribution of narrative categories in comic strips. *Cognitive Science*, *38*(7), 1317–1359. doi:10.1111/cogs.12116 PMID:24646175

Cohn, N., & Magliano, J. P. (2020). Editors' introduction and review: visual narrative research: an emerging field in cognitive science. *Topics in Cognitive Science*, *12*(1), 197–223. doi:10.1111/tops.12473 PMID:31865641

Colley, R. H. (1961). *Defining advertising goals for measured advertising results*. Association of National Advertisers.

Collins, W. A., & Wellman, H. M. (1982). Social scripts and developmental patterns in comprehension of televised narratives. *Communication Research*, *9*(3), 380–389. doi:10.1177/009365082009003003

Cooper, A. (2004). *The inmates are running the asulum.* Sams Publishing.

Copeland, M. E. (2002). *Wellness recovery action plan* (2nd ed.). Peach Press. doi:10.1300/J004v17n03_09

Corner, J. (2018). Documentary spectatorship and the navigation of "difficulty.". In C. Brylla & M. Kramer (Eds.), *Cognitive theory and documentary film* (pp. 59–73). Palgrave Macmillan. doi:10.1007/978-3-319-90332-3_4

Cortázar, J. (1991). *Hopscotch: A novel* (M. Heath, Trans.). Pantheon. (Original work published 1963)

Cosejo, D. G., Oesterreich, J., & Ohlsson, S. (2009). Re-categorization: Restructuring in categorization. In *Proceedings of the Annual Meeting of the Cognitive Science Society (CogSci 2009).* Red Hook, NY: Curran Associates.

Costello, F. J., & Keane, M. T. (2000). Efficient creativity: Constraint-guided conceptual combination. *Cognitive Science*, *24*(2), 299–349. doi:10.120715516709cog2402_4

Crask, M. R., & Laskey, H. A. (1990). A positioning-based decision model for selecting advertising messages. *Journal of Advertising Research*, *30*(4), 32–38.

Csikszentmihalyi, H., & Keith Sawyer, K. (1995). Creative insight: The social dimension of a solitary moment. In R. J. Sternberg & J. E. Davidson (Eds.), *The nature of insight* (pp. 329–364). MIT Press.

Cutting, J. E., Brunick, K. L., & Canden, A. (2012). Perceiving event dynamics and parsing Hollywood films. *Journal of Experimental Psychology*, *38*(6), 1476–1490. PMID:22449126

Cutting, J. E., & Iricinschi, C. (2014). Re-presentations of space in Hollywood movies: An event-indexing analysis. *Cognitive Science*, *35*, 1–23. PMID:25087776

Dainipponkoku hokkekyō genki. (1974). In *Ōjōden Hokkegenki* (pp. 43-219). Tokyo, Japan: Iwanami Shoten. (Original work published 1040–1043)

Damasio, A. R. (1998). Emotion in the perspective of an integrated nervous system. *Brain Research. Brain Research Reviews*, *26*(2-3), 83–86. doi:10.1016/S0165-0173(97)00064-7 PMID:9651488

Danto, A. C. (1965). *Analytical philosophy of history*. Cambridge University Press.

Darwin, C. (2016). Ningen no yurai (M. Hasegawa, Trans.). Tokyo, Japan: Kodansha.

Davis, F. (1979). *Yearning for yesterday: A sociology of nostalgia*. Free Press.

Dawkins, R. (1976). Rikotekina idensi (T. Hidaka, Trans.). Tokyo, Japan: Kinokuniyashoten.

de Maupassant, G. (2007). *The Short Stories of Guy de Maupassant*. Black's Readers Service.

Deahl, D. (2018), The EU plans to test an AI lie detector at border points. *The Verge*. https://www.theverge.com/2018/10/31/18049906/eu-artificial-intelligence-ai-lie-detector-border-points-immigration

Deleuze, G. (1983). *Cinema' Image-mouvement*. Paris, France: Les Editions de Minuit.

Deleuze, G. (1985). *Cinema' Image-temps*. Paris, France: Les Editions de Minuit.

Deleuze, G. (1985). *L' Image-temps*. Les Editions de Minuit.

Deleuze, G. (1990). *Pourparlers*. Les Editions de Minuit.

Denzin, N. K., & Lincoln, Y. S. (Eds.). (2000). *Handbook of qualitative research*. Sage Publications.

Derrida, J. (1978). Writing and difference. London: Routledge & Kegan. (Original work published 1967)

Descartes, R. (2006), A discourse on the method of correctly conducting one's reason and seeking truth in the sciences (I. MacLean, Trans.). Oxford University Press.

Dias, J., Mascarenhas, S., & Paiva, A. (2014). Fatima modular: Towards an agent architecture with a generic appraisal framework. In *Emotion Modeling* (pp. 44–56). Springer. doi:10.1007/978-3-319-12973-0_3

Diaz, L. (2016). *The woman who left* [Film]. L. Diaz (Producer).

Dilthey, W. (1900). *Die Entstehung der Hermeneutik*. Retrieved from https://archive.org/details/bub_gb_icZZAAAAcAAJ

Dog, N. (2013). *The Last of Us* [Computer video game]. Sony Interactive Entertainment.

Dōjōji engi. (1982). In *Muromachi jidai monogatari taisei, daiju, ten-hamo* [The Muromachi period narrative collection, No. 10, ten-hamo] (pp. 97–103). Tokyo, Japan: Kadokawa Shoten. (Original work published the Muromachi period)

Dōjōji monogatari. (1982). In *Muromachi jidai monogatari taisei, daiju, ten-hamo* [The Muromachi period narrative collection, No. 10, ten-hamo] (pp. 104–124). Tokyo, Japan: Kadokawa Shoten. (Original work published 1660, the Edo period)

Dōjōji. (1998). In Yōkyoku hyakuban, Shin nihon koten bungaku taikei, 57 [Nō 100 songs, New Japanese classic literature collection, 57] (pp. 52–57). Tokyo, Japan: Iwanami Shoten. (Original work published the Muromachi period)

Dormehl, L. (2018), A.I. border agents could use machine smarts to tell if travelers are lying. *Digital Trends*. https://www.digitaltrends.com/cool-tech/ai-border-airport-virtual-agent/

Dostoevsky, F. (2003). *The brothers Karamazov: A novel in four parts and an epilogue* (D. McDuff, Trans.). Penguin Classics. (Original work published 1880)

Dostoevsky, F. (2014). *Crime and punishment* (O. Ready, Trans.). Penguin Classics. (Original work published 1866)

Doyle, J. (1979). A truth maintenance system. *Artificial Intelligence*, *12*(3), 231–272. doi:10.1016/0004-3702(79)90008-0

Dream, Q. (2013). *Beyond Two Souls* [Computer video game]. Sony Interactive Entertainment.

Dreyer, C. T. (1928). *The passion of Joan of Arc* [Film]. France: Société générale des films.

Dyer, J., Gregersen, H., & Christensen, C. M. (2011). *The Innovator's DNA: Mastering the Five Skills of disruptive innovators*. Harvard Business School Press.

Eisenstein, S. M. (1942). *The film sense* (J. Leyda, Trans.). Harcourt Brace.

Eisenstein, S. M. (1949). *Film form* (J. Leyda, Trans.). Harcourt Brace.

Ekman, P. (1997). *What the face reveals: Basic and applied studies of spontaneous expression using the Facial Action Coding System (FACS)*. Oxford University Press.

Elliott, J. (2005). *Using narrative in social research: Qualitative and quantitative approaches*. Sage Publications. doi:10.4135/9780857020246

Fauconnier, G., & Turner, M. B. (1998). Conceptual integration networks. *Cognitive Science*, *22*(2), 133–187. doi:10.120715516709cog2202_1

Fauzi, F., & Belkhatir, M. (2014). Image understanding and the web: A state-of-the-art review. *Journal of Intelligent Information Systems*, *43*(2), 271–306. doi:10.100710844-014-0323-6

Finke, R. A., Ward, T. B., & Smith, S. M. (1990). *Creative cognition: Theory, research, and applications*. MIT Press.

Forster, E. M. (1927). *Aspects of the novel*. Edward Arnold.

Foucault, M. (1980). *Power/knowledge* (C. Gordon, L. Marshall, J. Mepham, & K. Soper, Trans.). Pantheon Books. (Original work published 1972, 1975, 1976, 1977)

Frederich, M. J., & Andrews, P. (2009). *Innovation passport: The IBM first-of-a-kind (FOAK) journey from research to realty*. IBM Press.

Fujii, S. (2004). *Monogatari riron kōgi* [Lectures on the theory of narrative]. Tokyo Daigaku Shuppankai.

Fukaya, M., & Tanaka, S. (1996). *The sense-making theory of <words>(<Kotoba> no Imizuke ron)*. Kinokuniya.

Fukushima, H. (2014). *Encyclopedia of Sake Terms*. Keio University.

Fukushima, H. (2020a in press). Cognitive Contents Generation: A Basic Introduction (Ninchiteki monogatari seisei heno shotai). In T. Ogata (Ed.), *Various Aspect of Post Narratology (Post Narratology no shoso)*. Shinyosya.

Fukushima, H. (2020b in press). Defining the aesthetic sake taste terms: A usage-based approach. In K. Toratani (Ed.), *The Language of Japanese Food: Cognitive Perspectives and Beyond*. John Benjamins.

Fukushima, K., Abe, A., & Kawada, R. (2019). How will sense of values change during art appreciation? *Proc. of 2019 IEEE International Conference on Data Mining Workshops (CDEC2019)*, 5—12. 10.1109/ICDMW.2019.00009

Furr, N., & Dyer, J. (2014). *The innovator's method: bringing the lean start-up into your organization.* Harvard Business Review Press.

Gauntner, J. (2011). Sake Handbook. North Clarendon: Tuttle Publishing.

Gaver, W. W., Beaver, J., & Benford, S. (2003). Ambiguity as a resource for design. In *Proceedings of the SIGCHI Conference on Human Factors in Computing Systems (pp.* 233-240*),* New York, NY: ACM.

Gavilances, J. M., Flatten, T. C., & Brettel, M. (2018). Content strategies for digital consumer engagement in social networks: Why advertising is an antecedent of engagement. *Journal of Advertising*, *47*(1), 4–23. doi:10.1080/00913367.2017.1405751

Genette, G. (1972). *Discours du récit, Essai de méthode, Figures III.* Paris: Seuil.

Genette, G. (1972). Discourse du recit: Essai de method. [Narrative discourse: An essay in method] In *Figures III.* Seuil.

Genette, G. (1980). *Narrative discourse: An essay in method* (J. E. Lewin, Trans.). Cornell University Press. (Original work published 1972)

Genji monogatari. (1993). *Shin nihon koten bungaku taikei, 19* [New Japanese classic literature collection, 19]. Tokyo, Japan: Iwanami Shoten. (Original work published early 11th century, the Heian period)

Genji monogatari. (1994). *Shin nihon koten bungaku taikei, 20* [New Japanese classic literature collection, 20]. Tokyo, Japan: Iwanami Shoten. (Original work published early 11th century, the Heian period)

Genji monogatari. (1995). *Shin nihon koten bungaku taikei, 21* [New Japanese classic literature collection, 21]. Tokyo, Japan: Iwanami Shoten. (Original work published early 11th century, the Heian period)

Genji monogatari. (1996). *Shin nihon koten bungaku taikei, 22* [New Japanese classic literature collection, 22]. Tokyo, Japan: Iwanami Shoten. (Original work published early 11th century, the Heian period)

Genji monogatari. (1997). *Shin nihon koten bungaku taikei, 23* [New Japanese classic literature collection, 23]. Tokyo, Japan: Iwanami Shoten. (Original work published early 11th century, the Heian period)

Genkyō shakusho. (1965). In *Shintei zōho kokushi taikei, 31* [New revision, supplement, national histories collection, 31]. Tokyo, Japan: Toshikawa Kobunkan. (Original work published 1322 (the Kamakura period)

Gergen, K. J. (1994a). *Toward transformation in social knowledge*. Sage Publications.

Gergen, K. J. (1994b). *Realities and relationships*. Harvard University Press.

Gergen, K. J. (1999). *An invitation to social construction*. Sage Publications.

Gervás, P. (2013). Propp's morphology of the folk tale as a grammar for generation. *OpenAccess Series in Informatics.*, *32*, 106–122. doi:10.4230/OASIcs.CMN.2013.106

Gibson, J. J. (1979). *The ecological approach to visual perception*. Houghton Mifflin.

Gick, M. L., & Holyoak, K. J. (1980). Analogical problem solving. *Cognitive Psychology*, *12*(3), 306–355. doi:10.1016/0010-0285(80)90013-4

Godard, J. L. (2018). *The image book* [Film].

Gratch, J., Marsella, S., Wang, N., & Stankovic, B. (2009). Assessing the validity of appraisal-based models of emotion. In *3rd International Conference on Affective Computing and Intelligent Interaction and Workshops* (pp. 1-8). IEEE. 10.1109/ACII.2009.5349443

Gratch, J., Okhmatovskaia, A., & Duncan, S. (2006, November). Virtual humans for the study of rapport in cross cultural settings. *25th Army Science Conference*, 27-30.

Gray, C. A. (2000). *The new social story book: Illustrated edition*. Future Horizons Inc.

Greenhalgh, T., & Hurwitz, B. (Eds.). (1998). *Narrative based medicine*. BMJ Books.

Guyatt, G. H. (1991). Evidence-based medicine. *American College of Physicians Journal Club*, *114*, A-16.

Haniya, Y. (1998). Shirei [Death Spirits]. Tokyo, Japan: Kōdansha (Original work publishued 1948-1995)

Hart, H. (2005). Do You See What I See? The Impact of Delsarte on Silent Film Acting. *Mime Journal*, *23*(1), 184–199. doi:10.5642/mimejournal.20052301.11

Hasumi, S. (2014). *Bovary fujin ron* [A Theory of "Madame Bovary"]. Chikumashobō.

Hattori, Y. (1970). Kabuki no kōzō [The structure of kabuki]. Tokyo, Japan: Iwanami Shoten.

Hayek, F. A. (2008). Kankakuchitujo (S. Akiyama, Trans.). Tokyo, Japan: Syunjuusha.

Hemingway, E. (1994). *For whom the bell tolls*. Arrow. (Original work published 1940)

Hesse, H. (2013). *Demian: The Story of Emil Sinclair's Youth* (D. Searls, Trans.). Penguin Random House. (Original work published 1919)

Hesse, H. (2015). *Steppenwolf* (D. Horrocks, Trans.). Picador USA. (Original work published 1927)

Hirota, A. (2016). The realization of prototypes in the product-design process - factors behind the realization of rapid prototyping. *Japan Society of Marketing & Distribution FY2016 Collection of Papers Presented at the National Research Conference, Japan Society of Marketing & Distribution*, (65), 6-98.

Hirota, A. (2017a). The prototype used in the implementation of the "Need-Solution Pairs" - The prototype use of single-user and multi-user to relate. In *Proceedings of 14th Open and User Innovation Society Meeting* (pp. 1-3). Innsbruck, Austria: Open and User Innovation Society Meeting.

Hirota, A. (2017b). The role of prototyping in creative moments (business insight). *Japan Society of Marketing & Distribution FY2017 Collection of Papers Presented at the National Research Conference, Japan Society of Marketing & Distribution*, *66*, 81-87.

Hirota, A. (2019). Pivot chain to create the pairing in Needs solutions pairs, Ferenze. *ISPIM Innovation Conference*, 1-17.

Hirota, A., Takemura, M., & Mizuno, M. (2017). Design prototyping in the "fuzzy front end" of product development – rapid prototyping at the stage of high uncertainty. *ISPIM Innovation Forum*, 1-7.

Hirota, A. (1999). Only one market -Development of "Cyber Zero" by ASICS, Tokyo. *Japan Marketing Journal*, *19*(2), 62–74.

Hirota, A. (2018). Effect of "prototyping stage" for "Need-Solution Pairs" in design thinking, Stockholm. *ISPIM Innovation Conference*, 1-16.

Hobbs, J. R., & Sticke, M. (1993). Interpretation as abduction. *Artificial Intelligence*, *63*(1-2), 69–142. doi:10.1016/0004-3702(93)90015-4

Holbrook, M. B., & Schindler, R. M. (1991). Echoes of the dear departed past: Some work in progress on nostalgia. In R. H. Holman & M. R. Solomon (Eds.), *Advances in Consumer Research, 18* (pp. 330–333). Association for Consumer Research.

Holyoak, K. J., & Thagard, P. (1995). *Mental leaps: Analogy in creative thought.* MIT Press.

Homer. (2006). The Iliad & The Odyssey (B. Williams, Trans.). TX: El Paso Norte Press. (Original work published the late eighth or early seventh century BC)

Hori, K. (2007). Sōzō katsudō shien system no riron to ōyō [Theories and applications of creativive activity aid]. Tokyo, Japan: Ohmsha.

Hori, K. (2007). *Creativity support, theory and application of AI approach toward creativity.* Ohmsha.

Hudlicka, E. (2003). Modeling effects of behavior moderators on performance: Evaluation of the MAMID methodology and architecture. *Proceedings of BRIMS*, 12.

Hugo, V. (1982). *Les miserable* (N. Denny, Trans.). Penguin Classics. (Original work published 1862)

Huhmann, B. A., & Albinsson, P. A. (2012). Does rhetoric impact advertising effectiveness with liking controlled? *European Journal of Marketing*, *46*(11/12), 1476–1500. doi:10.1108/03090561211259943

Hutson, J. P., Smith, T. J., Magliano, J. P., & Loschky, L. C. (2017). What is the role of the film viewer? The effects of narrative comprehension and viewing task on gaze control in film. *Cognitive Research: Principles and Implications*, *2*(1), 46. doi:10.118641235-017-0080-5 PMID:29214207

Ikeda, K. (2016). *Nihonjin no kangaekata sekai no hito no kangaekata—Sekaikachikanchousa kara mierumono* [Japanese way of thinking World way of thinking: What we can see from the World Values Survey?]. Keisoushobou.

Ikeo, K., Aoiki, K., Manami, C., & Inoue, T. (2010). *Matketing.* Yuibikaku.

Inoue, T. (2015). *Mishima Yukio "Hōjō no umi" vs Noma Hiroshi "Seinen no wa"—Sengo bungaku to zentai shōsetsu* [Mishima Yukio *The sea of fertility* vs. Noma Hirosi *The ring of young people*: Post-war literature and total novels]. Shintensha.

Ishii, J. (1993). *Marketing myths.* Iwanami Shoten.

Ishii, J. (2009). *Business insight – what is creation knowledge?* Iwanami Shoten.

Ito, T. (2000). Serufu herupu gurupu to kojin no monogatari [Self-help groups and personal stories]. *Japanese Sociological Review*, *51*(1), 88–103.

Iura, Y. (1990). "Sakusha shikihō kezairoku" kō (zen)—Mondaiten no shiteki to kaimei [Considering "Sakusha shikihō kezairoku" (the first part): Indication and elucidation of problems]. *Gakuen*, *610*, 2–13.

Iura, Y. (1991). "Sakusha shikihō kezairoku" kō (go)—Mondaiten no shiteki to kaimei [Considerating "Sakusha shikihō kezairoku" (the second part): Indication and elucidation of problems). *Gakuen*, *615*, 1–13.

Jauss, H. R. (1970). Literaturgeschichte als provokation. Frankfurt am Main: Suhrkamp Verlag.

Jenny, L., & Trezise, T. (1989). From Breton to Dali: The adventures of automatism. October, 51, 105-114.

Jhale, A., & Young, M. R. (2010). Cinematic visual discourse: Representation, generation, and evaluation. *IEEE Transactions on Computational Intelligence and AI in Games*, *2*(2), 69–81. doi:10.1109/TCIAIG.2010.2046486

Joyce, J. (2010). *Ulysses.* Wordsworth Editions. (Original work published 1922)

Kakas, A. C., Kowalski, R. A., & Toni, F. (1992). Abductive logic programming. *Journal of Logic and Computation*, *2*(6), 719–770. doi:10.1093/logcom/2.6.719

Kanai, A. (2016). Non-story, nostalgia, and film cognition: Nostalgia-based narrative rhetoric composition. In T. Ogata & T. Akimoto (Eds.), Computational and Cognitive Approaches to Narratology (pp. 376–390). Hershey, PA: Information Science Reference (IGI Global).

Kanai, A. (2018). Narrative simulation for film rhetoric composition with or without story and nostalgia effects. In T. Ogata & S. Asakawa (Eds.), Content generation through narrative communication and simulation (pp. 148–161). Hershey, PA: Information Science Reference (IGI Global). doi:10.4018/978-1-5225-4775-4.ch003

Kanai, A. (2019). Past, non-story narrative film, and nostalgia. In T. Ogata, & T. Akimoto (Eds.), Post-narratology through computational and cognitive approaches (pp. 277–291). Hershey, PA: Information Science Reference (IGI Global). doi:10.4018/978-1-5225-7979-3.ch006

Kanai, A. (2001a). The story as a constraint on film. *Proceedings of the Third International Conference on Cognitive Science*, 503–506.

Kanai, A. (2001b). Cognitive process model toward the rhetoric of the film. *Cognitive Studies*, *8*(2), 139–150.

Kanai, A. (2006). Reality of unfamiliar situations on moving image. *Proceedings of 19th Congress of the International Association of Empirical Aesthetics*, 800–803.

Kanai, A., & Kodama, K. (2010). Film edit design: Story and irrational cutting. *Cognitive Studies*, *17*(3), 444–458.

Kanai, A., & Ogata, T. (2004a). Non-story processing on the film rhetoric composition system. *Proceedings of 18th Congress of the International Association of Empirical Aesthetics*, 433–436.

Kanai, A., & Ogata, T. (2004b). Aspect of non-story processing and film rhetoric composition in the narrative generation mechanism. *Proceedings of the Ninth International Symposium on Artificial Life and Robotics*, 162–165.

Kandel, E. R. (2018). *Reductionism in art and brain science*. Columbia University Press.

Kano, F. (2015). Comprehending consumers: Computing complexity of cultures. In *Proceedings of the 29th Annual Conference of the Japanese Society for Artificial Intelligence*, Hokkaido. Japan: The Japanese Society for Artificial Intelligence.

Kato, S. (2010). Gendai seishinigaku niokeru rejiriansu no gainen no igi [Significance of the concept of the resilience in present psychiatry]. In Resilience: The new paradigm of present (pp. 1-23). Kanehara & Co., ltd.

Katō, T. (1997a). "Sekai kōmoku" no shohon—Sono ichizuke to tensha no keii [Various books of "sekai kōmoku": Their positioning and the process of transcription]. *Gakushūin Daigaku Kokugo Kokubungaku Kaishi*, *40*, 79–94.

Katō, T. (1997b). "Sekai kōmoku" no seiritsu [The establishment of "Sekai kōmoku"]. *Gakshūuin Diagaku Jinbun Kagaku Ronshū*, *6*, 77–100.

Katz, A. H., & Bender, E. I. (1976). Self-help groups in western society: History and prospects. *Journal of Behavioural Science*, *12*(3), 265–282. doi:10.1177/002188637601200302

Katz, S. D. (1991). *Film directing shot by shot visualizing from concept to screen*. Michael Wise Productions.

Kawai, M., Ono, J., & Ogata, T. (2020). Analyzing the stage performance structure of *kabuki*-dance "*Kyoganoko Musume Dojoji*." In *Proceedings of the 34th Annual Conference of the Japanese Society of Artificial Intelligence* (3D1-OS-22a-04). Tokyo, Japan: Japanese Society of Artificial Intelligence.

Kawai, H. (2001). Shinriryoho niokeru Monogatari no Igi [Significance of the "narrative" in psychotherapy]. *Japanese Journal of Psychotherapy*, *27*, 3–7.

Kawai, M., & Ogata, T. (2019). An analysis of "*Kyōganoko musume Dōjōji*" for the generation of the staging structures of *kabuki* from narrative plots. In *Proceedings of the 63th Special Interest Group on Language Sense Processing Engineering* (pp. 25–61). Tokyo, Japan: The Japanese Society for Artificial Intelligence.

Kawamura, Y., Kato, T., Tachibana, K., Okai, G., & Kinoshita, S. (2014). Kokoku kyanpen no jirei ruikei no kokoromi: Junkangata shohisha kodo moderu iDEACCycle no kanten kara [An attempt of case categorization of advertising campaigns: From the view point of cyclical consumer behavior model iDEACCycle]. *Bulletin of Nikkei Advertising Research Institute,* (273), 10-16.

Kawamura, Y. (2003). An analysis of the rhetoric of commercial film: Toward the building of a commercial film production support system based on image rhetoric. In *Proceedings of the 2003 IEEE International Conference on Systems, Man and Cybernetics* (pp.993-1000). Washington, DC: IEEE. 10.1109/ICSMC.2003.1243944

Kawamura, Y. (2004). An analysis of the rhetoric of commercial film: Toward the building of a commercial film production support information system. *Journal of Advertising Science*, *45*, 122–139.

Kawamura, Y. (2006). An analysis on rhetoric of the commercial film of beer and personal computer: A symbolic approach on rhetoric of commercial film. *Journal of Advertising Science*, *47*, 33–48.

Kawamura, Y. (2007). A study on the technique-rhetorics and effects of commercial film. *Cognitive Studies*, *14*(3), 409–423.

Kawamura, Y. (2009). An analysis on the story and editing techniques of commercial film: Toward the building of a commercial film production support information system. *Journal of Advertising Science*, *50*, 16–32.

Kawamura, Y. (2013). Junkangata shohisha kodo moderu iDEACCycle no teian: Shohisha nikki chosa/ intavu chosa wo motoni [A proposal of cyclical consumer behavior model iDEACCycle: Based on consumer dairy survey and interview survey]. *Japan Marketing Journal*, *33*(2), 95–109.

Kawamura, Y. (2018). Practice and modeling of advertising communication strategy: Sender-driven receiver-driven. In T. Ogata & S. Asakawa (Eds.), *Content generation through narrative communication and simulation* (pp. 358–379). IGI Global. doi:10.4018/978-1-5225-4775-4.ch013

Kawamura, Y. (2019). An attempt of the commercial film production support system based on the image rhetoric of commercial film. In T. Ogata & T. Akimoto (Eds.), *Post-narratology through computational and cognitive approaches* (pp. 292–317). IGI Global. doi:10.4018/978-1-5225-7979-3.ch007

Kawatake, T. (2003). Kabuki: Baroque fusion of the arts (Connel Hoff, F., & J., Trans.). Tokyo, Japan: The International House of Japan.

Kawatake, T. (1989). *Kabuki-bi ron* [On *kabuki*'s beauty]. Tokyo Daigaku Shuppankai.

Kawatake, T. (1992). *Kabuki: Its beauty and tradition* (C. Holmes, Trans.). Japan Arts Council.

Keller, K. (2013). *Strategic brand management* (4th ed.). Pearson.

Kelly, T., & Kelly, D. (2013). Creative confidence. New York, NY: Fletcher and Company.

Ki no kuni muro no kōri no ashiki on'na. (1974). In *Ōjōden Hokkegenki* (pp. 217–219). Tokyo, Japan: Iwanami Shoten. (Original work published 1040–1043)

Ki no kuni no Dōjōji no sō hōke wo utsushi hemi wo sukueru koto. (1993). In *Konjaku monogatari shū, 3* (pp. 286–290). Tokyo, Japan: Iwanami Shoten. (Original work published the end of the Heian period)

Kijima, N. (2014). *Cloninger no personality riron nyuumon* [An introduction to Croninger's personality theory]. Kitaoojishobou.

Klahr, D., & Dunbar, K. (1988). Dual space search during scientific reasoning. *Cognitive Science, 12*(1), 1–48. doi:10.120715516709cog1201_1

Kominami, T. (1998). "Kezairoku" no kōsatsu—Engeki ni okeru kankyaku ninshiki no shomondai [Consideration of "Kezairoku": Problems of audience recognition in a play]. *Otaru Joshi Tanki Daigaku Kenkyū Kiyō, 27*, 1–42.

Komori, Y., & Yamada, M. (2001). Seishinbunretsubyou no kazokushinrikyouiku niokeru naratibu apurochi [Narrative approach in family psychology education of schizophrenia]. *Family Therapy Research, 18*(2), 143–150.

Konjaku monogatari shū, 1. (1999a). In *Shin nihon koten bungaku taikei, 33* [New Japanese classic literature collection, 33]. Tokyo, Japan: Iwanami Shoten. (Original work published the end of the Heian period)

Konjaku monogatari shū, 2. (1999b). In *Shin nihon koten bungaku taikei, 34* [New Japanese classic literature collection, 34]. Tokyo, Japan: Iwanami Shoten. (Original work published the end of the Heian period)

Konjaku monogatari shū, 3. (1993). In *Shin nihon koten bungaku taikei, 35* [New Japanese classic literature collection, 35]. Tokyo, Japan: Iwanami Shoten. (Original work published the end of the Heian period)

Konjaku monogatari shū, 4. (1994). In *Shin nihon koten bungaku taikei, 36* [New Japanese classic literature collection, 36]. Tokyo, Japan: Iwanami Shoten. (Original work published the end of the Heian period)

Konjaku monogatari shū, 5. (1996). In *Shin nihon koten bungaku taikei, 37* [New Japanese classic literature collection, 37]. Tokyo, Japan: Iwanami Shoten. (Original work published the end of the Heian period)

Kotler, P., Kartajaya, H., & Setiawan, I. (2017). *Marketing 4.0: Moving from traditional to digital*. Gildan Media.

Kotler, P., & Keller, K. (2009). *Marketing management*. Pearson.

Kotler, P., & Takaoka, K. (2016). *Marketing no susume*. Chuoukouronsya.

Kristeva, J. (1980). *Desire in language: a semiotic approach to literature and art*. Columbia University Press.

Kurihara, K. (Ed.). (1886). *Kokon hauta taizen (Kaichū keikobon)*. Eirakudō.

Kuznetsov, O. P. (2013). Cognitive semantics and artificial intelligence. *Scientific and Technical Information Processing, 40*(5), 269–276. doi:10.3103/S0147688213050067

Kyōganoko musume Dōjōji. (1970). In *Meisaku kabuki zenshū, 19, Buyōgeki shū* [Masterpeace *kabuki* collection, 19, *Kabuki*-dance works] (pp. 15–30). Tokyo, Japan: Tokyo Sōgen Shinsha. (First performed in 1753. Written by Fujimoto Tobun and composed by Kineya Yasaburō)

Lakoff, G. (1987). *Women, fire and dangerous things: What categories reveal about the mind*. The University of Chicago Press. doi:10.7208/chicago/9780226471013.001.0001

Lakoff, G., & Johnson, M. (1980). *Metaphor we live by*. The University of Chicago Press.

Lane, H. C., Cahill, C., Foutz, S., Auerbach, D., Noren, D., Lussenhop, C., & Swartout, W. (2013). The Effects of a Pedagogical Agent for Informal Science Education on Learner Behaviors and Self-efficacy. In Artificial Intelligence in Education. AIED 2013. Lecture Notes in Computer Science, (vol. 7926). Springer.

Lascarides, A., & Oberlander, J. (1992). Abducing temporal discourse. *Proc. of 6th. int'l Workshop on Natural Language Generation (LNAI-587), 167—182.*

Laskey, H. A., Day, E., & Crask, M. R. (1989). Typology of main message strategies for television commercial. *Journal of Advertising, 18*(1), 36–41. doi:10.1080/00913367.1989.10673141

Laskey, H. A., Fox, R. J., & Crask, M. R. (1994). Investigation the impact of executional style on television commercial effectiveness. *Journal of Advertising Research, 34*(6), 9–16.

Laskey, H. A., Fox, R. J., & Crask, M. R. (1995). The relationship between advertising message strategy and television commercial effectiveness. *Journal of Advertising Research, 35*(2), 31–39.

Lazarus, R. S. (1991). Cognition and motivation in emotion. *The American Psychologist, 46*(4), 352–367. doi:10.1037/0003-066X.46.4.352 PMID:2048794

Lehrer, A. C. N.-P. L. (2009). Wine & conversation (2nd ed). Oxford: Oxford University Press.

Leong, C. W., Mihalcea, R., & Hassan, S. (2010). Text mining for automatic image tagging. In *Proceedings of the 23th International Conference on Computational Linguistics* (pp.647-655). Beijing, China: Academic Press.

Lester, R. K., & Piore, M. (2004). *Innovation the missing dimension.* Harvard University Press.

Levy-Strauss, C. (1976). Yasei no sikou (Y. Oohashi, Trans.). Tokyo, Japan: Misuzushobou.

Lisetti, C., & Hudlicka, E. (2015). Why and how to build emotion-based agent architectures. In R. A. Calvo, S. K. D'Mello, J. Gratch, & A. Kappas (Eds.), *Oxford Library of Psychology. The Oxford Handbook of Affective Computing* (pp. 94–109). Oxford University Press.

Longus. (2019). *Daphnis and Chloe* (E. Bowie, Trans.). Cambridge, UK: Cambridge University Press. (Original work published the second century)

López-Arroyo, B., & Roberts, R. P. (2014). English and Spanish descriptors in wine tasting terminology. *Terminology, 20*(1), 25–49. doi:10.1075/term.20.1.02lop

Loschky, L. C., Larson, A. M., Smith, J. T., & Magliano, J. P. (2020). The scene perception and event comprehension theory (SPECT) applied to visual narratives. *Topics in Cognitive Science, 12*(1), 311–351. doi:10.1111/tops.12455 PMID:31486277

Lusch, R. F., & Vargo, S. L. (2014). *Service dominant logic, premises, perspectives, possibilities.* Cambridge University Press.

Lutterbie, J. (2011). *Towards a general theory of acting: Cognitive science and performance.* Palgrave Macmillan. doi:10.1057/9780230119468

Lyotard, J. F. (1979). *La condition postmoderne* [The postmodern condition]. Minuit.

Maes, P. (1997). Intelligent Software. *IUI '97 Proceedings of the 2nd International Conference on Intelligent User Interfaces*, 41-43.

Magliano, J. P., & Zacks, J. M. (2011). The impact of continuity editing in narrative film on event segmentation. *Cognitive Science, 35*(8), 1489–1517. doi:10.1111/j.1551-6709.2011.01202.x PMID:21972849

Malatesta, L., Caridakis, G., Raouzaiou, A., & Karpouzis, K. (2007). Agent personality traits in virtual environments based on appraisal theory predictions. *Artificial and ambient intelligence, language, speech and gesture for expressive characters*, 7.

Malatesta, L., Raouzaiou, A., Karpouzis, K., & Kollias, S. (2009). Towards modeling embodied conversational agent character profiles using appraisal theory predictions in expression synthesis. *Applied Intelligence, 30*(1), 58–64. doi:10.100710489-007-0076-9

Malraux, A. (1990). *Man's fate* (H. M. Chevalier, Trans.). Vintage. (Original work published 1933)

Mann, T. (2005). *The magic mountain* (J. E. Woods, Trans.). Everyman. (Original work published 1924)

Manuel, J., Fang, L., Bellamy, J. L., & Bledsoe, S. E. (2008). Evaluating evidence. In R. M. Grinnell & Y. A. Unrau (Eds.), *Social work research and evaluation: Foundation of evidence-based practice* (8th ed., pp. 482–495). Oxford University Press.

Margolin, U. (1986). The doer and the deed: Action as a basis for characterization in narrative. *Poetics Today*, *7*(2), 205–225. doi:10.2307/1772759

Maslow, A. H. (1987). Ningensei no sinrigaku (T. Oguchi, Trans.). Tokyo, Japan: Kinokuniyashoten.

Mathematical Systems. (2011). *Text mining studio operating manual version 4.0*. Mathematical Systems.

Matsumoto, S. (1961). Warui yatsura [Bad Sorts]. Tokyo, Japan: Shinchōsha.

Matsumoto, S. (1962). Renkan [Linkage]. Tokyo, Japan: Kōdansha.

Matsumoto, S. (1980). Kurokawa no techō [Pocketbook of black leather]. Tokyo, Japan: Shinchōsha.

Matsumoto, S. (1981). Yakō no Kaidan [Stairs That Shine at Night]. Tokyo, Japan: Shinchōsha.

Matsuura, T. (1992). Terms for tasting Sake (Sake wo ajiwau kotoba). In I. Tajima & K. Niwa (Eds.), *Gendai Nihongo no kenkyū*. Izumi Shoin.

McAdams, D. P., & De St. Aubin, E. (Eds.) (1998). Generativity and adult development: How and why we care for the next generation. American Psychological Association.

McQuarrie, E. F., & Mick, D. G. (1996). Figures of rhetoric in advertising language. *The Journal of Consumer Research*, *22*(4), 424–438. doi:10.1086/209459

McQuarrie, E. F., & Mick, D. G. (1999). Visual rhetoric in advertising: Text-interpretive, experimental, and reader-response analysis. *The Journal of Consumer Research*, *26*(1), 37–54. doi:10.1086/209549

McQuarrie, E. F., & Mick, D. G. (2003). Visual and verbal rhetorical figures under directed processing versus incidental exposure to advertising. *The Journal of Consumer Research*, *29*(4), 579–587. doi:10.1086/346252

McQuarrie, E. F., & Phillips, B. J. (2005). Indirect persuasion in advertising: How consumers process metaphors presented in picture and words. *Journal of Advertising*, *34*(2), 7–19. doi:10.1080/00913367.2005.10639188

McRoy, S., & Hirst, G. (1993). Abductive explanation of dialogue misunderstandings. *Proc. of EACL93*, 277—286.

McRoy, S., & Hirst, G. (1995). The repair of speech act misunderstandings by abductive inference. *Computational Linguistics*, *21*(4), 435–478.

Mead, G. H. (1934). *Mind, self, and society*. University of Chicago Press.

Mehrabian, A. (1996). Pleasure-arousal-dominance: A general framework for describing and measuring individual differences in temperament. *Current Psychology (New Brunswick, N.J.)*, *14*(4), 261–292. doi:10.1007/BF02686918

Meisner, S. (1987). Sanford Meisner on acting. Vintage Random House.

Melville, H. 's (2009). Moby-Dick: Or, the whale. London, UK: Penguin Classics. (Original work published 1851)

Metz, C. (1968). *Essais sur la signification au cinéma* (Vol. 1). Klincksieck.

Miidera (1998). In *Yōkyoku hyakuban, Shin nihon koten bungaku taikei, 57* [Nō 100 songs, New Japanese classic literature collection, 57] (pp. 52–57). Tokyo, Japan: Iwanami Shoten. (Original work published the Muromachi period)

Mishima, Y. (1950). Ao no Jidai. Tokyo, Japan: Shinchōsha.

Mishima, Y. (1951, 1953). Kinjiki. Tokyo, Japan: Shinchōsha.

Mishima, Y. (1954a). Shiosai. Tokyo, Japan: Shinchōsha.

Mishima, Y. (1954b). Iwashiuri koi no hikiami [The sardine seller's net of love]. *Engekikai*, 11.

Mishima, Y. (1956). Kinkakuji. Tokyo, Japan: Shinchōsha.

Mishima, Y. (1962). Utsukushii hoshi. Tokyo, Japan: Shinchōsha.

Mishima, Y. (1968). Taiyō to tetsu. Tokyo, Japan: Kōdansha.

Mishima, Y. (1969a). Haru no yuki. Tokyo, Japan: Shinchōsha.

Mishima, Y. (1969b). Honba. Tokyo, Japan: Shinchōsha.

Mishima, Y. (1969c-1971). *Hōjō no umi* (4 books). Tokyo, Japan: Shinchōsha.

Mishima, Y. (1970). Akatsuki no tera. Tokyo, Japan: Shinchōsha.

Mishima, Y. (1971). Ten'nin gosui. Tokyo, Japan: Shinchōsha.

Mishima, Y. (2000–2006). *Ketteiban Mishima Yukio zenshū, 1–42* [Complete works of Mishima Yukio (definitive edition), Vols. 1–42]. Tokyo, Japan: Shinchōsha.

Mishler, E. G. (1995). Models of narrative analysis: A typology. *Journal of Narrative and Life History*, *5*(2), 87–123. doi:10.1075/jnlh.5.2.01mod

Mitchell, M. (2014). *Gone with the wind.* Pan Books. (Original work published 1936)

Miura, H. (2000). Naimaze. In Y. Hattori, T. Tomita, & T. Hirosue, (Eds.). Shinpan kabuki jiten [New edition kabuki dictionry] (p. 310). Tokyo, Japan: Heibonsha.

Momochidori musume Dōjōji. (1960). In T. Takano (Ed.), Nihon kayō shūsei, 9, Kinseihen (p. 86). Tokyo, Japan: Tokyodō Shuppan.

Monaco, J. (1977). *How to read a film: the art technology, language, history and theory of film and media.* Oxford University Press.

Morgan, A. (2000). *What is narrative therapy? An easy-to-read introduction.* Dulwich Center Publication.

Morgans, J. (2018). How video games cast actors just like movies do. *Vice*. https://www.vice.com/en_uk/article/gymde7/how-video-games-cast-actors-just-like-movies-do

Moriyama, S. (1993). Tsuruya Nanboku—Naimaze no sekai [Tsuruya Nanboku: Naimaze's world]. Tokyo, Japan: San'ichi Shobō.

Mothersbaugh, D. L., Huhmann, B. A., & Franke, G. R. (2002). Combinatory and separative effects of rhetorical figures on consumers' effort and focus in ad processing. *The Journal of Consumer Research*, *28*(4), 589–602. doi:10.1086/338211

Mukaiyachi, I. (2013) *Tojisha-kenkyu toha: Tojisha-kenkyu no rinen to kousei* [What is a Tojisha-Kenkyu? The concept and structure of a Tojisha-Kenkyu]. Tojisya-Kenkyu network (https://toukennet.jp/?page_id=56)

Murray, J. H. (1997). *Hamlet on the Holodeck: The future of narrative in cyberspace*. MIT Press.

Nagauta genpon shūsei kankōkai. (Ed.). (1927). *Momochidori musume Dōjōji (Nagauta genpon shūsei, Vol. 1)*. Tokyo, Japan: Nagauta Genpon Shūsei Kankōkai.

Naito, Y., & Kanai, A. (2012a). The narrative structure of nostalgia cognition and film. In *Proceedings of the International Conference on Cognitive Science '12*. Academic Press.

Naito, Y., & Kanai, A. (2012b). Nostalgia and film creation. In *Proceedings of the 26th Annual Conference of the Japanese Society for Artificial Intelligence (1N1-OS-1a-2)*. Yamaguchi, Japan: Academic Press.

Nakamura, J., & Ohsawa Y. (2008a). Effect of vacilliation on analogical thought process for concept creation. *Journal of Transdicsiplinary Federation of Science and technology*, *2*(1), 40-48.

Nakamura, J., & Ohsawa, Y. (2008b). *Kouji ninchi kinoo wo hikidasu kashikagiho* (in Japanese). Paper presented at the 25th Annual conference of Japanese Cognitive Science Society, Kyotanabe, Kyoto.

Nakamura, J., Takagi, S., & Ohsawa, Y. (2012). *Reframing method for deployment of business domain*. Paper presented at SIG-KST, Hiyoshi, Kanagawa.

Nakamura, J. (2014a). Discovering design concept for immovable objects. *Procedia Computer Science*, *35*, 954–960. doi:10.1016/j.procs.2014.08.175

Nakamura, J. (2014b). Creating essential hypothesis and rules in product planning: Introducing abductive inference model. In *Proceedings of International Conference on Data Mining Workshop (ICDM2014)* (pp. 471-476), New York: NY: IEEE. 10.1109/ICDMW.2014.63

Nakamura, J., & Ohsawa, Y. (2009). A shift of mind – introducing a concept creation model. *Information Sciences*, *179*(11), 1639–1646. doi:10.1016/j.ins.2008.11.036

Nakamura, J., & Teramoto, M. (2013). Frame as a clue to intention of data: Toward new product ideas with framed components. In *Proceedings of International Conference on Data Mining Workshop (ICDM2013)* (pp. 720-725). New York, NY: IEEE. 10.1109/ICDMW.2013.28

Natsume, S. (2007). Bungaku ron (I, II) [Literary theory (I, II)]. Tokyo, Japan: Iwanami Shoten. (Original work published 1907)

Natsume, S. (2014). Kōfu. Tokyo, Japan: Iwanami Shoten. (Original work published 1908)

Nelson, C. D. (2011). Narratives of classroom life: Changing conceptions of knowledge. *TESOL Quarterly*, *45*(3), 463–485. doi:10.5054/tq.2011.256799

Nixon, M., Pasquier, P., & El-Nasr, M. S. (2010). DelsArtMap: Applying delsarte's aesthetic system to virtual agents. In *International Conference on Intelligent Virtual Agents* (pp. 139-145). Springer. 10.1007/978-3-642-15892-6_15

Nobeoka, K., Kimura, M., & Osanai, A. (2015). Creation design value. *Hitotubashi Business Review*, *62*(4), 6–21.

Noguchi, Y. (Ed.). (2005). Naratibu no rinshoshakaigaku [Clinical sociology of narrative]. Keiso shobo.

Noma, H. (1983-84). *Seinen no wa, 1–5*. Tokyo, Japan: Iwanami Shoten. (Original work published 1949–1968)

Norman, D. (2013). *The Design of Everyday Things: Revised and expanded edition*. Basic Books.

O'Barr, W. M. (2010). The rise and fall of the TV commercial. *Advertising & Society Review*, *11*(2).

Ōe, K. (1979). Dōjidai game. Tokyo, Japan: Shinchōsha.

Ogata, T. (1999). Proposal of computational narratology. In *Proceedings of IPSJ SIG Computers and the Humanities* (pp. 53-60). Academic Press.

Ogata, T. (2002a). Expanded literary theory: Cognitive/computational expansion of literary theories and narratology. In *Proceedings of the 17th Congress of the International Association of Empirical Aesthetics* (pp. 163–166). Rome, Italy: University of Rome Tre.

Ogata, T. (2003a). Monogatari no tajūsei to kakuchō bungakuriron no gainen—System narratology ni mukete, I [Narrative multiplicity and the concept of expanded literary theory: Toward a system narratology, I]. In M. Yoshida (Ed.), Fukuzatsu-kei shakai riron no shin chihei [New paradigm of complex social system theory] (pp. 127–181). Tokyo, Japan: Senshu Daigaku Shuppankyoku.

Ogata, T. (2003b). Kakuchō bungakuriron no kokoromi—System narratology ni mukete, II [Attempts of expanded literary theory: Toward a system narratology II]. In M. Yoshida (Ed.), Fukuzatsu-kei shakai riron no shin chihei [New paradigm of complex social system theory] (pp. 309–356). Tokyo, Japan: Senshu Daigaku Shuppankyoku.

Ogata, T. (2010). Shōsetsu—Ryūdō to kotei, Sakuhin no hō he [Novels: Fluidity and fixation, toward works]. In T. Ogata & A. Kanai (Eds.), Monogatariron no jōhōgaku josetsu—Monogatari seisei no shisō to gijutsu wo megutte [An introduction to informatics of narratology: Around the thoughts and technologies of narrative generation] (pp. 130–169). Tokyo, Japan: Gakubunsha.

Ogata, T. (2016a). Narrative generation of *zukushi*—Towards its systematization through surveying *kabuki* and developing a proper noun conceptual dictionary in INGS. In *Proceedings of the 30th Annual Conference of the Japanese Society of Artificial Intelligence* (1K5-OS-06b-2). Tokyo, Japan: The Japanese Society for Artificial Intelligence.

Ogata, T. (2016b). Kabuki as multiple narrative structures. In T. Ogata & T. Akimoto (Eds.), Computational and cognitive approaches to narratology (pp. 391–422). Hershey, PA: Information Science Reference (IGI Global).

Ogata, T. (2018a). Kabuki ni mukete (1)—Shiisei to henshūsei no monogatari kara tajūsei to jitsuzon no monogatari he [Toward kabuki (1): From narratives of arbitrariness and editing to narratives of multiplicity and real existence]. In T. Ogata, Y. Kawamura, & A. Kanai (Eds.), Jōhō monogatariron—Jinkōchinō, ninchi, shakai katei to monogatari seisei [Informational narratology: Artificial intelligence/cognition/social process and narrative generation] (pp. 187–208). Tokyo, Japan: Hakutō Shobō.

Ogata, T. (2018b). Kabuki ni mukete (2)—Tajū monogatari kōzō no shosō [Toward kabuki (2): Aspects of multiple narrative structures]. In T. Ogata, Y. Kawamura, & A. Kanai (Eds.), Jōhō monogatariron—Jinkōchinō, ninchi, shakai katei to monogatari seisei [Informational narratology: Artificial intelligence/cognition/social process and narrative generation] (pp. 209–244). Tokyo, Japan: Hakutō Shobō.

Ogata, T. (2018c). Gaibu he no monogatari seisei mata ha geinō jōhō system ni mukete [Narrative generation to the outside or toward a geinō information system]. In T. Ogata, Y. Kawamura, & A. Kanai (Eds.), Jōhō monogatariron—Jinkōchinō, ninchi, shakai katei to monogatari seisei [Informational narratology: Artificial intelligence/cognition/social process and narrative generation] (pp. 327–353). Tokyo, Japan: Hakutō Shobō.

Ogata, T. (2018d). Naibu he no monogatari seisei mata ha watakushi monogatari ni mukete [Narrative generation to the inside or toward I narratives]. In T. Ogata, Y. Kawamura, & A. Kanai (Eds.), Jōhō monogatariron—Jinkōchinō, ninchi, shakai katei to monogatari seisei [Informational narratology: Artificial intelligence/cognition/social process and narrative generation] (pp. 355–370). Tokyo, Japan: Hakutō Shobō.

Ogata, T. (2018e). An integrated approach to narrative generation: From Mishima and kabuki to narrative generation systems. In T. Ogata & S. Asakawa (Eds.), Content generation through narrative communication and simulation (pp. 49–147). Hershey, PA: Information Science Reference (IGI Global)

Ogata, T. (2019a). A computational, cognitive, and narratological approach to narrative generation. In T. Ogata, & T. Akimoto (Eds.), Post-narratology through computational and cognitive approaches (pp.1-84). Hershey, PA: Information Science Reference (IGI Global). doi:10.4018/978-1-5225-7979-3.ch001

Ogata, T. (2019b). Toward a post-narratology or the narratology of narrative Generation. In T. Ogata, & T. Akimoto (Eds.), Post-narratology through computational and cognitive approaches (pp. 85–142). Hershey, PA: Information Science Reference (IGI Global). doi:10.4018/978-1-5225-7979-3.ch002

Ogata, T. (2019c). Kabuki as Multiple Narrative Structures and Narrative Generation. In T. Ogata & T. Akimoto (Eds.), Post-narratology through computational and cognitive approaches (pp. 192–275). Hershey, PA: Information Science Reference (IGI Global). doi:10.4018/978-1-5225-7979-3.ch005

Ogata, T. (2020a). Toward an integrated approach to narrative generation: Emerging eesearch and opportunities. Hershey, PA: Information Science Reference (IGI Global). doi:10.4018/978-1-5225-9693-6

Ogata, T. (2020b). Areas of narratives or narrative genres. In Toward an integrated approach to narrative generation: Emerging research and opportunities (pp. 59–161). Hershey, PA: Information Science Reference (IGI Global). doi:10.4018/978-1-5225-9693-6.ch002

Ogata, T. (2020c). Narratology and post-narratology. In Toward an integrated approach to narrative generation: Emerging research and opportunities (pp. 162–314). Hershey, PA: Information Science Reference (IGI Global). doi:10.4018/978-1-5225-9693-6.ch003

Ogata, T. (2020d). Theoretical or philosophical considerations for an integrated narrative generation approach. In Toward an integrated approach to narrative generation: Emerging research and opportunities (pp. 315–403). Hershey, PA: Information Science Reference (IGI Global). doi:10.4018/978-1-5225-9693-6.ch004

Ogata, T. (2020e). Internal and external narrative generation based on post-narratology: Emerging research and opportunities. Hershey, PA: Information Science Reference (IGI Global). doi:10.4018/978-1-5225-9943-2

Ogata, T. (2020f). An integrated narrative generation system: Synthesis and expansion. In Internal and external narrative generation based on post-narratology: Emerging research and opportunities (pp. 1–108). Hershey, PA: Information Science Reference (IGI Global).

Ogata, T. (2020g). Kabuki as a synthetic narrative: Synthesis and expansion. In Internal and external narrative generation based on post-narratology: Emerging research and opportunities (pp. 109–254). Hershey, PA: Information Science Reference (IGI Global).

Ogata, T. (2020h). Toward external and internal practices of narrative generation. In Internal and external narrative generation based on post-narratology: Emerging research and opportunities (pp. 328–427). Hershey, PA: Information Science Reference (IGI Global). doi:10.4018/978-1-5225-9943-2.ch004

Ogata, T., & Akimoto, T. (Eds.). (2019). Post-narratology through computational and cognitive approaches. Hershey, PA: Information Science Reference (IGI Global). doi:10.4018/978-1-5225-7979-3

Ogata, T., & Asakawa, S. (Eds.). (2018). Content generation through narrative communication and simulation. Hershey, PA: Information Science Reference (IGI Global). doi:10.4018/978-1-5225-4775-4

Ogata, T., & Kanai, A. (2010). Monogatariron no jōhōgaku josetsu—Monogatari seisei no shisō to gijutsu wo megutte [An introduction to informatics of narratology: Around the thoughts and technologies of narrative generation]. Tokyo, Japan: Gakubunsha.

Ogata, T., & Ono, J. (2019). Two elements and two techniques for the narrative generation of *kabuki*. In *Proceedings of the 33th Annual Conference of the Japanese Society of Artificial Intelligence* (1F2-NFC-1-03). Tokyo, Japan: Japanese Society of Artificial Intelligence.

Ogata, T., Fukuda, K., Ono, J., & Ito, K. (2020). Considering *naimaze* in *kabuki* narratives and a method. In *Proceedings of the 34th Annual Conference of the Japanese Society of Artificial Intelligence* (3P5-OS-16b-02). Tokyo, Japan: Japanese Society of Artificial Intelligence.

Ogata, T. (2002b). The concept of system narratology: From the viewpoint of expanded literary theory. In *Proceedings of the 3rd International Workshop of Literature in Cognition and Computer in held in PRICAI2002* (W7-1, pp. 1–10). Tokyo, Japan: Japanese Cognitive Science Society.

Ogata, T. (2007). To the rhetoric of story from Propp: Centering around the rhetoric of de-composition and re-composition. *Cognitive Studies, 14*(4), 532–558.

Ogata, T. (2010). "Monogatari seisei sisutemu" no haikei oyobi monogatari to bungaku no aida [Backgrounds of the "narrative generation system" and "relations between narrative and literature"]. In T. Ogata & A. Kanai (Eds.), *Monogatariron no jouhougaku josetsu: monogatari seisei no shisou to gijutsu wo megutte* [An introduction to the informatics of narratology: about the thought and technologies of narrative generation] (pp. 186–258). Gakubunsha.

Ogata, T. (2011). Narrative generation system as the practice of "informatics of narratology.". *Journal of Japan Society for Fuzzy Theory and Intelligent Informatics, 23*(5), 676–685. doi:10.3156/jsoft.23.5_676

Ogata, T. (2014). Expanded literary theory for automatic narrative generation. In *Proceedings of Joint 7th International Conference on Soft Computing and Intelligent Systems and 15th International Symposium on Advanced Intelligent Systems* (pp. 1558–1563). Tokyo, Japan: Japan Society of Fuzzy Theory and Intelligent Informatics.

Ogata, T. (2019d). A method of *naimaze* of narratives based on *kabuki* analyses and Propp's move techniques for an automated narrative generation system. *Journal of Robotics. Networking and Artificial Life, 6*(2), 71–78. doi:10.2991/jrnal.k.190829.001

Ogata, T. (2019e). Toward the incorporation of Japan's narratologies and literary theories into the narrative generation systems by the author. In *Proceedings of the 36th Annual Conference of the Japan Cognitive Science Society* (pp. 442–451). Kanagawa, Japan: Cognitive Science Society.

Ogata, T., & Amino, T. (2002). Tajū monogatari kōzō to *geinō* character no monogatari seisei system [A narrative generation system of multiple narrative structures and a *geinō* character]. In *Proceedings of the 19th Annual Meeting of the Japanese Cognitive Science Society* (pp. 130–131). Tokyo, Japan: Japanese Cognitive Science Society.

Ogata, T., Imabuchi, S., & Akimoto, T. (2014). Narratology and narrative generation: Expanded literary theory and the Integration as a narrative generation system (2). In *Proceedings of the 5th Augmented Human International Conference* (P-10). ACM Digital Library. 10.1145/2582051.2582077

Ogata, T., & Kanai, A. (2010). *Monogatariron no johogaku josetu—Monogatari seisei no gijutu wo megutte* [An introduction informatics of narratology: Around the thoughts and technologies of narrative generation]. Gakubunsha.

Ogata, T., & Kanai, A. (2010). *Monogatariron no jyohogaku jyosetsu: Monogatariseisei no shiso to gijyutsu wo megutte* [An introduction to informatics of narratology: Towards the thoughts and technologies of narrative generation]. Gakubun-sha.

Ohsawa, Y., & Nara, Y. (2002). Modeling the process of chance discovery by chance discovery on double helix. In *Proceedings of AAAI Technical Report FS-01-01*. Palo Alto, CA: AAAI Press.

Ohsawa, Y., Ito, T., & Kamata, M. I. (2008). Kamishibai KeyGraph as scenario map visualizer for detecting transient causes from sequential data. *Proc. of PAKDD 2008 Working Notes of Workshops on Data Mining for Decision Making and Risk Management*, 272—283.

Okamoto, K. (1969). Banchō sarayashiki. In Meisaku kabuki zenshū, 20, Shinkabuki shū [Masterpeace kabuki collection, 20, New kabuki works] (pp. 189–244). Tokyo, Japan: Tokyo Sōgen Shinsha. (First performed in 1916)

Olson, M. E., Seikkula, J., & Ziedonis, D. (2014). *The key elements of dialogic practice in open dialogue: Fidelity criteria.* https://www.umassmed.edu/globalassets/psychiatry/open-dialogue/keyelementsv1.109022014.pdf

Omicini, A., & Ossowski, S. (2003). Objective versus subjective coordination in the engineering of agent systems. In *Intelligent Information Agents* (pp. 179–202). Springer. doi:10.1007/3-540-36561-3_9

Oohira, H. (2010). *Kanjoo shinrigaku nyuumon* [Introduction to emotional psychology]. Yuibikaku.

Ortony, A., Clore, G. L., & Collins, A. (1990). *The cognitive structure of emotions.* Cambridge University Press.

Otokodate hatsugai soga. (1985). In *Kabuki daichō shūsei, 8* [*Kabuki* script collection, 8] (pp. 75–240). Tokyo, Japan: Benseisha. (First performed in 1759. Written by Fujimoto Tobun)

Otsuka H. (2004). *Analysis for Evaluated Expressions of Tasting Japanese Sake.* doi:10.11517/pjsai.JSAI04.0.147.0

Otsuka, H., Suwa, M., & Yamaguchi, K. (2015). *Studies of Expressions to Taste Japanese Sake by Creating Onomatopoeia.* Japanese Society for Artificial Intelligence. doi:10.11517/pjsai.JSAI2015.0_2N5OS16b5

Paradis, C., & Eeg-Olofsson, M. (2013). Describing Sensory Experience: The Genre of Wine Reviews. *Metaphor and Symbol, 28*(1), 22–40. doi:10.1080/10926488.2013.742838

Peirce, C. S. (1955). Philosophical writings of Peirce. Academic Press.

Phillips, B. J. (1997). Thinking into it: Consumer interpretation of complex advertising images. *Journal of Advertising, 26*(2), 77–87. doi:10.1080/00913367.1997.10673524

Picard, R. W., & Rosalind, W. (2000). Toward agents that recognize emotion. *Vivek-Bombay, 13*(1), 3–13.

Pingree, S. (1986). Children's activity and television comprehensibility. *Communication Research, 13*(2), 239–256. doi:10.1177/009365086013002005

Pinker, S. (1995). *Gengo wo umidasu honno* (N. Mukuda, Trans.). Tokyo, Japan: NHKbooks.

Polster, E. (1987). *Every person's life is worth a novel.* W. E. Norton & Company Inc.

Poole, D., Goebel, R., & Aleliunas, R. (1987). Theorist: A logical reasoning system for defaults and diagnosis. In The Knowledge Frontier: Essays in the Representation of Knowledge (pp. 331-352). Springer Verlag.

Pople, H. R. Jr. (1973). On the mechanization of abductive logic. *Proc. of IJCAI73*, 147—152.

Price, R. H., & Bouffard, D. L. (1974). Behavioral appropriateness and situational constraint as dimensions of social behavior. *Journal of Personality and Social Psychology, 30*(4), 579–586. doi:10.1037/h0037037

Prince, G. (2003). *A dictionary of narratology* (Revised ed.). University of Nebraska Press.

Propp, V. Y. (1968). *Morphology of the folktale* (L. Scott, Trans.). University of Texas Press. (Original work published 1928)

Propp, V. Y. (1968). *Morphology of the Folktale* (L. Scott, Trans.). University of Texas Press. (Original work published 1928)

Read, A. G. (1988). OuLiPo, architecture, and the practice of creative constraint. *Proc. of the 86th ACSA Annual Meeting and Technology Conference*, 190—194.

Rickel, J., & Johnson, W. L. (1997). Integrating pedagogical capabilities in a virtual environment agent. In *Proceedings of the First International Conference on Autonomous Agents* (pp. 30-38). 10.1145/267658.267664

Ries, E. (2011). *The lean startup: How today's entrepreneurs use continuous innovation to create radically successful businesses*. Currency.

Riessman, C. K. (1991). Beyond reductionism: Narrative genres in divorce accounts. *Journal of Narrative and Life History*, *1*(1), 41–68. doi:10.1075/jnlh.1.1.04bey

Riessman, C. K. (1993). *Narrative analysis*. Sage.

Rolland, R. (2007). *Jean-Christophe: Journey's end* (G. Cannan, Trans.). Echo Library.

Rolland, R. (2011). *Jean-Christophe* (Vol. I; G. Cannan, Trans.). Aeterna.

Rolland, R. (2018). *Jean-Christophe in Paris: The market-place, Antoinette, the house* (G. Cannan, Trans.). Sagwan Press.

Rooney, R., Benson, C., & Hennessy, E. (2012). The apparent reality of movies and emotional arousal: A study using physiological and self-report measures. *Poetics*, *40*(5), 405–422. doi:10.1016/j.poetic.2012.07.004

Rumelhart, D. E. (1975). Notes on a schema for stories. In D. G. Bobrow & A. Collins (Eds.), *Representation and understanding: Studies in cognitive science* (pp. 211–236). Academic Press. doi:10.1016/B978-0-12-108550-6.50013-6

Russell, J. A. (1980). A circumplex model of affect. *Journal of Personality and Social Psychology*, *39*(6), 1161–1178. doi:10.1037/h0077714

Saiki, S. (2008). Shounigan no kodomo no toubyoutaiken [Narrative based medicine and clinical intellect? Narrative of a pediatric cancer patient]. In Y. Yamada (Ed.), *Jinsei to yamai no katari* [Narrative of life and illness] (pp. 103–132). University of Tokyo Press.

Saio, T. (2014). Open dialogue approach to acute psychosis: Is its effectiveness well-established? *Clinical Evaluation*, *42*(2), 531–537.

Saito, S. (2003). Naratibubeisudomedisun to rinshochi [Narrative based medicine and clinical intellect]. In Y. Yamada (Ed.), *Jinsei to yamai no katari* [Narrative of the life and the illness] (pp. 133–163). University of Tokyo Press.

Sakamoto, K., Iwaoka, T., Ogawa, J., & Nakai, K. (2012). *Kentikukouseigaku, kentiku design no houhou*. Jikkyo Shuppan. (in Japanese)

Sake Competition. (2019). *The Official Guidbook for Sake Competition 2019*. Pia.

Sakusha shikihō kezairoku. (1972). In M. Gunji (Ed.), *Kinsei geidōron* (*Nihon shisō taikei, 61*) [Theories of early modern times' arts (Japanese classic phirosophical thoughts collection, Vol. 61)] (pp. 493–532). Tokyo, Japan: Iwanami Shoten. (Original work published 1801)

Sarasvathy, S. (2009). *Eectuation: Elements of entrepreneurial expertise*. Edward Elgar Pub.

Schank, R. C., & Abelson, R. P. (1977). *Scripts, plans, goals, and understanding: An inquiry into human knowledge structures*. Lawrence Erlbaum.

Schank, R., & Abelson, R. P. (1977). *Scripts, plans, goals, and understanding: An inquiry into human knowledge structure*. Lawrence Erlbaum Associates, Inc.

Schoen, D. A. (1983). *The reflective practitioner: How professionals think in action.* Routledge.

Schwan, S., & Garsoffky, B. (2004). The cognitive representation of filmic event summaries. *Applied Cognitive Psychology, 18*(1), 37–55. doi:10.1002/acp.940

Schwartz, S. H. (2012). An overview of the Schwartz theory of basic values. *Online Readings in Psychology and Culture, 2*(1). Advance online publication. doi:10.9707/2307-0919.1116

Scott, L. M. (1994). Images in advertising: The need for a theory of visual rhetoric. *The Journal of Consumer Research, 21*(2), 252–272. doi:10.1086/209396

Seino, K., Enomoto, Y., & Miyazawa, S. (2018). Narrative analysis of employment support for students with developmental disabilities: Using an objective analysis of free-expression answers. In T. Ogata & S. Asakawa (Eds.), Content generation through narrative communication and simulation (pp. 341-357). Information Science Reference (IGI Global).

Seino, K., Haruna, Y., & Ishizaki, S. (2016). Employment status and support needs of persons with disabilities in Japan: An analysis of narrative using narratology and text mining on a national survey. In T. Ogata, & T. Akimoto (Eds.), Computational and cognitive approaches to narratology (pp. 245-275). Information Science Reference (IGI Global).

Sekai kōmoku. (1916). In *Sekai kōmoku, shibai nenjū gyōji, gekikai chinwa* (*Chinsho kankōkai sōsyo, 9*) [Compendium of narrative worlds, annual events related to kabuki plays, rare tales in theatrical circles (Rare books publication society series, Vol. 9)] (pp. 6–46). Tokyo, Japan: Chinsyo kankōkai (Written in the late Edo period (before 1791))

Seki, K. (1978a). Nihon mukashibanashi taisei, 2—Honkaku mukashibanashi, 1 [Compilation of Japanese folktales, Vol. 2: Ordinary folktales, No. 1]. Tokyo, Japan: Kadokawa Shoten.

Seki, K. (1978b). Nihon mukashibanashi taisei, 3—Honkaku mukashibanashi, 2 [Compilation of Japanese folktales, Vol. 3: Ordinary folktales, No. 2]. Tokyo, Japan: Kadokawa Shoten.

Seki, K. (1978c). Nihon mukashibanashi taisei, 4—Honkaku mukashibanashi, 3 [Compilation of Japanese folktales, Vol. 4: Ordinary folktales, No. 3]. Tokyo, Japan: Kadokawa Shoten.

Seki, K. (1978d). Nihon mukashibanashi taisei, 5—Honkaku mukashibanashi, 4 [Compilation of Japanese folktales, Vol. 5: Ordinary folktales, No. 4]. Tokyo, Japan: Kadokawa Shoten.

Seki, K. (1978e). Nihon mukashibanashi taisei, 6—Honkaku mukashibanashi, 5 [Compilation of Japanese folktales, Vol. 6: Ordinary folktales, No. 5]. Tokyo, Japan: Kadokawa Shoten.

Seki, K. (1979a). Nihon mukashibanashi taisei, 1—Dōbutsu monogatari [Compilation of Japanese folktales, Vol. 1: Animal tales]. Tokyo, Japan: Kadokawa Shoten.

Seki, K. (1979b). Nihon mukashibanashi taisei, 7—Honkaku mukashibanashi, 6 [Compilation of Japanese folktales, Vol. 7: Ordinary folktales, No. 6]. Tokyo, Japan: Kadokawa Shoten.

Seki, K. (1979c). Nihon mukashibanashi taisei, 8—Warai banashi, 1 [Compilation of Japanese folktales, Vol. 8: Jokes and anecdotes, No. 1]. Tokyo, Japan: Kadokawa Shoten.

Seki, K. (1979d). Nihon mukashibanashi taisei, 9—Warai-banashi (Shōwa), 2 [Compilation of Japanese folktales, Vol. 9: Jokes and anecdotes, No. 2]. Tokyo, Japan: Kadokawa Shoten.

Seki, K. (1980). Nihon mukashibanashi taisei, 10—Warai-banashi (Shōwa), 3 [Compilation of Japanese folktales, Vol. 10: Jokes and anecdotes, No. 3]. Tokyo, Japan: Kadokawa Shoten.

Seki, K., Nomura, J., & Ōshima, H. (Eds.). (1979). Nihon mukashibanashi taisei, 12—Kenkyū hen [Compilation of Japanese folktales, 12: Researches]. Tokyo, Japan: Kadokawa Shoten.

Seki, K., Nomura, J., & Ōshima, H. (Eds.). (1980). Nihon mukashibanashi taisei, 11—Shiryō hen [Compilation of Japanese folktales, 11: Data]. Tokyo, Japan: Kadokawa Shoten.

Shimp, T. A. (1976). Methods of commercial presentation employed by national television advertisers. *Journal of Advertising*, *5*(4), 30–36. doi:10.1080/00913367.1976.10672661

Shklovsky, V. (1965). Art as technique. In *Russian formalist criticism: Four essays* (L. T. Lemon & M. J. Reis, Eds. & Trans.). University of Nebraska Press. (Original work published 1917)

Shojima, S. (2008). Multilayered contexts in which narrative is generated and interpreted: Toward the next narrative based inquiry. *Kyoto University Research Studies in Education*, *54*, 138–151.

Smith, G. M. (2003). *Film structure and the emotion system*. Cambridge University Press. doi:10.1017/CBO9780511497759

Smith, T. J. (2012). The attentional theory of cinematic continuity. *Projections*, *6*(1), 1–27. doi:10.3167/proj.2012.060102

Sophocles. (1972). Oedipus the king (A. Burgess, Trans.). Minneapolis, MN: University of Minnesota Press. (Original work first performed around 429 BC)

Spence, C. D. (1983). Narrative persuasion. *Psychoanalysis and Contemporary Thought*, *6*, 457–468.

Stanojevic, M., Vranes, S., & Velasevic, D. (1994). Using truth maintenance systems: A tutorial. *IEEE Expert: Intelligent Systems and Their Applications*, *9*(6), 46–56. doi:10.1109/64.363270

Stathakopoulos, V., Theodorakis, I. G., & Mastoridou, E. (2008). Visual and verbal rhetoric in advertising: The case of 'resonance'. *International Journal of Advertising*, *27*(4), 629–657. doi:10.2501/S0265048708080177

Steinert, M., & Larry, J. L. (2012). Finding one's way': Re-discovering a hunter-gatherer model based on wayfaring. *International Journal of Engineering Education*, *28*(2), 251–252.

Stein, N. L., & Nezworski, T. (1978). The effects of organization and instructional set on story memory. *Discourse Processes*, *1*(2), 177–193. doi:10.1080/01638537809544433

Stern, B. B. (1992). Historical and personal nostalgia in advertising text: The fin de siècle effect. *Journal of Advertising*, *21*(4), 11–22. doi:10.1080/00913367.1992.10673382

Stewart, D. W., & Furse, D. H. (1986). *Effective television advertising: A study of 1000 commercials*. Lexington Books.

Stickel, M. E. (1988). A Prolog-like system for computing minimum-cost abductive explanations in natural-language interpretation. *Proc. of Int'l Computer Science Conf.*, 343—350.

Stickel, M. E. (1989). Rationale and methods for abductive reasoning in natural language interpretation. *Proc. of Int'l Scientific Symp. "Natural Language and Logic" (LNAI-459), 233—252.*

Strong, E. K. (1925). Theories of selling. *The Journal of Applied Psychology*, *9*(9), 75–86. doi:10.1037/h0070123

Stuart, K. (2016). Video games where people matter? The strange future of emotional AI. *The Guardian*. https://www.theguardian.com/technology/2016/oct/12/video-game-characters-emotional-ai-developers

Sundström, P. (2005). *Exploring the affective*. University of Sweden. http://www.diva-portal.org/smash/record.jsf?pid=diva2:1041047

Suzuki, H. (2016b). Perspectives on projection science. In *Proceedings of the Annual Meeting of the Japanese Cognitive Science Society* (pp. 20-25). Japanese Cognitive Science Society.

Suzuki, H. (2016a). *Cognitive science as an education*. Tokyo University Publishing.

Tadaki, K., & Abe, A. (2020). The influence of story creating activities while appreciating abstract artworks. Advances in Artificial Intelligence, 1128, 126-135.

Tadaki, K., & Abe, A. (2017). Museum visitors' behavioral change caused by captions. *Proc. of the 2nd. Int'l Workshop on Language Sense on Computer in IJCAI2017*, 53—58.

Tadaki, K., & Abe, A. (2018). The influence of story writing worksheets on art appreciation. *Proc. of ICAROB2018*, 687—690. 10.5954/ICAROB.2018.OS17-7

Tadaki, K., & Abe, A. (2019). *The potentiality of first person views: Analysing narratives during art appreciation.* In T. Ogata & T. Akimoto (Eds.), *Post-Narratology Through Computational and Cognitive Approaches* (pp. 375–396). Information Science Reference. doi:10.4018/978-1-5225-7979-3.ch010

Tagaki, M. (2008). Experiences of people with acquired long-term physical disability: Life stories of persons with spinal cord injuries in Japan. *Qualitative Psychology Nexus*, *7*, 51–79.

Taikōki. (1996). In Shin nihon koten bungaku taikei, 60 [New Japanese classic literature collection, Vol. 60]. Tokyo, Japan: Iwanami Shoten (Original work published 1626)

Talmy, L. (2000). Toward a Cognitive Semantics: Vol. 2. *Typology and process in concept structuring*. The MIT Press. doi:10.7551/mitpress/6848.001.0001

Tanaka, K. (2010). Kigō to saiki—Kigōron no keishiki, Program no hitsuzen [Symbol and recursion: The forms of semiotics, the inevitability of programs]. Tokyo, Japan: Tokyo Daigaku Shuppankai (The University of Tokyo Press)

Tanaka, Y. (1981). Nantonaku kurisutaru. Tokyo, Japan: Chikuma Shobō Shinsha.

Tanaka, S., & Fukaya, M. (1998). *The Evolvement of the <Sense-Making Theory>(<Imizukeron> no Tenkai).* Kinokuniya.

Tanida, T., & HU, Y. (2013). Twitter, nan tweet areba kachikan ga wakarunoka? [Twitter, how many tweets do you understand your values?]. In *The 34th kotobakougakukenkyuukai* [*Japanese Conference for Language Engineering Research*]. Hokkaido, Japan: The Japanese Society for Artificial Intelligence.

Tanno, Y. (2003). *Seikaku no sinri—Big five to rinshou kara mita personality* [Personality psychology: Big Five and Clinical Personality]. Tokyo, Japan: Saiensusha.

Taura, T., Nagai, Y., & Tanaka, S. (2005). *Design space blending: A key for creative design.* Paper presented at the 15th International Conference on Engineering Design (ICED05), Melbourne, Australia.

Teramoto, Y., & Nakanishi, A. (1993). Chi no tenkansha tachi—Business trickstar [Transformers of knowledge: Business trickstars]. Tokyo, Japan: NTT Shuppan.

Theodorakis, I. G., Koritos, C., & Stathakopoulos, V. (2015). Rhetorical maneuvers in a controversial tide: Assessing the boundaries of advertising rhetoric. *Journal of Advertising*, *44*(1), 14–24. doi:10.1080/00913367.2014.930679

Thomas, P., & Braken, P. (2004). Critical psychiatry in practice. *Advances in Psychiatric Treatment*, *10*(5), 361–370. doi:10.1192/apt.10.5.361

Thorndyke, P. W. (1977). Cognitive structures in comprehension and memory of narrative discourse. *Cognitive Psychology*, *9*(1), 77–110. doi:10.1016/0010-0285(77)90005-6

Tolstoy, L. (2010). *War and peace* (A. Mandelker, Ed., MaudeL.MaudeA., Trans.). Oxford University Press. (Original work published 1969)

Torres, S. (2014), Video game characters modeled after real people. *Venture Beat*. https://venturebeat.com/2014/06/28/video-game-characters-modeled-after-real-people/

Trull, T. J., & Widiger, T. A. (2013). Dimensional models of personality: The five-factor model and the DSM-5. *Dialogues in Clinical Neuroscience, 15*(2), 135. PMID:24174888

Tsai, M. (2013). *Stray Dogs* [Film]. V. Wang (Producer).

Tsujii, J. (1987). *Chikishi no hyogen to riyoo*. Shokodo. (in Japanese)

Turing, A. (1950, October). Computing machinery and intelligence. *Mind, 59*(236), 433–460. doi:10.1093/mind/LIX.236.433

Uchida, Y., Kondō, H., & Funato, M. (1987). *"Sekai kōmoku"* sakuin—1985 nendo kyōdō kenkyu hōkoku [*"Sekai kōmoku"* index: 1985 joint study report]. *Kyōritsu Joshi Daigaku Bungei Gakubu Kiyō, 33*, 173–286.

Uejo, N. (1993). *Kokoku kopi gairon* [Advertising copy overview]. Sendenkaigi.

Urayama, M. (1974). Naimaze to sekai. *Geinō no kagaku, 5*, 103–120.

Utsunomiya, H., Isogai, A., Iwata, H., & Nakano, S. (2006). Flavor Terminology and Reference Standards for Sensory Analysis of Sake. *Journal of the Blewing Society of Japan, 101*(10), 730–739.

Vella, D. (2014). Modeling the semiotic structure of game characters. *Proceedings of DiGRA 2014*.

Vinyals, O., Toshev, A., Bengio, S., & Erhan, D. (2015). Show and tell: A neural image caption generator. In *Proceedings of 28th IEEE Conference on Computer Vision and Pattern Recognition* (pp. 3156-3164). Boston, MA: IEEE. 10.1109/CVPR.2015.7298935

Visch, V. T., & Tan, E. S. (2009). Categorizing moving objects into film genres: The effect of animacy attribution, emotional response, and the deviation from non-fiction. *Cognition, 110*(2), 265–272. doi:10.1016/j.cognition.2008.10.018 PMID:19118823

Von Hippel, E., & Von Krogh, G. (2016). CROSSROADS—Identifying Viable "Need–Solution Pairs"- Problem Solving Without Problem Formulation. *Organization Science, 27*(1), 207–221.

Vuokila-Oikkonen, P., Janhonen, S., & Vaisanen, L. (2004). "Shared-rhythm cooperation" in cooperative team meetings in acute psychiatric inpatient care. *Journal of Psychiatric and Mental Health Nursing, 11*(2), 129–140. doi:10.1111/j.1365-2850.2003.00679.x PMID:15009487

Watanabe, N. (2012). Nihon shōsetsu gijutsushi [A history of technologies of Japanese novels]. Tokyo, Japan: Shinchōsha.

Watanabe, T. (1986). Musume Dōjōji. Tokyo, Japan: Shinshindo Shuppan.

Watanabe, T. (1989). Kabuki—Kajō naru kigō no mori [Kabuki: The forest of excessive signs]. Tokyo, Japan: Shin'yōsha.

Watanabe, T. (2004). Kabuki—Kata no miryoku [Kabuki: Charm of forms]. Tokyo, Japan: Kadokawa Shoten.

Watanabe, T. (2012). Watashi no kabuki henreki—Aru gekihyō-ka no kokuhaku [My kabuki wandering: Confession of a drama critic]. Tokyo, Japan: Engeki Shuppansha.

Watanabe, T. (2018). Kudaime Danjūrō [Danjūrō IX]. Tokyo, Japan: Engeki Shuppansha.

Welles, O. (1958). *Touch of evil* [Film]. A. Zugsmith (Producer). USA: Universal Pictures.

Whitaker, R. (1970). *The language of film*. Prentice-Hall, Inc.

Wiley, N. (2003). Emotion and film theory. *Studies in Symbolic Interaction, 26*, 169–187. doi:10.1016/S0163-2396(02)26012-3

Wilkenfeld, M. J., & Ward, T. B. (2001). Similarity and emergence in conceptual combination. *Journal of Memory and Language, 45*(1), 21–38. doi:10.1006/jmla.2000.2772

Wiseman, F. (2017). *Ex libris: The New York public library* [Film].

Yamabe, S. (2012). Sosharuwaku niokeru ebidensu tohananika: Purorogu [What is the evidence in the social work?: Prologue]. In Japanese Society for the Study of Social Welfare (Ed.), Taironshakaihukushigaku [Discussion social welfare studies] (pp. 85-87). Chuohoki Publishing.

Yamada, Y. (2007a). Shitsutekikenkyuu niokeru taiwatekimoderukouseihou [Interactive model construction method in qualitative research]. *Japanese Journal of Qualitative Psychology, 6*, 174–194.

Yamada, Y. (2007b). *Shitsutekishinrigaku no houhou: Katari o kiku* [Method of qualitative psychology: Listening to narrative]. Shinyousha.

Yamamoto, Y. (1977). Sakazaki Dewanokami. In Yamamoto Yūzō zenshū, 1 [The complete works of Yamamoto Yūzō, 1]. (pp. 287–351). Tokyo, Japan: (Original work published 1921)

Yamashita, M. (2013). Innovation of next automobiles. *Journal of Society of Automotive Engineers of Japan, 67*(6), 12–17.

Yasuda, N. (1989). Dōjōji densetssu kara "Kyōganoko Musume Dōjōji" he [From Dōjōji legnds to *Kyōganoko Musume Dōjōji*]. *Shōtoku Gakuen Gifu Kyōiku Daigaku Kokugo Kokubungaku, 8*, 1–12.

Yik, M., Russell, J. A., & Steiger, J. H. (2011). A 12-point circumplex structure of core affect. *Emotion (Washington, D.C.), 11*(4), 705–731. doi:10.1037/a0023980 PMID:21707162

Yin, M., & Sun, Y. (2015). Human behavior models for virtual agents in repeated decision making under uncertainty. *Proceedings of the 14th International Conference on Autonomous Agents and Multiagent Systems.*

Yokomitsu, R. (1931). Kikai. Tokyo, Japan: Hakusuisha.

Yonemori, Y. (2007). *Abduction*. Keiso Shobo.

Yoshimoto, T. (1965). Gengo ni totte bi toha nanika [What is beauty for language?]. Tokyo, Japan: Keisō Shobō.

Yoshimoto, T. (1968). Kyōdō gensō ron [The theory of communal Illusion]. Tokyo, Japan: Kawade Shobō Shinsha.

Yoshimoto, T. (1971). Shinteki genshōron josetsu [A theory of mental phenomena: Introduction]. Tokyo, Japan: Hokuyōsha.

Yoshimoto, T. (2008). Shinteki genshōron honron [A theory of mental phenomena: Main issues]. Tokyo, Japan: Bunka Kagaku Kōtō Kenkyūin Shuppankyoku.

Yoshimoto, T. (1968). *Kyodo Gensoron* [The theory of communal fantasy]. Kawade Shobo Shinsha.

Yussen, S., Mathews, S., Huang, S. T., & Evans, R. (1988). The robustness and temporal course of the story schema's influence on recall. *Journal of Experimental Psychology. Learning, Memory, and Cognition, 14*(1), 173–179. doi:10.1037/0278-7393.14.1.173

Zackman, J. A. (1987). A framework for information systems architecture. *IBM Systems Journal, 26*(3), 276–292. doi:10.1147j.263.0276

Zeami, M. (1983). Sandō. In M. Yamazaki (Ed.), Zeami (pp. 253–268). Tokyo, Japan: Chūōkōronsha. (Original work published 1423)

Zhao, R., Papangelis, A., & Cassell, J. (2014). Towards a dyadic computational model of rapport management for human-virtual agent interaction. In Intelligent Virtual Agents. IVA 2014. Lecture Notes in Computer Science, (vol. 8637). Springer. doi:10.1007/978-3-319-09767-1_62

About the Contributors

Takashi Ogata, after he received his Bachelor in Social Sciences from Waseda University in 1983, he received his M.S. from Tsukuba University in 1992 and his Ph.D. from the University of Tokyo in 1995. He has attained industrial experience since 1983 at software development companies. Having been an Associate Professor in the faculty of Engineering at Yamanashi University since 1997, he is now a Professor in the faculty of Software and Information Science at Iwate Prefectural University since 2005. He is a member of the Japanese Society for Artificial Intelligence (JSAI), the Japanese Cognitive Science Society (JCSS), and the Japanese Association for Natural Language Processing. He was also the primary manager of the Literature, Cognition, and Computer (LCC) research group at the JCSS. He received the JSAI best paper award (1996), best paper award from the Japan Academy of Advertising (1996), and other academic awards. In addition to many papers, he has published academic books, including Monogatariron no Jōhogaku Josetu (An Introduction to the Informatics of Narratology) (T. Ogata & A. Kanai, 2010, Tokyo: Gakubunsha), Computational and Cognitive Approaches to Narratology (T. Ogata & T. Akimoto, 2016, IGI Global), Content Generation Through Narrative Communication and Simulation (T. Ogata & S. Asakawa, 2017, IGI Global), Jōho Monogatariron (Informational Narratology) (T. Ogata, Y. Kawamura, & A. Kanai, 2018, Tokyo: Hakutō Shobō), Post-Narratology Through Computational and Cognitive Approaches (T. Ogata & T. Akimoto, 2019, IGI Global), Toward an Integrated Approach to Narrative Generation: Emerging Research and Opportunities (T. Ogata, 2020, IGI Global), and Internal and External Narrative Generation Based on Post-Narratology: Emerging Research and Opportunities. (T. Ogata, 2020, IGI Global). Furthermore, he is currently writing and editing two books related to his concept of post-narratology by Japanese. His research interests include the following broad areas: (1) Computer science and information technologies (including narrative (story) generation (system), artificial intelligence, cognitive science, natural language processing and generation, ontology and conceptual dictionary), (2) Humanities (including narratology, literary theories, semiotics, semantics, discourse theory, philosophy, kabuki and Japanese arts and entertainments, Japanese narratologies and literary theories), and (3) Social sciences (including content business, marketing and advertising studies, sociology, social philosophy).

Jumpei Ono received his Bachelor in Faculty School of Software and Information Science from Iwate Prefectural University in 2010. He received his M.S. degree and his Ph.D. in Graduate School of Software and Information Science from Iwate Prefectural University in 2014 and 2018. He is having experienced instructor of information and communication technology at Vocational School of Digital Arts Sendai since 2018, he is Assistant Professor of the Faculty of Software and Information Technology at Aomori University since 2020.. He is a member of the Japanese Society for Artificial Intelligence

(JSAI), the Japanese Cognitive Science Society (JCSS), the Japanese Association for Natural Language Processing, Digital Games Research Association Japan (DiGRA Japan), and the Institute of Electronics, Information and Communication Engineers (IEICE). He received Excellent Student Award of the IEICE Tohoku Section (2017) and Iwate Prefectural University President's Award (2018). His main papers include "A Study on an Automatic Narrative Generation Game Based on Gaps and Surprise: An Approach Using Table-talk Role Playing Games and an Integrated Narrative Generation System"(Ono, J., 2018, Ph.D. dissertation of Iwate Prefectural University), "Surprise-Based Narrative Generation in an Automatic Narrative Generation Game" (Ono, J. & Ogata, T., 2018, IGI Global), "Toward developing a Mechanism for Automatic Narrative Generation based on a Story Generation Model of Table-top Role Playing Game" (Ono, J. & Ogata, T., 2017, Cognitive Study), and "A Design Plan of a Game System Including an Automatic Narrative Generation Mechanism: The Entire Structure and the World Settings" (Ono, J. & Ogata, T., 2016, Journal of Robotics, Networking and Artificial Life). His research interest includes narrative generation system, artificial intelligence, cognitive science, natural language processing, narratology, advertisement generation, rhetoric of surprise, and digital game and TRPG.

* * *

Akinori Abe received an M.E. and Doctor of Engineering (PhD) from the University of Tokyo in 1988 and 1991 respectively. His main research interests are abduction, analogical reasoning, data mining, Chance Discovery, curation and language sense processing (rather emotional aspects of language processing). He worked in NTT Communication Science Laboratories from 1991 to 2000, NTT MSC (Malaysia) from 2000 to 2002, and ATR From 2002 to 2009. He also worked as an associate professor of IREIIMS, Tokyo Women's Medical Univ. and a visiting associate professor of Kobe Univ. Graduate School (Cooperation Course). In 2009 he returned to NTT Communication Science Laboratories. Then he moved to Faculty of Letters, Chiba University in April 2012. Since then he is a full professor of Chiba University.

Hiroki Fukushima received a Master of Media and Governance and Ph.D. from Keio University in 2015 and 2018 respectively. His main research interests are semiotics, cognitive linguistics, and the theory of the representation-making of taste. He worked as an Assistant Professor of Keio University (2015-2018). He is now a Senior Lecturer at Kyushu Women's University.

Nobuyoshi Harada is a Special Researcher of Center for Frontier Medical Engineering, Chiba University, Japan.

Akihito Kanai is a professor of media and communication studies at Hosei university. His current research interests include film cognition, film rhetoric, film narrative, and computational film composition.

Yoji Kawamura is a professor of business administration at Kindai University. His majoring is modeling the various social (include business, advertising) phenomena based on the modeling of consumer and the system building for simulation. His published research has appeared in journals (in Japan) as Journal of Advertising Science, Japan Marketing Journal, Cognitive Studies, IPSJ Journal. He was received the 2005 Research Article Award given by Japan Academy of Advertising.

Jun Nakamura is a Professor at Faculty of Global Management, Chuo University, and a board member of Persol AVC Technology Co., Ltd., which is an engineering company to develop Audio & Video devices. He graduated from the University of Tokyo, Graduate School of Engineering, Department of Technology Management for Innovation with highest honors. The areas of his research are cognitive science, artificial information, and innovation management, particularly visualization of one's thinking process to be aware of novel ideas.

Sheldon Schiffer researches and experiments with "performance" capture of actors toward the "intelligent" behavior of autonomous animated characters in games and moving image production. Having taught acting for 14 years, and produced and directed both games and film, he works in performance media across multiple media genre. Schiffer is a professor at Georgia State University in the Creative Media Industries Institute, where he teaches courses in game development and design, animation, and sound design for games.

Index

T

U

V

W

www.ingramcontent.com/pod-product-compliance
Lightning Source LLC
LaVergne TN
LVHW060401300926
849162LV00006B/1317